Lecture Notes in Computer Science 13808

More information about this series at https://link.springer.com/bookseries/558

Leonid Karlinsky · Tomer Michaeli ·
Ko Nishino (Eds.)

Computer Vision – ECCV 2022 Workshops

Tel Aviv, Israel, October 23–27, 2022
Proceedings, Part VIII

 Springer

Editors
Leonid Karlinsky
IBM Research - MIT-IBM Watson AI Lab
Massachusetts, USA

Tomer Michaeli 🆔
Technion – Israel Institute of Technology
Haifa, Israel

Ko Nishino 🆔
Kyoto University
Kyoto, Japan

ISSN 0302-9743   ISSN 1611-3349 (electronic)
Lecture Notes in Computer Science
ISBN 978-3-031-25084-2   ISBN 978-3-031-25085-9 (eBook)
https://doi.org/10.1007/978-3-031-25085-9

This Springer imprint is published by the registered company Springer Nature Switzerland AG
The registered company address is: Gewerbestrasse 11, 6330 Cham, Switzerland

Foreword

Organizing the European Conference on Computer Vision (ECCV 2022) in Tel-Aviv during a global pandemic was no easy feat. The uncertainty level was extremely high, and decisions had to be postponed to the last minute. Still, we managed to plan things just in time for ECCV 2022 to be held in person. Participation in physical events is crucial to stimulating collaborations and nurturing the culture of the Computer Vision community.

There were many people who worked hard to ensure attendees enjoyed the best science at the 17th edition of ECCV. We are grateful to the Program Chairs Gabriel Brostow and Tal Hassner, who went above and beyond to ensure the ECCV reviewing process ran smoothly. The scientific program included dozens of workshops and tutorials in addition to the main conference and we would like to thank Leonid Karlinsky and Tomer Michaeli for their hard work. Finally, special thanks to the web chairs Lorenzo Baraldi and Kosta Derpanis, who put in extra hours to transfer information fast and efficiently to the ECCV community.

We would like to express gratitude to our generous sponsors and the Industry Chairs Dimosthenis Karatzas and Chen Sagiv, who oversaw industry relations and proposed new ways for academia-industry collaboration and technology transfer. It's great to see so much industrial interest in what we're doing!

Authors' draft versions of the papers appeared online with open access on both the Computer Vision Foundation (CVF) and the European Computer Vision Association (ECVA) websites as with previous ECCVs. Springer, the publisher of the proceedings, has arranged for archival publication. The final version of the papers is hosted by SpringerLink, with active references and supplementary materials. It benefits all potential readers that we offer both a free and citeable version for all researchers, as well as an authoritative, citeable version for SpringerLink readers. Our thanks go to Ronan Nugent from Springer, who helped us negotiate this agreement. Last but not least, we wish to thank Eric Mortensen, our publication chair, whose expertise made the process smooth.

October 2022

Rita Cucchiara
Jiří Matas
Amnon Shashua
Lihi Zelnik-Manor

Preface

Welcome to the workshop proceedings of the 17th European Conference on Computer Vision (ECCV 2022). This year, the main ECCV event was accompanied by 60 workshops, scheduled between October 23–24, 2022. We received 103 workshop proposals on diverse computer vision topics and unfortunately had to decline many valuable proposals because of space limitations. We strove to achieve a balance between topics, as well as between established and new series. Due to the uncertainty associated with the COVID-19 pandemic around the proposal submission deadline, we allowed two workshop formats: hybrid and purely online. Some proposers switched their preferred format as we drew near the conference dates. The final program included 30 hybrid workshops and 30 purely online workshops. Not all workshops published their papers in the ECCV workshop proceedings, or had papers at all. These volumes collect the edited papers from 38 out of the 60 workshops. We sincerely thank the ECCV general chairs for trusting us with the responsibility for the workshops, the workshop organizers for their hard work in putting together exciting programs, and the workshop presenters and authors for contributing to ECCV.

October 2022

Tomer Michaeli
Leonid Karlinsky
Ko Nishino

Organization

General Chairs

Rita Cucchiara — University of Modena and Reggio Emilia, Italy
Jiří Matas — Czech Technical University in Prague, Czech Republic
Amnon Shashua — Hebrew University of Jerusalem, Israel
Lihi Zelnik-Manor — Technion – Israel Institute of Technology, Israel

Program Chairs

Shai Avidan — Tel-Aviv University, Israel
Gabriel Brostow — University College London, UK
Giovanni Maria Farinella — University of Catania, Italy
Tal Hassner — Facebook AI, USA

Program Technical Chair

Pavel Lifshits — Technion – Israel Institute of Technology, Israel

Workshops Chairs

Leonid Karlinsky — IBM Research - MIT-IBM Watson AI Lab, USA
Tomer Michaeli — Technion – Israel Institute of Technology, Israel
Ko Nishino — Kyoto University, Japan

Tutorial Chairs

Thomas Pock — Graz University of Technology, Austria
Natalia Neverova — Facebook AI Research, UK

Demo Chair

Bohyung Han — Seoul National University, South Korea

Social and Student Activities Chairs

Tatiana Tommasi — Italian Institute of Technology, Italy
Sagie Benaim — University of Copenhagen, Denmark

Diversity and Inclusion Chairs

Xi Yin   Facebook AI Research, USA
Bryan Russell   Adobe, USA

Communications Chairs

Lorenzo Baraldi   University of Modena and Reggio Emilia, Italy
Kosta Derpanis   York University and Samsung AI Centre Toronto,
                                Canada

Industrial Liaison Chairs

Dimosthenis Karatzas   Universitat Autònoma de Barcelona, Spain
Chen Sagiv   SagivTech, Israel

Finance Chair

Gerard Medioni   University of Southern California and Amazon,
                                USA

Publication Chair

Eric Mortensen   MiCROTEC, USA

Workshops Organizers

W01 - AI for Space

Tat-Jun Chin   The University of Adelaide, Australia
Luca Carlone   Massachusetts Institute of Technology, USA
Djamila Aouada   University of Luxembourg, Luxembourg
Binfeng Pan   Northwestern Polytechnical University, China
Viorela Ila   The University of Sydney, Australia
Benjamin Morrell   NASA Jet Propulsion Lab, USA
Grzegorz Kakareko   Spire Global, USA

W02 - Vision for Art

Alessio Del Bue   Istituto Italiano di Tecnologia, Italy
Peter Bell   Philipps-Universität Marburg, Germany
Leonardo L. Impett   École Polytechnique Fédérale de Lausanne
                                (EPFL), Switzerland
Noa Garcia   Osaka University, Japan
Stuart James   Istituto Italiano di Tecnologia, Italy

W03 - Adversarial Robustness in the Real World

Angtian Wang	Johns Hopkins University, USA
Yutong Bai	Johns Hopkins University, USA
Adam Kortylewski	Max Planck Institute for Informatics, Germany
Cihang Xie	University of California, Santa Cruz, USA
Alan Yuille	Johns Hopkins University, USA
Xinyun Chen	University of California, Berkeley, USA
Judy Hoffman	Georgia Institute of Technology, USA
Wieland Brendel	University of Tübingen, Germany
Matthias Hein	University of Tübingen, Germany
Hang Su	Tsinghua University, China
Dawn Song	University of California, Berkeley, USA
Jun Zhu	Tsinghua University, China
Philippe Burlina	Johns Hopkins University, USA
Rama Chellappa	Johns Hopkins University, USA
Yinpeng Dong	Tsinghua University, China
Yingwei Li	Johns Hopkins University, USA
Ju He	Johns Hopkins University, USA
Alexander Robey	University of Pennsylvania, USA

W04 - Autonomous Vehicle Vision

Rui Fan	Tongji University, China
Nemanja Djuric	Aurora Innovation, USA
Wenshuo Wang	McGill University, Canada
Peter Ondruska	Toyota Woven Planet, UK
Jie Li	Toyota Research Institute, USA

W05 - Learning With Limited and Imperfect Data

Noel C. F. Codella	Microsoft, USA
Zsolt Kira	Georgia Institute of Technology, USA
Shuai Zheng	Cruise LLC, USA
Judy Hoffman	Georgia Institute of Technology, USA
Tatiana Tommasi	Politecnico di Torino, Italy
Xiaojuan Qi	The University of Hong Kong, China
Sadeep Jayasumana	University of Oxford, UK
Viraj Prabhu	Georgia Institute of Technology, USA
Yunhui Guo	University of Texas at Dallas, USA
Ming-Ming Cheng	Nankai University, China

W06 - Advances in Image Manipulation

Radu Timofte	University of Würzburg, Germany, and ETH Zurich, Switzerland
Andrey Ignatov	AI Benchmark and ETH Zurich, Switzerland
Ren Yang	ETH Zurich, Switzerland
Marcos V. Conde	University of Würzburg, Germany
Furkan Kınlı	Özyeğin University, Turkey

W07 - Medical Computer Vision

Tal Arbel	McGill University, Canada
Ayelet Akselrod-Ballin	Reichman University, Israel
Vasileios Belagiannis	Otto von Guericke University, Germany
Qi Dou	The Chinese University of Hong Kong, China
Moti Freiman	Technion, Israel
Nicolas Padoy	University of Strasbourg, France
Tammy Riklin Raviv	Ben Gurion University, Israel
Mathias Unberath	Johns Hopkins University, USA
Yuyin Zhou	University of California, Santa Cruz, USA

W08 - Computer Vision for Metaverse

Bichen Wu	Meta Reality Labs, USA
Peizhao Zhang	Facebook, USA
Xiaoliang Dai	Facebook, USA
Tao Xu	Facebook, USA
Hang Zhang	Meta, USA
Péter Vajda	Facebook, USA
Fernando de la Torre	Carnegie Mellon University, USA
Angela Dai	Technical University of Munich, Germany
Bryan Catanzaro	NVIDIA, USA

W09 - Self-Supervised Learning: What Is Next?

Yuki M. Asano	University of Amsterdam, The Netherlands
Christian Rupprecht	University of Oxford, UK
Diane Larlus	Naver Labs Europe, France
Andrew Zisserman	University of Oxford, UK

W10 - Self-Supervised Learning for Next-Generation Industry-Level Autonomous Driving

Xiaodan Liang	Sun Yat-sen University, China
Hang Xu	Huawei Noah's Ark Lab, China

Fisher Yu	ETH Zürich, Switzerland
Wei Zhang	Huawei Noah's Ark Lab, China
Michael C. Kampffmeyer	UiT The Arctic University of Norway, Norway
Ping Luo	The University of Hong Kong, China

W11 - ISIC Skin Image Analysis

M. Emre Celebi	University of Central Arkansas, USA
Catarina Barata	Instituto Superior Técnico, Portugal
Allan Halpern	Memorial Sloan Kettering Cancer Center, USA
Philipp Tschandl	Medical University of Vienna, Austria
Marc Combalia	Hospital Clínic of Barcelona, Spain
Yuan Liu	Google Health, USA

W12 - Cross-Modal Human-Robot Interaction

Fengda Zhu	Monash University, Australia
Yi Zhu	Huawei Noah's Ark Lab, China
Xiaodan Liang	Sun Yat-sen University, China
Liwei Wang	The Chinese University of Hong Kong, China
Xiaojun Chang	University of Technology Sydney, Australia
Nicu Sebe	University of Trento, Italy

W13 - Text in Everything

Ron Litman	Amazon AI Labs, Israel
Aviad Aberdam	Amazon AI Labs, Israel
Shai Mazor	Amazon AI Labs, Israel
Hadar Averbuch-Elor	Cornell University, USA
Dimosthenis Karatzas	Universitat Autònoma de Barcelona, Spain
R. Manmatha	Amazon AI Labs, USA

W14 - BioImage Computing

Jan Funke	HHMI Janelia Research Campus, USA
Alexander Krull	University of Birmingham, UK
Dagmar Kainmueller	Max Delbrück Center, Germany
Florian Jug	Human Technopole, Italy
Anna Kreshuk	EMBL-European Bioinformatics Institute, Germany
Martin Weigert	École Polytechnique Fédérale de Lausanne (EPFL), Switzerland
Virginie Uhlmann	EMBL-European Bioinformatics Institute, UK

Peter Bajcsy   National Institute of Standards and Technology,
                                     USA
Erik Meijering   University of New South Wales, Australia

W15 - Visual Object-Oriented Learning Meets Interaction: Discovery, Representations, and Applications

Kaichun Mo   Stanford University, USA
Yanchao Yang   Stanford University, USA
Jiayuan Gu   University of California, San Diego, USA
Shubham Tulsiani   Carnegie Mellon University, USA
Hongjing Lu   University of California, Los Angeles, USA
Leonidas Guibas   Stanford University, USA

W16 - AI for Creative Video Editing and Understanding

Fabian Caba   Adobe Research, USA
Anyi Rao   The Chinese University of Hong Kong, China
Alejandro Pardo   King Abdullah University of Science and
                                     Technology, Saudi Arabia
Linning Xu   The Chinese University of Hong Kong, China
Yu Xiong   The Chinese University of Hong Kong, China
Victor A. Escorcia   Samsung AI Center, UK
Ali Thabet   Reality Labs at Meta, USA
Dong Liu   Netflix Research, USA
Dahua Lin   The Chinese University of Hong Kong, China
Bernard Ghanem   King Abdullah University of Science and
                                     Technology, Saudi Arabia

W17 - Visual Inductive Priors for Data-Efficient Deep Learning

Jan C. van Gemert   Delft University of Technology, The Netherlands
Nergis Tömen   Delft University of Technology, The Netherlands
Ekin Dogus Cubuk   Google Brain, USA
Robert-Jan Bruintjes   Delft University of Technology, The Netherlands
Attila Lengyel   Delft University of Technology, The Netherlands
Osman Semih Kayhan   Bosch Security Systems, The Netherlands
Marcos Baptista Ríos   Alice Biometrics, Spain
Lorenzo Brigato   Sapienza University of Rome, Italy

W18 - Mobile Intelligent Photography and Imaging

Chongyi Li   Nanyang Technological University, Singapore
Shangchen Zhou   Nanyang Technological University, Singapore

Ruicheng Feng   Nanyang Technological University, Singapore
Jun Jiang   SenseBrain Research, USA
Wenxiu Sun   SenseTime Group Limited, China
Chen Change Loy   Nanyang Technological University, Singapore
Jinwei Gu   SenseBrain Research, USA

W19 - People Analysis: From Face, Body and Fashion to 3D Virtual Avatars

Alberto Del Bimbo   University of Florence, Italy
Mohamed Daoudi   IMT Nord Europe, France
Roberto Vezzani   University of Modena and Reggio Emilia, Italy
Xavier Alameda-Pineda   Inria Grenoble, France
Marcella Cornia   University of Modena and Reggio Emilia, Italy
Guido Borghi   University of Bologna, Italy
Claudio Ferrari   University of Parma, Italy
Federico Becattini   University of Florence, Italy
Andrea Pilzer   NVIDIA AI Technology Center, Italy
Zhiwen Chen   Alibaba Group, China
Xiangyu Zhu   Chinese Academy of Sciences, China
Ye Pan   Shanghai Jiao Tong University, China
Xiaoming Liu   Michigan State University, USA

W20 - Safe Artificial Intelligence for Automated Driving

Timo Saemann   Valeo, Germany
Oliver Wasenmüller   Hochschule Mannheim, Germany
Markus Enzweiler   Esslingen University of Applied Sciences,
                                   Germany
Peter Schlicht   CARIAD, Germany
Joachim Sicking   Fraunhofer IAIS, Germany
Stefan Milz   Spleenlab.ai and Technische Universität Ilmenau,
                                   Germany
Fabian Hüger   Volkswagen Group Research, Germany
Seyed Ghobadi   University of Applied Sciences Mittelhessen,
                                   Germany
Ruby Moritz   Volkswagen Group Research, Germany
Oliver Grau   Intel Labs, Germany
Frédérik Blank   Bosch, Germany
Thomas Stauner   BMW Group, Germany

W21 - Real-World Surveillance: Applications and Challenges

Kamal Nasrollahi   Aalborg University, Denmark
Sergio Escalera   Universitat Autònoma de Barcelona, Spain

Radu Tudor Ionescu	University of Bucharest, Romania
Fahad Shahbaz Khan	Mohamed bin Zayed University of Artificial Intelligence, United Arab Emirates
Thomas B. Moeslund	Aalborg University, Denmark
Anthony Hoogs	Kitware, USA
Shmuel Peleg	The Hebrew University, Israel
Mubarak Shah	University of Central Florida, USA

W22 - Affective Behavior Analysis In-the-Wild

Dimitrios Kollias	Queen Mary University of London, UK
Stefanos Zafeiriou	Imperial College London, UK
Elnar Hajiyev	Realeyes, UK
Viktoriia Sharmanska	University of Sussex, UK

W23 - Visual Perception for Navigation in Human Environments: The JackRabbot Human Body Pose Dataset and Benchmark

Hamid Rezatofighi	Monash University, Australia
Edward Vendrow	Stanford University, USA
Ian Reid	University of Adelaide, Australia
Silvio Savarese	Stanford University, USA

W24 - Distributed Smart Cameras

Niki Martinel	University of Udine, Italy
Ehsan Adeli	Stanford University, USA
Rita Pucci	University of Udine, Italy
Animashree Anandkumar	Caltech and NVIDIA, USA
Caifeng Shan	Shandong University of Science and Technology, China
Yue Gao	Tsinghua University, China
Christian Micheloni	University of Udine, Italy
Hamid Aghajan	Ghent University, Belgium
Li Fei-Fei	Stanford University, USA

W25 - Causality in Vision

Yulei Niu	Columbia University, USA
Hanwang Zhang	Nanyang Technological University, Singapore
Peng Cui	Tsinghua University, China
Song-Chun Zhu	University of California, Los Angeles, USA
Qianru Sun	Singapore Management University, Singapore
Mike Zheng Shou	National University of Singapore, Singapore
Kaihua Tang	Nanyang Technological University, Singapore

W26 - In-Vehicle Sensing and Monitorization

Jaime S. Cardoso	INESC TEC and Universidade do Porto, Portugal
Pedro M. Carvalho	INESC TEC and Polytechnic of Porto, Portugal
João Ribeiro Pinto	Bosch Car Multimedia and Universidade do Porto, Portugal
Paula Viana	INESC TEC and Polytechnic of Porto, Portugal
Christer Ahlström	Swedish National Road and Transport Research Institute, Sweden
Carolina Pinto	Bosch Car Multimedia, Portugal

W27 - Assistive Computer Vision and Robotics

Marco Leo	National Research Council of Italy, Italy
Giovanni Maria Farinella	University of Catania, Italy
Antonino Furnari	University of Catania, Italy
Mohan Trivedi	University of California, San Diego, USA
Gérard Medioni	Amazon, USA

W28 - Computational Aspects of Deep Learning

Iuri Frosio	NVIDIA, Italy
Sophia Shao	University of California, Berkeley, USA
Lorenzo Baraldi	University of Modena and Reggio Emilia, Italy
Claudio Baecchi	University of Florence, Italy
Frederic Pariente	NVIDIA, France
Giuseppe Fiameni	NVIDIA, Italy

W29 - Computer Vision for Civil and Infrastructure Engineering

Joakim Bruslund Haurum	Aalborg University, Denmark
Mingzhu Wang	Loughborough University, UK
Ajmal Mian	University of Western Australia, Australia
Thomas B. Moeslund	Aalborg University, Denmark

W30 - AI-Enabled Medical Image Analysis: Digital Pathology and Radiology/COVID-19

Jaime S. Cardoso	INESC TEC and Universidade do Porto, Portugal
Stefanos Kollias	National Technical University of Athens, Greece
Sara P. Oliveira	INESC TEC, Portugal
Mattias Rantalainen	Karolinska Institutet, Sweden
Jeroen van der Laak	Radboud University Medical Center, The Netherlands
Cameron Po-Hsuan Chen	Google Health, USA

Diana Felizardo	IMP Diagnostics, Portugal
Ana Monteiro	IMP Diagnostics, Portugal
Isabel M. Pinto	IMP Diagnostics, Portugal
Pedro C. Neto	INESC TEC, Portugal
Xujiong Ye	University of Lincoln, UK
Luc Bidaut	University of Lincoln, UK
Francesco Rundo	STMicroelectronics, Italy
Dimitrios Kollias	Queen Mary University of London, UK
Giuseppe Banna	Portsmouth Hospitals University, UK

W31 - Compositional and Multimodal Perception

Kazuki Kozuka	Panasonic Corporation, Japan
Zelun Luo	Stanford University, USA
Ehsan Adeli	Stanford University, USA
Ranjay Krishna	University of Washington, USA
Juan Carlos Niebles	Salesforce and Stanford University, USA
Li Fei-Fei	Stanford University, USA

W32 - Uncertainty Quantification for Computer Vision

Andrea Pilzer	NVIDIA, Italy
Martin Trapp	Aalto University, Finland
Arno Solin	Aalto University, Finland
Yingzhen Li	Imperial College London, UK
Neill D. F. Campbell	University of Bath, UK

W33 - Recovering 6D Object Pose

Martin Sundermeyer	DLR German Aerospace Center, Germany
Tomáš Hodaň	Reality Labs at Meta, USA
Yann Labbé	Inria Paris, France
Gu Wang	Tsinghua University, China
Lingni Ma	Reality Labs at Meta, USA
Eric Brachmann	Niantic, Germany
Bertram Drost	MVTec, Germany
Sindi Shkodrani	Reality Labs at Meta, USA
Rigas Kouskouridas	Scape Technologies, UK
Ales Leonardis	University of Birmingham, UK
Carsten Steger	Technical University of Munich and MVTec, Germany
Vincent Lepetit	École des Ponts ParisTech, France, and TU Graz, Austria
Jiří Matas	Czech Technical University in Prague, Czech Republic

W34 - Drawings and Abstract Imagery: Representation and Analysis

Diane Oyen	Los Alamos National Laboratory, USA
Kushal Kafle	Adobe Research, USA
Michal Kucer	Los Alamos National Laboratory, USA
Pradyumna Reddy	University College London, UK
Cory Scott	University of California, Irvine, USA

W35 - Sign Language Understanding

Liliane Momeni	University of Oxford, UK
Gül Varol	École des Ponts ParisTech, France
Hannah Bull	University of Paris-Saclay, France
Prajwal K. R.	University of Oxford, UK
Neil Fox	University College London, UK
Ben Saunders	University of Surrey, UK
Necati Cihan Camgöz	Meta Reality Labs, Switzerland
Richard Bowden	University of Surrey, UK
Andrew Zisserman	University of Oxford, UK
Bencie Woll	University College London, UK
Sergio Escalera	Universitat Autònoma de Barcelona, Spain
Jose L. Alba-Castro	Universidade de Vigo, Spain
Thomas B. Moeslund	Aalborg University, Denmark
Julio C. S. Jacques Junior	Universitat Autònoma de Barcelona, Spain
Manuel Vázquez Enríquez	Universidade de Vigo, Spain

W36 - A Challenge for Out-of-Distribution Generalization in Computer Vision

Adam Kortylewski	Max Planck Institute for Informatics, Germany
Bingchen Zhao	University of Edinburgh, UK
Jiahao Wang	Max Planck Institute for Informatics, Germany
Shaozuo Yu	The Chinese University of Hong Kong, China
Siwei Yang	Hong Kong University of Science and Technology, China
Dan Hendrycks	University of California, Berkeley, USA
Oliver Zendel	Austrian Institute of Technology, Austria
Dawn Song	University of California, Berkeley, USA
Alan Yuille	Johns Hopkins University, USA

W37 - Vision With Biased or Scarce Data

Kuan-Chuan Peng	Mitsubishi Electric Research Labs, USA
Ziyan Wu	United Imaging Intelligence, USA

W38 - Visual Object Tracking Challenge

Matej Kristan	University of Ljubljana, Slovenia
Aleš Leonardis	University of Birmingham, UK
Jiří Matas	Czech Technical University in Prague, Czech Republic
Hyung Jin Chang	University of Birmingham, UK
Joni-Kristian Kämäräinen	Tampere University, Finland
Roman Pflugfelder	Technical University of Munich, Germany, Technion, Israel, and Austrian Institute of Technology, Austria
Luka Čehovin Zajc	University of Ljubljana, Slovenia
Alan Lukežič	University of Ljubljana, Slovenia
Gustavo Fernández	Austrian Institute of Technology, Austria
Michael Felsberg	Linköping University, Sweden
Martin Danelljan	ETH Zurich, Switzerland

Contents – Part VIII

W36 - A Challenge for Out-of-Distribution Generalization in Computer Vision

W37 - Vision With Biased or Scarce Data

W38 - Visual Object Tracking Challenge

W31 - Challenge on Compositional and Multimodal Perception

W31 - Challenge on Compositional and Multimodal Perception

The International Challenge on Compositional and Multimodal Perception (CAMP) of this ECCV2022 workshop aims at gathering researchers who work on activity/scene recgnition, compositionality, multimodal perception and its applications.

People understand the world by breaking down into parts. Events are perceived as a series of actions, objects are composed of multiple parts, and this sentence can be decomposed into a sequence of words. Although our knowledge representation is naturally compositional, most approaches to computer vision tasks generate representations that are not compositional.

We also understand that people use a variety of sensing modalities. Vision is an essential modality, but it can be noisy and requires a direct line of sight to perceive objects. Other sensors (e.g. audio, smell) can combat these shortcomings. They may allow us to detect otherwise imperceptible information about a scene. Prior workshops focused on multimodal learning have focused primarily on audio, video, and text as sensor modalities, but we found that these sensor modalities may not be inclusive enough. Both these points present interesting components that can add structure to the task of activity/scene recognition yet appear to be underexplored. To help encourage further exploration in these areas, we believe a challenge with each of these aspects is appropriate.

October 2022

Kazuki Kozuka
Zelun Luo
Ehsan Adeli
Ranjay Krishna
Juan Carlos Niebles
Li Fei-Fei

YORO - Lightweight End to End Visual Grounding

Chih-Hui Ho[1]([✉]), Srikar Appalaraju[2], Bhavan Jasani[2], R. Manmatha[2], and Nuno Vasconcelos[1]

[1] UC San Diego, La Jolla, USA
{chh279,nvasconcelos}@ucsd.edu
[2] AWS AI Labs, San Diego, USA
{srikara,bjasani,manmatha}@amazon.com

Abstract. We present YORO - a multi-modal transformer encoder-only architecture for the Visual Grounding (VG) task. This task involves localizing, in an image, an object referred via natural language. Unlike the recent trend in the literature of using multi-stage approaches that sacrifice speed for accuracy, YORO seeks a better trade-off between speed an accuracy by embracing a single-stage design, without CNN backbone. YORO consumes natural language queries, image patches, and learnable detection tokens and predicts coordinates of the referred object, using a single transformer encoder. To assist the alignment between text and visual objects, a novel patch-text alignment loss is proposed. Extensive experiments are conducted on 5 different datasets with ablations on architecture design choices. YORO is shown to support real-time inference and outperform all approaches in this class (single-stage methods) by large margins. It is also the fastest VG model and achieves the best speed/accuracy trade-off in the literature. Code released (Code available at https://github.com/chihhuiho/yoro).

1 Introduction

There has been significant recent interest in Vision-Language (VL) learning and Visual Grounding (VG) [2,10,33,42,45,51,74,76,77,79,81,83,85,87,90]. This aims to localize, in an image, an object referred to by natural language, using a text query (see Fig. 1(a)). VG is potentially useful for many applications, ranging from cloud-based image retrieval from large corpora to resource constrained problems on devices of limited computational capacity. For example, in Fig. 1(a), a robot must use its limited computing resources to resolve a natural language query and locate the desired object. These applications currently pose an extreme challenge to VG, by compounding the difficulty of the task itself with

---

C.-H. Ho—Work done during an internship at Amazon.

Supplementary Information The online version contains supplementary material available at https://doi.org/10.1007/978-3-031-25085-9_1.

L. Karlinsky et al. (Eds.): ECCV 2022 Workshops, LNCS 13808, pp. 3–23, 2023.
https://doi.org/10.1007/978-3-031-25085-9_1

4   C.-H. Ho et al.

the need to solve it efficiently. When this is the case, computational efficiency becomes an unavoidable requirement of architecture design. Several examples exist in the vision literature, where branches of computationally efficient architectures have evolved for problems like object recognition [24] or detection [60], among others. The goal is not necessarily to develop the most powerful method in the literature, but to find the best trade-off [27] between task performance and some variable that reflects computation, e.g. speed or latency. This has led to the introduction of architectures, such as the MobileNet family [23,24,67] for recognition or the YOLO family [3,60–62] for object detection, that have become popular despite their weaker than state of the art accuracy. Given the obvious relationship between object detection and VG, it is not surprising that similar trends are starting to emerge for the latter. Like object detectors, VG methods can be broadly grouped in two categories: single [6,11,40,50,78,79] and multi-stage [7,19,45,69,81] methods. Multi-stage methods typically rely on a pre-trained visual backbone (such as the FasterRCNN [63]) as a visual encoder module. This simplifies the VG problem, reducing it to the selection of the object proposals that best match the query text. Like multi-stage object detectors, they are more accurate than single stage methods, but usually too complex for the applications of Fig. 1(a). In fact, the run-time complexity of many of these VG systems is lower-bounded by the run-time computation of the FasterRCNN [63] proposal stage, which is usually considered too expensive even for object detection, in domains like robotics. In these domains, single-stage detectors such as YOLO [3,60–62] are the architecture of choice. Unsurprisingly, the YOLO family has been the backbone of choice for various single-stage VG systems [6,79]. However, these and the other CNN-based single stage VG systems [6,40,50,78,79] previously proposed in the literature lack the powerful attention mechanisms that enable recent transformer architectures to excel at multi-modal data fusion [11]. This usually results in weak accuracy.

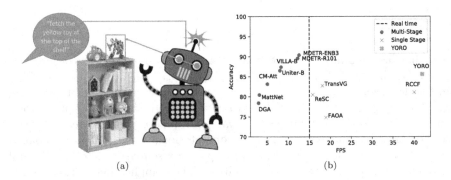

(a)   (b)

Fig. 1. (a) Example of VG with limited computational resources. (b) **Accuracy-Speed plot**: YORO surpasses real time FPS 15 and achieves better trade-off in the literature. (Color figure online)

In this work, we seek to endow single stage VG methods with transformer-style attention, so as to enable a better trade-off between accuracy and speed than currently available methods. For this, we propose an efficient and end-to-end differentiable VG architecture, denoted as *You Only Refer Once* (YORO). YORO is inspired by a transformer architecture, ViT [14], that has achieved success for image classification. ViT has also recently been extended to a transformer architecture, ViLT [33], for vision-language tasks such as visual reasoning [70] or visual question answering [20]. We investigate the more challenging extension of ViLT for tasks that require object localization, such as VG. The main challenge is how to establish explicit connections between words and bounding boxes without the addition of an object detector. Inspired by recent approaches to the design of transformers for object detection [4,17,30], we augment the ViLT architecture with a set of detection tokens that enable this localization. This makes YORO an encoder-only multi-modal transformer architecture that consumes three data streams: referred text tokens, an image in the form of flattened patches, and learnable detection tokens. YORO then uses self-attention to fuse information from all these streams in order to localize the referred object.

Similarly to transformer-based detectors [4,17], YORO is trained by Hungarian matching between the localization hypotheses derived from its tokens and the ground-truth bounding boxes. This combines a bounding box regression loss that minimizes the difference between predicted and ground truth boxes and a classification loss that associates the referred text to the bounding box. The alignment between language and vision representations is further strengthened by an object-text alignment loss that operates in feature space, encouraging the visual features of the predicted object to match the text features of the corresponding text. However, this must be done carefully. In the early stages of training, when object predictions tend to be incorrect, this object-text alignment loss can hamper learning. To both address this issue and encourage more fine-grain alignment, we propose a novel patch-text alignment loss that aligns features of input patches to their corresponding text. Extensive experiments on five datasets show that YORO establishes the new SOTA for single stage VG models. Overall, as illustrated in Fig. 1(b), YORO (red) achieves the best accuracy/speed trade-off among the methods in the literature. It has substantially better accuracy than previous single stage VG methods (green) and is substantially faster than multi-stage VG models (blue). Ablation studies compare different variants of the proposed model and demonstrate the effectiveness of the proposed patch-text alignment loss. With this we summarize our contributions:

- A novel multi-modal visual grounding architecture YORO is proposed using an encoder-only transformer (Sect. 3). YORO is a simple single-stage design and is end-to-end trainable.
- A novel patch-text alignment loss is introduced for enabling fine-grained visual language alignment (Sect. 4).
- YORO achieves the best accuracy/speed trade-off in the literature, outperforming existing real-time VG methods on five datasets (Sect. 5).

2   Related Work

Prior work in visual grounding (VG) may be mainly classified into multi-stage and single-stage approaches based on the design of the visual branch.

Multi-stage VG approaches [7,9,15,19,48,49,81,84] usually leverage a region proposal network (RPN) [63] to handle the selection of candidate object locations. In a first-stage, the RPN proposes a set of object candidates. In a second-stage, the VG model selects the region proposal that best matches the query text. This reduces VG to a retrieval problem [18,25,26,65], where the model only needs to search within the proposed object candidates. Similarly to two-stage object detectors [29,44], these methods are powerful for VG. However, like two-stage detectors, they tend to be computationally heavy, making them impractical for real-time operation on complexity constrained devices, as illustrated in Fig. 1. When compared to the object detection literature [29,44], the optimization of the trade-off between speed and performance has received much less attention in VG. On the contrary, recent VG methods [9,30] pursue higher accuracy by adopting more powerful language models [46,57] and encoder-decoders transformer architectures [37,55,57,58,72,86,88] with more parameters and slower speeds. Instead, YORO pursues a better trade-off between accuracy and speed. In this context, it is not clear that a multi-stage architecture is the best solution. Since there is no way to recover objects missed by the RPN, strong performance can require a large number of proposals [52,85], which is inherently expensive. In fact, the use of a preliminary object selection prevents the exploitation of contextual clues that may be available in the query text (e.g. the component "on the top of the shelf" of the query of Fig. 1) and could be critical for solving the task with less proposals. The integration of language and vision is, after all, the difference between object detection and VG. This suggests that single stage architectures, like YORO, could be more effective when computation has a cost. Our experimental results support this hypothesis.

Single-stage VG approaches [1,6,11,38,40,50,78,79] achieve greater inference speeds by discarding the proposal generation stage. Most of these methods are based on CNN object detectors, namely the single-stage detector YOLOV3 [62]. This is, for example, adopted to extract visual features by SSG [6] and FAOA [79]. These features are then fused with text features to regress the bounding boxes. ReSC [78] improves FAOA [79] with a recursive sub-query construction module. MCN [50] further extends YOLOV3 [62] for the joint solution of VG and segmentation, using multitask losses. These models lack the powerful attention mechanisms of the transformer architecture, which are critical for the integration of information both spatially, across image regions, and modally, across the vision and language information streams. When compared to them, YORO has the benefit of leveraging the transformer to implement this powerful integration of information. This enables substantially higher accuracy than previous single-stage approaches. Recently, TransVG [11] has also adopted a transformer model [72] to fuse language and vision features. While TransVG does not use a two-stage object detector, it uses two stages of transformers. The vision and language streams are first processed by independent transformers, whose outputs are then fed to a multi-modal transformer that integrates the two modalities. This makes it substantially slower than YORO, which performs all operations with a single transformer, and could give rise to loss of information needed to reason jointly about vision and language, as discussed above for the RPN. The fact that TransVG is both slower and less accurate than YOLO suggests that there is no benefit to the two transformer stage architecture.

Vision Transformer. YORO is based on the Vision transformer (ViT), introduced in [14] for image classification by direct consumption of image patches, i.e. without a preliminary CNN backbone. Beyond classification [14,56,71], the vision transformer has been applied to detection [4,17,89] and segmentation [59,68], as discussed in the recent surveys of [21,32]. Recently, ViLT [33] has extended the ViT for various visual-language tasks that do not require object localization, such as visual question answering. The extension to tasks, such as VG, that require object localization is non-trivial. This is due to the need to establish explicit connections between words and bounding boxes, which requires much more fine-grained understanding of both language and images. YORO addresses this limitation by the addition of learnable detection tokens and Hungarian matching between token-based object predictions and ground truth bounding boxes, a component of modern transformer-based object detectors [4,17,89]. To the best of our knowledge, it is the first architecture to perform VG tasks with an encoder-only transformer without a visual backbone. This is critical for speed. YORO is shown to achieve the best trade-off between speed and accuracy in the VG literature, as illustrated in Fig. 1(b).

3   YORO Architecture

The YORO architecture is summarized in Fig. 2. The input query sentence and image are first converted into text and visual embeddings respectively. These embeddings are augmented by a set of learnable detection tokens and fed into a transformer encoder. Within the transformer, learning happens via self-attention [72]. The transformer predicted features corresponding to the detection tokens are finally fed into several heads, implemented with a multi-layer perceptron, which produce class and bounding box predictions. The model is trained end-to-end, using a combination of losses. Each model component is discussed next.

3.1   Multi-modal Inputs

YORO consumes multi-modal inputs: language, vision and detection tokens.

The Language Modality is depicted in the blue block of Fig. 2. The referred text is converted into sequences of tokens with a BERT tokenizer [12]. Assume the sequence has length m and a dictionary size of v for the pre-trained tokenizer. Each text token $t_i \in \mathbb{R}^v$ in the sequence $\{t_1; \ldots; t_m\}$ is then projected into a d dimensional feature vector $f_i^l = \mathcal{W}^l t_i$, where $\mathcal{W}^l \in \mathbb{R}^{d \times v}$ is a projection layer.

The m projected token vectors are then augmented with a text classification token $f_{cls}^l \in \mathbb{R}^d$. Similar to [33], each of these tokens is summed with a text positional embedding $f_{pos}^l \in \mathbb{R}^{(m+1) \times d}$ and a modality type embedding $f_{type}^l \in \mathbb{R}^{(m+1) \times d}$. The positional embedding specifies the location of each token, while the modality type embedding tells apart the text input from visual input. The overall text input embedding may be written as

$$f^l = [f_{cls}^l; f_1^l; \ldots; f_m^l] + f_{pos}^l + f_{type}^l. \tag{1}$$

The Vision Modality is shown in the green block of Fig. 2. The input image $I \in \mathbb{R}^{H \times W \times 3}$ is partitioned into n visual patches $\{p_1, \ldots, p_n\}$ of size s, where $p_i \in \mathbb{R}^{s^2 \times 3}$ and $n = \frac{HW}{s^2}$. Each visual patch is then projected, with a linear layer of weight matrix $\mathcal{W}^v \in \mathbb{R}^{d \times (s^2 \times 3)}$, into a d dimensional visual patch feature $f_i^v = \mathcal{W}^v p_i$. Similar to the language input, the n visual patch features are augmented with a visual classification token $f_{cls}^v \in \mathbb{R}^d$ and summed with a visual positional embedding $f_{pos}^v \in \mathbb{R}^{(n+1) \times d}$ and a modality type embedding $f_{type}^v \in \mathbb{R}^{(n+1) \times d}$, to produce an overall visual input embedding

$$f^v = [f_{cls}^v; f_1^v; \ldots; f_n^v] + f_{pos}^v + f_{type}^v. \tag{2}$$

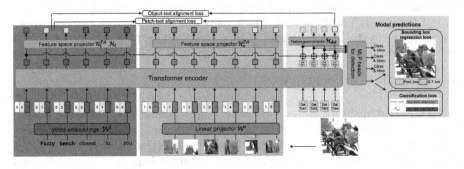

Fig. 2. YORO architecture. Blue, green, yellow and pink blocks represent language, vision, detection and prediction branches respectively. YORO does **not** use a large pre-trained visual backbone. Input image is divided into patches. Those of IOU ≥ 0.5 with ground truth box (red) are marked in light green. (Color figure online)

Learnable Detection Tokens. The vision and language models above are similar to those implemented by ViT [14] and ViLT [33]. These models were proposed for tasks, such as image classification or visual question-answering that do not require object localization. In this case, the use of holistic classifications tokens f_{cls}^l, f_{cls}^v is sufficient to encode the image identity. However, our experiments (see ablations in Sect. 5.3) show that this is not the case for VG, which depends critically on object localization. In the object detection literature, this problem is addressed by introducing learnable detection tokens, which are individually associated with bounding boxes [4]. YORO leverages such tokens, as shown in the yellow block of Fig. 2. The text f^l and visual f^v embeddings are complemented by q learnable detection tokens $f^{det} = [f_1^{det}; \ldots; f_q^{det}]$. Each token represents an object and different tokens may represent objects of different sizes.

3.2 Transformer Encoder

YORO transformer encoder consists of D stacked layers, each including a multi-headed self-attention (MSA) network and a feed forward network (FFN). Denoting x_d as the input of layer d, the transformer output is computed as follows:

$$x_0 = [f^l; f^v; f^{det}] \tag{3}$$

$$x_d' = x_{d-1} + \text{MSA}(\text{LN}(x_{d-1})), \quad d = 1 \ldots D \tag{4}$$

$$x_d = x_d' + \text{FFN}(\text{LN}(x_d'))), \quad d = 1 \ldots D \tag{5}$$

where LN denotes layer normalization [72]. D is set to 12 in our experiments.

3.3 Multi-modal Transformer Outputs

The transformer input $[f^l; f^v; f^{det}]$ is a sequence of length $n + m + q + 2$. The output is a sequence $[o^l; o^v; o^{det}]$ of the same length, where $o^l = [o_{cls}^l; o_1^l; \ldots; o_m^l]$, $o^v = [o_{cls}^v; o_1^v; \ldots; o_n^v]$ and $o^{det} = [o_1^{det}; \ldots; o_q^{det}]$, are the outputs of the language, vision and detection branches respectively. The transformed text class token o_{cls}^l is added to the transformed detection tokens o_i^{det} before being passed to the detection heads. This strengthens the dependence of the bounding box predictions on the input text, beyond what would be possible by the use of attention alone.

3.4 Feature Projector and Detection Heads

To optimize YORO with the proposed losses, several modules are added to the transformer output, including a detection head for the bounding box and classification loss (red block in Fig. 2), and a projection head to enable the object-text alignment and the patch-text alignment loss discussed in Sect. 4. All these heads are stacked fully connected layers. We next discuss the details of all losses.

4 Training Losses

YORO is trained with 4 different losses: (1) a bounding box regression loss to regress bounding box coordinates, (2) a classification loss to classify which text tokens the bounding box is associated to, (3) an object-text alignment loss that aligns detection token features with corresponding text features and (4) a novel patch-text alignment loss that aligns image patches and text tokens. During training, Hungarian matching [35] computes the optimal bipartite matching between predicted and ground truth boxes. During inference, the bounding box of highest classification score is selected.

Bounding Box Regression Loss. Two losses are used for bounding box regression: the L_1 and the generalized IoU loss [64] L_{giou}. These losses are applied to ground truth bounding box $B_{g.t.} \in \mathbb{R}^4$ and predicted box $B_{pred} \in \mathbb{R}^4$ pairs. This results in bounding box regression loss

$$L^{bbox} = \lambda_1 L_1(B_{pred}, B_{g.t.}) + \lambda_2 L_{giou}(B_{pred}, B_{g.t.}), \qquad (6)$$

where λ_1 and λ_2 are experimentally set to 2 and 5 respectively.

Classification Loss. Unlike the classification losses of object detection, where a single object class is predicted, YORO predicts the set of language tokens to be associated with a bounding box. This is needed to encourage the association of the text phrase with the corresponding objects. For example, in Fig. 2 the red box in the image is associated with the words "fuzzy" and "bench" with probabilities 0.5 and 0.5, respectively. This is implemented by generating predictions with a softmax output of dimension equal to the maximum token length and using a soft cross entropy loss L_{sce} based on the ground-truth text/object assignments. This is the classification loss $L^{cls} = L_{sce}(P_{pred}, P_{g.t.})$, where P_{pred} is the predicted probability distribution and $P_{g.t.}$ the ground truth distribution.

Object-Text Alignment Loss. Inspired by recent uses of region-text alignment for fusing words and image regions [5,7,13,30,36,39], YORO further aligns the language transformer output o^l and the detection transformer output o^{det} in a suitable feature space. o^l and o^{det} are first mapped to a common feature space $\mathcal{F}$, with projection heads $\mathcal{H}_l$ and $\mathcal{H}_{det}$, which implement the mappings $\hat{o}^l = \mathcal{H}_l(o^l)$ and $\hat{o}^{det} = \mathcal{H}_{det}(o^{det})$, where $\mathcal{H}_{det}, \mathcal{H}_{det}$ are linear projections. Consider an image with g ground truth bounding boxes and let $\mathcal{T}_i \subseteq \{j : 1 \le j \le m\}$, where m is the length of the language token sequence, be the subset of language token indices that a projected detection feature $\hat{o}_i^{det}$ should be aligned to. This alignment is encouraged with the object-to-text alignment (OTA) loss

$$L^{OTA} = \sum_{i=1}^{g} \frac{1}{|\mathcal{T}_i|} \sum_{j \in \mathcal{T}_i} -\log \left(\frac{\exp(\hat{o}_i^{det\mathsf{T}} \hat{o}_j^l / \tau)}{\sum_{k=1}^{m} \exp(\hat{o}_i^{det\mathsf{T}} \hat{o}_k^l / \tau)} \right), \qquad (7)$$

where τ is experimentally set to 0.07. Consider Fig. 2 and assume, for example, that detection token 4 is selected to match the groundtruth (red) bounding box by the Hungarian matching algorithm. In this case, L^{OTA} encourages the embeddings $\hat{o}_0^l$ and $\hat{o}_1^l$ of the words "fuzzy" and "bench" to be closer to the embedding of $\hat{o}_4^{det}$ than those of the words "closest," "to," and "you".

Similarly, let $\mathcal{O}_i \subseteq \{j : 1 \le j \le g\}$ be the set of object indices that a projected language token feature $\hat{o}_i^l$ should be aligned to. This alignment is encouraged by the text-object alignment loss L^{TOA}

$$L^{TOA} = \sum_{i=1}^{m} \frac{1}{|\mathcal{O}_i|} \sum_{j \in \mathcal{O}_i} -\log \left(\frac{\exp(\hat{o}_i^{l\mathsf{T}} \hat{o}_j^{det} / \tau)}{\sum_{k=1}^{g} \exp(\hat{o}_i^{l\mathsf{T}} \hat{o}_k^{det} / \tau)} \right), \qquad (8)$$

which is a symmetric version of L^{OTA}. The overall object alignment loss is $L^{OA} = \frac{1}{2}L^{OTA} + \frac{1}{2}L^{TOA}$.

Patch-Text Alignment Loss. While the object-text alignment loss is desirable to encourage feature space alignment between *predicted* boxes and corresponding text tokens, this assumes that the predicted boxes are correct. However, as illustrated in Fig. 3, this assumption is frequently violated during training, especially in the early epochs. In this case, the object-text alignment loss can provide ambiguous supervision (e.g. supervision from background instead of the object in the early epochs of Fig. 3). To address this limitation, we propose to align the features extracted from the *ground truth patches at the input of the network* (green patches of Fig. 2) with the features of the associated text tokens. More specifically, the indices of input patches (p_5 and p_6 in Fig. 2) that have at least 0.5 IOU with the ground truth box (red box in Fig. 2) are first identified and the corresponding features aligned with those of the ground truth language tokens ("fuzzy" and "bench" in Fig. 2). As shown in Fig. 3, while each patch feature may not provide full information about the object, the patch features never provide contradictory supervision. This helps stabilize the learning.

Epoch 0   Epoch 5   Epoch 20

Fig. 3. Visualization of predicted box through training epochs. Red, green and blue boxes indicate ground truth box, ground truth patch and predicted box respectively. Patch-Text Alignment loss helps localization. Best viewed in color digitally. (Color figure online)

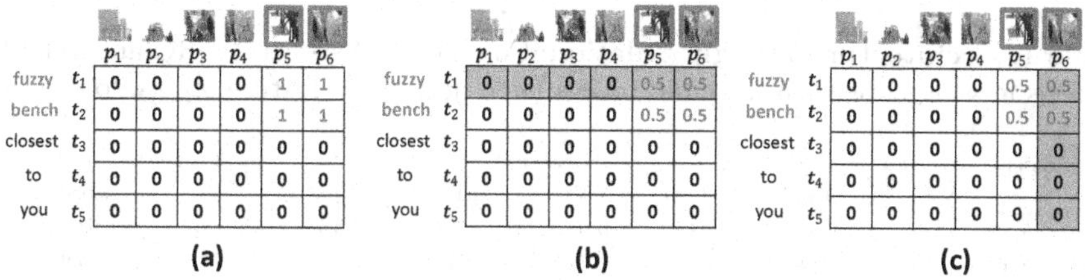

(a)   **(b)**   **(c)**

Fig. 4. Patch-Text Alignment Loss: (a) Ground truth for patch-text alignment. (b) Column wise normalization (c) Row wise normalization.

To align text and patch features, the text transformer output o^l and vision transformer outputs o^v are first projected to a shared feature space using the patch alignment projection heads $\mathcal{H}_l^{PA}$ and $\mathcal{H}_v^{PA}$ respectively, which leads to $\tilde{o}^l$ and $\tilde{o}^v$. Given m text tokens and n visual patches, a ground truth table $A \in \mathbb{R}^{m \times n}$ is then created, where A_{ij} is 1 if text token i and visual patch j should be aligned and 0 otherwise, as shown in Fig. 4(a). This matrix is column normalized into matrix A^C (Fig. 4(b)) using

$$A_{ij}^c = \begin{cases} \frac{A_{ij}}{\sum_i A_{ij}} & \text{if } i \in \mathcal{S}^{TPA} \\ 0 & \text{otherwise} \end{cases}, \qquad (9)$$

where $\mathcal{S}^{TPA} = \{i | \sum_i A_{ij} > 0\}$ is the set of token indices associated with at least 1 patch. For example, $\mathcal{S}^{TPA} = \{1, 2\}$ for Fig. 4. The row-wise normalization of A into matrix A^r is defined in a similar fashion, as shown in Fig. 4(c).

To encourage the fine-grained alignment between patches and text, a text-patch alignment loss L^{TPA} is defined as

$$L^{TPA} = \sum_i \sum_j p_{ij}^c ln \frac{p_{ij}^c}{A_{ij}^c} \quad \forall i \in \mathcal{S}^{TPA}, \qquad (10)$$

where $p_{ij}^c = \frac{exp(\tilde{o}_i^{l\mathsf{T}} \tilde{o}_j^v)/\tau}{\sum_k exp(\tilde{o}_i^{l\mathsf{T}} \tilde{o}_k^v)/\tau}$ is a contrastive distribution defined over the language and visual embedding. For example, the word "fuzzy" (t_1) of Fig. 4(b) should be aligned with patch p_5 and p_6 according to the ground truth distribution $A_1^c = [0; 0; 0; 0; 0.5; 0.5]$. (10) minimizes the KL divergence between this and the predicted distribution $p_1^c = [p_{11}^c; \ldots; p_{1n}^c]$. This encourages the alignment of the feature vector $\tilde{o}_1^l$ extracted from word t_1 with the feature vectors $\tilde{o}_5^v$ and $\tilde{o}_6^v$ extracted from patches p_5 and p_6.

Similar to the L^{TPA} of (10), the patch-text alignment loss L^{PTA} aligns each patch to its associated tokens. As shown in Fig. 4(c), p_6 should be aligned to t_1 and t_2. The final patch alignment loss L^{PA} is $L^{PA} = \frac{1}{2}L^{TPA} + \frac{1}{2}L^{PTA}$.

Combined Loss. The final loss combines the bounding box regression loss, the classification loss, the object alignment loss and the patch alignment loss with $L = L^{bbox} + L^{cls} + L^{OA} + L^{PA}$.

Table 1. Architecture Design Ablation: Comparison between YORO and ablated versions missing either learnable *det* tokens (w/o det) or *cls* feature o_{cls}^l (w/o cls).

Model variant	Refcoco			Refcoco+			Refcocog	
	Val	TestA	TestB	Val	TestA	TestB	Val	Test
YORO w/o det.	74.9 (−8.0)	79.4 (−6.2)	68.8 (−8.6)	66.6 (−6.9)	73.2 (−5.4)	59.2 (−5.7)	69.7 (−3.7)	68.9 (−5.4)
YORO w/o cls.	82.4 (−0.5)	84.8 (−0.8)	76.7 (−0.7)	72.8 (−0.7)	77.7 (−0.9)	64.0 (−0.9)	72.4 (−1.0)	73.9 (−0.4)
YORO	82.9	85.6	77.4	73.5	78.6	64.9	73.4	74.3

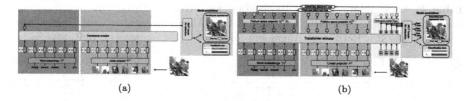

Fig. 5. (a) YORO w/o det. token. (b) YORO w/o cls. feature o_{cls}^l.

5   Experiments

5.1   Dataset

YORO is evaluated on five VG datasets using the accuracy metric, where the predicted box is correct when the IOU with the ground truth box is at least 0.5.

RefCoco/RefCoco+/RefCocog [31,82] are curated from MSCOCO [41]. Ref-COCO consists of 142,209 referred expressions for 50,000 objects in 19,994 images. We adopt the split of [7,10,43,69,74,80,87] into train, val, testA, and testB datasets with 120,624, 10,834, 5,657 and 5,095 expressions respectively. RefCOCO+ contains 141,564 expressions for 49,856 objects in 19,992 images. We use the split of [6,19,42,50,80,90], where train, val, testA and testB datasets contain 120,191, 10,758, 5,726 and 4,889 image text pairs, respectively. RefCOCOg consists of 85,474 image text pairs of 54,822 objects across 26,711 images. We adopt the UMD split of [42,45,50,77,81], into 80,512, 4,896 and 9,602 pairs for train, val and test dataset, respectively.

ReferItGame [31] contains 19,997 images from the SAIAPR-12 [16] dataset. It has 130,363 text expressions for 99,296 objects. We follow the split of [25,51,73, 81,87] for training, testing and validation set.

CopsRef [8], derived from GQA [28], contains 508 object classes and average expression length 14.4. When compared to RefCOCO and RefCOCOg (80 object classes and average length around 6), CopsRef has longer expression and more diverse object classes. We adopt the `WithoutDist` version of the dataset [8,15], where train, val and test contains 119,628, 12,586 and 16,524 pairs respectively.

5.2   Training and Evaluation Details

We summarize some implementation details for reproducibility. YORO uses `bert-base-uncase` [12] pre-trained tokenizer from HuggingFace [75] with a maximum text length of 40. The pre-trained checkpoint from [33] is used to obtain initial weights. Five detection tokens are used (see Table 3 for a token number ablation) and randomly initialized using a normal distribution of zero mean and standard deviation 0.02. The AdamW optimizer [47] with initial learning rate of 10^{-4} and weight decay of 10^{-2} is adopted. This warm up setting is used for the first 10% training epochs and the learning rate is then linearly decayed

Table 2. Loss Ablation. CL: Classification loss, RE: Bounding box regression loss, OA: Object-text Alignment Loss and PA: Patch-text Alignment loss.

Loss				CopRef		ReferItGame	
CL	RE	OA	PA	Val.	Test	Val.	Test
✓	✓			67.05	70.78	71.55	69.86
✓	✓	✓		67.73 (+0.68)	71.0 (+0.22)	71.89 (+0.34)	70.39 (+0.53)
✓	✓	✓	✓	**68.08** (+1.03)	**71.3** (+0.52)	**72.67**(+1.12)	**71.9** (+2.04)

Table 3. Ablation of number of detection tokens on RefCoco+.

Det. Tok. #	RefCoco+		
	Val.	TestA	TestB
1	73.2	**79.1**	**64.9**
5	**73.5**	78.6	**64.9**
10	72.8	78.78	62.8

to 0. YORO is pre-trained on the concatenated detection dataset curated by [30] for 40 epochs. Since the VG task contains a single referred object, image-text pairs with a single bounding box are sampled. This pre-training allows a good initialization of the detection branch. The pre-trained model is fine-tuned on downstream datasets for 40 epochs. All experiments are conducted using PyTorch [53] with batch size 128. No VL augmentation like MixGen [22] was used.

5.3 Ablation Study

Architecture Design. The performance of YORO is compared to the two variants of Fig. 5 on RefCoco/RefCoco+/RefCocog, with the results of Table 1.

YORO w/o Det. Token. Figure 5(a) shows a simple extension of the ViLT [33] architecture to VG. In this case, the classification feature of the text modality o_{cls}^l (See Sect. 3.3) is forwarded through the detection head to regress a bounding box, without the use of detection branch or any detection tokens. The first row of Table 1 shows that, when compared to YORO, this is between 3.7% and 8.6% worse. For almost all splits, the performance loss is larger than 5%, demonstrating a clear benefit in using detection tokens for the VG task.

YORO w/o Cls. Feature o_{cls}^l. Conversely, the use of detection tokens without the holistic classification feature o_{cls}^l is implemented with the architecture of Fig. 5(b). This performs slightly worse (0.4% to 1%) than YORO. The holistic feature contributes to YORO's performance but is much less critical than the detection tokens. Altogether, these ablations support the proposed design.

Loss Function. Table 2 studies the importance of the various losses in Sect. 4. The baseline consists of classification (CL) and bounding-box regression (RE) losses. Adding object-text (OA) alignment loss improves accuracy by +0.22% and +0.53%. When the PA loss is optimized, the gain over baseline increases to +0.52% and +2.04% on the test set of CopsRef and ReferItGame respectively. This indicates both OA and PA loss improve detection accuracy on multiple datasets. The improved performance of PA validates the importance of accounting for the inaccuracy of bounding box predictions in the early training epochs.

Number of Detection Tokens. Table 3 presents an ablation of the number of detection tokens. On average, using 1 or 5 tokens has comparable results, with a

Table 4. Refcoco/Refcoco+/Refcocog. Compared with single-stage models, YORO achieves the best accuracy. YORO is one of the smallest models and is also the fastest.

Method	Backbone		Refcoco			Refcoco+			Refcocog		Param.	FPS
	Lang.	Visual	Val	TestA	TestB	Val	TestA	TestB	Val	Test	(M)	
FAOA [79]	Bert	Darknet53	72.05	74.81	67.91	-	-	-	-	-	182.9	18.9
RCCF [40]	BiLSTM	DLA34	-	81.06	71.85	-	70.35	56.32	-	-	-	40
SSG [6]	BiLSTM	Darknet53	-	76.51	67.50	-	62.14	49.27	58.80	-	-	-
MCN [50]	BiGRU	Darknet53	80.08	82.29	74.98	67.16	72.86	57.31	66.46	66.01	-	-
ReSC [78]	Bert	Darknet53	76.59	78.22	73.25	63.23	66.64	55.53	64.87	64.87	179.9	15.8
TransVG [11]	Bert	Res101	81.02	82.72	**78.35**	64.82	70.70	56.94	68.67	67.73	149.7	18.1
YORO	Bert	Linear	**82.9**	**85.6**	77.4	**73.5**	**78.6**	**64.9**	**73.4**	**74.3**	114.3	41.9

Table 5. Compared to single/two-stage methods, YORO achieves SOTA on **ReferItGame** [31].

Two-stage	Acc.	Single-stage (*)	Acc.
CMN [25]	28.33	RCCF* [40]	63.79
VC [87]	31.13	SSG* [6]	54.24
Luo [51]	31.85	ReSC-B* [78]	64.33
MAttNet [81]	29.04	ReSC-L* [78]	64.60
Sim. Net [73]	34.54	FAOA* [79]	59.30
CITE [54]	35.07	ZSGNet* [66]	58.63
PIRC [34]	59.13	TransVG* [11]	70.73
DDPN [85]	63.00	**YORO***	**71.90**

Table 6. YORO achieves SOTA on **CopsRef** [8] w/o using g.t. box.

Method (* indicates single stage)	Proposal	Acc.
GroundeR [65]	g.t.	75.7
Shuffle-GroundeR	g.t.	58.5
Obj-Attr-GroundeR	g.t.	68.8
MattNet [81]	g.t.	77.9
CM-Att-Erase [45]	g.t.	80.4
MattNet-Mine [8]	g.t.	78.4
VGTR-ResNet50 [15]	Pred.	66.73
VGTR-ResNet101 [15]	Pred.	67.75
ReSC-B* [78]	Pred.	64.49
ReSC-L* [78]	Pred.	65.32
YORO*	Pred.	**71.3**

slight decrease for 10 tokens. Since the VG task only contains a single referent, the lack of benefit in using a large number of detection tokens is sensible.

5.4 Quantitative Comparison

RefCoco/RefCoco+/RefCocog. Table 4 summarizes performance of single-stage methods on RefCoco, RefCoco+ and RefCocog for different splits (val, testA and testB). These are the methods that achieve real time speed (FPS > 15) and thus directly comparable to YORO. YORO outperforms all single stage methods on seven splits (between abs +1.88% and +8.6%). The only exception is Refcoco testB (−0.95% lower than TransVG [11]). The most competitive method in this class is TransVG, which has 1.3× the size, is 2.3× slower than YORO, and achieves significantly lower accuracies on most splits. The only method of speed comparable to YORO (RCCF) has significantly lower accuracy on all splits (up to a 14 point drop on Refcocog Val). A more extensive comparison, including the much heavier two stage models, is presented in the appendix. Overall, as summarized in Fig. 1(b) and Fig. 6, YORO has a significantly better trade-off between speed, parameter size and accuracy than all methods in the literature.

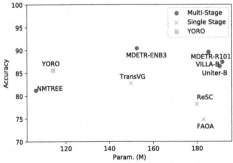

(a) big cow   (b) blue colored mattress

Fig. 6. YORO beats the prior Pareto front for the memory cost/accuracy trade-off.

Fig. 7. Top: Predicted (blue) and ground truth (red) box. Bottom: Attention from the selected head. (Color figure online)

ReferItGame. Table 5 compares YORO to both two-stage (left) and single-stage (right) baselines. YORO outperforms the best two-stage (DDPN [85]) and single-stage (TransVG [11]) method by 8.9% and 1.2% respectively. For roughly the same inference speed, YORO also beats RCCF [40] by 8.1%. Note that YORO achieves SOTA on ReferItGame for both speed and performance.

CopsRef. Table 6 summarizes the accuracy of YORO and the baselines on CopsRef [8]. The lower part of the table contains baselines that either detect the referent or select the referent from object proposals. YORO outperforms the previous SOTA by 3.4%, without using proposals. The upper part of the table refers to models that use *ground truth* bounding boxes. YORO fares well even under this unfair comparison, outperforming some of these baselines. Overall, these results demonstrate that YORO can generalize to VG datasets containing more object classes and longer input text.

5.5 Comparison of Trade-Offs Between Size, Speed, and Accuracy

To quantify the efficiency of YORO, we compare YORO with other VG baselines in terms of parameter size and inference speed.

Inference speed (FPS) is measured by forwarding 100 images (batch size 1) from the Refcoco testA dataset through the VG model. The results are shown in the rightmost column of Table 4 and Fig. 1(b). All FPS measurements besides RCCF [40], MattNet [81] and DGA [77] were conducted by ourselves, using a single Titan Xp GPU and Intel Xeon CPU E5-2630 v4 @ 2.20G[1]. The FPS of [40,77,81] are copied from [40], which measures speed on a Titan Xp GPU (identical to ours) and Intel Xeon CPU E5-2680v4 @ 2.4G (superior to ours).

---

[1] Note that [79] reported a FPS of 26.3 for FAOA, using an NVIDIA 1080TI and Intel Core i9-9900K @ 3.60G set-up, while we measured a FPS of 18.9 under our set-up.

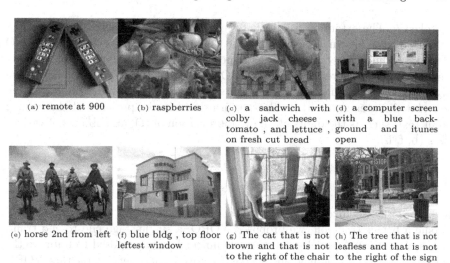

Fig. 8. YORO Qualitative: Visualization of the predicted box (**blue**) and ground truth box (red) in Refcoco+ (a–b), Refcocog (c–d), ReferItGame (e–f) and CopsRef (g–h). The input text for the referent is listed below each image. (Color figure online)

YORO achieves 41.9 FPS and is the fastest VG model in the literature. It is between 3.3× and 14× faster than multi-stage models and at least twice as fast than all single-stage models other than RCCF, which has much weaker accuracy.

Figure 1(b) summarizes the trade-off between speed and accuracy of several methods on Refcoco testA. Single-stage methods tend to have significantly lower accuracy than multi-stage methods. The major exception is YORO, which is substantially more accurate than the other single-stage methods, bridging a significant portion of the accuracy gap between the two types of models. Note that YORO is well above the line connecting MDETR and RCCF, which determines the Pareto front for the VG problem. It also has a large gain in inference speed (close to 3.5× faster) over MDETR. More plots are shown in the supplemental.

We also breakdown the percentage of inference time consumed per YOLO module. Multi-modal VG inputs (Sect. 3.1), transformer encoder (Sect. 3.2) and detection head (Sect. 3.4) consume 33.5%, 62.8% and 3.7% of the inference time, respectively. This suggests that future research should focus on optimizing the design of the transformer layers for both speed and accuracy.

Parameter sizes are compared in the second column from the right of Table 4 and Fig. 6. YORO is smaller than all baselines besides NMTREE [42], which uses a LSTM based language model. Note that, in particular, YORO is the smallest of the single-stage methods (See green × in Fig. 6).

5.6 Qualitative Results

Figure 7 shows the patches of higher attention weight for two example images, computed as follows. Let the detection token corresponding to the predicted box be token k. The attention weight between each patch and detection token k is converted to an heatmap and overlaid on the input image. It is clear that YORO attends to the referred objects. See appendix for other examples. Figure 8 and the appendix show predictions by YORO on all 5 datasets. YORO can perform high quality detection not only on short sentences (a–b), but also on longer and convoluted sentences (c–h). It also performs well in the presence of challenging input text, such as the digits of (a, e), abbreviation of (f), and objects of various sizes (b, f, h).

6   Conclusion

In this work, we investigate the speed accuracy trade-off of VG models. To pursue a better trade-off, we proposed YORO, a single-stage end-to-end trainable architecture. A novel patch-text alignment loss is proposed to improve the alignment between language and image features. Experiments show that YORO outperforms existing single-stage methods. It is the fastest VG and achieves the best trade-off between size, speed, and accuracy in the literature. We believe this is of interest for many resource constrained VG applications.

References

1. Appalaraju, S., Jasani, B., Kota, B.U., Xie, Y., Manmatha, R.: DocFormer: end-to-end transformer for document understanding. In: Proceedings of the IEEE/CVF International Conference on Computer Vision (ICCV), pp. 993–1003, October 2021
2. Biten, A.F., Litman, R., Xie, Y., Appalaraju, S., Manmatha, R.: LaTr: layout-aware transformer for scene-text VQA. In: Proceedings of the IEEE/CVF Conference on Computer Vision and Pattern Recognition (CVPR), pp. 16548–16558, June 2022
3. Bochkovskiy, A., Wang, C.Y., Liao, H.Y.M.: YOLOv4: optimal speed and accuracy of object detection. arXiv preprint arXiv:2004.10934 (2020)
4. Carion, N., Massa, F., Synnaeve, G., Usunier, N., Kirillov, A., Zagoruyko, S.: End-to-end object detection with transformers. ArXiv abs/2005.12872 (2020)
5. Chen, T., Deng, J., Luo, J.: Adaptive offline quintuplet loss for image-text matching. arXiv preprint arXiv:2003.03669 (2020)
6. Chen, X., Ma, L., Chen, J., Jie, Z., Liu, W., Luo, J.: Real-time referring expression comprehension by single-stage grounding network. ArXiv abs/1812.03426 (2018)
7. Chen, Y.-C., et al.: UNITER: UNiversal image-TExt representation learning. In: Vedaldi, A., Bischof, H., Brox, T., Frahm, J.-M. (eds.) ECCV 2020. LNCS, vol. 12375, pp. 104–120. Springer, Cham (2020). https://doi.org/10.1007/978-3-030-58577-8_7
8. Chen, Z., Wang, P., Ma, L., Wong, K.Y.K., Wu, Q.: Cops-Ref: a new dataset and task on compositional referring expression comprehension. In: Proceedings of the IEEE/CVF Conference on Computer Vision and Pattern Recognition (CVPR), June 2020
9. Cho, J., Lei, J., Tan, H., Bansal, M.: Unifying vision-and-language tasks via text generation. In: ICML (2021)

10. Deng, C., Wu, Q., Wu, Q., Hu, F., Lyu, F., Tan, M.: Visual grounding via accumulated attention. In: Proceedings of the IEEE Conference on Computer Vision and Pattern Recognition (CVPR), June 2018
11. Deng, J., Yang, Z., Chen, T., Zhou, W., Li, H.: TransVG: end-to-end visual grounding with transformers. In: Proceedings of the IEEE/CVF International Conference on Computer Vision (ICCV), pp. 1769–1779, October 2021
12. Devlin, J., Chang, M.W., Lee, K., Toutanova, K.: BERT: pre-training of deep bidirectional transformers for language understanding. In: NAACL (2019)
13. Diao, H., Zhang, Y., Ma, L., Lu, H.: Similarity reasoning and filtration for image-text matching. In: AAAI (2021)
14. Dosovitskiy, A., et al.: An image is worth 16×16 words: transformers for image recognition at scale. In: International Conference on Learning Representations (2021). https://openreview.net/forum?id=YicbFdNTTy
15. Du, Y., Fu, Z., Liu, Q., Wang, Y.: Visual grounding with transformers. CoRR abs/2105.04281 (2021). https://arxiv.org/abs/2105.04281
16. Escalante, H.J., et al.: The segmented and annotated IAPR TC-12 benchmark. Comput. Vis. Image Underst. **114**(4), 419–428 (2010). https://doi.org/10.1016/j.cviu.2009.03.008
17. Fang, Y., et al.: You only look at one sequence: rethinking transformer in vision through object detection. arXiv preprint arXiv:2106.00666 (2021)
18. Fukui, A., Park, D.H., Yang, D., Rohrbach, A., Darrell, T., Rohrbach, M.: Multimodal compact bilinear pooling for visual question answering and visual grounding. In: Proceedings of the 2016 Conference on Empirical Methods in Natural Language Processing, Austin, Texas, pp. 457–468. Association for Computational Linguistics, November 2016. https://doi.org/10.18653/v1/D16-1044. https://aclanthology.org/D16-1044
19. Gan, Z., Chen, Y.C., Li, L., Zhu, C., Cheng, Y., Liu, J.: Large-scale adversarial training for vision-and-language representation learning. In: NeurIPS (2020)
20. Goyal, Y., Khot, T., Summers-Stay, D., Batra, D., Parikh, D.: Making the V in VQA matter: elevating the role of image understanding in visual question answering. In: 2017 IEEE Conference on Computer Vision and Pattern Recognition (CVPR), pp. 6325–6334 (2017)
21. Han, K., et al.: A survey on visual transformer. ArXiv abs/2012.12556 (2020)
22. Hao, X., et al.: MixGen: a new multi-modal data augmentation (2022). https://doi.org/10.48550/ARXIV.2206.08358. https://arxiv.org/abs/2206.08358
23. Howard, A.G., et al.: Searching for MobileNetV3. In: 2019 IEEE/CVF International Conference on Computer Vision (ICCV), pp. 1314–1324 (2019)
24. Howard, A.G., et al.: MobileNets: efficient convolutional neural networks for mobile vision applications. CoRR abs/1704.04861 (2017). http://arxiv.org/abs/1704.04861
25. Hu, R., Rohrbach, M., Andreas, J., Darrell, T., Saenko, K.: Modeling relationships in referential expressions with compositional modular networks. In: 2017 IEEE Conference on Computer Vision and Pattern Recognition (CVPR), pp. 4418–4427 (2017)
26. Hu, R., Xu, H., Rohrbach, M., Feng, J., Saenko, K., Darrell, T.: Natural language object retrieval. In: Proceedings of the IEEE Conference on Computer Vision and Pattern Recognition (2016)
27. Huang, J., et al.: Speed/accuracy trade-offs for modern convolutional object detectors. In: 2017 IEEE Conference on Computer Vision and Pattern Recognition (CVPR), pp. 3296–3297 (2017)

28. Hudson, D.A., Manning, C.D.: GQA: a new dataset for real-world visual reasoning and compositional question answering. In: 2019 IEEE/CVF Conference on Computer Vision and Pattern Recognition (CVPR), pp. 6693–6702 (2019)

29. Jiao, L., et al.: A survey of deep learning-based object detection. IEEE Access **7**, 128837–128868 (2019). https://doi.org/10.1109/ACCESS.2019.2939201

30. Kamath, A., Singh, M., LeCun, Y., Misra, I., Synnaeve, G., Carion, N.: MDETR-modulated detection for end-to-end multi-modal understanding. arXiv preprint arXiv:2104.12763 (2021)

31. Kazemzadeh, S., Ordonez, V., Matten, M.A., Berg, T.L.: ReferitGame: referring to objects in photographs of natural scenes. In: EMNLP (2014)

32. Khan, S.H., Naseer, M., Hayat, M., Zamir, S.W., Khan, F.S., Shah, M.: Transformers in vision: a survey. ArXiv abs/2101.01169 (2021)

33. Kim, W., Son, B., Kim, I.: ViLT: vision-and-language transformer without convolution or region supervision. In: Meila, M., Zhang, T. (eds.) Proceedings of the 38th International Conference on Machine Learning. Proceedings of Machine Learning Research, vol. 139, pp. 5583–5594. PMLR, 18–24 July 2021. http://proceedings.mlr.press/v139/kim21k.html

34. Kovvuri, R., Nevatia, R.: PIRC Net: using proposal indexing, relationships and context for phrase grounding. CoRR abs/1812.03213 (2018). http://arxiv.org/abs/1812.03213

35. Kuhn, H.W.: The Hungarian method for the assignment problem. Naval Res. Logist. Q. **2**(1–2), 83–97 (1955). https://doi.org/10.1002/nav.3800020109

36. Lee, K.H., Chen, X., Hua, G., Hu, H., He, X.: Stacked cross attention for image-text matching. arXiv preprint arXiv:1803.08024 (2018)

37. Lewis, M., et al.: BART: denoising sequence-to-sequence pre-training for natural language generation, translation, and comprehension. In: Proceedings of the 58th Annual Meeting of the Association for Computational Linguistics, pp. 7871–7880. Association for Computational Linguistics, July 2020. https://doi.org/10.18653/v1/2020.acl-main.703. https://aclanthology.org/2020.acl-main.703

38. Li, C., Fehérvári, I., Zhao, X., Macedo, I., Appalaraju, S.: SeeTek: very large-scale open-set logo recognition with text-aware metric learning. In: Proceedings of the IEEE/CVF Winter Conference on Applications of Computer Vision (WACV), pp. 2544–2553, January 2022

39. Li, K., Zhang, Y., Li, K., Li, Y., Fu, Y.: Visual semantic reasoning for image-text matching. In: ICCV (2019)

40. Liao, Y., et al.: A real-time cross-modality correlation filtering method for referring expression comprehension. In: 2020 IEEE/CVF Conference on Computer Vision and Pattern Recognition (CVPR), pp. 10877–10886 (2020)

41. Lin, T.-Y., et al.: Microsoft COCO: common objects in context. In: Fleet, D., Pajdla, T., Schiele, B., Tuytelaars, T. (eds.) ECCV 2014. LNCS, vol. 8693, pp. 740–755. Springer, Cham (2014). https://doi.org/10.1007/978-3-319-10602-1_48

42. Liu, D., Zhang, H., Wu, F., Zha, Z.J.: Learning to assemble neural module tree networks for visual grounding. In: Proceedings of the IEEE/CVF International Conference on Computer Vision (ICCV), October 2019

43. Liu, J., Wang, L., Yang, M.H.: Referring expression generation and comprehension via attributes. In: 2017 IEEE International Conference on Computer Vision (ICCV), pp. 4866–4874 (2017). https://doi.org/10.1109/ICCV.2017.520

44. Liu, L., et al.: Deep learning for generic object detection: a survey. Int. J. Comput. Vis. **128**, 261–318 (2019). https://doi.org/10.1007/s11263-019-01247-4

45. Liu, X., Wang, Z., Shao, J., Wang, X., Li, H.: Improving referring expression grounding with cross-modal attention-guided erasing. In: Proceedings of the IEEE/CVF Conference on Computer Vision and Pattern Recognition (CVPR), June 2019
46. Liu, Y., et al.: RoBERTa: a robustly optimized BERT pretraining approach. ArXiv abs/1907.11692 (2019)
47. Loshchilov, I., Hutter, F.: Decoupled weight decay regularization. In: International Conference on Learning Representations (2019). https://openreview.net/forum?id=Bkg6RiCqY7
48. Lu, J., Batra, D., Parikh, D., Lee, S.: ViLBERT: pretraining task-agnostic visiolinguistic representations for vision-and-language tasks. In: Advances in Neural Information Processing Systems, pp. 13–23 (2019)
49. Lu, J., Goswami, V., Rohrbach, M., Parikh, D., Lee, S.: 12-in-1: multi-task vision and language representation learning. In: The IEEE/CVF Conference on Computer Vision and Pattern Recognition (CVPR), June 2020
50. Luo, G., et al.: Multi-task collaborative network for joint referring expression comprehension and segmentation. In: 2020 IEEE/CVF Conference on Computer Vision and Pattern Recognition (CVPR), pp. 10031–10040 (2020)
51. Luo, R., Shakhnarovich, G.: Comprehension-guided referring expressions. In: Proceedings of the IEEE Conference on Computer Vision and Pattern Recognition (CVPR), July 2017
52. Nagaraja, V.K., Morariu, V.I., Davis, L.S.: Modeling context between objects for referring expression understanding. In: Leibe, B., Matas, J., Sebe, N., Welling, M. (eds.) ECCV 2016. LNCS, vol. 9908, pp. 792–807. Springer, Cham (2016). https://doi.org/10.1007/978-3-319-46493-0_48
53. Paszke, A., et al.: PyTorch: an imperative style, high-performance deep learning library. ArXiv abs/1912.01703 (2019)
54. Plummer, B.A., Kordas, P., Kiapour, M.H., Zheng, S., Piramuthu, R., Lazebnik, S.: Conditional image-text embedding networks. In: Ferrari, V., Hebert, M., Sminchisescu, C., Weiss, Y. (eds.) ECCV 2018. LNCS, vol. 11216, pp. 258–274. Springer, Cham (2018). https://doi.org/10.1007/978-3-030-01258-8_16
55. Qi, W., et al.: ProphetNet: predicting future N-gram for sequence-to-sequence pretraining. In: Proceedings of the 2020 Conference on Empirical Methods in Natural Language Processing: Findings, pp. 2401–2410 (2020)
56. Radford, A., et al.: Learning transferable visual models from natural language supervision. CoRR abs/2103.00020 (2021). https://arxiv.org/abs/2103.00020
57. Raffel, C., et al.: Exploring the limits of transfer learning with a unified text-to-text transformer. J. Mach. Learn. Res. **21**(140), 1–67 (2020). http://jmlr.org/papers/v21/20-074.html
58. Raffel, C., et al.: Exploring the limits of transfer learning with a unified text-to-text transformer. ArXiv abs/1910.10683 (2020)
59. Ranftl, R., Bochkovskiy, A., Koltun, V.: Vision transformers for dense prediction. ArXiv preprint (2021)
60. Redmon, J., Divvala, S., Girshick, R., Farhadi, A.: You only look once: unified, real-time object detection. In: 2016 IEEE Conference on Computer Vision and Pattern Recognition (CVPR), pp. 779–788 (2016). https://doi.org/10.1109/CVPR.2016.91
61. Redmon, J., Farhadi, A.: YOLO9000: better, faster, stronger. In: 2017 IEEE Conference on Computer Vision and Pattern Recognition (CVPR), pp. 6517–6525 (2017)
62. Redmon, J., Farhadi, A.: YOLOv3: an incremental improvement. ArXiv abs/1804.02767 (2018)

63. Ren, S., He, K., Girshick, R., Sun, J.: Faster R-CNN: towards real-time object detection with region proposal networks. In: Proceedings of the 28th International Conference on Neural Information Processing Systems, NIPS 2015, vol. 1, pp. 91–99. MIT Press, Cambridge (2015)
64. Rezatofighi, S.H., Tsoi, N., Gwak, J., Sadeghian, A., Reid, I.D., Savarese, S.: Generalized intersection over union: a metric and a loss for bounding box regression. In: 2019 IEEE/CVF Conference on Computer Vision and Pattern Recognition (CVPR), pp. 658–666 (2019)
65. Rohrbach, A., Rohrbach, M., Hu, R., Darrell, T., Schiele, B.: Grounding of textual phrases in images by reconstruction. ArXiv abs/1511.03745 (2016)
66. Sadhu, A., Chen, K., Nevatia, R.: Zero-shot grounding of objects from natural language queries. In: The IEEE International Conference on Computer Vision (ICCV), October 2019
67. Sandler, M., Howard, A.G., Zhu, M., Zhmoginov, A., Chen, L.C.: MobileNetV2: inverted residuals and linear bottlenecks. In: 2018 IEEE/CVF Conference on Computer Vision and Pattern Recognition, pp. 4510–4520 (2018)
68. Strudel, R., Garcia, R., Laptev, I., Schmid, C.: Segmenter: transformer for semantic segmentation. In: Proceedings of the IEEE/CVF International Conference on Computer Vision (ICCV), pp. 7262–7272, October 2021
69. Su, W., et al.: VL-BERT: pre-training of generic visual-linguistic representations. In: International Conference on Learning Representations (2020). https://openreview.net/forum?id=SygXPaEYvH
70. Suhr, A., Zhou, S., Zhang, A., Zhang, I., Bai, H., Artzi, Y.: A corpus for reasoning about natural language grounded in photographs. In: Proceedings of the 57th Annual Meeting of the Association for Computational Linguistics, Florence, Italy, pp. 6418–6428. Association for Computational Linguistics, July 2019. https://doi.org/10.18653/v1/P19-1644. https://aclanthology.org/P19-1644
71. Touvron, H., Cord, M., Douze, M., Massa, F., Sablayrolles, A., Jegou, H.: Training data-efficient image transformers & distillation through attention. In: International Conference on Machine Learning, vol. 139, pp. 10347–10357, July 2021
72. Vaswani, A., et al.: Attention is all you need. In: Guyon, I., et al. (eds.) Advances in Neural Information Processing Systems, vol. 30. Curran Associates, Inc. (2017). https://proceedings.neurips.cc/paper/2017/file/3f5ee243547dee91fbd053c1c4a845aa-Paper.pdf
73. Wang, L., Li, Y., Huang, J., Lazebnik, S.: Learning two-branch neural networks for image-text matching tasks. TPAMI 41(2), 394–407 (2019)
74. Wang, P., Wu, Q., Cao, J., Shen, C., Gao, L., van den Hengel, A.: Neighbourhood watch: referring expression comprehension via language-guided graph attention networks. In: Proceedings of the IEEE/CVF Conference on Computer Vision and Pattern Recognition (CVPR), June 2019
75. Wolf, T., et al.: Transformers: state-of-the-art natural language processing. In: Proceedings of the 2020 Conference on Empirical Methods in Natural Language Processing: System Demonstrations, pp. 38–45. Association for Computational Linguistics, October 2020. https://www.aclweb.org/anthology/2020.emnlp-demos.6
76. Yang, S., Li, G., Yu, Y.: Cross-modal relationship inference for grounding referring expressions. In: Proceedings of the IEEE/CVF Conference on Computer Vision and Pattern Recognition (CVPR), June 2019
77. Yang, S., Li, G., Yu, Y.: Dynamic graph attention for referring expression comprehension. In: Proceedings of the IEEE/CVF International Conference on Computer Vision (ICCV), October 2019

78. Yang, Z., Chen, T., Wang, L., Luo, J.: Improving one-stage visual grounding by recursive sub-query construction. In: Vedaldi, A., Bischof, H., Brox, T., Frahm, J.-M. (eds.) ECCV 2020. LNCS, vol. 12359, pp. 387–404. Springer, Cham (2020). https://doi.org/10.1007/978-3-030-58568-6_23

79. Yang, Z., Gong, B., Wang, L., Huang, W., Yu, D., Luo, J.: A fast and accurate one-stage approach to visual grounding. In: Proceedings of the IEEE/CVF International Conference on Computer Vision (ICCV), October 2019

80. Yu, F., et al.: ERNIE-ViL: knowledge enhanced vision-language representations through scene graphs. In: Proceedings of the AAAI Conference on Artificial Intelligence, vol. 35, no. 4, pp. 3208–3216, May 2021. https://ojs.aaai.org/index.php/AAAI/article/view/16431

81. Yu, L., et al.: MAttNet: modular attention network for referring expression comprehension. In: Proceedings of the IEEE Conference on Computer Vision and Pattern Recognition (CVPR), June 2018

82. Yu, L., Poirson, P., Yang, S., Berg, A.C., Berg, T.L.: Modeling context in referring expressions. ArXiv abs/1608.00272 (2016)

83. Yu, L., Tan, H., Bansal, M., Berg, T.L.: A joint speaker-listener-reinforcer model for referring expressions. In: 2017 IEEE Conference on Computer Vision and Pattern Recognition (CVPR), pp. 3521–3529 (2017)

84. Yu, L., Tan, H., Bansal, M., Berg, T.L.: A joint speaker-listener-reinforcer model for referring expressions. In: 2017 IEEE Conference on Computer Vision and Pattern Recognition (CVPR), pp. 3521–3529 (2017). https://doi.org/10.1109/CVPR.2017.375

85. Yu, Z., Yu, J., Xiang, C., Zhao, Z., Tian, Q., Tao, D.: Rethinking diversified and discriminative proposal generation for visual grounding. ArXiv abs/1805.03508 (2018)

86. Zaheer, M., et al.: Big bird: transformers for longer sequences. In: Advances in Neural Information Processing Systems, vol. 33 (2020)

87. Zhang, H., Niu, Y., Chang, S.F.: Grounding referring expressions in images by variational context. In: Proceedings of the IEEE Conference on Computer Vision and Pattern Recognition (CVPR), June 2018

88. Zhang, J., Zhao, Y., Saleh, M., Liu, P.J.: PEGASUS: pre-training with extracted gap-sentences for abstractive summarization (2019)

89. Zhu, X., Su, W., Lu, L., Li, B., Wang, X., Dai, J.: Deformable DETR: deformable transformers for end-to-end object detection. In: International Conference on Learning Representations (2021). https://openreview.net/forum?id=gZ9hCDWe6ke

90. Zhuang, B., Wu, Q., Shen, C., Reid, I., van den Hengel, A.: Parallel attention: a unified framework for visual object discovery through dialogs and queries. In: Proceedings of the IEEE Conference on Computer Vision and Pattern Recognition (CVPR), June 2018

W32 - Uncertainty Quantification
for Computer Vision

W32 - Uncertainty Quantification for Computer Vision

Nowadays, machine learning and deep learning approaches continually demonstrate their viability to solve vision challenges with models deployed to solve practical tasks. While performance (in terms of accuracy) is good, these models are predominately used as black boxes, and it is difficult to ascertain whether or not their outputs are reasonable. Even manual data set inspection, to discriminate between well predicted simple samples and errors on hard samples, may not be feasible. Uncertainty quantification and calibration are powerful tools that may be employed by engineers and researchers to better understand model outputs reliability which is hugely beneficial to safe decision making. The machine learning community has placed great effort in developing novel techniques (e.g., Bayesian methods, post-hoc calibration and distribution-free approaches) and bench-marking them with classic research data sets. Our goal for this workshop is twofold. Firstly, we are interested in extending uncertainty quantification methods to more challenging computer vision data sets or practical use cases from an industrial perspective. Secondly, we wish to stimulate debate in the community about how to best integrate uncertainty in a community that often aims at 100

October 2022

Andrea Pilzer
Martin Trapp
Arno Solin
Yingzhen Li
Neill Campbell

Localization Uncertainty Estimation
for Anchor-Free Object Detection

Youngwan Lee[1,2]($\boxtimes$), Joong-Won Hwang[1], Hyung-Il Kim[1], Kimin Yun[1],
Yongjin Kwon[1], Yuseok Bae[1], and Sung Ju Hwang[2]

[1] Electronics and Telecommunications Research Institute (ETRI),
Daejeon, South Korea
{yw.lee,jwhwang,hikim,kimin.yun,scocso,ysbae}@etri.re.kr
[2] Korea Advanced Institute of Science and Technology (KAIST),
Daejeon, South Korea
sjhwang82@kaist.ac.kr

Abstract. Since many safety-critical systems, such as surgical robots
and autonomous driving cars operate in unstable environments with sensor noise and incomplete data, it is desirable for object detectors to take
the localization uncertainty into account. However, there are several limitations of the existing uncertainty estimation methods for anchor-based
object detection. 1) They model the uncertainty of the heterogeneous
object properties with different characteristics and scales, such as location (center point) and scale (width, height), which could be difficult to
estimate. 2) They model box offsets as Gaussian distributions, which is
not compatible with the ground truth bounding boxes that follow the
Dirac delta distribution. 3) Since anchor-based methods are sensitive
to anchor hyper-parameters, their localization uncertainty could also be
highly sensitive to the choice of hyper-parameters. To tackle these limitations, we propose a new localization uncertainty estimation method
called UAD for anchor-free object detection. Our method captures the
uncertainty in four directions of box offsets (left, right, top, bottom)
that are homogeneous, so that it can tell which direction is uncertain,
and provide a quantitative value of uncertainty in [0, 1]. To enable such
uncertainty estimation, we design a new uncertainty loss, negative power
log-likelihood loss, to measure the localization uncertainty by weighting
the likelihood loss by its IoU, which alleviates the model misspecification problem. Furthermore, we propose an uncertainty-aware focal loss
for reflecting the estimated uncertainty to the classification score. Experimental results on COCO datasets demonstrate that our method significantly improves FCOS [32], by up to 1.8 points, without sacrificing computational efficiency. We hope that the proposed uncertainty estimation
method can serve as a crucial component for the safety-critical object
detection tasks.

1 Introduction

CNN-based object detection models are widely used in many safety-critical systems such as autonomous vehicles and surgical robots [29]. For such safety-

© The Author(s), under exclusive license to Springer Nature Switzerland AG 2023
L. Karlinsky et al. (Eds.): ECCV 2022 Workshops, LNCS 13808, pp. 27–42, 2023.
https://doi.org/10.1007/978-3-031-25085-9_2

Fig. 1. Example of 4-directional uncertainty for anchor-free object detection. C_L, C_R, C_T, and C_B denote the estimated certainty in $[0, 1]$ value with respect to left, right, top, and bottom. For example, the proposed UAD estimates low top-directional certainty due to its ambiguous head boundary of the cat wearing a hat. This demonstrates that our method enables the detection network to quantify which direction is uncertain due to unclear or obvious objects.

critical systems, it is essential to measure how reliable the estimated output from the object detection is, in addition to achieving good performance. Although object detection is a task that requires both object localization and classification, most of the state-of-the-art methods [1,3,21,26,37] utilize the classification scores as detection scores without considering the localization uncertainty. As a result, these methods output mislocalized detection boxes that are highly over-confident [10]. Thus the confidence of the bounding box localization should also be taken into consideration when estimating the uncertainty of the detection.

Recently, there have been many attempts [8,10,15,18] that try to model localization uncertainty for object detection. All of these efforts model the uncertainty of location (center point) and scale (width, height) by modeling their distributions as Gaussian distributions [23,26] with extra channels in the regression output. However, since the center points, widths, and heights have semantically different characteristics [15], this approach which considers each value equally is inappropriate for modeling localization uncertainty. For example, the estimated distributions of the center point and scale are largely different according to [15]. In terms of the loss function for uncertainty modeling, conventional methods [8,15,18] use the negative log-likelihood loss to regress outputs as Gaussian distributions. He *et al.* [10] introduce KL divergence loss by modeling the box prediction as Gaussian distribution and the ground-truth box as Dirac delta function. In the perspective of cross-entropy, however, these methods face the model misspecification problem [14] in that the Dirac delta function cannot be exactly represented with Gaussian distributions, *i.e.*, for any μ and σ, $\delta(x) \neq N(x|\mu, \sigma^2)$.

Recently, anchor-free methods [3,17,32,39,40] that do not require heuristic anchor-box tuning (*e.g.*, scale, aspect ratio) have outperformed conventional anchor-box based methods such as Faster R-CNN [26], RetinaNet [21], and their variants [1,38]. As a representative anchor-free method, FCOS [32] adopts the concept of centerness as an *implicit* localization uncertainty that measures how well the center of the predicted bounding boxes fits the ground-truth boxes. The centerness score can be multiplied by the classification score to calibrate the box quality score during the test phase. However, the centerness faces the problem of inconsistency [19,36] between the training and test phase as it is trained separately with the classification network. Besides, the centerness does not fully account for localization uncertainty of bounding boxes (*e.g.*, scale or direction).

To deal with these limitations, in this paper, we propose Uncertainty-Aware Detection (UAD), which *explicitly* estimates the localization uncertainty for an anchor-free object detection model. The proposed method estimates the uncertainty of the four values that define the box offsets (left, right, top, bottom) to fully describe the localization uncertainty. It is advantageous to estimate the uncertainty of the four box offsets having a similar semantic characteristic compared to conventional algorithms that estimate localization uncertainty for anchor-based detection [8,10,15,18]. Our method can also capture richer information for localization uncertainty than just the centerness of FCOS.

The proposed method enables to inform which direction of a box boundary is uncertain as a quantitative value in [0,1] independently from the overall box uncertainty as shown in Fig. 1 (please refer to more examples in Fig. 3). To this end, we model the box offset and its uncertainty as Gaussian distributions by introducing a newly designed uncertainty loss and an uncertainty network. To resolve the aforementioned model misspecification [14] between Dirac delta and Gaussian distribution, we design a novel uncertainty loss, *negative power log-likelihood loss (NPLL)*, inspired by Power likelihood [7,11,24,28,31], to enable the uncertainty network to learn to estimate localization uncertainty by weighing the log-likelihood loss by Intersection-over-Union (IoU).

To handle the inconsistency between the training and test phase, we also propose an uncertainty-aware classification by reflecting the estimated uncertainty into the classification score in both the training/inference phase. To this end, we introduce a Certainty-aware representation Network (CRN) to represent features with the classification network jointly. We also define an uncertainty-aware focal loss (UFL) that adjusts the loss contributions of examples differently based on their estimated uncertainties. UFL focuses on the high-quality examples obtaining lower uncertainty (*i.e.*, higher certainty) by weighting the estimated certainty, which enables to generate uncertainty-reflected localization score. The difference from other focal losses such as QFL [19] and VFL [36] is that we use the estimated uncertainty as a weighting factor instead of IoU. The 4-directions uncertainty captures the object localization quality better than IoU (*i.e.*, *scale*), and the uncertainty-based methods empirically yield better performance.

By introducing the uncertainty network with NPLL and uncertainty-aware classification with UFL to FCOS [33], we build an uncertainty-aware detector, UAD. We validate UAD on the challenging COCO [22] benchmarks. Through extensive experiments, we find that the Gaussian modeling with the proposed NPLL and the newly defined focal loss, UFL, yields better performance over baseline methods. Besides, UAD improves the FCOS [33] baseline by 1–1.8 gains in AP using different backbone networks without additional computation burden. In addition to detection performance, the proposed UAD accurately estimates the uncertainty in 4-directions as well as the detection performance as shown in Fig. 1, 3.

The main contributions of our work can be summarized as follows:

- We propose a simple and effective method to measure the localization uncertainty for anchor-free object detectors that can serve as a detection quality measure and provide confidences in [0, 1] for four directions (left, right, top, bottom) from the center of the object.
- We propose a novel uncertainty-aware loss function, inspired by *power likelihood* that weighs the negative log-likelihood loss by the IoU, which resolves the model misspecification problem.
- We introduce an uncertainty-aware classification scheme with the proposed *uncertainty-aware focal loss* by leveraging the estimated uncertainty during both the training and the test phase in a consistent manner.

2   Related Works

2.1   Anchor-Free Object Detection

Recently, anchor-free object detectors [3,17,32,39,40] have attracted attention beyond anchor-based methods [1,21,26,38] that need to tune sensitive hyper-parameters related to anchor box (*e.g.*, scale, aspect ratio, etc.). CornerNet [17] predicts an object location as a pair of keypoints (top-left and bottom-right). CenterNet [3] extends CornerNet as a triplet instead of a pair of key points to boost performance. This idea is extended by CenterNet [3] that utilizes a triplet instead of a pair of key points to boost performance. ExtremeNet [40] locates four extreme points (top, bottom, left, right) and one center point to generate the object box. Zhu *et al.* [39] utilizes keypoint estimation to predict center point objects and regresses to other attributes, including size, orientation, pose, and 3D location. FCOS [32] views all points (or locations) inside the ground-truth box as positive samples and regresses four distances (left, right, top, bottom) from the points. We propose to endow FCOS with localization uncertainty due to its simplicity and performance.

2.2   Uncertainty Estimation

Uncertainty in deep neural networks can be estimated in two types [4,13,18]: epistemic (sampling-based) and aleatoric (sampling-free) uncertainty. Epistemic

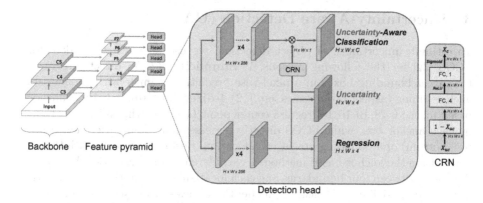

Fig. 2. Architecture of the proposed UAD. Differently from FCOS [32], our UAD can estimate localization uncertainty from a separate *uncertainty* network that outputs the uncertainty on four values that describes a bounding box (left, right, top, and bottom). In addition, the estimated uncertainty is utilized for uncertainty-aware classification as a box-quality confidence.

uncertainty measures the model uncertainty in the models' parameters through Bayesian neural networks [30], Monte Carlo dropout [5], and Bootstrap Ensemble [16]. As they need to be re-evaluated several times and store several sets of weights for each network, it is hard to apply them for real-time applications. Aleatoric uncertainty is data and problems inherent such as sensor noise and ambiguities in data. It can be estimated by explicitly modeling it as model output.

Recent works [8,10,16,18] have adopted uncertainty estimation for object detection. Lakshminarayanan *et al.* [16] and Harakeh *et al.* [8] use Monte Carlo dropout in Epistemic based methods. As described above, since epistemic uncertainty needs to be inferred several times, it is not suitable for real-time object detection. Le *et al.* [18] and Choi *et al.* [2] are aleatoric based methods and jointly estimate the uncertainties of four parameters of bounding box from SSD [23] and YOLOv3 [25]. He [10] estimates the uncertainty of the bounding box by minimizing the KL-divergence loss for the Gaussian distribution of the predicted box and Dirac delta distribution of the ground-truth box on the Faster R-CNN [26] (anchor-based method). From the cross-entropy perspective, however, Dirac delta distribution cannot be represented as a Gaussian distribution, which results in a misspecification problem [14]. To overcome this problem, we design a new uncertainty loss function, *negative power log-likelihood loss*, inspired by the power likelihood concept [7,11,24,28,31]. The latest concurrent work is the Generalized Focal loss (GFocal) [19] that represents jointly localization quality, classification, and model bounding box as arbitrary distribution. The distinct difference from GFocal [19] is that our method estimates 4-directions uncertainties as quantitative values in the range [0, 1] thus, these estimated values can be used as an informative cue for decision-making.

3   Uncertainty-Aware Detection (UAD)

To estimate uncertainty of object detection, we use an anchor-free detector, FCOS [32] for the following reasons: 1) **Simplicity**. FCOS directly regresses the target bounding boxes in a pixel-wise prediction manner without heuristic anchor tuning (aspect ratio, scales, etc.). 2) **4-directions uncertainty**. anchor-based methods [8,10,16,18] regress center point (x,y), width, and height based on each anchor box, while FCOS directly regresses four boundaries (left, right, top, bottom) of a bounding box at each location. Besides, the center, width, and height from the anchor-based methods have different characteristics, whereas the distances between four boundaries and each location are semantically symmetric. In terms of modeling, it is easier to model the values sharing semantic meanings that have similar properties. Furthermore, it enables to notify *which direction of a box boundary is uncertain* separately from the overall box uncertainty.

FCOS also adopts centerness to suppress the low-quality predicted boxes in the inference stage, which estimates how well center of the predicted box fits any of the objects. However, centerness is an *implicit* score of uncertainty which is insufficient in measuring localization uncertainty, since it does not fully capture the box quality such as scale or directions. It also has the inconsistency problem, since centerness as localization uncertainty is independently obtained regardless of the classification score and multiplied to the classification score only at the inference phase.

To overcome such limitations of the centerness-based uncertainty estimation, we propose a novel method, Uncertainty-Aware Detection (UAD), to endow FCOS with a localization uncertainty estimator that reflects the box quality along the 4-directions of the bounding box. This approach allows the network to estimate not only a single measure of object localization uncertainty but also explicitly output the uncertainty in each of the four directions. To this end, firstly, we introduce an uncertainty network to FCOS with a newly proposed uncertainty loss, *negative power log-likelihood loss* (NPLL). Next, we propose an uncertainty-aware focal loss (UFL) to jointly represent the estimated uncertainty with the classification network during both training and test step. The overview of UAD is illustrated in Fig. 2.

3.1   Power Likelihood

In FCOS [32], if the location belongs to the ground-truth box area, it is regarded as a positive sample and as a negative sample otherwise. On each location (x, y), the box offsets are regressed as 4D vector $B_{x,y} = [l, r, t, b]^\top$ that is the distances from the location to four sides of the bounding box (*i.e.*, left, right, top, and bottom). The regression targets $B_{x,y}^g = [l^g, r^g, t^g, b^g]^\top$ are computed as,

$$l^g = x - x_{lt}^g, \ r^g = x_{rb}^g - x, \ t^g = y - y_{lt}^g, \ b^g = y_{rb}^g - y, \tag{1}$$

where (x_{lt}^g, y_{lt}^g) and (x_{rb}^g, y_{rb}^g) denote the coordinates of the left-top and right-bottom corners of the ground-truth box, respectively. Then, for all locations of positive samples, the IoU loss [35] is measured between the predicted $B_{x,y}$ and the ground truth $B_{x,y}^g$ for the regression loss.

To better estimate localization uncertainty than centerness, it is necessary to consider the four directions that represent the object boundary. Therefore, we introduce an uncertainty network that estimates the localization uncertainty of the box based on the regressed box offsets (l, r, t, b). To predict the uncertainties of four box offsets, we model the box offsets as Gaussian distributions and train the network to estimate their uncertainties (standard deviation). Assuming that each instance of box offsets is independent, we use multivariate Gaussian distribution of output B^* with diagonal covariance matrix Σ_B to model each box offset B:

$$P_\Theta(B^*|B) = \mathcal{N}(B^*; \boldsymbol{\mu}_B, \boldsymbol{\Sigma}_B), \tag{2}$$

where Θ is the learnable network parameters. $\boldsymbol{\mu}_B = [\mu_l, \mu_r, \mu_t, \mu_b]^\top$ and $\boldsymbol{\Sigma}_B = \mathrm{diag}(\sigma_l^2, \sigma_r^2, \sigma_t^2, \sigma_b^2)$ denote the predicted box offset and its *uncertainty*, respectively.

Existing works [2,10,15,18] model ground-truth as Dirac delta distribution and box offset as Gaussian estimation, respectively. [2,15,18] adopt negative log-likelihood loss (NLL) and [10] use KL-divergence loss (KL-Loss). In cross-entropy perspective, minimizing NLL and KL-loss is equivalent as below:

$$\mathcal{L} = -\frac{1}{N} \sum_x P_D(x) \cdot \log P_\Theta(x), \tag{3}$$

where P_D and P_Θ are Dirac delta function and Gaussian probability density function, respectively. When the box offset is located in a ground-truth box, the P_D is 1 then Eq. 3 becomes the negative log-likelihood loss. However, this is problematic since the Dirac delta distribution does not belong to the family of Gaussian distributions, which results in the model misspecification problem [14]. In a number of statistical literature [7,11,24,28,31], to estimate parameters of interest in a robust way when the model is misspecified, the Power likelihood $(P_\Theta(\cdot)^w)$ is often used to replace the original likelihood, which raises the likelihood $(P_\Theta(\cdot))$ to a power (w) to reflect how influential a data instance is. To fill the gaps between Dirac delta distribution and Gaussian distribution, we utilize this Power likelihood and introduce a novel uncertainty loss, *negative power log-likelihood loss (NPLL)*, that exploits Intersection-over-Union (IoU) as the power term since the offset that has higher IoU should be more influential. By the property of the logarithm, the log-likelihood is multiplied by the IoU. Thus, the new uncertainty loss is defined as:

$$\mathcal{L}_{uc} = - \sum_{k \in \{l,r,t,b\}} IoU \cdot \log P_\Theta\left(B_k^g | \mu_k, \sigma_k^2\right) \tag{4}$$

$$= IoU \times \left[\sum_{k \in \{l,r,t,b\}} \left\{ \frac{(B_k^g - \mu_k)^2}{2\sigma_k^2} + \frac{1}{2}\log\sigma_k^2 \right\} + 2\log 2\pi \right], \tag{5}$$

where IoU is the intersection-over-union between the predicted box and the ground-truth box and k is $\in \{l, r, t, b\}$. With this uncertainty loss, when the predicted coordinate μ_k from the regression branch is inaccurate, the network will output larger uncertainty σ_k. Note that unlike centerness, our network is trained to directly estimate the localization uncertainties for the four directions $\sigma_u = (\sigma_l, \sigma_r, \sigma_t, \sigma_b)$ that define the box offsets. This allows to estimate which direction of a box boundary is uncertain, separately from the overall box uncertainty.

To realize this idea, we add a 3×3 Conv layer with 4 channels as an uncertainty network for FCOS [32] (w/o centerness) as shown in Fig. 2. Our network predicts a probability distribution in addition to the box coordinates. The mean values μ_k of each box offset are predicted from the regression branch, and the new uncertainty network with the sigmoid function outputs four uncertainty values $\sigma_k \in [0, 1]$. The regression and the uncertainty networks share the same feature (4 Conv layers) as their inputs to estimate the mean μ_k and the standard deviation σ_k. Note that the computation burden of adding the uncertainty network is negligible.

3.2 Uncertainty-Aware Classification

Dense object detectors including FCOS generate several predicted boxes and then utilize non-maximum suppression (NMS) to filter out redundant boxes by ranking the boxes with classification confidence. However, as the classification score does not account for the quality of the detected bounding box, centerness [32] or IoU [34] is often used to weigh the classification confidence at the inference phase to calibrate the detection score. To address the misalignment problem between confidence and localization quality, GFL [19] or VFL [36] suggests a method to reflect the box quality (e.g., IoU) into the classification score in both the training and test phase.

Certainty Representation Network. As the localization uncertainty is negatively correlated to the detected box quality, it is natural to apply the estimated localization uncertainty to the classification network for uncertainty-aware classification. The higher the box quality is, the closer the predicted probability is to 1, so that the uncertainty is taken in reverse. Thus, we convert the uncertainty features $X_{uc} \in \mathbb{R}^{4 \times W \times H}$ from the uncertainty network into certainty features $X_c \in \mathbb{R}^{1 \times W \times H}$ through the certainty representation network (CRN), to combine the confidence scores with the classification features. The CRN consists of two fully-connected layers, $W_1 \in \mathbb{R}^{4 \times 4}$ and $W_2 \in \mathbb{R}^{1 \times 4}$. Specific process is described as below:

$$X_c = \phi(W_2(\delta(W_1(1 - X_{uc})))), \tag{6}$$

where ϕ and δ denote the Sigmoid and ReLU activation function, respectively. The certainty feature is multiplied by the classification features $X_{cls} \in \mathbb{R}^{256 \times W \times H}$ to obtain a representation that accounts for both certainty and the classification score.

Table 1. Comparison of various focal loss. y is target IoU between the predicted box and the ground-truth as a soft label instead of one-hot category. p denotes the predicted classification score, α is a weighting factor, and $\mathcal{L}_{BCE}$ means binary cross-entropy loss. UFL denotes the proposed uncertainty-aware focal loss. $f(\sigma_u)$ and σ_u denote a certainty function and the estimated localization uncertainties from the uncertainty branch, respectively.

Loss type	$y > 0$	$y = 0$
QFL [19]	$\|y - p\|^{\gamma} \cdot \mathcal{L}_{BCE}$	$-p^{\gamma} \log(1 - p)$
VFL [36]	$y \cdot \mathcal{L}_{BCE}$	$-\alpha p^{\gamma} \log(1 - p)$
UFL (ours)	$f(\sigma_u) \cdot \mathcal{L}_{BCE}$	$-\alpha p^{\gamma} \log(1 - p)$

Uncertainty-Aware Focal Loss. Both GFL [19] and VarifocalNet [36] introduce IoU-classification representation and new classification losses, Quality focal loss (QFL) and Varifocal loss (VFL), that inherit Focal loss (FL) [21] for addressing class imbalance between positive/negative examples. Table 1 describes the detail definitions of focal losses. QFL and VFL utilize IoU score between the predicted box and the ground-truth as a soft label (*e.g.*, $y \in [0, 1]$), instead of one-hot category (*e.g.*, $y \in \{0, 1\}$). α is a weighting factor and p denotes the predicted classification score. Following FL, QFL focuses on hard positive examples by down-weighting the loss contribution of easy examples with a modulating factor ($|y - p|^{\gamma}$), while VFL pays more attention to easy examples (*i.e.*, higher-IoU) by weighting the target IoU (*i.e.*, y) and only reduces the loss contribution from negative examples by scaling the loss with a weighting factor of αp^{γ}. That is, QFL and VFL adjust the loss contribution with respect to training samples by weighting box quality measure (*e.g.*, IoU).

Unlike GFL and VFL based on deterministic detection results (*e.g.*, only box boundaries), our method can estimate object boundaries as probabilistic distributions. Specifically, our method estimates the means (μ_k) and uncertainties as standard deviations (σ_k), of the four offset distributions. In this way, our method utilizes the estimated uncertainty as a box quality measure instead of IoU as in QFL and VFL. From this perspective, we design a new focal loss, *uncertainty-aware focal loss* (**UFL**) for learning uncertainty-aware classification as defined in the third row of Table 1. p is the estimated probability from the uncertainty-aware classification network. $f(\sigma_u)$ and σ_u denote certainty function and the estimated localization uncertainties from the uncertainty branch, respectively. Instead of weighting the IoU, we use the certainty score obtained from the certainty function $f(\sigma_u)$ to give more attention to high quality examples. In this paper, $f(\sigma_u)$ is defined as below:

$$f(\sigma_u) = \frac{1}{4} \sum_{k \in \{l, r, t, b\}} (1 - \sigma_k), \qquad (7)$$

which averages $(1 - \sigma_k)$ for all $k \in \{l, r, t, b\}$. Thus, we weigh the positive examples with the estimated certainty score instead of the target IoU (y). For

Table 2. Comparison with data representations of box regression on FCOS. NPLL denotes the proposed negative power log-likelihood loss.

Distribution	AP	AP_{75}	AP_S	AP_M	AP_L
Dirac delta [32]	37.8	40.8	21.2	42.1	48.2
Gaussian w/NLL [2,10,15]	38.3	42.2	21.2	42.5	49.4
General w/DFL [19]	39.0	42.3	**22.6**	43.0	50.6
Gaussian w/NPLL (ours)	**39.0**	**42.9**	21.8	**43.2**	**50.7**

negative samples, we find that down-weighting the loss contribution yields better performance with a smaller scale factor of α than VFL (*e.g.,* 0.25 vs. 0.75) in Table 3 (right). Also, following QFL or VFL, we set γ to 2.

3.3   Training and Inference

We define the total loss $\mathcal{L}$ of UADet as below:

$$\mathcal{L} = \frac{1}{N_{pos}} \sum_i \mathcal{L}_{uac} + \frac{1}{N_{pos}} \sum_i \mathbb{1}_{\{c_i^* > 0\}} (\mathcal{L}_{bbox} + \lambda \mathcal{L}_{uc}), \qquad (8)$$

where $\mathcal{L}_{uc}$ is NPLL (Sect. 3.1), $\mathcal{L}_{bbox}$ is the GIoU loss [27], and $\mathcal{L}_{uac}$ is UFL (Sect. 3.2). N_{pos} denotes the number of positive samples, $\mathbb{1}$ is the indicator function, c_i^* is class label of the location i, and λ is to balance weight for $\mathcal{L}_{uc}$. The summation is calculated over all positive locations i on the feature maps. Since our baseline is FCOS, We follow the sampling strategy of FCOS. With the proposed NPLL and UFL, UADet is learned to estimate uncertainties in four directions as well as the uncertainty-aware classification score. In the test phase, the predicted uncertainty-aware classification score is utilized in the NMS post-processing step for ranking the detected boxes.

4   Experiments

Experimental Setup. In this section, we evaluate the effectiveness of UADet on the challenging COCO [22] dataset which has 80 object categories. We use COCO `train2017` set for training and `val2017` set for ablation studies. Final results are evaluated on `test-dev2017` in the evaluation server for comparison with state-of-the-art. Since FCOS [32] without centerness is our baseline, we use the default hyper-parameters of FCOS. We train UADet by stochastic gradient descent algorithm with a mini-batch of size 16 and the initial learning rate is 0.01. For the ablation study, we use ResNet-50 backbone with ImageNet pre-trained weights and 1× schedule [9] without multi-scale training. For comparison with state-of-the-art methods, we adopt 2× schedule with multi-scale augmentation where the shorter image side is randomly sampled from [640, 800] pixels. We implement the proposed UAD based on `Detectron2` [6].

Table 3. Varying λ and α for NPLL and UFL, respectively.

λ	AP	α	AP
0.100	39.5	1.00	39.4
0.075	39.5	0.75	39.4
0.050	**39.8**	0.50	39.6
0.025	39.3	**0.25**	**39.8**
0.010	39.3	0.10	39.4

Table 4. Comparison between box quality estimation methods on FCOS.

Box quality methods	AP	AP_{75}	AP_S	AP_M	AP_L
FCOS w/o centerness	37.8	40.8	21.2	42.1	48.2
centerness-branch [32]	38.5	41.6	**22.4**	42.4	49.1
IoU-branch [12,34]	38.7	42.0	21.6	43.0	50.3
QFL [19]	39.0	41.9	22.0	43.1	51.0
VFL [36]	39.0	41.9	21.9	42.6	51.0
UFL (ours)	**39.5**	**42.6**	22.1	**43.4**	**51.8**

4.1 Ablation Study

Power Likelihood. We investigate the effectiveness of the proposed Gaussian modeling with the proposed uncertainty loss (*i.e.*, NPLL) for localization uncertainty. Table 2 shows different data representation methods. The Dirac delta distribution in the first row is the baseline which is FCOS without centerness. Compared to naive Gaussian distribution methods [2,10,15], our method obtains more accuracy gain (+1.2% vs. +0.5%), which shows the proposed IoU power term effectively overcomes the model misspecification problem [14]. Besides, our method shows similar performance compared to general distribution with DFL [19], demonstrating that Gaussian distribution with NPLL effectively models the underlying distribution of the object localization as well. Meanwhile, we test the hyper-parameter λ of NPLL in Table 3 (left) and 0.05 shows the best performance, thus we use the value in the rest experiments. It is also worth noting that the proposed method allows the network to well estimate which direction is uncertain as a quantified value $\in [0,1]$ as shown in Fig. 3.

Uncertainty-Aware Classification. We compare the proposed uncertainty-aware classification using UFL with other box quality estimation methods in Table 4. We can find that the box-quality based representation methods including QFL [19], VFL [36], and UFL (ours) surpass the centerness [32] and IoU branch [12,34] methods which combine quality score only during the test phase. The proposed UFL (39.5%) consistently achieves better performance than both QFL(39.0%) and VFL(39.0%). We also investigate the effects of various focal

Table 5. Comparison of different focal losses on UAD (ours).

Focal loss type	AP	AP_{75}	AP_S	AP_M	AP_L
QFL [19]	39.1	42.9	21.9	43.2	50.8
VFL [36]	39.3	42.5	21.6	43.2	51.1
UFL (ours)	**39.8**	**43.0**	**22.0**	**44.0**	**51.4**

Table 6. The effect of each component on UAD (ours). The baseline is FCOS without centerness. The inference time is measured at V100 GPU with a batch size of 1. CRN denotes the certainty-aware representation network in the Fig. 2.

NPLL	UFL	CRN	AP	Inference time (s)
			37.8	0.037
✓			39.0	0.038
✓	✓		39.5	0.038
✓	✓	✓	39.8	0.038

losses on UAD as shown in Table 5. UFL still outperforms IoU-based focal losses such as QFL and VFL, which demonstrates our uncertainty modeling strategy on 4-directions effectively captures more degrees (l, r, t, b) of the box quality compared to only an overall quality IoU.

Components of UAD. As shown in Table 6, we investigate the effect of each component of the proposed UAD. In addition, we measure the GPU inference time at NVIDIA V100 GPU with a batch size of 1. We start FCOS without centerness as a baseline (37.8 AP). Adding the uncertainty network and learning with NPLL improve the baseline to 39.0 AP (+1.2). Replacing focal loss with the proposed UFL obtains 0.5 AP gain, and adding CRN boosts further performance gain (+0.3). These results demonstrate that the proposed method not only effectively estimates 4-directions localization uncertainty but also boosts detection performance. We also emphasize that all these components require negligible computational overhead as shown in Table 6.

4.2   Comparison with Other Methods

Using different backbones, we compare the proposed UAD with FCOS and other methods on COCO [22] test-dev2017. Table 7 summarizes the results. Compared to FCOS [33], UAD achieves consistent performance gains based on various backbones, such as ResNet-50/101 and ResNeXt-101-32x8d/64x4d. We note that UAD further boosts performance on deeper and more complex backbone ResNeXt than ResNet. Specifically, ResNeXt-101-32x8d/64x4d/32x8d-DCN obtains +1.3/+1.4/+1.8 AP gains while ResNet-R-50/101 get 1.0 AP gain, respectively. These results show the uncertainty representation of UAD makes the network result in a synergy effect with the deeper feature representation.

Table 7. UAD performance on COCO test-dev2017. These results are tested with single-model and single-scale. Note that the results of FCOS is the latest update version in [33]. FCOS and UAD are trained with the same training protocols such as multi-scale augmentation and 2× schedule in [9]. DCN:Deformable Convolutional Network v2.

Method	Backbone	Epoch	AP	AP_{50}	AP_{75}	AP_S	AP_M	AP_L
Anchor-based								
FPN [20]	ResNet-101	24	36.2	59.1	39.0	18.2	39.0	48.2
RetinaNet [21]	ResNet-101	18	39.1	59.1	42.3	21.8	42.7	50.2
ATSS [37]	ResNet-101	24	43.6	62.1	47.4	26.1	47.0	53.6
GFL [19]	ResNet-101	24	45.0	63.7	48.9	27.2	48.8	54.5
GFL [19]	ResNeXt-101-32x8d-DCN	24	48.2	67.4	52.6	29.2	51.7	60.2
Anchor-free								
CornerNet [17]	Hourglass-104	200	40.6	56.4	43.2	19.1	42.8	54.3
CenterNet [3]	Hourglass-104	190	44.9	62.4	48.1	25.6	47.4	57.4
SAPD [41]	ResNet-101	24	43.5	63.6	46.5	24.9	46.8	54.6
FCOS [33]	ResNet-50	24	41.4	60.6	44.9	25.1	44.2	50.9
UAD	ResNet-50	24	**42.4** +1.0	60.6	46.1	24.5	45.5	53.1
FCOS [33]	ResNet-101	24	43.2	64.5	46.8	26.1	46.2	52.8
UAD	ResNet-101	24	**44.2** +1.0	62.6	48.0	25.6	47.4	55.2
FCOS [33]	ResNeXt-101-32x8d	24	44.1	66.0	47.9	27.4	46.8	53.7
UAD	ResNeXt-101-32x8d	24	**45.4** +1.3	64.1	49.3	27.7	48.6	56.0
FCOS [33]	ResNeXt-101-64x4d	24	44.8	66.4	48.5	27.7	47.4	55.0
UAD	ResNeXt-101-64x4d	24	**46.2** +1.4	64.9	50.3	28.2	49.6	57.2
FCOS [33]	ResNeXt-101-32x8d-DCN	24	46.6	65.9	50.8	28.6	49.1	58.6
UAD	ResNeXt-101-32x8d-DCN	24	**48.4** +1.8	67.1	52.7	29.5	51.6	61.0

Compared to GFL [19] that utilizes better sampling method (*e.g.*, ATSS [37]) based on anchor-based detection, UAD shows lower AP on ResNet-101. However, UAD achieves 48.4 AP, which is higher than GFL (48.2) when using the ResNeXt-101-32x8d-DCN backbone.

4.3 Discussion

GFL [19] also estimates the general distributions along with four directions like UAD. GFL only *qualitatively* interprets the directional uncertainty according to the shape of the distribution (*e.g.*, sharp or flatten). However, different from GFL, our UAD not only can capture the overall uncertainty of the objects, but also estimate the four-directional uncertainties as *quantitative* values in [0, 1]. Specifically, as shown in Fig. 3, UAD can estimate lower certainty values on unclear and ambiguous boundaries due to occlusion and shadow. For examples, UAD estimates the lower bottom-directional certainty value (*e.g.*, C_B: 34%) of the dog in the center-bottom image due to the its shadow. For the upper-mid image, the left-directional certainty value (C_L) of the bird is estimated by only 25% because the body of the bird is occluded by the tree branch. Besides, as the giraffe in the upper-rightmost image is occluded by another giraffe, it has the lower right- and bottom-directional certainties (57% and 36%). Hence, UAD can captures lower certainties on unclear or cloaked sides. From these results, we might expect that the estimated 4-directional quantitative values can be used as crucial information for the safety-critical application or decision-making system.

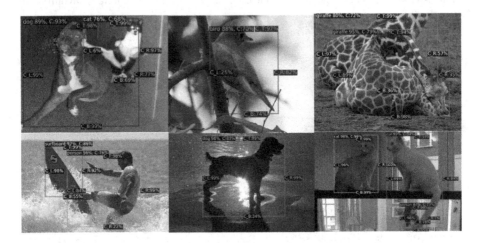

Fig. 3. Estimated uncertainty examples of the proposed UAD. Since there is no supervision of uncertainty, we analyze the estimated uncertainty qualitatively. UAD estimates lower certainties on unclear and ambiguous boundaries due to occlusion and shadow. For example, Both the surfboard and the person in the left-bottom image have much lower bottom-directional certainties (*i.e.*, C_B: 55% and 22%) as their shapes are quite unclear due to the water. Also, the right-directional certainty (C_R) of the woman in the right-bottom image is estimated only by 1% because it is covered by the tail of a cat on the TV.

5   Conclusion

We have proposed UAD that estimates 4-directions uncertainty for anchor-free object detectors. To this end, we design the new uncertainty loss, negative Power log-likelihood loss, to train the network that produces the localization uncertainty and enables accurate localization. Our uncertainty estimation method captures not only the quality of the detected box but also which direction is uncertain as a quantified value in [0, 1]. Furthermore, we also propose an uncertainty-aware focal loss and the certainty representation network for uncertainty-aware classification. It helps to correctly rank detected objects in the NMS step. Experiments on challenging COCO datasets demonstrate that UAD improves our baseline, FCOS, without the additional computational overhead. We hope the proposed UAD can serve as a component providing localization uncertainty as an essential cue and improving the performance of the anchor-free object detection methods.

Acknowledgement. This work was supported by Institute of Information & Communications Technology Planning & Evaluation(IITP) grant funded by the Korea government(MSIT) (No.2014-3-00123, Development of High Performance Visual BigData Discovery Platform for Large-Scale Realtime Data Analysis, No.2022-0-00124, Development of Artificial Intelligence Technology for Self-Improving Competency-Aware Learning Capabilities).

References

1. Cai, Z., Vasconcelos, N.: Cascade R-CNN: delving into high quality object detection. In: CVPR (2018)
2. Choi, J., Chun, D., Kim, H., Lee, H.J.: Gaussian YOLOv3: an accurate and fast object detector using localization uncertainty for autonomous driving. In: ICCV (2019)
3. Duan, K., Bai, S., Xie, L., Qi, H., Huang, Q., Tian, Q.: CenterNet: keypoint triplets for object detection. In: ICCV (2019)
4. Gal, Y.: Uncertainty in deep learning. Ph.D. thesis (2016)
5. Gal, Y., Ghahramani, Z.: Dropout as a Bayesian approximation: representing model uncertainty in deep learning. In: ICML (2016)
6. Girshick, R., Radosavovic, I., Gkioxari, G., Dollár, P., He, K.: Detectron (2018). https://github.com/facebookresearch/detectron
7. Grünwald, P., van Ommen, T.: Inconsistency of Bayesian inference for misspecified linear models, and a proposal for repairing it. Bayesian Anal. **12**(4), 1069–1103 (2017)
8. Harakeh, A., Smart, M., Waslander, S.L.: BayesOD: a Bayesian approach for uncertainty estimation in deep object detectors. arXiv:1903.03838 (2019)
9. He, K., Girshick, R., Dollar, P.: Rethinking ImageNet pre-training. In: ICCV (2019)
10. He, Y., Zhu, C., Wang, J., Savvides, M., Zhang, X.: Bounding box regression with uncertainty for accurate object detection. In: CVPR (2019)
11. Holmes, C.C., Walker, S.G.: Assigning a value to a power likelihood in a general Bayesian model. Biometrika **104**(2), 497–503 (2017)
12. Jiang, B., Luo, R., Mao, J., Xiao, T., Jiang, Y.: Acquisition of localization confidence for accurate object detection. In: Ferrari, V., Hebert, M., Sminchisescu, C., Weiss, Y. (eds.) Computer Vision – ECCV 2018. LNCS, vol. 11218, pp. 816–832. Springer, Cham (2018). https://doi.org/10.1007/978-3-030-01264-9_48
13. Kendall, A., Gal, Y.: What uncertainties do we need in Bayesian deep learning for computer vision? In: NeurIPS (2017)
14. Kleijn, B.J.K., van der Vaart, A.W.: The Bernstein-Von-Mises theorem under misspecification. Electron. J. Stat. **6**, 354–381 (2012)
15. Kraus, F., Dietmayer, K.: Uncertainty estimation in one-stage object detection. In: ITSC (2019)
16. Lakshminarayanan, B., Pritzel, A., Blundell, C.: Simple and scalable predictive uncertainty estimation using deep ensembles. In: NeurIPS (2017)
17. Law, H., Deng, J.: CornerNet: detecting objects as paired keypoints. In: ECCV, pp. 734–750 (2018)
18. Le, M.T., Diehl, F., Brunner, T., Knol, A.: Uncertainty estimation for deep neural object detectors in safety-critical applications. In: ITSC (2018)
19. Li, X., et al.: Generalized focal loss: learning qualified and distributed bounding boxes for dense object detection. In: NeurIPS (2020)

20. Lin, T.Y., Dollár, P., Girshick, R.B., He, K., Hariharan, B., Belongie, S.J.: Feature pyramid networks for object detection. In: CVPR (2017)
21. Lin, T.Y., Goyal, P., Girshick, R., He, K., Dollár, P.: Focal loss for dense object detection. In: ICCV (2017)
22. Lin, T.-Y., et al.: Microsoft COCO: common objects in context. In: Fleet, D., Pajdla, T., Schiele, B., Tuytelaars, T. (eds.) ECCV 2014. LNCS, vol. 8693, pp. 740–755. Springer, Cham (2014). https://doi.org/10.1007/978-3-319-10602-1_48
23. Liu, W., et al.: SSD: single shot multibox detector. In: Leibe, B., Matas, J., Sebe, N., Welling, M. (eds.) ECCV 2016. LNCS, vol. 9905, pp. 21–37. Springer, Cham (2016). https://doi.org/10.1007/978-3-319-46448-0_2
24. Lyddon, S.P., Holmes, C.C., Walker, S.G.: General Bayesian updating and the loss-likelihood bootstrap. Biometrika **106**(2), 465–478 (2019)
25. Redmon, J., Farhadi, A.: YOLOv3: an incremental improvement. arXiv:1804.02767 (2018)
26. Ren, S., He, K., Girshick, R., Sun, J.: Faster R-CNN: towards real-time object detection with region proposal networks. In: NeurIPS (2015)
27. Rezatofighi, H., Tsoi, N., Gwak, J., Sadeghian, A., Reid, I., Savarese, S.: Generalized intersection over union (2019)
28. Royall, R., Tsou, T.S.: Interpreting statistical evidence by using imperfect models: robust adjusted likelihood functions. J. Roy. Stat. Soc. Ser. B (Stat. Methodol.) **65**(2), 391–404 (2003)
29. Sarikaya, D., Corso, J.J., Guru, K.A.: Detection and localization of robotic tools in robot-assisted surgery videos using deep neural networks for region proposal and detection. IEEE Trans. Med. Imaging **36**, 1542–1549 (2017)
30. Shridhar, K., Laumann, F., Liwicki, M.: A comprehensive guide to Bayesian convolutional neural network with variational inference. arXiv:1901.02731 (2019)
31. Syring, N., Martin, R.: Calibrating general posterior credible regions. Biometrika **106**(2), 479–486 (2019)
32. Tian, Z., Shen, C., Chen, H., He, T.: FCOS: fully convolutional one-stage object detection. In: ICCV (2019)
33. Tian, Z., Shen, C., Chen, H., He, T.: FCOS: a simple and strong anchor-free object detector. IEEE Trans. Pattern Anal. Mach. Intell. (TPAMI) (2021)
34. Wu, S., Li, X., Wang, X.: IoU-aware single-stage object detector for accurate localization. Image Vision Comput. **97**, 103911 (2020)
35. Yu, J., Jiang, Y., Wang, Z., Cao, Z., Huang, T.: Unitbox: an advanced object detection network. In: MM (2016)
36. Zhang, H., Wang, Y., Dayoub, F., Sünderhauf, N.: VarifocalNet: an IoU-aware dense object detector. arXiv preprint arXiv:2008.13367 (2020)
37. Zhang, S., Chi, C., Yao, Y., Lei, Z., Li, S.Z.: Bridging the gap between anchor-based and anchor-free detection via adaptive training sample selection. In: CVPR (2020)
38. Zhang, S., Wen, L., Bian, X., Lei, Z., Li, S.Z.: Single-shot refinement neural network for object detection. In: CVPR (2018)
39. Zhou, X., Wang, D., Krähenbühl, P.: Objects as points. arXiv:1904.07850 (2019)
40. Zhou, X., Zhuo, J., Krahenbuhl, P.: Bottom-up object detection by grouping extreme and center points. In: CVPR (2019)
41. Zhu, C., Chen, F., Shen, Z., Savvides, M.: Soft anchor-point object detection. arXiv preprint arXiv:1911.12448 (2019)

Variational Depth Networks: Uncertainty-Aware Monocular Self-supervised Depth Estimation

Georgi Dikov[⊠] and Joris van Vugt

Qualcomm Technologies Netherlands B.V., Nijmegen, The Netherlands
gdikov@qti.qualcomm.com

Abstract. Using self-supervised learning, neural networks are trained to predict depth from a single image without requiring ground-truth annotations. However, they are susceptible to input ambiguities and it is therefore important to express the corresponding depth uncertainty. While there are a few truly monocular and self-supervised methods modelling uncertainty, none correlates well with errors in depth. To this end we present Variational Depth Networks (VDN): a probabilistic extension of the established monocular depth estimation framework, MonoDepth2, in which we leverage variational inference to learn a parametric, continuous distribution over depth, whose variance is interpreted as uncertainty. The utility of the obtained uncertainty is then assessed quantitatively in a 3D reconstruction task, using the ScanNet dataset, showing that the accuracy of the reconstructed 3D meshes highly correlates with the precision of the predicted distribution. Finally, we benchmark our results using 2D depth evaluation metrics on the KITTI dataset.

Keywords: Self-supervised learning · Depth estimation · Variational inference

1 Introduction

Depth estimation is an important task in computer vision, since it forms the basis of many algorithms in applications such as 3D scene reconstruction [2,38, 39,47,55] or autonomous driving [52,57,60] among others. Inferring depth from a single image is an inherently ill-posed problem due to a scale ambiguity: an object in an image will appear the same if it were twice as large and placed twice as far away [20]. Nevertheless, deep neural networks are able to provide reliable, dense depth estimates by learning relative object sizes from data [10]. To this end, there are two main learning paradigms: *supervised* training from

J. van Vugt—Work done while at Qualcomm Technologies Netherlands B.V.

Supplementary Information The online version contains supplementary material available at https://doi.org/10.1007/978-3-031-25085-9_3.

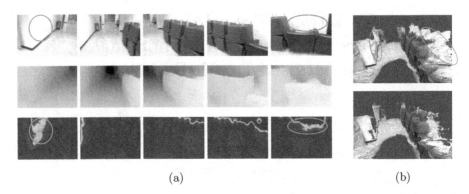

(a)   (b)

Fig. 1. (a) Top to bottom: input images from ScanNet [7] (scene 0019_00), predicted depths (far ■■■ close), predicted uncertainties (low ■■■ high); (b) Top: reconstructed scene from all dense depth predictions, bottom: reconstructed scene from filtered depths. Notice how the predicted uncertainty highlights regions (circled in red) for which the network would not have received meaningful error signal during self-supervised training, and is therefore susceptible to mistakes. Those can then be filtered using thresholded uncertainties as masks, leading to a sparser but more accurate scene reconstruction (Color figure online)

dense [12, 49, 51] or sparse [34] ground truth depth maps, e. g.obtained by a Time-of-Flight [22, 43] sensor such as LiDAR [6], and *self-supervised* training which exploits 3D geometric constraints to construct an auxiliary task of photometric consistency between different views of the same scene [13, 16, 17, 62]. The latter approach is particularly useful as it does not require ground truth depth images, and can be applied on sequences of frames taken by an ordinary, off-the-shelf monocular camera.

To reliably make use of the estimated depth in downstream tasks, a dense quantification of the uncertainty associated to the predictions is essential [28]. Consider the example given in Fig. 1 where the depth predictions for the overexposed, blank white walls are compromised (see red markings in Fig. 1a), leading to a noisy scene reconstruction as shown in the top part of Fig. 1b. To mitigate this, one can use the uncertainty maps to filter the potentially erroneous depth pixels and produce a sparser but more accurate mesh, cf.bottom of Fig. 1b. However, obtaining meaningful confidence values from a single image in a fully self-supervised learning setting is an especially challenging task, as the depth is only indirectly learnt. Consequently the majority of existing uncertainty-aware methods are either trained in a supervised fashion [4, 31, 35, 50], assume that multiple views are available at test time [26] or model other types of uncertainty, e. g.on the photometric error [58].

The goal of our work is to extend self-supervised depth training with principled uncertainty estimation. To that end, we present Variational Depth Networks (VDN)—an entirely monocular, probabilistically motivated approach to depth uncertainty. It builds upon established self-supervised methods and leverages advancements in approximate Bayesian learning. Specifically, VDN extends MonoDepth2 [17] to model the depth as a continuous distribution, whose parameters are optimised using the framework of variational inference [18, 30]. As a result, the network learns to assign high uncertainty to regions for which the depth can vary a lot without significantly increasing the photometric error, and low uncertainty otherwise. Building up on this idea, in Sect. 4 we also present a new method to quantitatively evaluate the utility of the uncertainty maps in a 3D reconstruction task using the ScanNet dataset [7], and benchmark the quality of the 2D depth predictions on the KITTI dataset [14]. In summary, our main contributions are as follows:

- We propose VDN as a novel probabilistic framework for monocular, self-supervised depth estimation, which uses approximate Bayesian inference to learn a continuous, parametric distribution over the depth. The uncertainty is then expressed as the variance of this distribution.
- We show qualitatively that the obtained uncertainty is more interpretable as it highlights regions in the image which are difficult to learn in a self-supervised setting.
- We also demonstrate that high confidence predictions are more likely to be accurate. For that, we propose an evaluation scheme based on the task of 3D scene reconstruction, where the depth uncertainty is used to filter unreliable predictions before fusion.

2   Related Work

Self-supervised Uncertainty. Self-supervised learning for monocular depth estimation was originally proposed by Zhou et al. [62]. Their core idea is that a network that predicts the depth and relative pose of a video frame can be optimised by using the photometric consistency with warped neighbouring frames as a loss function. They also include an *explainability mask* in their network to account for moving objects and non-Lambertian surfaces, which can be interpreted as a form of uncertainty estimation. Later, Godard et al. [17] consolidated several improvements into a conceptually simple method called MonoDepth2, which did not include the explainability mask since it did not have a significant impact on the accuracy of the estimated depth in practice.

Klodt and Vedaldi [31] were the first to probabilistically model the depth, pose and photometric error and use the estimated uncertainties to down-weight regions in the image that violate the colour constancy assumption made by the photometric objective function. The depth and poses are modelled through Laplacian distributions where the likelihoods of target depth and pose, obtained from a classical *Structure-from-Motion* system [36], are maximised. In contrast to their method, ours is self-contained, i.e.it does not rely on external teachers and therefore its performance is not bounded by the quality of those. In an analogous way, Yang et al. [58] also model the photometric error as a Laplacian distribution, and show that its variance can be used to improve the downstream task of visual odometry [59].

Alternatively, depth estimation can be reframed as a discrete classification problem, as proposed by Johnston and Carneiro [24], which allows for computing the variance without any additional prediction head in the network. However, their approach does not have strong guarantees on the quality of the output distribution [19] and in practice the variance appears to mostly inversely correlate with the predicted disparity except for the furthest regions in the image. On the other hand, Poggi et al. [42] present a comprehensive summary of various depth uncertainty estimation techniques for self-supervised learning and propose a combination of ensembling and self-teaching methods as an effective way to improve depth accuracy. They also propose evaluation metrics based on sparsification, which can be used to assess the quality of the predicted uncertainty. In our work we will compare to baselines from both [24] and [42].

Last, the shortcomings of photometric uncertainty estimation in the context of Multi-View Stereo [44] are addressed by Xu et al. [56] with the goal of directly improving the predicted depth. In contrast, we aim for a monocular method with interpretable uncertainty values.

Supervised Uncertainty. A fully supervised probabilistic approach is taken by Liu et al. [35], where the authors update a discrete depth probability volume (DPV) for each image, by fusing information from consecutive frames in an iterative Bayesian filtering fashion. Due to the discrete nature of the DPV, arbitrary distributions can be expressed, however to obtain an initial estimate for it, one needs to compute a cost volume from a number of frames in a video sequence. Moreover, their confidence maps show banding artefacts originating from the discrete depth representation in the cost volume.

Whereas most prior work uses a Laplacian or Gaussian distribution to model the depth and its uncertainty, ProbDepthNet from Brickwedde et al. [4] uses a Gaussian mixture model (GMM). The main benefit of GMMs is that they can represent multi-modal distributions, which can occur in foreground-background ambiguity. Walz et al. [50] propose a method for depth estimation on gated images and model the aleatoric depth uncertainty. Ke et al. [26] have the goal to improve scene reconstruction using depth uncertainty in a two-stage method: (i) predict a rough depth and uncertainty estimates using optical flow and triangulation from multiple frames; and (ii) refine the outputs of the first stage in an iterative procedure based on recurrent neural networks.

3   Methods

3.1   Background and Motivation

Fundamentals. Let $\mathcal{D} = \{I_t\}_{t=1...N}$ be a sequence of image frames and $T_{t\to s}$ the corresponding 3D camera motion from a target frame t to a source frame s. Further, let K denote the camera intrinsic matrix, projecting from 3D camera coordinates to 2D pixel coordinates $x \in \mathcal{X}$. Then, by exploiting 3D geometric constraints, one can cast the task of learning a depth map D_t for a frame I_t as a photometric consistency optimisation problem between the target and the warped source frames [13,17,62]:

$$\mathcal{L}_{\text{photo}}(I_t, D_t) = \sum_{x \in \mathcal{X}} \|I_s \langle K T_{t\to s} D_t(x) K^{-1} x \rangle - I_t(x)\|, \qquad (1)$$

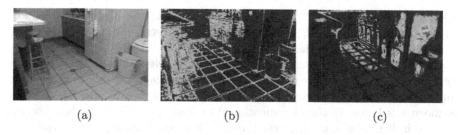

     (a)   (b)   (c)

Fig. 2. (a) A sample input image from ScanNet [7] (scene 0000_00); (b) Photometric uncertainty; (c) Variational depth uncertainty; (low ▬▬▬ high)

where $I_s\langle\cdot\rangle$ stands for a (bilinear) interpolation on the source frame I_s, following the notation of [17]. For the sake of notational brevity, here and throughout the rest of the paper we omit the dependencies on K as well as on I_s and $T_{t\to s}$ in the losses.

The estimated depth D_t is usually expressed as the inverse disparity output of a deterministic convolutional neural network μ_θ, parametrised by weights θ:

$$D_t = \mu_\theta(I_t)^{-1}. \tag{2}$$

For numerical reasons, the disparity output is activated by a sigmoid non-linearity and stretched to a predefined $\left[d_{\max}^{-1}, d_{\min}^{-1}\right]$ range. In practice, the loss from Eq. (1) is also extended to account for multiple source frames (e. g. using the *minimum reprojection error* [17]) and combined with other terms such as structural similarity [53] or smoothness regularisation [16,17]. In this paper we will refer to the extended loss as $\mathcal{L}_{\mathrm{photo}}$ and to the full model as MonoDepth2. Importantly, this will serve us as a basis framework for monocular, self-supervised depth learning upon which we will introduce a probabilistic extension in Sect. 3.2.

Uncertainty Estimation. Despite its wide-spread popularity, MonoDepth2 is not designed to account for the uncertainty associated with D_t. Following a paradigm of modelling the aleatoric uncertainty explicitly [28] one can reframe the loss from Eq. (1) into an exponential family likelihood with a learnable variance $\hat{\sigma}_\theta$:

$$p(I_t \mid D_t) \propto \frac{1}{\hat{\sigma}_\theta(I_t)} \exp\left(-\frac{\mathcal{L}_{\mathrm{photo}}(I_t, D_t)}{\hat{\sigma}_\theta(I_t)}\right), \tag{3}$$

where we abuse the notation for the weights θ and the neural network $\hat{\sigma}$, which may share only some of its parameters with μ. At this point, it is important to clarify that $\hat{\sigma}_\theta(I_t)$, as used in Eq. (3), accounts merely for the variance in the photometric error, $\mathcal{L}_{\mathrm{photo}}$, and not the predicted depth D_t. To give and intuitive explanation why the two uncertainties are not interchangeable, consider the following thought experiment: let all pixels in I_t and I_s have the same colour value. Then, for any predicted D_t and arbitrary $T_{t\to s}$ we have that

$I_t(x) = I_s \langle KT_{t \to s} D_t(x) K^{-1} x \rangle = I_s(x), \forall x \in \mathcal{X}$ and the likelihood from Eq. (3) is maximised with $\hat{\sigma}_\theta(I_t) \to 0$. Thus the photometric variance will collapse, while the actual depth variance is large.

In reality, this scenario can occur at large textureless surfaces, such as walls or overexposed regions close to light sources. Figures 2a and 2b show an example input and the corresponding photometric uncertainty. Notice how the network confidence is lowest in the aforementioned regions and highest on their boundaries or in high-frequency patterned areas, where small changes in D_t can substantially increase $\mathcal{L}_{\text{photo}}$. Thus, the photometric variance does not necessarily correlate with the uncertainty in the depth estimate, and in some cases it is even complementary to the latter. On the other hand, VDN is able to assign high depth variance to those regions, cf. Fig. 2c.

Despite that, the photometric uncertainty has been reported to quantitatively improve the depth estimates [42,58]. We hypothesise that this can be attributed to the effect of loss attenuation as the supervisory signal is not dominated by noise stemming from the difficult, depth sensitive areas such as non-Lambertian objects, similarly to the observations made by [28] in a supervised depth regression setup. Nevertheless, there are real-world applications, such as 3D scene reconstruction where proper depth uncertainty estimation is of greater importance, as we will show experimentally in Sect. 4.

3.2   Variational Depth Networks

Objective.   In the following, we will introduce a probabilistic extension to the self-supervised depth learning pipeline, in which the variance of the predicted depth maps can be reliably estimated. Intuitively speaking, we will assume that D_t is a random variable following some conditional distribution and we will make the image warping transformation in Eq. (1) aware of the probabilistic nature of D_t. We find this intuition to fit well into the Bayesian framework of reasoning and we will leverage approximate variational inference [18,25,30] to optimise a parametric distribution over D_t.

In essence, it requires that we specify a likelihood $p(I_t \mid D_t)$, a prior $p(D_t)$ and a posterior distribution $p(D_t \mid I_t)$ to which a tractable approximation, $q_\theta(D_t \mid I_t)$ is fit. Then, using q_θ we can derive a lower bound on the marginal log-likelihood:

$$\mathbb{E}_{p_D}[\log p_\theta(I_t)] = \mathbb{E}_{p_D}\left[\log \mathbb{E}_{q_\theta}\left[\frac{p(I_t \mid D_t)p(D_t)}{q_\theta(D_t \mid I_t)} \right] \right] \tag{4}$$

$$\geq \mathbb{E}_{p_D, q_\theta}\left[\log \frac{p(I_t \mid D_t)p(D_t)}{q_\theta(D_t \mid I_t)} \right]. \tag{5}$$

This can be further decomposed into a log-likelihood and a KL-divergence term, into the so-called *evidence lower bound*:

$$\mathcal{L}_{\text{ELBO}}(I_t, D_t) = \mathbb{E}_{p_D, q_\theta}[\log p(I_t \mid D_t)] \\ - \mathbb{E}_{p_D}[\text{KL}\left(q_\theta(D_t \mid I_t) \parallel p(D_t) \right)]. \tag{6}$$

One can show that maximising $\mathcal{L}_{\text{ELBO}}$ w.r.t. θ is equivalent to minimising $\mathbb{E}_{p_\mathcal{D}}[\text{KL}\left(q_\theta(D_t \mid I_t) \parallel p(D_t \mid I_t)\right)]$ thus closing the gap between the approximation and the underlying true posterior [18,25]. For the likelihood of VDN we choose an unnormalised density as in Eq. (3), however, throughout this work we will not model both the photometric and depth uncertainty, so as to isolate the effects of our contribution. In the subsequent sections we will specify the exact form of $q_\theta(D_t \mid I_t)$ and $p(D_t)$.

Approximate Posterior. In the context of depth estimation, one has to take into account two considerations when choosing a suitable family of variational distributions. First, it has to have a positive, bounded support over $[d_{\min}, d_{\max}]$ and, second, it has to allow for reparametrisation so that the weights θ can be learnt with backpropagation. One such candidate distribution is given by the truncated normal distribution [5], constrained to the aforementioned interval, whose location parameter is defined by the output of the neural network μ_θ and the scale by σ_θ, similarly to Eq. (3). Unlike the photometric variance $\hat{\sigma}_\theta$, σ_θ will have a direct relation to the variance of the estimated depth. For numerical reasons, however, it may be beneficial to express the approximate posterior over disparity instead of depth [17], and convert disparity samples to depth as per Eq. (2):

$$q_\theta(D_t^{-1} I_t) = \mathcal{N}_{\text{tr}}\left(D_t^{-1} \mid \mu_\theta(I_t), \sigma_\theta(I_t), d_{\max}^{-1}, d_{\min}^{-1}\right). \tag{7}$$

Backpropagating to μ_θ and σ_θ is possible through a reparametrisation using the inverse CDF function, which is readily implemented in TensorFlow [1,11] and in third-party packages [40] for Pytorch [41].

Here we assume that q_θ is a pixelwise factorised distribution and we obtain a disparity prediction using the mode, $\mu_\theta(I_t)$. Since we have defined a distribution over the disparity, it is not straightforward to obtain the mode of the transformed distribution over the depth, q_θ^{-1}. Fortunately however, for the given truncated normal parametrisation and the reciprocal transformation one can compute it analytically from the density of q_θ^{-1} using the change of variables trick, see Appendix A.2 for details:

$$\text{mode}\left(q_\theta^{-1}(D_t \mid I_t)\right) = \min(\max(m, d_{\min}), d_{\max}),$$
$$\text{where} \quad m = \frac{\sqrt{\mu_\theta(I_t)^2 + 8\sigma_\theta(I_t)^2} - \mu_\theta(I_t)}{4\sigma_\theta(I_t)^2}. \tag{8}$$

Finally, to obtain the estimated pixelwise depth uncertainty, one can compute the sample variance of q_θ^{-1}.

Prior. The choice of depth prior is particularly important for us because it can adversely bias the shape of the variational posterior. To understand the reason for that, one has to compare the VDN model with a regular VAE [30]: while both models encode the input image in a latent representation, a VDN does not use

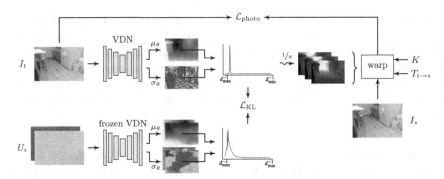

Fig. 3. Model overview of VDN with an example input from ScanNet [7] (scene 0000_00). Given a target image I_t, the subnetworks μ_θ and σ_θ predict the pixelswise location and scale parameters of the approximate posterior, resulting in a factorised distribution over disparities. Then, multiple samples are drawn and the reciprocal of each is used independently in a warping transformation of a source image I_s, assuming known intrinsics K and pose $T_{t \to s}$. The warped and interpolated source frames are used to compute the likelihood. The prior is given by the predicted location and scale parameters from a set of pseudo-inputs U_i as per [48]. The arrow $\rightsquigarrow$ denotes a sampling operation

a learnable decoder to form the likelihood but rather a fixed warping transformation. This means that a bias in the latent space, cannot be compensated for during decoding, resulting in hindered weight optimisation. For this reason we opt for a learnable prior given by the aggregated approximate posterior, which is provably the optimal prior for that task [46,48], see Appendix A.1 for details:

$$p^*(D_t) = \sum_{I_t \in \mathcal{D}} q_\theta(D_t \mid I_t) p_\mathcal{D}(I_t). \tag{9}$$

Unfortunately, however, the estimation of the aggregate posterior is computationally prohibitive for large, high-dimensional datasets. Therefore, we employ an approximation by Tomczak et al. [48], called *VampPrior*, where the prior is given as a mixture of the variational posteriors computed on a set of learnable pseudo inputs $\{U_i\}_{i=1...k}$:

$$p(D_t) \approx \frac{1}{k} \sum_{i=1}^{k} q_\theta(D_t \mid U_i). \tag{10}$$

Earlier, we expressed the approximate posterior in disparity- rather than depth-space and consequently the prior becomes a mixture distribution over disparities too. Since the KL-divergence is invariant to continuous, invertible transformations [33] (such as the reciprocal relation of depth and disparity), one can compute $\mathrm{KL}\left(q_\theta(D_t^{-1} \mid I_t) \| p(D_t^{-1})\right)$ instead. In summary, all of the components of VDN are presented in Fig. 3.

4 Experiments

4.1 Setup

Datasets

ScanNet. The ScanNet [7] dataset contains 1513 video sequences collected in indoor environments, annotated with 3D poses, dense depth maps and reconstructed meshes. The reason to use this dataset is to evaluate the per-image depth and uncertainty estimation and to assess the utility of uncertainty in 3D reconstruction. Consequently, we use the ground-truth poses to compute the photometric error instead of predicting them. For training we only consider every 10th frame as target to reduce redundancy and for each, we find a source frame both backwards and forwards in time with a relative translation of 5–10cm and a relative rotation of at most 5°. All images are resized to 384 × 256 pixels. We use the ground-truth poses to compute the photometric error and do not train a network to predict the pose since we want to focus our analysis in this experiment on the depth and uncertainty estimates only.

KITTI. The KITTI dataset [14] is an established benchmark dataset for depth estimation research and consists of 61 sequences collected from a vehicle. Following [17], we use the Eigen split [12], resize the input images to 640 × 192 and evaluate against LiDAR ground-truth capped at 80 m. Unlike the ScanNet experiments, here the camera poses are learnt the same way as in [17] so as to allow for fair comparison.

Metrics

3D. Previous works on 3D reconstruction [3,37,45] use point-to-point distances as the basis for comparing to ground-truth meshes. They convert each mesh to a point cloud by only considering its vertices, or by sampling points on the faces, essentially discarding the surface information of the mesh. However, if a predicted point lies on the surface of the ground-truth mesh it can still incur a non-zero error since only the distance to the closest vertex is considered. To mitigate this, we propose to use a cloud-to-mesh (c → m) distance as a basis for our 3D reconstruction error computation, which is readily available in open-source software like CloudCompare [15]. Given a mesh $\mathcal{M} = (\mathcal{V}, \mathcal{F})$, where $\mathcal{V}$ denotes the vertices and $\mathcal{F}$ the faces, we compute the accuracy as the fraction of vertices for which the Euclidean distance to the closest face $f' \in \mathcal{F}'$ in another mesh $\mathcal{M}'$ is smaller than a threshold ϵ:

$$\mathrm{acc}_{c \to m}(\mathcal{M}, \mathcal{M}') = \frac{1}{|\mathcal{V}|} \sum_{v \in \mathcal{V}} \mathbb{1} \left[\min_{f' \in \mathcal{F}'} \mathrm{dist}(v, f') < \epsilon \right]. \tag{11}$$

Here $\mathbb{1}[\cdot]$ denotes the indicator function. Given predicted and ground-truth meshes, $\mathcal{M}_{\mathrm{pred}}$ and $\mathcal{M}_{\mathrm{gt}}$ respectively, we define the *precision* as $\mathrm{acc}_{c \to m}(\mathcal{M}_{\mathrm{pred}},$

$\mathcal{M}_{gt}$) and the *recall* as $\text{acc}_{c \to m}(\mathcal{M}_{gt}, \mathcal{M}_{pred})$. The F-score is the harmonic mean of the precision and recall [32]. Following standard practices in 3D reconstruction literature [37,45] we use a threshold of $\epsilon = 5\,\text{cm}$ in all our evaluations.

2D. For the evaluation of the 2D predicted depth maps we compute the widely used metrics proposed by Eigen et al. [12]. Uncertainty is evaluated using sparsification curves [23] and the Area Under the Sparsification Error (AUSE) and Area Under the Random Gain (AURG) as proposed by Poggi et al. [42]. Note AURG and AUSE are computed w.r.t.another 2D depth metric and therefore comparison among different models is fair only when they perform similarly on that metric too.

Implementation Details

Network Architectures and Training Details. Even though our model architecture strictly follows [17] there are a couple of deviations. In particular, to accommodate the prediction of the distribution location and scale parameters, we duplicate the original disparity decoder architecture and, for the scale parameter only, change the output activation to linear. To avoid numerical instability issues with the scale, we clip it to the $[10^{-6}, 3]$ interval. In all our experiments we use a ResNet-18 encoder [21], pretrained on ImageNet [9], the Adam optimiser [29] with an initial learning rate of 10^{-4} which we reduce by a factor of 10 after 30 epochs, for a total of 40 epochs. The VampPrior for our VDN models is computed as described in Sect. 3.2 with 20 pseudo-inputs, which we initialise by broadcasting a random colour value over the height and width dimensions. To estimate the loss $\mathcal{L}_{ELBO}$ from Eq. (6) the approximate posterior is sampled 10 times.

3D Reconstruction. We use the TSDF-fusion algorithm implemented in Open3D [61] to reconstruct ScanNet [7] scenes. To speed up reconstruction, we only integrate every 10th frame and, during fusion, we use a sample size of $5\,\text{cm}$ and a truncation distance of $20\,\text{cm}$. For evaluation we use the ground-truth meshes provided with the dataset.

4.2   ScanNet: Uncertainty-Aware Reconstruction

To evaluate the usefulness of the predicted uncertainty we use the task of 3D reconstruction on ScanNet [7] scenes. In this experiment we leverage the depth uncertainty for measurement selection by masking out pixels with uncertainty above a preselected threshold during the integration process. We compare our method to several other recently proposed depth uncertainty estimation methods, all implemented on top of the same MonoDepth2 framework. *Photometric uncertainty* refers to Eq. (3), which is used by D3VO [58] to improve visual odometry. *Self-teaching* refers to the method proposed by Poggi et al. [42], where we use the model without uncertainty as a teacher for training the student network

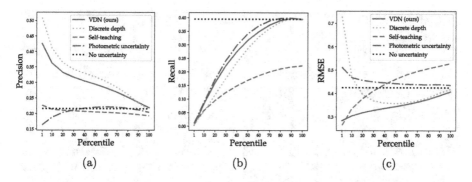

Fig. 4. ScanNet: mean reconstruction precision (a) and recall (b) as well as 2D depth RMSE (c) curves on the validation set for various filtering thresholds on the uncertainty. A monotonically decreasing precision curve indicates that the uncertainty correlates well with the errors in the depth maps used for fusion while a higher recall means that smaller portions of the geometry are being removed

in a supervised way. *Discrete depth* predicts a discrete disparity volume [24], from which continuous depth and variance can be derived. Each of these methods constitute a fair baseline as all are fully self-supervised and monocular.

Table 1. ScanNet: 2D depth, 2D uncertainty and 3D reconstruction metrics. All methods are based on the same MonoDepth2 architecture and are our own (re)implementations. ↑ and ↓ denote if higher or lower score is better

Method	Abs Rel ↓	Sq Rel ↓	RMSE ↓	RMSE log ↓	δ < 1.25 ↑	δ < 1.25² ↑	δ < 1.25³ ↑	AUSE[a] ↓	AURG[a] ↑	Precision ↑	Recall ↑	F-score ↑
No uncertainty (MonoDepth2)	0.146	0.088	0.425	0.204	0.800	0.948	0.985	-	-	0.216	0.395	0.276
Photometric uncertainty	0.154	0.098	0.426	0.215	0.787	0.940	0.979	0.102	-0.008	0.204	0.395	0.266
Self-teaching [42]	0.170	0.115	0.529	0.246	0.690	0.914	0.975	0.056	0.034	0.194	0.223	0.204
Discrete depth [24]	0.147	0.086	0.419	0.202	0.796	0.946	0.984	0.091	-0.001	0.212	0.392	0.272
VDN (fixed prior)	0.148	0.093	0.416	0.211	0.797	0.944	0.981	0.083	0.006	0.217	0.392	0.272
VDN (VampPrior)	0.144	0.085	0.402	0.194	0.801	0.948	0.985	0.083	0.003	0.219	0.395	0.279
VDN (fixed prior, 10 scenes)	0.414	0.495	1.036	0.787	0.318	0.546	0.675	0.125	0.094	0.096	0.262	0.138
VDN (VampPrior, 10 scenes)	0.287	0.238	0.680	0.348	0.494	0.797	0.931	0.152	0.008	0.103	0.261	0.144

[a] Measured on Abs Rel.

Table 1 summarises the results for the standard 2D depth and 3D reconstruction metrics. First, we note that photometric uncertainty performs considerably worse than the other methods. Discrete depth performs generally on par with the No uncertainty baseline. VDN slightly outperforms all baselines on most metrics. Figure 4a shows the mean reconstruction precision when increasing the uncertainty threshold at which predictions are considered valid. We expect to see a downwards trend, as using more uncertain predictions should decrease the accuracy of the reconstructed mesh. Here, the photometric uncertainty does not show this behaviour, whereas the variational and discrete uncertainty do show it, with discrete generally having a higher precision everywhere except when using more than 90% of the pixels. Conversely, Fig. 4b shows the mean reconstruction recall where a rapid increase signifies that larger pieces of the scene geometry are being cut out. For the sake of completeness, in Fig. 4c we also a provide

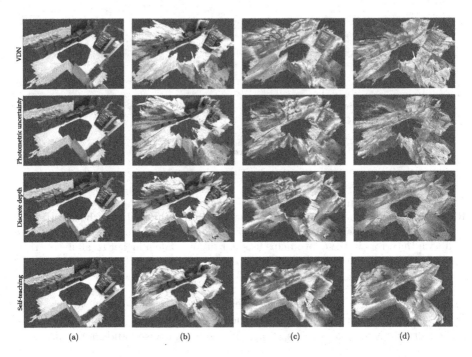

Fig. 5. (a) Meshes constructed using the ground-truth depth maps from ScanNet [7] (scene 0019_00); (b) Coloured meshes using the predicted depths; (c) Meshes from predicted depth, coloured by the cloud-to-mesh distances from the ground-truth; (d) Meshes from predicted depth, coloured by the depth uncertainty; (low ▬▬▬ high)

similar plots for the mean RMSE as measured on 2D depth images. Figures 5a and 5b show reconstructions from ground-truth and predicted depths for all uncertainty-aware baselines, and Figs. 5c and 5d depict the corresponding cloud-to-mesh distances and uncertainties. Notice how the photometric uncertainty anti-correlates with the precision, while the discrete depth merely increases the uncertainty with the distance from the camera. The output of the self-teaching model is not very interpretable either as it models the aleatoric noise in the teacher network. More qualitative examples are disclosed in Appendix B.

4.3 ScanNet: Prior Ablation Study

To investigate the adverse effects of naively specifying a prior distribution over the disparity, we compare the VampPrior against a truncated normal distribution with fixed location and scale parameters at 0.5 and 2.0 respectively, in two training scenarios: on the full training data and on a subset of 10 scenes only. The latter setup is especially interesting because it exacerbates any undesirable influence the prior might have onto the approximate posterior due to the lack of sufficient training data. While both priors are capable of regularising the spread of the variational posterior, the VampPrior shows superior results as presented

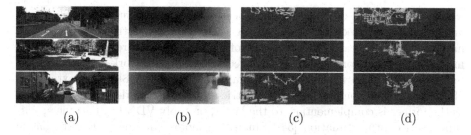

(a)	(b)	(c)	(d)

Fig. 6. (a) Sample input images from the Eigen test split [12] in KITTI [14]; (b) Predicted disparities; (c) Predicted disparity variance; (d) Estimated depth variance using 100 samples

Table 2. KITTI: 2D depth and uncertainty evaluation results on the Eigen test split [12] with raw LiDAR ground truth (80 m)

Method	Abs Rel ↓	Sq Rel ↓	RMSE ↓	RMSE log ↓	$\delta < 1.25$ ↑	$\delta < 1.25^2$ ↑	$\delta < 1.25^3$ ↑	Abs Rel AUSE ↓	Abs Rel AURG ↑	RMSE AUSE ↓	RMSE AURG ↑	$\delta < 1.25$ AUSE ↓	$\delta < 1.25$ AURG ↑
No uncertainty (MonoDepth2 [17])	0.115	0.882	4.791	0.190	0.879	0.961	0.982	-	-	-	-	-	-
Boot+Self [42]ᵃ	0.111	0.826	4.667	0.184	0.880	0.961	0.983	0.033	0.040	2.124	1.857	0.033	0.077
Photometric uncertainty [42]ᵃ	0.113	0.928	4.919	0.192	0.876	0.958	0.981	0.051	0.027	3.097	1.188	0.060	0.056
VDN (ours)	0.117	0.882	4.815	0.195	0.873	0.959	0.981	0.058	0.018	1.942	2.140	0.085	0.030

ᵃ The scores are taken from Tables 10 and 13 in the supplementary material of [42].

in the bottom half of Table 1. In particular, in the low data regime, it achieves significantly better scores on most metrics.

4.4 KITTI: 2D Depth Evaluation

In order to benchmark VDN on the KITTI dataset [14] against comparable prior work, we have selected as baselines the original MonoDepth2 [17], referred to as *No uncertainty*, the MonoDepth2 (*Boot+Self*) from Poggi et al. [42], which does account for depth uncertainty through self-teaching and bootstrapped ensemble learning, and the *Photometric uncertainty* baseline also presented in [42] under the name *MonoDepth2-Log*. Table 2 shows the depth and uncertainty results for the VDN and the baselines. Our model performs slightly worse than the baselines except for the RMSE-AUSE and AURG metrics, which we attribute to the increased amount of noise during training, stemming from the stochastic sampling operations. Figure 6a shows three example inputs from the test set with their corresponding disparity location and scale predicted parameters in Figs. 6b and 6c. The resulting depth uncertainty is illustrated in Fig. 6d, which highlights the depth ambiguity of the sky and distant, indistinguishable objects.

5 Conclusions

We have presented a probabilistic extension of MonoDepth2, which learns a parametric posterior distribution over depth. The method yields useful uncertainty, which correlates well with the error in the depth predictions and consequently, we have shown that one can use the uncertainty to mask out unreliable pixels and improve the precision of meshes in a 3D scene reconstruction task. Such masking, however, can come at a cost of decreased recall, resulting in sparser meshes. It is therefore a promising direction for future work to combine our method with a disparity [27] or mesh completion algorithm [8]. Other extensions of our work could combine the photometric and variational depth uncertainties, as the former is complementary to the latter, or apply VDN to multi-view, self-supervised depth estimation [54]. Finally, we note that due to the stochastic nature of our method, it is moderately demanding on computation and memory resources during training, as an additional forward-pass is needed for the VampPrior, and multiple samples are drawn from the approximate posterior to estimate the likelihood and KL-divergence terms of the loss. In addition, the depth uncertainty is computed from samples of the transformed disparity posterior. For the training and evaluation of all models we have used a single NVIDIA RTX A5000 GPU with 24 GB of memory.

Acknowledgements. We highly appreciate the constructive feedback and suggestions from our colleagues Mohsen Ghafoorian and Alex Bailo as well as the consistent support from Gerhard Reitmayr and Eduardo Esteves.

References

1. Abadi, M., et al.: TensorFlow: large-scale machine learning on heterogeneous systems (2015). https://www.tensorflow.org/
2. Bloesch, M., Czarnowski, J., Clark, R., Leutenegger, S., Davison, A.J.: CodeSLAM-learning a compact, optimisable representation for dense visual slam. In: Proceedings of the IEEE Conference on Computer Vision and Pattern Recognition, pp. 2560–2568 (2018)
3. Božič, A., Palafox, P., Thies, J., Dai, A., Nießner, M.: TransformerFusion: monocular RGB scene reconstruction using transformers. arXiv preprint arXiv:2107.02191 (2021)
4. Brickwedde, F., Abraham, S., Mester, R.: Mono-SF: multi-view geometry meets single-view depth for monocular scene flow estimation of dynamic traffic scenes. In: Proceedings of the IEEE/CVF International Conference on Computer Vision, pp. 2780–2790 (2019)
5. Burkardt, J.: The truncated normal distribution. Department of Scientific Computing Website, Florida State University, pp. 1–35 (2014). https://people.sc.fsu.edu/jburkardt/presentations/truncated_normal.pdf
6. Christian, J.A., Cryan, S.: A survey of lidar technology and its use in spacecraft relative navigation. In: AIAA Guidance, Navigation, and Control (GNC) Conference, p. 4641 (2013)
7. Dai, A., Chang, A.X., Savva, M., Halber, M., Funkhouser, T., Nießner, M.: ScanNet: richly-annotated 3D reconstructions of indoor scenes. In: Proceedings of the Computer Vision and Pattern Recognition (CVPR). IEEE (2017)
8. Dai, A., Ruizhongtai Qi, C., Nießner, M.: Shape completion using 3D-encoder-predictor CNNs and shape synthesis. In: Proceedings of the IEEE Conference on Computer Vision and Pattern Recognition, pp. 5868–5877 (2017)

9. Deng, J., Dong, W., Socher, R., Li, L.J., Li, K., Fei-Fei, L.: ImageNet: a large-scale hierarchical image database. In: 2009 IEEE Conference on Computer Vision and Pattern Recognition, pp. 248–255. IEEE (2009)

10. Dijk, T.V., Croon, G.D.: How do neural networks see depth in single images? In: Proceedings of the IEEE/CVF International Conference on Computer Vision, pp. 2183–2191 (2019)

11. Dillon, J.V., et al.: Tensorflow distributions. arXiv preprint arXiv:1711.10604 (2017)

12. Eigen, D., Puhrsch, C., Fergus, R.: Depth map prediction from a single image using a multi-scale deep network. arXiv preprint arXiv:1406.2283 (2014)

13. Garg, R., B.G., V.K., Carneiro, G., Reid, I.: Unsupervised CNN for single view depth estimation: geometry to the rescue. In: Leibe, B., Matas, J., Sebe, N., Welling, M. (eds.) ECCV 2016. LNCS, vol. 9912, pp. 740–756. Springer, Cham (2016). https://doi.org/10.1007/978-3-319-46484-8_45

14. Geiger, A., Lenz, P., Stiller, C., Urtasun, R.: Vision meets robotics: the KITTI dataset. Int. J. Robot. Res. (IJRR) **32**, 1231–1237 (2013)

15. Girardeau-Montaut, D.: Cloudcompare. France: EDF R&D Telecom ParisTech (2016). https://www.cloudcompare.org/

16. Godard, C., Mac Aodha, O., Brostow, G.J.: Unsupervised monocular depth estimation with left-right consistency. In: Proceedings of the IEEE Conference on Computer Vision and Pattern Recognition, pp. 270–279 (2017)

17. Godard, C., Mac Aodha, O., Firman, M., Brostow, G.J.: Digging into self-supervised monocular depth estimation. In: Proceedings of the IEEE/CVF International Conference on Computer Vision, pp. 3828–3838 (2019)

18. Graves, A.: Practical variational inference for neural networks. Adv. Neural Inf. Process. Syst. **24** (2011)

19. Guo, C., Pleiss, G., Sun, Y., Weinberger, K.Q.: On calibration of modern neural networks. In: International Conference on Machine Learning, pp. 1321–1330. PMLR (2017)

20. Hartley, R., Zisserman, A.: Multiple View Geometry in Computer Vision. Cambridge University Press (2004). https://doi.org/10.1017/cbo9780511811685

21. He, K., Zhang, X., Ren, S., Sun, J.: Deep residual learning for image recognition. In: Proceedings of the IEEE Conference on Computer Vision and Pattern Recognition, pp. 770–778 (2016)

22. Horaud, R., Hansard, M., Evangelidis, G., Ménier, C.: An overview of depth cameras and range scanners based on time-of-flight technologies. Mach. Vis. Appl. **27**(7), 1005–1020 (2016). https://doi.org/10.1007/s00138-016-0784-4

23. Ilg, E., et al.: Uncertainty estimates and multi-hypotheses networks for optical flow. In: Proceedings of the European Conference on Computer Vision (ECCV), pp. 652–667 (2018)

24. Johnston, A., Carneiro, G.: Self-supervised monocular trained depth estimation using self-attention and discrete disparity volume. In: Proceedings of the IEEE/CVF Conference on Computer Vision and Pattern Recognition, pp. 4756–4765 (2020)

25. Jordan, M.I., Ghahramani, Z., Jaakkola, T.S., Saul, L.K.: An introduction to variational methods for graphical models. Mach. Learn. **37**(2), 183–233 (1999)

26. Ke, T., Do, T., Vuong, K., Sartipi, K., Roumeliotis, S.I.: Deep multi-view depth estimation with predicted uncertainty. In: 2021 IEEE International Conference on Robotics and Automation (ICRA), pp. 9235–9241. IEEE (2021)

27. Keltjens, B., van Dijk, T., de Croon, G.: Self-supervised monocular depth estimation of untextured indoor rotated scenes. arXiv preprint arXiv:2106.12958 (2021)

28. Kendall, A., Gal, Y.: What uncertainties do we need in Bayesian deep learning for computer vision? arXiv preprint arXiv:1703.04977 (2017)
29. Kingma, D.P., Ba, J.L.: Adam: a method for stochastic gradient descent. In: ICLR: International Conference on Learning Representations, pp. 1–15 (2015)
30. Kingma, D.P., Welling, M.: Auto-encoding variational bayes. In: Bengio, Y., LeCun, Y. (eds.) 2nd International Conference on Learning Representations, ICLR 2014, Banff, AB, Canada, 14–16 April 2014, Conference Track Proceedings (2014). http://arxiv.org/abs/1312.6114
31. Klodt, M., Vedaldi, A.: Supervising the new with the old: learning SFM from SFM. In: Proceedings of the European Conference on Computer Vision (ECCV), pp. 698–713 (2018)
32. Knapitsch, A., Park, J., Zhou, Q.Y., Koltun, V.: Tanks and temples: Benchmarking large-scale scene reconstruction. ACM Trans. Graph. (ToG) 36(4), 1–13 (2017)
33. Kullback, S., Leibler, R.A.: On information and sufficiency. Ann. Math. Stat. 22(1), 79–86 (1951)
34. Kuznietsov, Y., Stuckler, J., Leibe, B.: Semi-supervised deep learning for monocular depth map prediction. In: Proceedings of the IEEE Conference on Computer Vision and Pattern Recognition, pp. 6647–6655 (2017)
35. Liu, C., Gu, J., Kim, K., Narasimhan, S.G., Kautz, J.: Neural RGB (R) D sensing: depth and uncertainty from a video camera. In: Proceedings of the IEEE/CVF Conference on Computer Vision and Pattern Recognition, pp. 10986–10995 (2019)
36. Mur-Artal, R., Montiel, J.M.M., Tardos, J.D.: Orb-slam: a versatile and accurate monocular slam system. IEEE Trans. Rob. 31(5), 1147–1163 (2015)
37. Murez, Z., van As, T., Bartolozzi, J., Sinha, A., Badrinarayanan, V., Rabinovich, A.: Atlas: end-to-end 3D scene reconstruction from posed images. In: Vedaldi, A., Bischof, H., Brox, T., Frahm, J.-M. (eds.) ECCV 2020, Part VII. LNCS, vol. 12352, pp. 414–431. Springer, Cham (2020). https://doi.org/10.1007/978-3-030-58571-6_25
38. Newcombe, R.A., et al.: KinectFusion: real-time dense surface mapping and tracking. In: 2011 10th IEEE International Symposium on Mixed and Augmented Reality, pp. 127–136. IEEE (2011)
39. Nießner, M., Zollhöfer, M., Izadi, S., Stamminger, M.: Real-time 3D reconstruction at scale using voxel hashing. ACM Trans. Graph. (ToG) 32(6), 1–11 (2013)
40. Obukhov, A.: Truncated normal distribution in PyTorch (2020), https://github.com/toshas/torch_truncnorm
41. Paszke, A., et al.: Pytorch: An imperative style, high-performance deep learning library. In: Wallach, H., Larochelle, H., Beygelzimer, A., d' Alché-Buc, F., Fox, E., Garnett, R. (eds.) Advances in Neural Information Processing Systems, vol. 32, pp. 8024–8035. Curran Associates, Inc. (2019). http://papers.neurips.cc/paper/9015-pytorch-an-imperative-style-high-performance-deep-learning-library.pdf
42. Poggi, M., Aleotti, F., Tosi, F., Mattoccia, S.: On the uncertainty of self-supervised monocular depth estimation. In: Proceedings of the IEEE/CVF Conference on Computer Vision and Pattern Recognition, pp. 3227–3237 (2020)
43. Remondino, F., Stoppa, D.: TOF Range-imaging Cameras, vol. 68121. Springer, Heidelberg (2013). https://doi.org/10.1007/978-3-642-27523-4
44. Seitz, S.M., Curless, B., Diebel, J., Scharstein, D., Szeliski, R.: A comparison and evaluation of multi-view stereo reconstruction algorithms. In: 2006 IEEE Computer Society Conference on Computer Vision and Pattern Recognition (CVPR 2006), vol. 1, pp. 519–528. IEEE (2006)

45. Sun, J., Xie, Y., Chen, L., Zhou, X., Bao, H.: NeuralRecon: real-time coherent 3D reconstruction from monocular video. In: Proceedings of the IEEE/CVF Conference on Computer Vision and Pattern Recognition, pp. 15598–15607 (2021)
46. Takahashi, H., Iwata, T., Yamanaka, Y., Yamada, M., Yagi, S.: Variational autoencoder with implicit optimal priors. In: Proceedings of the AAAI Conference on Artificial Intelligence, vol. 33, pp. 5066–5073 (2019)
47. Tateno, K., Tombari, F., Laina, I., Navab, N.: CNN-SLAM: real-time dense monocular SLAM with learned depth prediction. In: Proceedings of the IEEE Conference on Computer Vision and Pattern Recognition, pp. 6243–6252 (2017)
48. Tomczak, J., Welling, M.: VAE with a VampPrior. In: International Conference on Artificial Intelligence and Statistics, pp. 1214–1223. PMLR (2018)
49. Ummenhofer, B., et al.: DeMoN: depth and motion network for learning monocular stereo. In: Proceedings of the IEEE Conference on Computer Vision and Pattern Recognition, pp. 5038–5047 (2017)
50. Walz, S., Gruber, T., Ritter, W., Dietmayer, K.: Uncertainty depth estimation with gated images for 3D reconstruction. In: 2020 IEEE 23rd International Conference on Intelligent Transportation Systems (ITSC), pp. 1–8. IEEE (2020)
51. Wang, C., Buenaposada, J.M., Zhu, R., Lucey, S.: Learning depth from monocular videos using direct methods. In: Proceedings of the IEEE Conference on Computer Vision and Pattern Recognition, pp. 2022–2030 (2018)
52. Wang, Y., Chao, W.L., Garg, D., Hariharan, B., Campbell, M., Weinberger, K.Q.: Pseudo-lidar from visual depth estimation: bridging the gap in 3D object detection for autonomous driving. In: Proceedings of the IEEE/CVF Conference on Computer Vision and Pattern Recognition, pp. 8445–8453 (2019)
53. Wang, Z., Bovik, A.C., Sheikh, H.R., Simoncelli, E.P.: Image quality assessment: from error visibility to structural similarity. IEEE Trans. Image Process. **13**(4), 600–612 (2004)
54. Watson, J., Mac Aodha, O., Prisacariu, V., Brostow, G., Firman, M.: The temporal opportunist: self-supervised multi-frame monocular depth. In: Proceedings of the IEEE/CVF Conference on Computer Vision and Pattern Recognition, pp. 1164–1174 (2021)
55. Whelan, T., Leutenegger, S., Salas-Moreno, R., Glocker, B., Davison, A.: ElasticFusion: dense slam without a pose graph. In: Robotics: Science and Systems (2015)
56. Xu, H., et al.: Digging into uncertainty in self-supervised multi-view stereo. In: Proceedings of the IEEE/CVF International Conference on Computer Vision, pp. 6078–6087 (2021)
57. Xu, Y., Zhu, X., Shi, J., Zhang, G., Bao, H., Li, H.: Depth completion from sparse LiDAR data with depth-normal constraints. In: Proceedings of the IEEE/CVF International Conference on Computer Vision, pp. 2811–2820 (2019)
58. Yang, N., Stumberg, L.v., Wang, R., Cremers, D.: D3VO: deep depth, deep pose and deep uncertainty for monocular visual odometry. In: Proceedings of the IEEE/CVF Conference on Computer Vision and Pattern Recognition, pp. 1281–1292 (2020)
59. Yang, N., Wang, R., Stuckler, J., Cremers, D.: Deep virtual stereo odometry: leveraging deep depth prediction for monocular direct sparse odometry. In: Proceedings of the European Conference on Computer Vision (ECCV), pp. 817–833 (2018)
60. Yin, Z., Shi, J.: Geonet: Unsupervised learning of dense depth, optical flow and camera pose. In: Proceedings of the IEEE Conference on Computer Vision and Pattern Recognition, pp. 1983–1992 (2018)

Unsupervised Joint Image Transfer and Uncertainty Quantification Using Patch Invariant Networks

Christoph Angermann$^{(\boxtimes)}$ [iD], Markus Haltmeier [iD], and Ahsan Raza Siyal

Department of Mathematics, University of Innsbruck,
Technikerstraße 13, 6020 Innsbruck, Austria
christoph.angermann@uibk.ac.at
http://applied-math.uibk.ac.at/

Abstract. Unsupervised image transfer enables intra- and inter-modality image translation in applications where a large amount of paired training data is not abundant. To ensure a structure-preserving mapping from the input to the target domain, existing methods for unpaired image transfer are commonly based on cycle-consistency, causing additional computational resources and instability due to the learning of an inverse mapping. This paper presents a novel method for uni-directional domain mapping that does not rely on any paired training data. A proper transfer is achieved by using a GAN architecture and a novel generator loss based on patch invariance. To be more specific, the generator outputs are evaluated and compared at different scales, also leading to an increased focus on high-frequency details as well as an implicit data augmentation. This novel patch loss also offers the possibility to accurately predict aleatoric uncertainty by modeling an input-dependent scale map for the patch residuals. The proposed method is comprehensively evaluated on three well-established medical databases. As compared to four state-of-the-art methods, we observe significantly higher accuracy on these datasets, indicating great potential of the proposed method for unpaired image transfer with uncertainty taken into account. Implementation of the proposed framework is released here: https://github.com/anger-man/unsupervised-image-transfer-and-uq.

Keywords: Unsupervised image transfer · Uncertainty quantification · Generative adversarial network · Patch invariance · Modality propagation · Accelerated MRI · Radiotherapy

1 Introduction

Image transfer, e.g., within and across medical imaging modalities, has gained a lot of popularity in the last decade of research [1]. The application range of inter- and intra-modality image translation is multifaceted and can help to overcome key weaknesses of an acquisition method. For example, modality propagation in

L. Karlinsky et al. (Eds.): ECCV 2022 Workshops, LNCS 13808, pp. 61–77, 2023.
https://doi.org/10.1007/978-3-031-25085-9_4

magnetic resonance imaging (MRI) is of high interest since acquisition of multiple contrasts is crucial for better diagnosis for many clinical protocols [2]. Especially acquiring T1-weighted (T1w) and T2-weighted (T2w) contrasts increases scanning time significantly and thus is often formulated as a translation task from T1w to T2w. Another example is accelerated MRI, where medical costs and patient stress are minimized by decreasing the amount of k-space measurements [3,4]. Such methods for MRI reconstruction of undersampled measurements allow the use of MRI in applications where it is currently too time and resource intensive. A third application to mention here is automated computed tomography (CT) synthesis based on MRI images, which allows for MRI-only treatment planning in radiation therapy. A MRI-to-CT synthesizing method can eliminate the need for CT simulation and therefore improves the treatment workflow and reduces radiation exposure for the patient during radiotherapy [1,5–8].

Just discussed applications correspond to (un)supervised image transfer, which targets at translating an image from one domain to another. Supervised approaches exploit the inter-domain correspondence between input and output data [9]. These methods rely on large amounts of paired data and perfectly registered scans of the same patient, which are not always abundant in medical applications [5]. Unsupervised approaches are commonly built on a generative adversarial network (GAN) [10] that assimilates the distribution of the generated samples to the real distribution of the target domain by employing an adversarial discriminator network. To ensure that the synthesized output does not become irrelevant to the input, additional constraints may be added to the generator loss [11–14]. Especially cycle-consistency is a well-received method for structure preservation in fully unsupervised medical image transfer [6–9]. However, a cycle-consistent GAN (cycleGAN) requires the parallel learning of an inverse mapping. As a result, the training time is significantly increased and the final performance depends on the inverse transfer function. Furthermore, cycle-consistent GANs compare the reconstruction on a whole-image base and therefore pay less attention to fine structures and high-frequency details.

Although current methods provide powerful tools for high-dimensional image transfer, the generator loss calculation usually assumes that the learned mappings are correct. This represents a significant source for instability and erroneous predictions when considering out-of-distribution (OOD) data during training [9,15]. Tackling data-dependent uncertainty in deep computer vision has raised a lot of interest in the last years of research and provided effective tools to check the reliability of a model's predictions in supervised applications. However, research on modeling uncertainty inherent from data in a completely unsupervised setting is still limited [9,16] and needs a deeper investigation.

This paper presents a novel GAN approach to fully unpaired medical image transfer, including prediction of data-dependent uncertainty and invariance over patches. To be more exact, a Wasserstein generative adversarial network (WGAN) [17] is leveraged to an uni-directional image transfer model. Structural correspondence between input and target modality is guaranteed using a novel generator loss that enforces invariance over image patches. Furthermore, the patch-based residuals are assumed to follow a zero-mean Laplace distribution with the scale parameter being a function of the input. As a consequence, the generator is allowed to predict uncertainty that operates as a learned loss attenuation and can be used to indicate the quality of a transferred image in the absence of ground truth data. The proposed model and training strategy is evaluated in three different unsupervised scenarios: modality propagation using T1w and T2w brain MRI from the IXI [18] database; accelerated MRI enhancement using emulated single-coil knee MRI from the FastMRI [3] database; MRI-to-CT synthesis using head CT scans from the CQ500 [19] database. The proposed framework is benchmarked against state-of-the-art works for unidirectional and bi-directional image translation [9,11,12,14]. We not only evaluate accuracy on unseen test data but further investigate robustness to perturbed inputs.

Contributions:

- We present an unidirectional framework that enables fully unsupervised image transmission of medical data while preserving fine structures.
- Structural correspondence between different characteristics and modalities is ensured by an improved generator loss based on patch invariance. This also yields implicit data augmentation for the critic and generator networks.
- In addition to the image transmission, the model provides an uncertainty map that correlates with the prediction error, indicating the quality of a mapped instance.

2   Related Work

2.1   Generative Adversarial Networks

The GAN architecture [10] is composed of a generator network $G : \mathcal{Z} \to \mathcal{X}$ and an adversarial part $f : \mathcal{X} \to [0, 1]$. The generator maps from a latent space $\mathcal{Z}$ to image space $\mathcal{X}$, where the parameters of G are adapted such that the distribution of the synthesized examples assimilates to the real data distribution on $\mathcal{X}$. Simultaneously, the adversarial f is trained to distinguish between synthesized and real instances. In a two-player min-max game, generator parameters are updated to fool a steadily improving discriminator [20]. Improving the joint loss functional of the generating and the adversarial part yielded improved modifications of the initial GAN framework, like WGAN [17], improved WGAN [21], LSGAN [22] or SNGAN [23]. While GANs reach outstanding performance in image synthesis [24,25], they are also well accepted in improving prediction quality in supervised image applications such as super-resolution [26], paired image translation [27,28], and medical image enhancement [5,29].

2.2   Unpaired Image Transfer and Domain Mapping

Unpaired image translation maps an image from input to target domain where corresponding samples from both spaces are hard to obtain or applied registration methods yield too much misalignment. In these cases, cycleGAN [11] has become the gold standard since it learns an inverse mapping from target domain back to input space. The core idea of cycleGAN is that the synthesized image must retain enough detail of the input instance in the target domain to allow for reconstruction. Especially in the medical sector, a structure-preserving transfer function is of high priority. Wolterink et al. [7] utilized cycle-consistency for MRI-only treatment planning in radiotherapy. Hiasa et al. [8] improved the cycleGAN architecture in this application area by adding a gradient consistency loss to pay more attention to the edges in the image. Yang et al. [6] added to cycleGAN a structure-consistency loss based on a modality independent neighborhood descriptor.

Learning an inverse GAN framework simultaneously (bi-directional) in order to ensure input-output consistency increases hardware requirements and introduces an additional instability if the inverse generator is not trained sufficiently or if the transfer mapping is not injective. Fu et al. [12] investigated geometry-consistent GAN (gcGAN), an uni-directional approach that enforces consistency when applying geometric transformations (rotation, flipping) before and after propagation through the generator network. Benaim et Wolf [13] considered GAN in combination with a distance constraint, where the distance between two samples from the input domain should be preserved after mapping to the target domain. A very recent and successful uni-directional approach to unpaired domain mapping was made by Park et al. [14] that uses contrastive unpaired translation (CUT), i.e., structure-consistency is preserved by matching patches of the input and the synthesized instance using an additional classification step.

2.3 Uncertainty Quantification

Uncertainty quantification methods have been applied to solve a variety of real-world problems in computer vision where in addition to the model's response also a measure on its confidence is provided [30]. In general, two broad categories of uncertainty are considered: aleatoric uncertainty captures noise inherent in the data and epistemic/model uncertainty describes uncertainty in the model parameters [15]. The latter type of uncertainty occurs in finite data settings and thus can be explained away providing a sufficient amount of data. Bayesian models provide a mathematically grounded framework that can account for model uncertainty in combination with Bayesian inference techniques. Gal et Ghahramani [31] set up a theoretical framework that casts the dropout technique as approximate Bayesian inference, enabling a rather simple calculation of epistemic uncertainty by multiple network forward passes. The works of Saatci et Wilson [16] as well as Palakkadavath et Srijith [32] leverage this framework to Bayesian GANs and show that considering Bayesian learning principles can address mode collapse in image synthesis. Kendall et Gal [15] have explored the benefits of modeling aleatoric and epistemic uncertainty simultaneously in image segmentation and regression and concluded that the two types of uncertainty are not mutually exclusive, but in fact complementary in different data scenarios. Upadhyay et al. modeled aleatoric uncertainty for MRI image enhancement [2] and unsupervised image transfer [9] by introducing uncertainty-aware general-ized adaptive cycleGAN (UGAC). Therefore, the latter work will also serve as a benchmark method for the proposed uncertainty-aware uni-directional image transfer approach.

3 Method

3.1 Preliminaries and GAN Architecture

The underlying structure of the proposed uncertainty-aware domain mapping is a GAN combined with a patch invariant generator term. Let $\mathcal{X} \subset \mathbb{R}^{d \times d \times c_{\text{in}}}$ and $\mathcal{Y} \subset \mathbb{R}^{d \times d \times c_{\text{out}}}$ denote the input and the target domain, respectively. For simplicity we consider quadratic instances with the number of image pixels equal to d^2. Furthermore, let $X := \{x_1, \ldots, x_M\}$ be the set of M given input images and $Y := \{y_1, \ldots, y_N\}$ the set of N available but unaligned target images. $P_{\mathcal{X}}$ and $P_{\mathcal{Y}}$ denote the distributions of the images in both domains. The proposed image transfer is built on a generator function $G_\theta \colon \mathcal{X} \to \mathcal{Y}$, which aims to map an input sample to a corresponding instance in the target domain. The generator function is approximated by a convolutional neural network (CNN), which is parameterized by a weight vector θ. By adjusting θ, the distribution P_θ of generator outputs may be brought closer to the real data distribution $P_{\mathcal{Y}}$ in the target domain. The distance between the generator distribution and the real distribution is estimated with the help of the critic $f_\omega \colon \mathcal{Y} \to \mathbb{R}$, which is parameterized by weight vector ω and is trained simultaneously with the generator network since P_θ changes after each update to the generator weights θ [20].

We choose a network critic based on the Wasserstein-1 distance [17,20,33]. The Wasserstein-1 distance between two distributions P_1 and P_2 is defined as $\mathcal{W}_1(P_1, P_2) := \inf_{J \in \mathcal{J}(P_1, P_2)} \mathbb{E}_{(x,y) \sim J} \|x - y\|$, where the infimum is taken over the set of all joint probability distributions that have marginal distributions P_1 and P_2. The Kantorovich-Rubinstein duality [33] yields

$$\mathcal{W}_1(P_1, P_2) = \sup_{\|f\|_L \leq 1} \left[\mathbb{E}_{y \sim P_1} f(y) - \mathbb{E}_{y \sim P_2} f(y) \right], \tag{1}$$

where $\|\cdot\|_L \leq C$ denotes that a function is C-Lipschitz. Equation (1) indicates that a good approximation to $\mathcal{W}_1(P_{\mathcal{Y}}, P_\theta)$ is found by maximizing $\mathbb{E}_{y \sim P_{\mathcal{Y}}} f_\omega(y) - \mathbb{E}_{y \sim P_\theta} f_\omega(y)$ over the set of CNN weights $\{\omega \mid f_\omega \colon \mathcal{Y} \to \mathbb{R} \text{ 1-Lipschitz}\}$, where the Lipschitz continuity of f_ω can be enhanced via a gradient penalty [21]. Given training batches $\mathbf{y} = \{y_n\}_{n=1}^b$, $y_n \overset{\text{iid}}{\sim} P_{\mathcal{Y}}$ and $\mathbf{x} = \{x_n\}_{n=1}^b$, $x_n \overset{\text{iid}}{\sim} P_{\mathcal{X}}$, this yields the following empirical risk for critic f_ω:

$$\ell_{\text{cri}}(\omega, \theta, \mathbf{y}, \mathbf{x}, p) := \frac{1}{b} \sum_{n=1}^b f_\omega(G_\theta(x_n)) - f_\omega(y_n) + p \cdot \left(\left(\|\nabla_{\tilde{y}_n} f_\omega(\tilde{y}_n)\|_2 - 1 \right)_+ \right)^2, \tag{2}$$

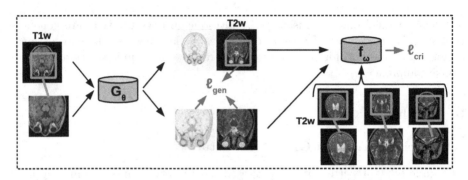

Fig. 1. Utilizing patch invariance for unsupervised MRI propagation and uncertainty quantification. The T1w input and a corresponding random patch on a finer scale are fed to the generator G_θ, which outputs the synthesized T2w counterparts and corresponding scale maps (red). The synthesized patch and the corresponding patch of the full-size output are compared. Loss attenuation is introduced by the scale map of the synthesized patch. The generator is additionally updated using the Wasserstein-1 distance, estimated with the help of f_ω. (Color figure online)

where p denotes the influence of the gradient penalty, $(\cdot)_+ := \max(\{0, \cdot\})$ and $\tilde{y}_n := \epsilon_n \cdot G_\theta(x_n) + (1 - \epsilon_n) \cdot y_n$ for $\epsilon_n \overset{\text{iid}}{\sim} \mathcal{U}[0, 1]$. Since only the first term of the functional in (2) depends on θ and the goal for the generator is to minimize the Wasserstein-1 distance, the adversarial empirical risk for generator G_θ simplifies as follows:

$$\ell_{\text{gen}}(\theta, \omega, \mathbf{x}) := -\frac{1}{b} \sum_{n=1}^{b} f_\omega(G_\theta(x_n)). \tag{3}$$

3.2   Patch Invariance

In the frame of medical image translation, it is not sufficient to ensure that the output samples lie in the target domain. Great attention should be paid that a model also preserves global structure as well as fine local details. Let $x \in \mathbb{R}^{d \times d \times c}$ and $\Phi := \{(\rho, j_1, j_2) \in [0.7, 1] \times [0, d]^2 \mid j_1 + \rho d \leq d \wedge j_2 + \rho d \leq d\}$. We define the patch operator $\mathcal{P} : \Phi \times \mathbb{R}^{d \times d \times c} \to \mathbb{R}^{d \times d \times c}$ as follows:

$$\mathcal{P}(\rho, j_1, j_2)(x) := \mathcal{R}_{d \times d \times c}(x[j_1 : j_1 + \rho d, j_2 : j_2 + \rho d, :]), \tag{4}$$

where $(\rho, j_1, j_2) \in \Phi$ and $\mathcal{R}_{d \times d \times c}$ resizes the patch to original image size $d \times d \times c$. The patch operator $\mathcal{P}$ chooses a quadratic patch of 70% to 100% the input size and conducts resampling to original size (cf. Fig. 1). For resampling, we use bicubic interpolation[1].

The basic intuition now is: if we take a patch of the input image and propagate it through the generator, than it should be equal to the corresponding patch

---

[1] https://www.tensorflow.org/api_docs/python/tf/image/resize.

of the transferred full-size image. We choose the 1-norm for comparing the corresponding patches and ensure realistic synthesized patches by adding the patch operator also to the Wasserstein-1 critic. This yields the following improvements for the critic and the generator risks:

$$\ell_{\text{cri}}(\omega, \theta, \mathbf{y}, \mathbf{x}, p, \vec{\phi}) := \frac{1}{b} \sum_{n=1}^{b} \left[f_\omega(G_\theta(x_n)) - f_\omega(y_n) \right.$$

$$\left. + f_\omega\left(\mathcal{P}_{\phi_n}(G_\theta(x_n))\right) - f_\omega(\mathcal{P}_{\phi_n}(y_n)) + p \cdot (\ldots)^2 \right], \tag{5}$$

$$\ell_{\text{gen}}(\theta, \omega, \mathbf{x}, \vec{\phi}, \lambda) := \frac{1}{b} \sum_{n=1}^{b} \left[-f_\omega\left(G_\theta(x_n)\right) - f_\omega\left(\mathcal{P}_{\phi_n}\left(G_\theta(x_n)\right)\right) \right.$$

$$\left. + \lambda \cdot \underbrace{d^{-2} \left\| G_\theta(\mathcal{P}_{\phi_n}(x_n)) - \mathcal{P}_{\phi_n}(G_\theta(x_n)) \right\|_1}_{\ell_{\text{patch}}(\theta, x_n, \phi_n)} \right], \tag{6}$$

where λ controls the influence of the patch loss and the patch extraction settings $\phi = \{\phi_n\}_{n=1}^{b}$, $\phi_n \in \Phi$ are chosen randomly at each risk calculation.

This approach yields some practical advantages: The generator is forced to be consistent over smaller patches, which prevents the network to generate modes with highest similarity to the real data (mode collapse). Furthermore, the generator is prevented from learning arbitrary mappings between input and target domain (e.g., mapping a T1w MRI of an old lady to a T2w MRI of a young boy), because this memorized mapping would then also have to be fulfilled for all smaller patches, i.e., the transfer would also have to be memorized on arbitrary scales. The patch extractor can also be viewed as a magnification and cropping operation. This yields a higher penalty when comparing fine structures that may not have much effect on the loss function when compared on full image scale. Finally, patch extraction causes implicit data augmentation and can help to avoid critic overfitting where the critic is tempted to memorize training samples.

3.3   Uncertainty by Loss Attenuation

We consider now Eq. (6) from a probabilistic point of view. For $x \in \mathcal{X}$, let $a = \mathcal{P}_\phi(G_\theta(x)))$ and $b = G_\theta(\mathcal{P}_\phi(x))$ for a patch configuration $\phi \in \Phi$. If we force the patch invariance via the 1-norm, the underlying assumption is that every pixel of the residual $\epsilon := a - b$ should follow a zero-mean and fixed-scale Laplace distribution [9]. Consider residual pixel $\epsilon_j \sim \text{Laplace}(0, \sigma)(\epsilon_j) = \frac{1}{2\sigma} \exp(-|\epsilon_j|/\sigma)$ where σ represents the scale parameter of the distribution. Maximum likelihood (ML) optimization on the full image (note a and b are functions of θ) yields

$$\max_\theta \prod_{j=1}^{d^2} \frac{1}{2\sigma} \exp\left(-|a_j - b_j|/\sigma\right). \tag{7}$$

Applying the negative logarithm and dividing by factor d^2 results in

$$\min_\theta \frac{1}{d^2} \sum_{j=1}^{d^2} |a_j - b_j|/\sigma + \log(2\sigma), \tag{8}$$

which is equivalent to minimizing $\ell_{\mathrm{patch}}(\theta, x, \phi)$ in Eq. (6) when considering a fixed scale σ. The assumption of a fixed scale for the pixel-wise residuals is quite strong and may not hold in the presence of OOD data. The idea now is to consider individual scales for every pixel. Inspired by [9,15], we make the scale σ a function of input x, i.e., we split the generator $G_\theta(x) = [G_\theta^I(x), G_\theta^\sigma(x)]$ in the output branch and return two images, the transferred image $G_\theta^I(x)$ and the corresponding pixel-wise scale map $G_\theta^\sigma(x)$ for the residuals. This results in

$$\ell_{\mathrm{patch}}(\theta, x, \phi) = \frac{1}{d^2} \sum_{j=1}^{d^2} \frac{|G_\theta^I(\mathcal{P}_\phi(x))_j - \mathcal{P}_\phi(G_\theta^I(x))_j|}{G_\theta^\sigma(\mathcal{P}_\phi(x))_j} + \log\left(2 \cdot G_\theta^\sigma(\mathcal{P}_\phi(x))_j\right). \tag{9}$$

This can be seen as a loss attenuation as we get high values in $G_\theta^\sigma(x)$ for image regions with high absolute residuals. At the same time, the logarithmic term discourages the model to predict high uncertainty for all pixels. The proposed generator loss for patch invariant and uncertainty-aware image transfer is obtained by inserting ℓ_{patch} (9) into ℓ_{gen} (6).

3.4   Implementation Details

In this work, the generator is a U-net [34] with five downsampling operations and approximately 10.7×10^6 parameters. After the last upsampling operation, the U-Net is split into two branches to generate two responses, the transferred image $G_\theta^I(\cdot)$ and the corresponding uncertainty map $G_\theta^\sigma(\cdot)$, c.f. (9). A non-negative scale map is enforced by applying the softplus activation function $softplus(x) := \log(\exp(x) + 1)$ to the latter output branch. A decoding network for the Wasserstein critic is built following the DCGAN critic [35] with 5 downsampling steps and approximately 4.7×10^6 parameters. Detailed information on critic and generator implementation can be found in the github repository. All models are trained using the Adam optimizer [36] with $\beta_1 = 0$, $\beta_2 = 0.9$ and minibatch size 8. The learning rate is set to 5×10^{-5} for the generator and 2×10^{-5} for the critic network. No learning rate scheduler or further data augmentation techniques are applied. The total amount of generator updates is 15k and we iterate between 15 critic updates and 1 generator update. Gradient penalty parameter p equals 10, the influence λ of the patch constraint is chosen for each data set individually by a grid search.

4   Experiments

4.1   Datasets

We consider three different tasks in medical image-to-image translation.

Modality Propagation: The IXI [18] database consists of registered T1w and T2w scans of 577 patients. We want to demonstrate the plausibility of our model for unpaired modality propagation and thus build a model for T1w to T2w transfer. To do so, we remove 10% of the patients for evaluation. The remaining patients are randomly split into input and target data, where no patient contributes to both domains at the same time. This is done to simulate a scenario where no paired slices are available throughout the entire training process. We only use the core 60% of all axial slices, which yields approximately 20k train slices for input, 20k train slices for target, and 4k pairs for evaluation. The spatial dimension d equals 256.

Accelerated MRI Enhancement: The FastMRI [3] database consists of more than 1500 multi-coil diagnostic knee MRI scans and corresponding emulated single-coil data. Our experiments are based on a subset of nearly 800 coronal proton-density weighted scans without fat-suppression from the official train and validation single-coil releases. We remove 10% of the patients for evaluation and split the remaining patients into input and target data. While the target domain consists of slices from fully-sampled MRI scans, we consider 4x acceleration (only 25% of k-space measurements) for the slices by using the subsampling scheme discussed in [3,4]. This yields 7.1k train slices for input and 6.9k for target. We are aware that enhancement of accelerated MRI can also be considered as a supervised task since generation of paired instances is feasible. However, this experiment should demonstrate possible applicability of the proposed framework to inverse problems in general.

MRI-to-CT Synthesis: The CQ500 [19] consists of CT scans of nearly 500 patients, where we use a randomly selected subset of 80 patients for the target domain. Furthermore, we make use of T1w MRI scans of 144 randomly selected patients taken from IXI [18] as input data. For each dataset we use the core 60% of all axial slices, which yields 11.4k train slices for input and 10.4k train slices for target. All slices are subsampled to spatial size $d = 256$. Note that in this experiment input and target data is coming from completely separated datasets (inconsistent head orientations, different brain areas, resolution, etc.). An additional challenge are artifacts outside the skull caused by the CT table and the measurement equipment in CQ500. We step away from any preprocessing here and investigate how the method reacts to this kind of artifacts. Since no ground truth data is available for the two separated databases only qualitative evaluation is conducted.

4.2 Compared Methods and Scenarios

We compare our approach to a variety of state-of-the-art methods for unsupervised image transfer that have already been introduced in Sect. 2.2 and Sect. 2.3: cycle-consistent GAN (cycleGAN) [11], uncertainty-aware generalized adaptive cycleGAN (UGAC) [9], geometry-consistent GAN (gcGAN) [12] with horizontal flip and contrastive unpaired translation (CUT) [14]. We test two versions of our approach: the first version utilizes only patch invariance (PI, cf. (6)) and the second version utilizes uncertainty-aware patch invariance (UAPI, cf. (9)). For cycleGAN, UGAC and gcGAN we make use of the same generator and critic architecture and training configurations as described in Sect. 3.4 to guarantee a fair comparison. For CUT, we use the publicly available github repository[2]. All slices are normalized to range $[0, 255]$ and handled as grayscale images. During optimization, the images are scaled to $[-1, 1]$ to speed up training while evaluation metrics, the structural similarity index (SSIM) [37] and the peak-signal-to-noise-ratio (PSNR), are calculated on original image scale.

For each of the three applications, we want to test not only performance on unseen assessment data, but also robustness to different types of perturbations. All approaches are trained on unaffected images (without additional noise) and evaluated in the following scenarios: GN0 (original test images); GN5 (adding Gaussian noise, deviation 5% of image range); GN10 (adding Gaussian noise, deviation 10%); GN20 (adding Gaussian noise, deviation 20%); IP2 (impulse perturbation, 2% of pixels replaced by random values); IP5 (5% of pixels replaced); IP10 (10% of pixels replaced).

4.3 Quantitative Evaluation

Table 1. Quantitative evaluation of our approach and four compared methods on two datasets under seven different noise scenarios. The reported metrics are the structural similarity index and the peak-signal-to-noise-ratio (SSIM/PSNR, higher is better).

Methods	GN0	GN5	GN10	GN20	IP2	IP5	IP10
	IXI test data						
cycleGAN	78.99/20.90	77.04/20.71	72.53/20.14	63.87/19.19	69.81/20.39	64.51/19.99	60.08/19.50
UGAC	79.31/21.21	77.24/21.05	72.53/20.43	**64.03**/19.37	70.15/20.68	64.88/20.22	61.01/19.78
gcGAN	81.00/21.52	76.51/20.66	69.96/19.50	57.75/17.90	73.79/20.30	69.24/19.55	63.91/18.78
CUT	78.67/21.15	76.10//21.22	71.43/20.58	61.27/19.67	74.11/21.19	71.78/20.94	68.97/.65
PI(ours)	**82.53**/22.18	69.82/21.49	51.90/20.64	36.96/19.26	54.85/20.92	44.10/20.24	38.73/19.70
UAPI(ours)	79.99/**22.62**	**78.76/22.20**	74.49/21.70	62.24/**20.34**	**77.44/21.93**	74.29/21.56	**69.35/21.14**
	FastMRI test data						
cycleGAN	81.60/21.48	70.71/19.45	59.59/17.89	40.87/15.80	62.95/18.35	57.50/17.74	50.80/17.02
UGAC	85.29/22.62	72.17/19.91	59.64/17.99	43.21/16.24	72.01/20.03	63.97/18.73	56.36/17.71
gcGAN	80.73/20.76	68.67/18.13	59.23/16.60	48.31/15.26	71.51/18.61	65.22/17.54	58.66/16.61
CUT	89.36/23.37	**85.67/22.71**	**83.32/ 22.15**	76.63/ 20.58	**86.61/23.02**	**84.45/ 22.45**	**82.59 /21.86**
PI(ours)	85.85/23.37	76.22/20.85	66.44/18.94	53.46/17.19	75.77/20.80	67.61/19.28	59.44/18.04
UAPI(ours)	**90.37/ 24.74**	81.08/21.37	69.97/18.41	55.72/16.07	79.18/19.81	71.69/17.97	65.44/16.98

The quantitative metrics in Table 1 obtained on unseen test data indicate superior performance of our approach for the modality propagation task on IXI. Considering evaluation on clean test data, usage of patch invariance gives an increase in SSIM and PSNR metrics compared to the bi-directional (cycleGAN, UGAC) and uni-directional (gcGAN, CUT) benchmarks. As compared to the benchmark methods, consideration of uncertainty-aware patch invariance even

---

[2] https://github.com/taesungp/contrastive-unpaired-translation.

improves results, also for the scenarios with perturbed test data (GN5 to IP10). Robustness of UAPI is also established visually in Fig. 2. Especially for scenarios with high perturbations (GN10, GN20, IP5, IP10), we observe a performance advandtage when using the uncertainty-aware method UAPI. Interestingly, usage of PI without uncertainty awareness on perturbation scenarios GN5 to IP10 yields a significant decrease for the SSIM but not for the PSNR metric. This rather unexpected observation needs further investigation in future research.

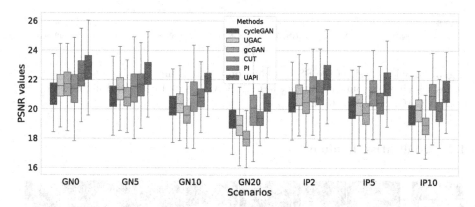

Fig. 2. Visual analysis of the PSNR values on IXI under seven different test scenarios, obtained by our two approaches (PI, UAPI) and 4 compared methods (cycleGAN, UGAC, gcGAN, CUT).

In Table 1 we observe for the accelerated MRI enhancement task on FastMRI that our method UAPI significantly outperforms other benchmark on unaffected test data but gives modest accuracy when additional noise and perturbed pixels are added. Figure 3 shows the superior performance of CUT in terms of the PSNR metric for noisy data. This is quite interesting since that has not been the case for the previous modality propagation application. The task of accelerated MRI enhancement strongly differs from the other two applications. While the goal of modality propagation and MRI-to-CT synthesis is to come up with a completely new image, the aim of accelerated MRI enhancement is to improve quality of a already existing image. In fact, the methods cycleGAN, UGAC, gcGAN, PI and UAPI depend on a rather simple U-Net [34] implementation and a standard DCGAN critic [35] with the aim to demonstrate plausibility of different transfer approaches on easy-to-implement frameworks. The CUT method is a benchmark where the publicly available source code had to be used, consisting of a ResNet-based generator [11] and built-in data augmentation techniques that may better compensate for noisy input data. Nevertheless, our methods PI and UAPI seemingly achieve better results compared to the U-Net based benchmarks. We will take up investigation of robustness of our methods in combination with different network architectures as a future goal.

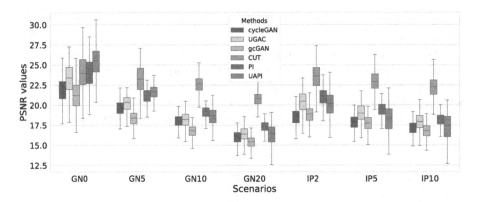

Fig. 3. Visual analysis of the PSNR values on FastMRI under seven different test scenarios, obtained by our two approaches (PI, UAPI) and 4 compared methods (cycleGAN, UGAC, gcGAN, CUT).

4.4   Qualitative Evaluation

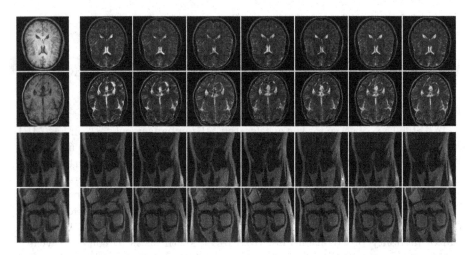

Fig. 4. Evaluation samples of our approach and four compared methods on IXI and FastMRI for scenario GN0 (test data without perturbations). From left to right: input, images transferred by cycleGAN, UGAC, gcGAN, CUT, PI, UAPI and ground truth.

In Fig. 4 we analyze the prediction quality of our and compared approaches in a qualitative way. Considering modality propagation in MRI, we see that usage of uncertainty-aware patch invariance (UAPI) gives a better detailed weighting of the cerebrospinal fluid in the middle of the brain. In general, employing patch invariance yields better preservation of fine structures. This observation also applies to accelerated MRI enhancement. In particular, CUT and UAPI provide comparatively sharper knee images with more high-frequency details than the other methods.

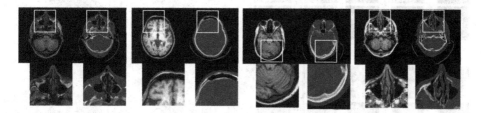

Fig. 5. Evaluation samples of the UAPI method on unseen brain MRI slices. For every data pair, the input slice and the corresponding UAPI prediction are visualized on the left and the right side, respectively. The first row contains the images on original scale, the second row selected patches to visualize the prediction quality for detailed structures.

Qualitative evaluation plays an important role for the third investigated application, namely MRI-to-CT synthesis, where quantitative comparison is not possible due to lack of ground truth data. Satisfying results were obtained with the UAPI method, which are visualized in Fig. 5. Cavities and brain shapes are well preserved by our method although we used two completely independent and unaligned head datasets for this experiment. UAPI synthesizes brain table artifacts that are also visible in CQ500. A proper evaluation on cleaned CT data is necessary and thus will be considered as a future working step.

4.5   Uncertainty Scores

Additional to improved accuracy we demonstrate the efficacy of estimating the scale maps with the proposed method. The input-dependent non-negative scale maps are derived from the second output branch G_θ^σ, see (9). Indeed, the predicted scale maps are able to model uncertainty inherent from data. This can be observed in Fig. 6, where in addition to the transferred images also the predicted scale maps and the absolute residuals between predicted and ground truth images are displayed. Obviously, uncertainty is relatively greater in regions with higher residual values. From the scale maps it can be deduced for which positions the generator is comparatively uncertain in its prediction, such as the cerebral cortex and eye sockets in head MRI or the lateral knee ligaments in knee MRI.

The correspondence between residual and scale maps suggests that the latter can be used as an approximation to a prediction's residuals that are not available due to the lack of ground truth data in unsupervised learning. In order to quantitatively study this relationship we visualize mean absolute residual score and mean uncertainty maps for 512 randomly selected unseen test images in a scatter plot (see Fig. 7). Moreover, we compare our uni-directional method UAPI also to the relations observed by UGAC that models uncertainty with the help

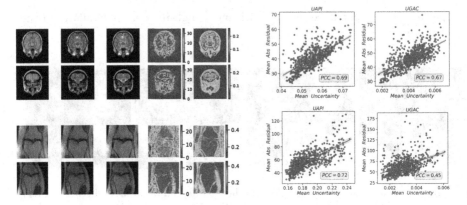

Fig. 6. Position-based relation between abs. residuals and predicted scale maps on IXI (top) and FastMRI (bottom). Left to right: input, ground truth, prediction by UAPI, abs. residuals and predicted scale map.

Fig. 7. Scatter plot between abs. residual and scale map values on IXI (top) and FastMRI (bottom). The predictions are generated by UAPI (left) and UGAC (right).

of a bi-directional cycleGAN [9]. For modality propagation as well as accelerated MRI enhancement we visually observe an approximate positive linear correlation between mean absolute residual scores and mean uncertainty scores. We calculate the Pearson correlation coefficient (PCC) to obtain a quality estimate for the linear correlation and compare between UAPI and UGAC. Our method returns a slightly higher PCC on IXI (UAPI: 0.69, UGAC: 0.67). The discrepancy between both methods even increases on FastMRI (UAPI: 0.72, UGAC: 0.45). This further encourages the idea that scale maps derived from our approach can be used to indicate the overall quality of a transferred image.

5 Conclusions

In this paper we proposed a WGAN-based approach using patch invariance to employ joint image transfer and uncertainty quantification in an fully unsupervised manner. We demonstrate superior performance of our uni-directional method for modality propagation and accelerated MRI enhancement compared to four state-of-the-art benchmarks in unpaired image translation. Moreover, the method reaches qualitatively satisfying results for MRI-to-CT synthesis using completely unaligned databases during training. The predicted uncertainty can be representative of the residual maps and thus indicate the quality of a transferred image in the absence of ground truth data. Further investigation of the network architecture and improvement in robustness represents an important goal for future research. Future work will also include the application of the uncertainty-aware and patch invariant network to other unpaired image-to-image applications outside the medical sector.

References

1. Wang, T., et al.: A review on medical imaging synthesis using deep learning and its clinical applications. J. Appl. Clin. Med. Phys. **22**(1), 11–36 (2021), https://aapm.onlinelibrary.wiley.com/doi/abs/10.1002/acm2.13121

2. Upadhyay, U., Sudarshan, V.P., Awate, S.P.: Uncertainty-aware GAN with adaptive loss for robust MRI image enhancement. In: 2021 IEEE/CVF International Conference on Computer Vision Workshops (ICCVW), pp. 3248–3257 (2021)

3. Zbontar, J., Knoll, F., et al.: fastMRI: an open dataset and benchmarks for accelerated MRI (2018). https://arxiv.org/abs/1811.08839

4. Hyun, C.M., Kim, H.P., Lee, S.M., Lee, S., Seo, J.K.: Deep learning for undersampled MRI reconstruction. Phys. Med. Biol. **63**(13), 135007 (2018). https://doi.org/10.1088/1361-6560/aac71a

5. Lei, Y., et al.: MRI-only based synthetic CT generation using dense cycle consistent generative adversarial networks. Med. Phys. **46**(8), 3565–3581 (2019)

6. Yang, H., et al.: Unpaired brain MR-to-CT synthesis using a structure-constrained CycleGAN. In: Stoyanov, D., et al. (eds.) DLMIA/ML-CDS -2018. LNCS, vol. 11045, pp. 174–182. Springer, Cham (2018). https://doi.org/10.1007/978-3-030-00889-5_20

7. Wolterink, J.M., Dinkla, A.M., Savenije, M.H.F., Seevinck, P.R., van den Berg, C.A.T., Išgum, I.: Deep MR to CT synthesis using unpaired data. In: Tsaftaris, S.A., Gooya, A., Frangi, A.F., Prince, J.L. (eds.) SASHIMI 2017. LNCS, vol. 10557, pp. 14–23. Springer, Cham (2017). https://doi.org/10.1007/978-3-319-68127-6_2

8. Hiasa, Y., et al.: Cross-modality image synthesis from unpaired data using Cycle-GAN. In: Gooya, A., Goksel, O., Oguz, I., Burgos, N. (eds.) SASHIMI 2018. LNCS, vol. 11037, pp. 31–41. Springer, Cham (2018). https://doi.org/10.1007/978-3-030-00536-8_4

9. Upadhyay, U., Chen, Y., Akata, Z.: Robustness via uncertainty-aware cycle consistency. In: Ranzato, M., Beygelzimer, A., Dauphin, Y., Liang, P., Vaughan, J.W. (eds.) Advances in Neural Information Processing Systems, vol. 34, pp. 28261–28273. Curran Associates, Inc. (2021). https://proceedings.neurips.cc/paper/2021/file/ede7e2b6d13a41ddf9f4bdef84fdc737-Paper.pdf

10. Goodfellow, I., et al.: Generative adversarial nets. In: Ghahramani, Z., Welling, M., Cortes, C., Lawrence, N., Weinberger, K. (eds.) Advances in Neural Information Processing Systems, vol. 27. Curran Associates, Inc. (2014). https://proceedings.neurips.cc/paper/2014/file/5ca3e9b122f61f8f06494c97b1afccf3-Paper.pdf

11. Zhu, J., Park, T., Isola, P., Efros, A.A.: Unpaired image-to-image translation using cycle-consistent adversarial networks. In: 2017 IEEE International Conference on Computer Vision (ICCV), pp. 2242–2251 (2017)

12. Fu, H., Gong, M., Wang, C., Batmanghelich, K., Zhang, K., Tao, D.: Geometry-consistent generative adversarial networks for one-sided unsupervised domain mapping. In: Proceedings of the IEEE/CVF Conference on Computer Vision and Pattern Recognition, pp. 2427–2436 (2019)

13. Benaim, S., Wolf, L.: One-sided unsupervised domain mapping. Adv. Neural Inf. Process. Syst. **30** (2017)

14. Park, T., Efros, A.A., Zhang, R., Zhu, J.-Y.: Contrastive learning for unpaired image-to-image translation. In: Vedaldi, A., Bischof, H., Brox, T., Frahm, J.-M. (eds.) ECCV 2020. LNCS, vol. 12354, pp. 319–345. Springer, Cham (2020). https://doi.org/10.1007/978-3-030-58545-7_19

15. Kendall, A., Gal, Y.: What uncertainties do we need in Bayesian deep learning for computer vision? Adv. Neural Inf. Process. Syst. **30** (2017)
16. Saatci, Y., Wilson, A.G.: Bayesian GAN. Adv. Neural Inf. Process. Syst. **30** (2017)
17. Arjovsky, M., Chintala, S., Bottou, L.: Wasserstein generative adversarial networks. In: Proceedings of the 34th International Conference on Machine Learning, vol. 70, pp. 214–223. PMLR (2017)
18. Robinson, E.C., Hammers, A., Ericsson, A., Edwards, A.D., Rueckert, D.: Identifying population differences in whole-brain structural networks: a machine learning approach. Neuroimage **50**(3), 910–919 (2010)
19. Chilamkurthy, S., et al.: Deep learning algorithms for detection of critical findings in head CT scans: a retrospective study. Lancet **392**(10162), 2388–2396 (2018)
20. Angermann, C., Schwab, M., Haltmeier, M., Laubichler, C., Jónsson, S.: Unsupervised single-shot depth estimation using perceptual reconstruction. CoRR abs/2201.12170 (2022). https://arxiv.org/abs/2201.12170
21. Gulrajani, I., Ahmed, F., Arjovsky, M., Dumoulin, V., Courville, A.C.: Improved training of wasserstein GANs. In: Guyon, I., et al. (eds.) Advances in Neural Information Processing Systems, vol. 30. Curran Associates, Inc. (2017). https://proceedings.neurips.cc/paper/2017/file/892c3b1c6dccd52936e27cbd0ff683d6-Paper.pdf
22. Mao, X., Li, Q., Xie, H., Lau, R.Y., Wang, Z., Paul Smolley, S.: Least squares generative adversarial networks. In: Proceedings of the IEEE International Conference on Computer Vision, pp. 2794–2802 (2017)
23. Miyato, T., Kataoka, T., Koyama, M., Yoshida, Y.: Spectral normalization for generative adversarial networks. In: International Conference on Learning Representations (2018)
24. Karras, T., Laine, S., Aila, T.: A style-based generator architecture for generative adversarial networks. In: Proceedings of the IEEE/CVF Conference on Computer Vision and Pattern Recognition, pp. 4401–4410 (2019)
25. Karras, T., Aila, T., Laine, S., Lehtinen, J.: Progressive growing of GANs for improved quality, stability, and variation. In: International Conference on Learning Representations (2018)
26. Ledig, C., et al.: Photo-realistic single image super-resolution using a generative adversarial network. In: Proceedings of the IEEE Conference on Computer Vision and Pattern Recognition, pp. 4681–4690 (2017)
27. Isola, P., Zhu, J.Y., Zhou, T., Efros, A.A.: Image-to-image translation with conditional adversarial networks. In: Proceedings of the IEEE Conference on Computer Vision and Pattern Recognition, pp. 1125–1134 (2017)
28. Arslan, A.T., Seke, E.: Face depth estimation with conditional generative adversarial networks. IEEE Access **7**, 23222–23231 (2019)
29. Armanious, K., et al.: MedGAN: medical image translation using GANs. Comput. Med. Imaging Graph. **79**, 101684 (2020)
30. Abdar, M., et al.: A review of uncertainty quantification in deep learning: techniques, applications and challenges. Inf. Fusion **76**, 243–297 (2021)
31. Gal, Y., Ghahramani, Z.: Dropout as a Bayesian approximation: representing model uncertainty in deep learning. In: Balcan, M.F., Weinberger, K.Q. (eds.) Proceedings of The 33rd International Conference on Machine Learning. Proceedings of Machine Learning Research, vol. 48, pp. 1050–1059. PMLR, New York (2016). https://proceedings.mlr.press/v48/gal16.html
32. Palakkadavath, R., Srijith, P.: Bayesian generative adversarial nets with dropout inference. In: 8th ACM IKDD CODS and 26th COMAD, pp. 92–100 (2021)

33. Villani, C.: Optimal Transport: Old and New, vol. 338. Springer, Heidelberg (2008). https://doi.org/10.1007/978-3-540-71050-9
34. Ronneberger, O., Fischer, P., Brox, T.: U-net: convolutional networks for biomedical image segmentation. In: Navab, N., Hornegger, J., Wells, W.M., Frangi, A.F. (eds.) MICCAI 2015. LNCS, vol. 9351, pp. 234–241. Springer, Cham (2015). https://doi.org/10.1007/978-3-319-24574-4_28
35. Radford, A., Metz, L., Chintala, S.: Unsupervised representation learning with deep convolutional generative adversarial networks (2015)
36. Kingma, D.P., Ba, J.: Adam: a method for stochastic optimization (2014). https://arxiv.org/abs/1412.6980
37. Wang, Z., Bovik, A.C., Sheikh, H.R., Simoncelli, E.P.: Image quality assessment: from error visibility to structural similarity. IEEE Trans. Image Process. **13**(4), 600–612 (2004)

Uncertainty Quantification Using Query-Based Object Detectors

Meet P. Vadera[✉], Colin Samplawski, and Benjamin M. Marlin

University of Massachusetts, Amherst, USA
{mvadera,csamplawski,marlin}@cs.umass.edu

Abstract. Recently, a new paradigm of query-based object detection has gained popularity. In this paper, we study the problem of quantifying the uncertainty in the predictions of these models that derive from model uncertainty. Such uncertainty quantification is vital for many high-stakes applications that need to avoid making overconfident errors. We focus on quantifying multiple aspects of detection uncertainty based on a deep ensembles representation. We perform extensive experiments on two representative models in this space: DETR and AdaMixer. We show that deep ensembles of these query-based detectors result in improved performance with respect to three types of uncertainty: location uncertainty, class uncertainty, and objectness uncertainty (Code available at: https://github.com/colinski/uq-query-object-detectors).

Keywords: Uncertainty quantification · Deep ensembles · Query-based object detection · Transformer · Mixer

1 Introduction

The advent of deep learning has led to the deployment of computer vision systems in countless domains. Of particular interest is the problem of object detection. However, there is considerable risk involved when deploying object detectors for high-stakes applications, such as in autonomous vehicles or medical imaging. This is due to the well-known property that deep networks often make over-confident errors. This can be mitigated by leveraging approaches that quantify aspects of detection uncertainty, such as deep ensembles or approximate Bayesian methods [2,6,8,16,32].

Recently a new paradigm of query-based object detection has generated significant interest. These models learn a set of query embeddings that can be directly decoded into bounding box predictions. The query embeddings learn to attend to any part of the image, allowing prediction at any location and scale. This offers significant benefits over the earlier dense detectors that make predictions for a set of anchor points. These query-based models are learned end-to-end with less hand engineering, require less post-processing, and show

Supplementary Information The online version contains supplementary material available at https://doi.org/10.1007/978-3-031-25085-9_5.

L. Karlinsky et al. (Eds.): ECCV 2022 Workshops, LNCS 13808, pp. 78–93, 2023.
https://doi.org/10.1007/978-3-031-25085-9_5

equal or better detection performance than previous approaches. Therefore, it is likely that query-based detectors will become a popular choice for real-world deployment. However, to the best of our knowledge, no prior work has studied how uncertainty in model parameters impacts uncertainty in detection outputs for these models.

In this paper, we take the first step in answering that question by studying detection uncertainty derived from deep ensemble of query-based detectors. We focus on three different aspects of detection uncertainty: a) the uncertainty in the location of a bounding box prediction, b) class uncertainty from the classification output of the detector, and c) objectness uncertainty, which considers the uncertainty in the model's prediction of where there is an object at all in a given area in the scene. We show that using a deep ensemble of query-based detectors results in improved performance with respect to all three types of output uncertainty. In Fig. 1 we show a example of how a deep ensemble can be used to correct an overconfident error from a single detector.

Fig. 1. Example of an overconfident error corrected by a deep ensemble. Far left image shows ground-truth boxes. Center image shows detections from a single DETR model, with overconfident error marked in blue. Far right image shows boxes from merged clusters from the ensemble. We see that the overconfident error has been removed. (Color figure online)

2 Background and Related Work

2.1 Object Detection

Until recently, the dominant paradigm for detector design was dense object detection. In this framework, a bounding box predictions are made at a dense grid of anchor points [3,10,13,25,26] or even every pixel in the anchor-free setting [7,17,28]. This allows the model to more easily predict boxes at every location and scale. However, this approach can often lead to an over-prediction of boxes, necessitating the use of non-maximal suppression (NMS) and other post-processing steps.

Recently, a new paradigm of object detection, called query-based detection, has been gaining in popularity. In this setting, a set of query embeddings are learned that can be decoded into bounding box predictions directly. The first work proposing this approach is the Detection Transformer (DETR) model [4]. This model uses a transformer encoder-decoder [30] pair to predict a set of detections from learned query embeddings. DETR is trained using a set-based objective that uses bipartite matching to match query-generated predictions to ground-truth boxes. A number of follow-on works to DETR have been proposed that improve performance and speed-up convergence [20,31,33]. Most recently, the AdaMixer model [9] replaces the transformers of the DETR family with a mixer model [29]. This leads to performance and training time that is superior to earlier transformer-based approaches. A major benefit of query-based detectors is that they no longer need to make dense predictions as the query embeddings can naturally learn to attend to relevant objects in the image. This reduces the need for post-processing and hand-crafted model components, such as NMS and anchor boxes.

2.2   Uncertainty Quantification in Object Detection

There has been some prior work on approximate Bayesian inference for deep CNN-based object detection models. Much of this work looks at estimating uncertainty in the prediction of bounding-box properties such as location and size. [15] was one of the earliest works in this space that proposed using log-likelihood loss to estimate heteroscedastic aleatoric uncertainty for bounding-box regression. [18] builds upon this to have the neural network model explicitly produce an estimated output variance like in heteroscedastic regression. This requires a single forward pass through the neural network to estimate the output variance. [22] use MC Dropout and then compute the mean and variance in the context of bounding-box regression by computing statistics over spatially correlated outputs. The BayesOD approach [12] builds upon [18] to get rid of the non-maximal suppression with Bayesian inference and provides a multivariate extension instead of the diagonal covariance of the standard log-likelihood for bounding box regression.

In terms of evaluating detection uncertainty, the widely used metric for evaluating object detection models is the mean average precision (mAP) score, but the mAP score only evaluates the accuracy of the hard predictions. It does not account for the uncertainty in model predictions. Previous work by [11] proposed *Probability-based Detection Quality (PDQ)*, a metric to evaluate the probabilistic aspect of the outputs. PDQ consists of two components: spatial quality, and label quality. However, there are certain issues with the PDQ evaluation framework. First, the label quality is measured as the probability of the correct class, with the unmatched ground-truth and predicted bounding boxes contributing a score of 0. This is problematic because, as we highlight later in Sect. 3.3, a more reasonable way to assess the class uncertainty on unmatched predicted bounding boxes is to consider the NLL of the background class prediction. Second, while looking at spatial quality, the metric looks at two components: foreground loss and background loss. Foreground loss is the negative log-likelihood of the pixels in the ground-truth bounding box under the predicted distribution that assigns a probability value on whether a given pixel in the scene is a part of the detection. Background loss penalizes pixels by using the complement of the predicted distribution, by computing the NLL of the pixels present in the predicted bounding box, but not in the ground-truth bounding box. This is again problematic, as it composes location uncertainty with evaluating how well the ground-truth bounding box area is covered in a way that is hard to decompose. Finally, while matching ground-truth bounding boxes to the predicted bounding boxes, the authors propose computing PDQ scores for every combination of the predicted bounding boxes and ground-truth bounding boxes. This can lead to leakage of label information at the time of matching, which is not desirable.

3   Uncertainty Estimation in Query-Based Detectors

In this section we first describe the specific query-based detectors we use in more detail. We then explain our proposed approach to merge the output of an ensemble of these detectors. Finally, we describe how we quantify different kinds of uncertainty derived from the ensemble.

3.1   Detection Transformer and AdaMixer

At the most abstract level, query-based detectors can be thought of as learning a set of query embeddings in a latent embedding space that are decoded into bounding box predictions. A decoder function is then learned which takes as input the image features from a CNN backbone and the query embeddings. The output of the decoder is a new set of transformed query embeddings of the same shape. These can then be transformed into class and bounding box predictions using MLP output heads.

In the case of DETR and AdaMixer, a set of 100 256-dimensional query embedding vectors is used. Furthermore, both models use ResNet50 as the backbone [14]. In the case of DETR, the decoder is a transformer that computes cross-attention between image features and query embeddings. AdaMixer replaces the transformer with an MLP mixer model. This decoder uses position-aware adaptive mixing between the query embeddings and image features. This has been shown to result in faster convergence and better performance than DETR.

For each bounding box at index i in the training dataset, the ground-truth can be denoted as $y_i = (c_i, b_i)$, where c_i denotes the class label, and $b_i \in [0,1]^4$ denotes the x, y coordinates of the center of the box, as well as the height and width in normalized coordinates. In general, there are more queries than ground-truth objects in an image. To assign predictions to ground-truth boxes a bipartite matching problem is solved using the Hungarian algorithm. This results in the following training loss:

$$\mathcal{L}_{\text{Hungarian}}(y, \hat{y}) = -\log \hat{p}_{\hat{\sigma}(i)}(c_i) + \mathbb{1}_{\{c_i \neq \text{"background"}\}} \mathcal{L}_{\text{box}}(b_i, \hat{b}_{\hat{\sigma}}(i)) \quad (1)$$

Here, $\hat{\sigma}(i)$ denotes the permutation of the predictions that minimizes the matching loss between predictions and the ground-truth. Furthermore, $\hat{p}_{\hat{\sigma}(i)}(c_i)$ denotes the predicted probability for class c_i under the $\hat{\sigma}(i)$ permutation, and $\hat{b}_{\hat{\sigma}}(i)$ denotes the predicted bounding box coordinates under the $\hat{\sigma}(i)$. It's important to note that this matching computation is only required during training. $\mathcal{L}_{\text{box}}$

is constructed using two components: the IoU loss ($\mathcal{L}_{\text{iou}}$) and the L1 regression loss:

$$\mathcal{L}_{\text{box}}(b_i, \hat{b}_{\hat{\sigma}}(i)) = \lambda_{\text{iou}}\mathcal{L}_{\text{iou}}(b_i, \hat{b}_{\hat{\sigma}}(i)) + \lambda_{\text{L1}}\left\|b_i - \hat{b}_{\sigma(i)}\right\|_1 \tag{2}$$

Here, λ_{iou} and λ_{L1} denote the hyperparameters for weighing the two components of the loss function.

3.2   Merging Bounding Boxes

When using an ensemble of classifiers, it is straightforward to merge the outputs of the ensemble members by computing the ensemble average class distribution. However, in our case, we have multiple bounding box candidates generated by each ensemble member. Different ensemble members may predict different numbers of boxes and boxes in different locations as well as different class distributions for similar boxes. This variability across the ensemble is exactly the manifestation of model uncertainty that we aim to quantify.

To summarize class and location uncertainty, we begin by merging the bounding box predictions from different members of the ensemble using the Basic Sequential Algorithmic Scheme (BSAS) clustering method [21, 27]. For a given input, denote the set of outputs of each of the members of the ensemble as $D_k = \cup\{(b_{ik}, s_{ik})\}_{i=1}^{N}$, where i denotes the index of the output bounding box, k denotes the index of the model in the ensemble, b denotes the predicted output corresponding to the location of the bounding box, and s denotes the predicted class distribution for the bounding box. Furthermore, let $\mathcal{O}_m$ denote the mth cluster. Detection are sequentially assigned to clusters if the maximum IoU between detection D_i and any existing $\mathcal{O}_m$ is greater than or equal to a set threshold ω. Otherwise, the detection is assigned as a cluster of its own. Once the clusters are formed, we produce the posterior predictive class distribution as follows:

$$\hat{s}_i = \frac{1}{K}\sum_{k=1}^{K} s_k, \text{ where } K = |\mathcal{O}_i| \tag{3}$$

For merging the bounding box location predictions, there are several options. First, we could simply compute the "average" bounding box, formed by taking the average of the top-left and bottom-right coordinates of boxes in the same cluster. In fact, the average merging strategy has been used in the past to merge bounding boxes [23]. In addition to the average bounding box, we could also generate a "maximal" bounding box or a "minimal" bounding box. A maximal bounding box, by definition, is the tightest bounding box that encapsulates all the bounding boxes within a cluster. On the other hand, a minimal bounding box is the intersection of all the bounding boxes within a cluster. Consider the bounding box $b_j = (x_{1j}, y_{1j}, x_{2j}, y_{2j})$ in a cluster $\mathcal{O}_i$, where (x_{1j}, y_{1j}) indicate the coordinates of the top-left corner, and (x_{2j}, y_{2j}) indicate the coordinates of the bottom-right corner, then the merged bounding is obtained by:

$$\hat{b}_i^{\mathrm{avg}} = \frac{1}{J} \sum_{j=1}^{J} b_j, \text{ where } J = |\mathcal{O}_i| \text{ (average bounding box)} \tag{4}$$

$$\hat{b}_i^{\min} = (\min(\mathbf{x_1}), \min(\mathbf{y_1}), \max(\mathbf{x_2}), \max(\mathbf{y_2})) \text{ (maximal bounding box)} \tag{5}$$

$$\hat{b}_i^{\max} = (\max(\mathbf{x_1}), \max(\mathbf{y_1}), \min(\mathbf{x_2}), \min(\mathbf{y_2})) \text{ (minimal bounding box)} \tag{6}$$

In the above set of equations, $\mathbf{x_1} = [x_{11}, x_{12}, ..., x_{1J}]$, $\mathbf{y_1} = [y_{11}, y_{12}, ..., y_{1J}]$, $\mathbf{x_2} = [x_{21}, x_{22}, ..., x_{2J}]$, and $\mathbf{y_2} = [y_{21}, y_{22}, ..., y_{2J}]$. An illustration depicting individual bounding box predictions of different members of the ensemble and merging strategy has been given in Fig. 2. Note that the clustering method involves representing the bounding box by its top-left and bottom-right coordinates, instead of the representation using center coordinates and bounding box dimensions that DETR typically outputs.

Fig. 2. (Left) Image with ground-truth bounding boxes. (Middle) Clustered bounding boxes from the full ensemble for one object. (Right) Average bounding box post merging for one object.

3.3   Uncertainty Quantification

Location Uncertainty. The first aspect of predictive uncertainty that we evaluate is the uncertainty in the location of the object. Specifically, we look at the uncertainty in the center location of the output bounding box. To evaluate the location uncertainty, we compute the negative log likelihood of the true location of the center $(x_{\mathrm{center}}, y_{\mathrm{center}})$ under a Gaussian distribution $\mathcal{N}(\hat{\mu}, \hat{\sigma}^2)$, where $\hat{\mu}$ is the predicted center location, that is,

$$\hat{\mu} = (\hat{x}_{\mathrm{center}}, \hat{y}_{\mathrm{center}}) \tag{7}$$

$$\hat{\sigma}^2 = \left(\frac{1}{N} \sum_{n=1}^{N} (\hat{x}_{\mathrm{n,center}} - x_{\mathrm{n,center}})^2, \frac{1}{N} \sum_{n=1}^{N} (\hat{y}_{\mathrm{n,center}} - y_{\mathrm{n,center}})^2 \right) \tag{8}$$

In the above set of equations, $(x_{\mathrm{n,center}}, y_{\mathrm{n,center}}) \in \mathcal{D}_{\mathrm{boxes,train}}$, where $\mathcal{D}_{\mathrm{boxes,train}}$ denotes the collection of the bounding boxes from the entire training set. The variance computation is equivalent to computing the average of the squared residuals of the model predictions. A prerequisite to computing the variance is that we first need to match the ground-truth bounding boxes with the predicted bounding boxes on the training data. Predicted bounding boxes are assigned to ground-truth bounding boxes by solving the minimum weight matching problem, where each matching has a weight of (1 - IoU). Furthermore, we set a threshold on minimum IoU for the matching between the predicted bounding box and the ground-truth bounding box to be valid during evaluation. An important point to note here is that we can only compute the variance using the matched bounding boxes on the training data. Although the equations mentioned above look at the predictions from a single model, this can be extended in a straightforward way for ensembles by using a Gaussian mixture model likelihood, where the mixture is the cluster forming the detections, and thus each member of the mixture is an individual detection in the cluster. For every member of the mixture, the mean parameter is the center of the detection, and the variance is the average of the variances across all members of the ensemble (each computed using the equations given above).

Class Uncertainty. Along with each bounding box, the DETR model outputs a $(K + 1)$-dimensional softmax probability distribution, where K denotes the number of classes and the final dimension is reserved for the "background" class. If the bounding box has maximum probability on the background class, it is filtered out during evaluation. In order to compute any metrics for the classification performance, we need to assign predicted bounding boxes to the ground-truth bounding boxes. This aligns a predicted softmax distribution with a ground-truth class label.

We use the assignment described in the previous subsection for assigning the predicted bounding boxes to ground-truth bounding boxes. For the predicted bounding boxes that do not get assigned to any ground-truth bounding box, we set their respective ground-truth class to background. On the other hand, if after assignment there exist ground-truth bounding boxes that do not have any assigned predicted bounding boxes, we assert a uniform distribution over classes as the predicted distribution. This penalizes the missed bounding box targets. Once this assignment is complete, we evaluate the negative log-likelihood (NLL), expected calibration error (ECE), and accuracy for this multi-class classification task. NLL and ECE are especially useful as they account for the probabilistic nature of the output.

Objectness Uncertainty. In safety-critical applications of object detection models, it is often important to understand how well the detector performs when it comes to identifying the presence of "any" object in an area of interest in the scene. For example, in the case of autonomous driving, it may be sufficient to identify that there is are one or more objects in a given area in the scene for the car to avoid it, without identifying the correct classes of the objects.

As we described earlier, among other object categories, DETR and AdaMixer also output the probability that the bounding box belongs to the background class. This information can be used to predict whether the bounding box belongs to any actual object or to the background. We can consider this a kind of binary classification problem, where the probability of a bounding box belonging to an actual object is $p(\text{"Objectness"}|\cdot) = 1 - p(\text{"Background"}|\cdot)$. Next, we assign this objectness probability to every pixel belonging to the predicted bounding box. In the case where a pixel belongs to multiple bounding boxes, we take the maximum of the objectness probabilities across the boxes, to reflect the model's highest objectness confidence value. Generating the ground-truth objectness map is very straightforward. Every pixel in an image is labelled as belonging to an object if it belongs to even a single ground-truth bounding box, or background otherwise.

Next, we downsample the prediction and ground-truth grids using max pooling, as processing numerous images with a high number of pixels in each of them becomes computationally prohibitive. For each image, we obtain the objectness probability and ground-truth vectors by flattening the grids, which are then finally concatenated to compute performance metrics on the objectness classification problem. The final concatenation step is important because different images in the dataset can have different sizes, and thus it is important not to bias the performance computation by assigning the same weight to metrics from different image sizes.

4 Experiments

In this section, we present our experimental framework and our experimental results. First, we begin by describing the experimental protocols involved, followed by different experiments and results, each looking at a specific aspect of uncertainty estimation, as described in the previous section.

4.1 Experimental Protocols

For our empirical evaluations, we use the DETR and AdaMixer models with the COCO 2017 dataset [19]. COCO 2017 is a widely used large-scale dataset for evaluating object detection models, containing 80 object categories. The COCO training dataset contains 118,287 images and a total of 860,001 bounding boxes. The COCO validation dataset contains 5,000 images, and a total of 36,781 bounding boxes.

For uncertainty estimation in DETR and AdaMixer, we generate two variants of deep ensembles: a) full ensembles and b) decoder ensembles. In full ensembles, we train each member from scratch on the training data with different random initialization. For the decoder ensembles, we freeze the backbone and encoder to the pre-trained model parameters, and only unfreeze and retrain the parameters of the decoder and output heads after new random initialization. Each of these ensembles has 9 models, unless otherwise specified. Deep ensembles have shown promising performance at uncertainty estimation with smaller ensemble size, and that motivates our choice of using 9 models [1]. During merging, we remove clusters that have fewer than 3 members [23]. Note that the AdaMixer model as published uses 80 binary classifiers rather than a single softmax classifier as in

DETR. To make our results comparable, we replace the classifier in AdaMixer with the same (80 + 1)-way softmax classifier of DETR. We then train using the same loss as DETR, replacing AdaMixer's focal loss with standard cross entropy. We notice a drop of ~0.2 average precision when making this change. To train our models, we use the PyTorch-based MMDetection library [5,24].

4.2   Experiment 1: Evaluating Location Uncertainty

In the first experiment, we evaluate the location uncertainty aspect of the different methods. We use the evaluation procedure reported in Sect. 3.3, and set the matching IoU threshold to 0.5. We also note that the location uncertainty is only computed for the matched bounding boxes, as setting penalties for unmatched bounding boxes is not straightforward. Furthermore, given that the dataset consists of images of different sizes, we transform the coordinates to normalized coordinates ranging between 0 and 1. We present the negative log-likelihood results for the center location prediction in Table 1.

While aggregating the bounding boxes for ensemble-based methods, we use the average merging strategy. From the results presented in Table 1, we observe that while the full ensemble has the lowest negative log-likelihood of the methods presented for each model architecture, the performance improvement is not very significant. While comparing against models and methods, DETR full ensemble shows the best performance. Next, we turn our focus towards evaluating the class uncertainty in the model predictions.

Table 1. Location uncertainty results.

	NLL ($\downarrow$)
Base DETR	−5.68
Base AdaMixer	−5.59
DETR decoder ensemble	−5.88
DETR full ensemble	−6.03
AdaMixer full ensemble	−5.91
AdaMixer decoder ensemble	−5.87

4.3   Experiment 2: Evaluating Classification Performance

In this experiment, we evaluate and compare the classification performance across different methods and merging strategies. In Fig. 3, we present the results comparing classification performance across different methods. We focus on the average merging strategy for aggregating bounding boxes for deep ensembles as it results in better overall performance when we consider the entire validation dataset, as well as leads to a higher fraction of the ground-truth bounding boxes matched (shown in Appendix A.1). We observe that deep ensembles help us improve over the performance of a single model across all three performance metrics we measure. However, especially for NLL, the performance gap closes between a single model and the ensemble as we set a stricter matching IoU threshold. This is again due to the dominance of penalty terms from unmatched bounding boxes. We defer the additional results from this experiment evaluating different merging strategies, along with the results on only the matched bounding boxes to the Appendix A.1.

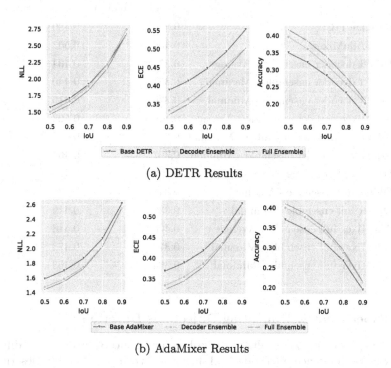

(a) DETR Results

(b) AdaMixer Results

Fig. 3. A comparison of classification performance metrics across different methods on the entire validation data set. We present negative log-likelihood (NLL, ↓), expected calibration error (ECE, ↓), and accuracy (↑) for different matching IoU thresholds. For deep ensembles, we use the average bounding box merging strategy.

Additionally, we present the average precision and recall results using the COCO evaluation API [19] in Table 2. The COCO evaluation API has been widely used in the existing literature to evaluate object detectors' performance. However, we must note that the average precision and recall results do not evaluate the probabilistic aspect of the classification performance, unlike the NLL and ECE metrics shown earlier in this section. From the results in Table 2, we observe that the full ensemble with average merging performs the best in terms of average precision and recall, however, the gains are still somewhat modest as compared to the respective base models. Finally, while comparing against models and methods, we observe that Adamixer full ensemble with average merging strategy tends to outperform the rest on average precision and recall. We observe a similar trend when looking at NLL, ECE, and accuracy on the entire validation set.

Table 2. Comparison of average precision and recall results using COCO evaluation API.

	Average precision	Average recall
Base DETR	0.386	0.480
Base AdaMixer	0.400	0.503
DETR decoder ensemble - Average	0.375	0.464
DETR decoder ensemble - Maximal	0.355	0.440
DETR decoder ensemble - Minimal	0.336	0.416
AdaMixer decoder ensemble - Average	0.399	0.503
AdaMixer decoder ensemble - Maximal	0.375	0.475
AdaMixer decoder ensemble - Minimal	0.367	0.465
DETR full ensemble - Average	0.399	0.494
DETR full ensemble - Maximal	0.349	0.439
DETR full ensemble - Minimal	0.332	0.419
AdaMixer full ensemble - Average	0.408	0.518
AdaMixer full ensemble - Maximal	0.373	0.473
AdaMixer full ensemble - Minimal	0.364	0.463

4.4   Experiment 3: Evaluating Objectness Uncertainty

In this experiment, we evaluate the objectness uncertainty across different approximate inference methods and merging strategies. As we describe in Sect. 3.3, our goal is to evaluate how the models perform on the binary classification task of identifying whether a local region belongs to any object or not. We introduce additional methods for this task that look at composing the objectness information from deep ensembles without clustering. To do this, we simply generate an objectness probability map for each member of the ensemble, and while composing the maps, for every pixel, we average the objectness probabilities across all the members of the ensemble. The results of this experiment are presented in Table 3. Note that for the final downsampling step of the image grids using max-pooling, we use a kernel of size (4, 4).

There are a few key takeaways from this experiment. First, when evaluating the probabilistic aspect of the methods using NLL and AUROC, we observe that the full ensemble without merging predicted bounding boxes performs much better than the rest of the methods. Specifically, we see that both decoder ensemble and full ensemble without merging perform better than their counterparts which include merging. This is intuitive, as the merging step can lead to filtering information if a cluster does not satisfy the minimum number of detections threshold. Averaging the most confident objectness probability scores across multiple members of the ensemble can lead to small objectness probability mass assigned to areas of the image where an individual model does not assign. Furthermore, the reverse is also true. In cases where an individual model has an overconfident

Table 3. Objectness Uncertainty Results.

Method	Merging strategy	NLL	Accuracy	AUROC	Precision	Recall	F1
Base DETR	–	0.313	0.916	0.959	0.865	0.958	0.909
Base AdaMixer	–	0.369	0.913	0.955	0.864	0.952	0.906
DETR decoder ensemble	Average	0.317	0.914	0.954	0.866	0.953	0.907
	Maximal	0.304	0.909	0.956	0.848	0.969	0.904
	Minimal	0.383	0.910	0.942	0.878	0.923	0.900
AdaMixer decoder ensemble	Average	0.319	0.911	0.951	0.861	0.951	0.904
	Maximal	0.298	0.904	0.954	0.839	0.969	0.899
	Minimal	0.411	0.904	0.935	0.875	0.912	0.893
DETR full ensemble	Average	0.289	0.917	0.958	0.865	0.961	0.911
	Maximal	0.287	0.907	0.959	0.837	0.980	0.903
	Minimal	0.390	0.909	0.940	0.881	0.917	0.899
AdaMixer full ensemble	Average	0.305	0.913	0.954	0.863	0.954	0.906
	Maximal	0.286	0.904	0.956	0.834	0.976	0.899
	Minimal	0.418	0.904	0.934	0.878	0.907	0.893
DETR decoder ensemble	No merging	0.261	0.921	0.968	0.878	0.953	0.914
DETR full ensemble	No merging	0.205	0.928	0.977	0.888	0.956	0.921
AdaMixer decoder ensemble	No merging	0.238	0.921	0.969	0.884	0.945	0.913
AdaMixer full ensemble	No merging	0.214	0.926	0.974	0.884	0.951	0.919

prediction, averaging the objectness scores can also help reduce the prediction probability to a more moderate level. An example demonstrating this is shown in Fig. 4.

Next, when comparing between merging strategies, we find that the minimal merging strategy achieves the lowest performance among other merging techniques. Again, this is intuitive, as the minimal bounding-box-generating strategy would assign objectness probability to fewer pixels in the image. Comparing the full ensemble and the decoder ensemble, except for the minimal merging strategy, the full ensemble tends to perform better for a given merging strategy. Finally, when comparing across models and methods, we note that DETR full ensemble with no merging performs the best on all the metrics (except recall).

4.5   Experiment 4: Runtime-Aware Performance Comparison

In this experiment, we evaluate the prediction latency of the methods introduced to assess the cost of using the proposed deep ensembles approach to quantify detection uncertainty. In the previous experiments, we used an ensemble size of 9 models for both the decoder and full ensembles. However, the decoder ensemble backbone is fixed, thus we can cache the outputs from the encoder and run them

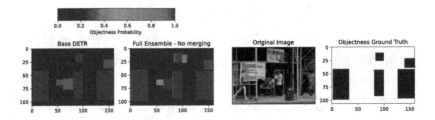

Fig. 4. (First) Objectness probability map using full ensemble (without merging). (Second) Objectness probability map using Base DETR. (Third) Original image with ground-truth bounding boxes. (Fourth) Ground-truth objectness map. The blue area corresponds to objects, and the white area corresponds to the background. Note that the base DETR model assigns a strong objectness probability to the shopping cart in the middle of the image, which does not have a ground-truth bounding box assigned to it. However, the full ensemble helps moderate the confidence due to model averaging. (Color figure online)

Table 4. Runtime latency results. Median runtime latency numbers are presented as average milliseconds per image on the validation dataset. The number of models in the ensemble is indicated in parentheses. We use an NVIDIA 3080Ti GPU to evaluate the runtime latency for all models.

	Backbone	Encoder	Decoder	Clustering	Total	FPS
Base DETR	15.5	4.28	5.25	–	25.0	40.0
Base AdaMixer	17.7	–	16.4	–	34.1	29.7
DETR decoder ensemble (9)	15.5	4.28	47.3	7.1	74.1	13.5
DETR decoder ensemble (18)	15.5	4.3	94.5	28.3	142.6	7.0
DETR full ensemble (5)	77.4	21.4	26.3	2.9	127.9	7.8
DETR full ensemble (9)	139.3	38.5	47.3	6.3	231.4	4.3
AdaMixer decoder ensemble (9)	17.7	–	147.6	7.2	172.5	5.8
AdaMixer full ensemble (9)	159.3	–	147.6	7.2	314.1	3.2

through all 9 decoders. This helps us to reduce the overall prediction latency. We give a detailed breakdown of the latency numbers in Table 4. We note that the results presented in Table 4 are approximate runtime numbers and that they do not account for additional overheads such as the time required to load models in the memory. Furthermore, the runtime numbers for ensembles are obtained using linear extrapolation over the results presented for a single model. We also provide frames/sec results based on the total runtime latency.

For this experimental section, we compare classification performance results from a full ensemble consisting of 5 models and a decoder ensemble that consists of 18 models using the DETR architecture. As noted in Table 4, the runtime of these are very similar, and thus they make for a more fair runtime-aware comparison. We focus specifically on the average bounding box merging strategy. For merging, we use the same IoU threshold of 0.75 for grouping and ensure the same fraction of the total ensemble as the threshold of the minimum number of detections by setting the minimum number of detections for a cluster to 2 for

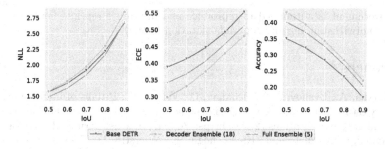

Fig. 5. A comparison of classification performance metrics across different methods on DETR on the entire validation data set. We present negative log-likelihood (NLL, ↓), expected calibration error (ECE, ↓), and accuracy (↑) for different matching IoU thresholds. For deep ensembles, we use the average bounding box merging strategy.

the 5-member full ensemble and 7 for the 18-member decoder ensemble [23]. The results of the classification performance are presented in Fig. 5.

We observe that the full ensemble still performs better than the decoder ensemble on the NLL metric. Looking at ECE and accuracy on the entire validation dataset, we observe that the decoder ensemble has better performance, while on the matched bounding boxes, the full ensemble tends to yield better accuracy and ECE when compared with the decoder ensemble. This is especially an interesting observation, as the accuracy and ECE improvements on the larger decoder ensemble on the full validation set do not necessarily translate to an improvement in NLL. Additional results on the matched bounding boxes is presented in Appendix A.2.

5 Conclusions

In this paper, we have explored the application of deep ensemble-based uncertainty quantification methods to query-based object detectors. Through our empirical analysis, we have shown how this approach can help us improve on class uncertainty, location uncertainty, and objectness uncertainty estimates. Furthermore, our runtime latency results show how a larger decoder ensemble with similar latency to a smaller full ensemble can lead to interesting tradeoffs among different performance metrics. Several important directions for future work emerge from this paper, including studying other methods for deriving and representing model uncertainty, exploring other detection merging techniques, and investigating the application of these methods to other detection architectures.

Acknowledgements. Research reported in this paper was sponsored in part by the CCDC Army Research Laboratory under Cooperative Agreement W911NF-17-2-0196 (ARL IoBT CRA). The views and conclusions contained in this document are those of the authors and should not be interpreted as representing the official policies, either expressed or implied, of the Army Research Laboratory or the U.S. Government. The

U.S. Government is authorized to reproduce and distribute reprints for Government purposes notwithstanding any copyright notation herein.

References

1. Ashukha, A., Lyzhov, A., Molchanov, D., Vetrov, D.: Pitfalls of in-domain uncertainty estimation and ensembling in deep learning. In: International Conference on Learning Representations (2020). https://openreview.net/forum?id=BJxI5gHKDr
2. Blundell, C., Cornebise, J., Kavukcuoglu, K., Wierstra, D.: Weight uncertainty in neural network. In: Bach, F., Blei, D. (eds.) Proceedings of the 32nd International Conference on Machine Learning. Proceedings of Machine Learning Research, vol. 37, pp. 1613–1622. PMLR, Lille (2015). https://proceedings.mlr.press/v37/blundell15.html
3. Bochkovskiy, A., Wang, C.Y., Liao, H.Y.M.: Yolov4: optimal speed and accuracy of object detection. arXiv preprint arXiv:2004.10934 (2020)
4. Carion, N., Massa, F., Synnaeve, G., Usunier, N., Kirillov, A., Zagoruyko, S.: End-to-end object detection with transformers. In: Vedaldi, A., Bischof, H., Brox, T., Frahm, J.-M. (eds.) ECCV 2020. LNCS, vol. 12346, pp. 213–229. Springer, Cham (2020). https://doi.org/10.1007/978-3-030-58452-8_13
5. Chen, K., et al.: MMDetection: open mmlab detection toolbox and benchmark. arXiv preprint arXiv:1906.07155 (2019)
6. Chen, T., Fox, E., Guestrin, C.: Stochastic gradient hamiltonian monte carlo. In: International Conference on Machine Learning, pp. 1683–1691. PMLR (2014)
7. Duan, K., Bai, S., Xie, L., Qi, H., Huang, Q., Tian, Q.: Centernet: keypoint triplets for object detection. In: ICCV (2019)
8. Gal, Y., Ghahramani, Z.: Dropout as a bayesian approximation: representing model uncertainty in deep learning. In: International Conference on Machine Learning, pp. 1050–1059 (2016)
9. Gao, Z., Wang, L., Han, B., Guo, S.: Adamixer: a fast-converging query-based object detector. In: CVPR (2022)
10. Girshick, R., Donahue, J., Darrell, T., Malik, J.: Rich feature hierarchies for accurate object detection and semantic segmentation. In: Proceedings of the IEEE Conference on Computer Vision and Pattern Recognition (CVPR) (2014)
11. Hall, D., et al.: Probabilistic object detection: definition and evaluation. In: Proceedings of the IEEE/CVF Winter Conference on Applications of Computer Vision, pp. 1031–1040 (2020)
12. Harakeh, A., Smart, M., Waslander, S.L.: Bayesod: a bayesian approach for uncertainty estimation in deep object detectors. In: 2020 IEEE International Conference on Robotics and Automation (ICRA), pp. 87–93. IEEE (2020)
13. He, K., Gkioxari, G., Dollár, P., Girshick, R.: Mask R-CNN. In: ICCV, pp. 2980–2988 (2017)
14. He, K., Zhang, X., Ren, S., Sun, J.: Deep residual learning for image recognition. In: Proceedings of the IEEE Conference on Computer Vision and Pattern Recognition, pp. 770–778 (2016)
15. Kendall, A., Gal, Y.: What uncertainties do we need in bayesian deep learning for computer vision? In: Guyon, I., et al. (eds.) Advances in Neural Information Processing Systems 30: Annual Conference on Neural Information Processing Systems 2017(December), Long Beach, CA, USA, pp. 4–9, 2017, pp. 5574–5584 (2017). https://proceedings.neurips.cc/paper/2017/hash/2650d6089a6d640c5e85b2b88265dc2b-Abstract.html

16. Lakshminarayanan, B., Pritzel, A., Blundell, C.: Simple and scalable predictive uncertainty estimation using deep ensembles. In: Advances in Neural Information Processing Systems, pp. 6402–6413 (2017)
17. Law, H., Deng, J.: Cornernet: detecting objects as paired keypoints. In: ECCV (2018)
18. Le, M.T., Diehl, F., Brunner, T., Knol, A.: Uncertainty estimation for deep neural object detectors in safety-critical applications. In: 2018 21st International Conference on Intelligent Transportation Systems (ITSC), pp. 3873–3878. IEEE (2018)
19. Lin, T.-Y., et al.: Microsoft COCO: common objects in context. In: Fleet, D., Pajdla, T., Schiele, B., Tuytelaars, T. (eds.) ECCV 2014. LNCS, vol. 8693, pp. 740–755. Springer, Cham (2014). https://doi.org/10.1007/978-3-319-10602-1_48
20. Meng, D., et al.: Conditional detr for fast training convergence. In: Proceedings of the IEEE International Conference on Computer Vision (ICCV) (2021)
21. Miller, D., Dayoub, F., Milford, M., Sünderhauf, N.: Evaluating merging strategies for sampling-based uncertainty techniques in object detection. In: 2019 International Conference on Robotics and Automation (ICRA), pp. 2348–2354. IEEE (2019)
22. Miller, D., Nicholson, L., Dayoub, F., Sünderhauf, N.: Dropout sampling for robust object detection in open-set conditions. In: 2018 IEEE International Conference on Robotics and Automation (ICRA), pp. 3243–3249. IEEE (2018)
23. Miller, D., Sünderhauf, N., Milford, M., Dayoub, F.: Uncertainty for identifying open-set errors in visual object detection. IEEE Rob. Autom. Lett. (2021). https://doi.org/10.1109/LRA.2021.3123374
24. Paszke, A., et al.: Pytorch: an imperative style, high-performance deep learning library. In: Wallach, H., Larochelle, H., Beygelzimer, A., d' Alché-Buc, F., Fox, E., Garnett, R. (eds.) Advances in Neural Information Processing Systems, vol. 32, pp. 8024–8035. Curran Associates, Inc. (2019)
25. Redmon, J., Divvala, S., Girshick, R., Farhadi, A.: You only look once: unified, real-time object detection. In: Proceedings of the IEEE Conference on Computer Vision and Pattern Recognition, pp. 779–788 (2016)
26. Ren, S., He, K., Girshick, R., Sun, J.: Faster r-cnn: towards real-time object detection with region proposal networks. Adv. Neural Inf. Process. Syst. 28 (2015)
27. Theodoridis, S., Koutroumbas, K.: Pattern Recognition. Elsevier, Amsterdam (2006)
28. Tian, Z., Shen, C., Chen, H., He, T.: FCOS: fully convolutional one-stage object detection. In: ICCV (2019)
29. Tolstikhin, I., et al.: MLP-mixer: an all-mlp architecture for vision. In: NeurIPS (2021)
30. Vaswani, A., et al.: Attention is all you need. In: NeurIPS (2017)
31. Wang, Y., Zhang, X., Yang, T., Sun, J.: Anchor detr: query design for transformer-based detector. In: AAAI (2022)
32. Welling, M., Teh, Y.W.: Bayesian learning via stochastic gradient langevin dynamics. In: Proceedings of the 28th International Conference on Machine Learning (ICML-2011), pp. 681–688 (2011)
33. Zhu, X., Su, W., Lu, L., Li, B., Wang, X., Dai, J.: Deformable DETR: deformable transformers for end-to-end object detection. In: 9th International Conference on Learning Representations, ICLR 2021, Virtual Event, Austria, 3–7 May 2021. OpenReview.net (2021). https://openreview.net/forum?id=gZ9hCDWe6ke

W33 - Recovering 6D Object Pose

W33 - Recovering 6D Object Pose

This workshop covers topics related to 6DoF object pose estimation, which is of major importance to many higher level applications such as robotic manipulation and augmented/virtual reality. The workshop features four invited talks by experts in the field, discussion on open problems, and presentations of accepted workshop papers and of relevant papers invited from the main conference. In conjunction with the workshop, we organize the BOP Challenge 2022 to determine state-of-the-art object pose estimation methods.

October 2022

Martin Sundermeyer
Tomas Hodan
Yann Labbé
Gu Wang
Lingni Ma
Eric Brachmann
Bertram Drost
Sindi Shkodrani
Rigas Kouskouridas
Ales Leonardis
Carsten Steger
Vincent Lepetit
Jiří Matas

CenDerNet: Center and Curvature Representations for Render-and-Compare 6D Pose Estimation

Peter De Roovere[1,2(✉)] [iD], Rembert Daems[1,3,4] [iD], Jonathan Croenen[2] [iD], Taoufik Bourgana[5] [iD], Joris de Hoog[5] [iD], and Francis wyffels[1] [iD]

[1] IDLab-AIRO, Ghent University – imec, Ghent, Belgium
peter.deroovere@ugent.be
[2] RoboJob, Heist-op-den-Berg, Belgium
[3] Department of Electromechanical, Systems and Metal Engineering, Ghent University, Ghent, Belgium
[4] EEDT-DC, Flanders Make, Gent, Belgium
[5] DecisionS, Flanders Make, Leuven, Belgium

Abstract. We introduce CenDerNet, a framework for 6D pose estimation from multi-view images based on center and curvature representations. Finding precise poses for reflective, textureless objects is a key challenge for industrial robotics. Our approach consists of three stages: First, a fully convolutional neural network predicts center and curvature heatmaps for each view; Second, center heatmaps are used to detect object instances and find their 3D centers; Third, 6D object poses are estimated using 3D centers and curvature heatmaps. By jointly optimizing poses across views using a render-and-compare approach, our method naturally handles occlusions and object symmetries. We show that CenDerNet outperforms previous methods on two industry-relevant datasets: DIMO and T-LESS.

Keywords: Object detection · 6D object pose estimation · Industrial robotics · Render-and-Compare · Curvature maps

1 Introduction

1.1 Context

6D pose estimation is an essential aspect of industrial robotics. Today's high-volume production lines are powered by robots reliably executing repetitive movements. However, as the manufacturing industry shifts towards high-mix, low-volume production, there is a growing need for robots that can handle more variability [2]. Estimating 6D poses for diverse sets of objects is crucial to that goal.

Supplementary Information The online version contains supplementary material available at https://doi.org/10.1007/978-3-031-25085-9_6.

Manufacturing use cases present unique challenges. Many industrial objects are reflective and textureless, with scratches or saw patterns affecting their appearance [4,31]. Parts are often stacked in dense compositions, with many occlusions. These densely stacked, reflective parts are problematic for existing depth sensors. In addition, object shapes vary greatly, often exhibiting symmetries leading to ambiguous poses. Many applications require sub-millimeter precision and the ability to integrate new, unseen parts swiftly.

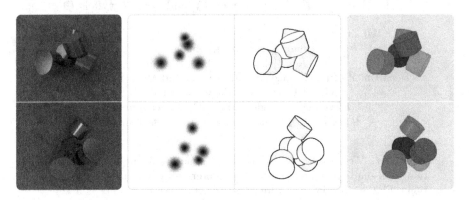

Fig. 1. Multi-view input images (left) are converted to center and curvature heatmaps (center), used to estimate 6D object poses (right).

This work presents a framework for 6D pose estimation targeted to these conditions. Our approach predicts object poses for known textureless parts from RGB images with known camera intrinsics and extrinsics. We use multi-view data as monocular images suffer from ambiguities in appearance and depth. In practice, images from multiple viewpoints are easily collected using multi-camera or hand-in-eye setups. This setup reflects many real-world industrial use-cases.

1.2   Related Work

Recent progress in 6D pose estimation from RGB images was demonstrated in the 2020 BOP challenge [10]. Convolutional neural networks (CNNs) trained on large amounts of synthetic data are at the core of this success. Many recent methods consist of three stages: (1) object detection, (2) pose estimation, and (3) refinement, using separate neural networks for each stage. While most approaches operate on monocular RGB images, some focus on multi-view data.

Object Detection. In the first stage, CNN-based neural networks are used to detect object instances. Although these detectors typically represent objects by 2D bounding boxes [6,7,17,22], 2D center points can be a simple and efficient alternative [33].

Pose Estimation. In the second stage, the 6D pose of each detected object is predicted. Classical methods based on local features [1] or template matching [8] have been replaced by learning systems. CNNs are used to detect local features [14, 25, 26, 30] or find 2D-3D correspondences [18, 19, 29, 32]. Crucial aspects are the parameterization of 6D poses [34] and how symmetries are handled [20].

Refinement. In the final stage, the estimated poses are iteratively refined by comparing object renders to the original image. This comparison is not trivial, as real-world images are affected by lighting, texture, and background changes, not captured in the corresponding renders. However, CNNs can be trained to perform this task [14, 16, 32].

Multi-view. Approaches for multi-view 6D object pose estimation extend existing single-view methods. First, poses hypotheses are generated from individual images. Next, these estimates are fused across views [14, 15].

1.3   Contributions

We present CenDerNet, a framework for 6D pose estimation from multi-view images based on center and curvature representations. First, a convolutional neural network is trained to predict center and curvature heatmaps. Second, center heatmaps are used to detect object instances and find their 3D centers. These centers initialize and constrain the pose esimates. Third, curvature heatmaps are used to optimize these poses further using a render-and-compare approach (Fig. 1).

Our system is conceptually simple and easy to use. Many existing methods consist of multiple stages, each with different training and tuning requirements. Our framework is more straightforward. We use a single, fully-convolutional neural network to convert RGB images to interpretable representations. Next, we use classical optimization techniques that require little tuning.

Using a render-and-compare approach, we jointly estimate poses for all objects in a scene across all viewpoints. As a result, our method naturally handles occlusions and object symmetries. We provide a GPU implementation of our render-and-compare method that allows evaluating over 2,000 scene pose estimates per second.

We evaluate CenDerNet using DIMO and T-LESS, two challenging, industry-relevant datasets. On DIMO, our method outperforms PVNet by a large margin. On T-LESS, CenDerNet outperforms the 2020 ECCV results of CosyPose, the leading multi-view method.

2   CenDerNet

Our system consists of three stages:

1. A convolutional neural network predicts center and curvature heatmaps for multi-view input images.
2. The predicted center heatmaps are converted to 3D center points. These 3D centers initialize and constrain the set of predicted object poses.
3. Object poses are optimized by comparing curvature renders to the predicted curvature heatmaps.

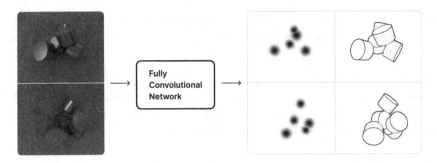

Fig. 2. Step 1: multi-view RGB images are converted to center and curvature heatmaps by a single, fully convolutional network.

2.1 From Images to Center and Curvature Heatmaps

This step eliminates task-irrelevant variations by converting images into center and curvature representations (Fig. 2). RGB images can vary significantly due to lighting, background, and texture changes. These changes, however, do not affect object poses. This step eliminates these effects by transforming images into representations that simplify pose estimation. As we want our system to apply to a wide range of objects, these representations should be category-independent. We identify centers and curvatures as suitable representations with complementary properties.

Centers. We use center heatmaps—modeling the probability of object center points—to detect objects and roughly estimate their locations. Previous work has shown that detecting objects as center points is simple and efficient [33]. Moreover, 2D center points can be triangulated to 3D, initializing object poses and enabling geometric reasoning. For example, center predictions located at impossible locations can be discarded. When predicting center points, there is a trade-off between spatial precision and generalization. Precise center locations can differ subtly between similar objects. This makes it difficult for a learning system to predict precise centers for unseen objects. As we want our system to generalize to unseen categories, we relax spatial precision requirements, by training our model to predict gaussian blobs at center locations. We use a single center heatmap for all object categories.

Curvature. We use curvature heatmaps to highlight local geometry and enable comparison between images and renders. Representations based on 3D geometry are robust to changes in lighting, texture, and background and can be created from textureless CAD files. Representations based on global geometry [29] or category-level semantics [5,19] do not generalize to unseen object types. Therefore, we focus on local geometry. Previous work has shown that geometric edges can be used for accurately estimating 6D poses [3,12,13]. We base our representation on view-space curvature. To obtain these view-space curvatures, we first render normals in view-space. Next, we approximate the gradients for each pixel using the Prewitt operator [21]. Finally, we calculate the 2-norm of these gradients to obtain a per-pixel curvature value. Areas with high curvatures correspond to geometric edges or object boundaries and are visually distinct. Similarly, visually similar areas, like overlapping parallel planes, contain no curvatures values, as shown in Fig. 3b.

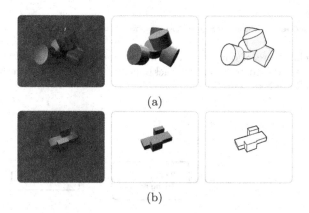

(a)

(b)

Fig. 3. (a) Curvatures are calculated by rendering normals in view space, approximating per-pixel gradients using the Prewitt operator, and calculating the 2-norm. (b) Visually similar areas, like overlapping parallel planes, exhibit no curvature values.

Model and Training. We use a fully convolutional network to predict center and curvature heatmaps. The same weights are applied to images from different viewpoints. The architecture is based on U-net [23] and shown in Fig. 4. A shared backbone outputs feature maps with a spatial resolution equal to the input images. Separate heads are used for predicting center and curvature heatmaps. Ground-truth center heatmaps are created by projecting 3D object centers to each image and splatting the resulting points using a Gaussian kernel, with standard deviation adapted by object size and distance. Curvature heatmaps are created as explained in Sect. 2.1. Binary cross entropy loss is used for both outputs.

2.2 From Center Heatmaps to 3D Centers

This step converts multi-view center heatmaps to 3D center points. First, local maxima are found in every heatmap using a peak local max filter [28] (Fig. 5). Each of these 2D maxima represents a 3D ray, defined by the respective camera intrinsics and extrinsics. Next, for each pair of 3D rays, the shortest mutual distance and midpoint are calculated [24]. When this distance is below a threshold d_t, the midpoint is added to the set of candidates. Within this set, points that are closer to each other than a distance d_c are merged. Finally, the remaining points are refined by maximizing their reprojection score across views, using

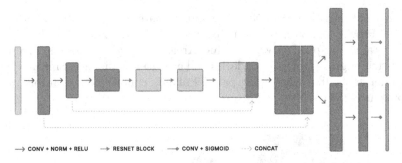

Fig. 4. The architecture of our fully convolutional network is based on U-net. Input images are first processed by a shared backbone. Afterwards, separate heads output center and curvature heatmaps.

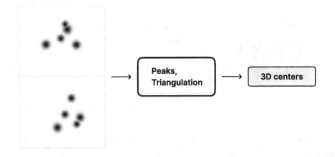

Fig. 5. Step 2: multi-view 2D center heatmaps are converted to 3D center points.

Scipy's Nelder-Mead optimizer [27]. This leads to a set of 3D points, each with per-view heatmap scores. If information about the number of objects in the scene is available, this set is further pruned. 3D centers are sorted by their aggregated heatmap score (accumulated for all views) and removed if they are closer than a distance d_o to higher-scoring points.

2.3 6D Pose Estimation

This step optimizes the 6D poses for all detected objects, using a render-and-compare approach based on curvatures (Fig. 6). Object CAD models, camera intrinsics, and extrinsics are available. Consequently, curvature maps can be rendered for each set of 6D object pose candidates. We define a cost function that compares such curvature renders to the predicted curvature heatmaps. This cost function is used for optimizing a set of object poses, initialized by the previously detected 3D centers.

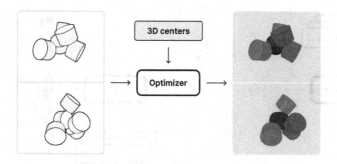

Fig. 6. Step 3: Multi-view curvature heatmaps and 3D object centers are used to find 6D object poses.

Cost Function. Predicted curvature heatmaps are converted to binary images with threshold t_b. Next, for each binary image, a distance map is created where each pixel contains the distance to the closest non-zero (true) pixel, using scikit-image's distance transform [28]. The resulting distance maps have to be calculated only once and can be reused throughout optimization (Fig. 7).

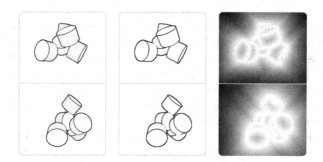

Fig. 7. Target curvature heatmaps (left) are converted to binary images (center). Next, distance maps (right) are calculated, where each pixel contains the distance to the closest non-zero (true) pixel.

Using these distance maps, curvature renders can be efficiently compared to the target heatmaps. Pixel-wise multiplication of a render to a distance map returns an image where each pixel contains the distance to the closest true curvature pixel, weighted by its curvature value. As a result, regions with high curvature weigh more, and regions with zero curvature do not contribute. The final cost value is obtained by summing the resulting image, and dividing by the sum of the rendered curvature map. This is done for each viewpoint. The resulting costs are weighed by view-specific weights w_v and summed, resulting in a final scalar cost value.

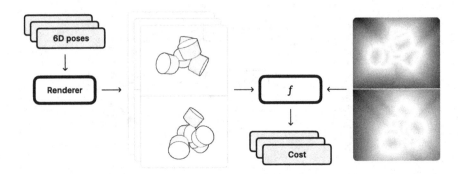

Fig. 8. Overview of the cost function. First, curvature maps are rendered for each set of 6D object pose estimates. These curvature maps are compared to a pre-computed distance map based on the target curvatures. This results in a cost value for each sample, calculated in parallel.

Figure 8 shows an overview of the cost function. We implement this function—including curvature rendering—on GPU. Our implementation runs at 2,000 calls per second on an NVIDIA RTX3090 Ti for six 256×320 images per call.

Optimization. We sequentially optimize 6D object poses by evaluating pose candidates anchored by the detected 3D centers. As scenes can consist of densely stacked objects with many occlusions, we optimize objects sequentially. We argue highly visible objects are easier to optimize and—crucially—should be taken into account when optimizing objects they occlude. For each 3D center, we use per-view center heatmap scores as a proxy for visibility. We optimize objects in order of decreasing visibility, and weigh the contribution of each view to the cost according to these scores. When estimating an object pose, we first evaluate a set of 2,000 pose candidates with random rotations and translations normally distributed around the 3D center. Afterwards, the best candidates are further optimized using a bounded Nelder-Mead optimizer [27].

3 Experiments

We evaluate our method on the 6D localization task as defined in the BOP challenge [10] on DIMO [4] and T-LESS [9], two industry-relevant datasets. On DIMO, we show our method significantly outperforms PVNet [19], a strong single-view baseline. On T-LESS, CenDerNet improves upon the 2020 ECCV results of CosyPose [14], the leading multi-view method.

3.1 DIMO

Dataset. The dataset of industrial metal objects (DIMO) [4] reflects real-world industrial conditions. Scenes consist of symmetric, textureless and highly reflective metal objects, stacked in dense compositions.

Baseline. We compare our method to PVNet with normalized pose rotations. PVNet [19] is based on estimating 2D keypoints followed by perspective n-point optimization. This method is robust to occlusion and truncation, but suffers from pose ambiguities caused by object symmetries. This is amended by mapping rotations to canonical rotations before training [20].

Experiments. We use a subset of the dataset, focusing on the most reflective parts, and scenes with objects of the same category. For training, only synthetic images are used. We report results on both synthetic and real-world test data. When predicting center and curvature heatmaps, the same output channel is used for all object categories.

Evaluation. We report average recall (AR) for two different error functions:

- MSPD, the maximum symmetry-aware projection distance, as it is relevant for evaluating RGB-only methods.
- MSSD, the maximum symmetry-aware surface distance, as it is relevant for robotic manipulation.

Strict thresholds of correctness θ_e are chosen, as manufacturing use-cases require high-precision poses. θ_{MSSD} is reported for 5% of the object diameter, θ_{MSPD} for 5px.

Our system predicts poses in world frame, whereas PVNet—a single-view method—predicts per-image poses that are relative to the camera. For better comparison, we transform our estimated poses to camera frames and compute per-image metrics.

Table 1. Results on DIMO for synthetic test images. CenDerNet significantly outperforms the PVNet baseline. The difference between MSSD and MSPD shows the importance of multi-view images for robotics applications where spatial precision is important.

	$\mathrm{AR}_{\mathrm{MSPD}} < 5\mathrm{px}$	$\mathrm{AR}_{\mathrm{MSSD}} < 5\%$
PVNet	0.577	0.079
CenDerNet (ours)	**0.721**	**0.639**

Results. As shown in Table 1 and Table 2, CenDerNet significantly outperforms the PVNet baseline. Despite rotation normalizations, PVNet still suffers from pose ambiguities, struggling to predict high-precision poses.

Table 2. Results on DIMO for real-world test images. Both methods are affected by the sim-to-real gap. While the results of PVNet plummet, CenDerNet is more robust.

	AR$_{\text{MSPD}}$ < 5px	AR$_{\text{MSSD}}$ < 5%
PVNet	0.016	0.000
CenDerNet (ours)	**0.516**	**0.403**

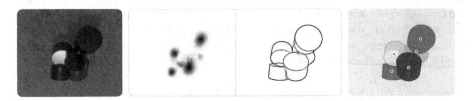

Fig. 9. Results on synthetic test image of the DIMO dataset. Despite severe reflections and shadows, our system recovers qualitative center and curvature representations and precise 6D poses.

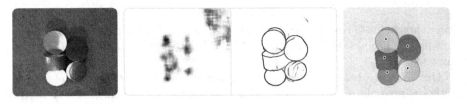

Fig. 10. Results on real-world test image of the DIMO dataset. The lower quality of the center and curvature heatmaps is due to the sim-to-real gap. Nevertheless, the estimated 6D poses are accurate.

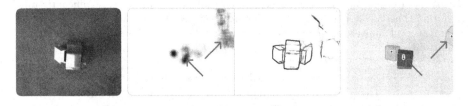

Fig. 11. Failure case on a real-world test image of the DIMO dataset. The unexpected output in the center heatmap (orange arrow) could be due to the sim-to-real gap. Objects cannot be differentiated when object centers align (red arrow). (Color figure online)

The large gap in MSPD en MSSD scores for PVNet shows the importance of multi-view images for robotics applications. While PVNet's MSPD score (in image space) is reasonable, its MSSD score (in world/robot space) is subpar.

Reflections and shadows significantly affect many test images, leading to challenging conditions, as shown in Fig. 9 and Fig. 10. Nevertheless, center and curvature predictions are often sufficient to find accurate object poses. The difference in results between Table 1 and Table 2 shows the effect of the sim-to-real gap. Here, many of the keypoints predicted by PVNet are too far off to meet the strict accuracy requirements, causing results to plummet. CenDerNet is more robust.

An important advantage of our method is the interpretability of the predicted center and curvature representations. The top-right area of the center heatmap in Fig. 10 and Fig. 11 could be due to the sim-to-real gap. Adding more background variations while training could improve the robustness of the model.

Figure 11 shows a failure case. When object centers align, the system fails to differentiate between them.

3.2   T-LESS

Dataset. The T-LESS dataset provides multi-view images of non-reflective, colorless industrial objects. As part of the BOP benchmark, it allows easy comparison to state-of-the-art methods. For the 2020 BOP challenge [11], photorealistic synthetic training images (PBR) were provided.

Baseline. We compare our method to CosyPose [14], the multi-view method that won the 2020 BOP challenge. We use the ECCV 2020 BOP evaluation results for 8 views made available by the authors.

Experiments. We use only PBR images for training. As each scene contains a mix of object types, we extend our model to predict a separate center heatmap for each object category. This reduces the generality of our method, but allows us to select the correct CAD model when optimizing poses. As training images contain distractor objects, we take object visibility into account when generating ground-truth center and curvature heatmaps.

Evaluation. In addition to reporting $AR_{MSPD} < 5px$ and $AR_{MSSD} < 5\%$, we report the default BOP [10] evaluation metrics (VSD, MSSD and MSPD). We transform poses to camera-frames and calculate metrics for all images.

Results. As shown in Table 3, our method outperforms CosyPose in high-precision scenarios. Table 4 shows we also outperform the provided CosyPose predictions on the default BOP metrics. However, the CosyPose-variant that is currently at the top of the BOP leaderboard still performs significantly better.

Table 3. Results on TLESS. CenDerNet outperforms CosyPose on both metrics. The stark difference in MSSD score is relevant for high-precision robotics applications.

	$AR_{MSPD} < 5px$	$AR_{MSSD} < 5\%$
CosyPose (ECCV 2020)	0.499	0.250
CenDerNet (ours)	**0.543**	**0.544**

Table 4. Results on TLESS for the default BOP evaluation metrics. CenDerNet outperforms the provided CosyPose results. However, the scores reported on the BOP leaderboard are still significantly higher.

	AR	AR$_{MSPD}$	AR$_{MSSD}$	AR$_{VSD}$
CosyPose (ECCV 2020)	0.617	0.686	0.610	0.557
CenDerNet (ours)	0.713	0.715	0.717	0.707
CosyPose (BOP leaderboard)	**0.839**	**0.907**	**0.836**	**0.773**

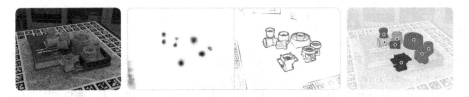

Fig. 12. Results on T-LESS. Our model is able to eliminate irrelevant scene elements, like backgrounds and distractor objects.

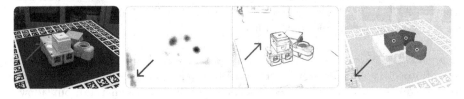

Fig. 13. Failure case on T-LESS. The model is confused about the charuco tags around the scene, leading to an incorrect 3D center. The pose optimization process cannot recover from this error.

Figure 12 shows predicted heatmaps, object centers and poses. In many situations, our method is capable of eliminating irrelevant backgrounds and objects, despite being trained only on synthetic data. There is, however, still a sim-to-real gap, as shown in Fig. 13. Again, the predicted representations provide hints on how to improve the system. In this case, the model is confused by charuco tags on the edge of the table.

4   Conclusions

We present CenDerNet, a system for multi-view 6D pose estimation based on center and curvature representations. Our system is conceptually simple and therefore easy-to-use. First, a single neural network converts images into interpretable representations. Next, a render-and-compare approach, with GPU-optimized cost function, allows for jointly optimizing object poses across multiple viewpoints, thereby naturally handling occlusions and object symmetries. We demonstrate our system on two challenging, industry-relevant datasets and show it outperforms PVNet, a strong single-view baseline, and CosyPose, the leading multi-view approach. In future work, we plan to explore ways to further decrease the processing time of our method. In addition, we will investigate improvements to the robustness of our pipeline. For example, fusing information from multiple viewpoints could improve center and curvature predictions. We also look forward to evaluating our method in a real-world setup.

Acknowledgements. The authors wish to thank everybody who contributed to this work. The authors thank RoboJob (https://www.robojob.eu) for their support and the members of the *Keypoints Gang* for many insightfull discussions. This research was supported by VLAIO Baekeland Mandate HBC.2019.2162 and by Flanders Make (https://www.flandersmake.be) through the PILS SBO project (Project Inspection with Little Supervision) and the CADAIVISION SBO project. Furthermore this research received funding from the Flemish Government under the "Onderzoeksprogramma Artificiële Intelligentie (AI) Vlaanderen" programme.

References

1. Bay, H., Ess, A., Tuytelaars, T., Gool, L.V.: Speeded-up robust features (SURF). Comput. Vis. Image Underst. **110**(3), 346–359 (2008). https://doi.org/10.1016/j.cviu.2007.09.014
2. Commission, E.: Directorate-General for Internal Market, Industry, E., SMEs: a vision for the European industry until 2030 : final report of the Industry 2030 high level industrial roundtable. Publications Office (2019). https://doi.org/10.2873/34695
3. Daems, R., Victor, J., Baets, P.D., Onsem, S.V., Verstraete, M.: Validation of three-dimensional total knee replacement kinematics measurement using single-plane fluoroscopy. Int. J. Sustain. Constr. Des. **7**(1), 14 (2016). https://doi.org/10.21825/scad.v7i1.3634
4. De Roovere, P., Moonen, S., Michiels, N., Wyffels, F.: Dataset of industrial metal objects. arXiv preprint arXiv:2208.04052 (2022)
5. Florence, P.R., Manuelli, L., Tedrake, R.: Dense object nets: learning dense visual object descriptors by and for robotic manipulation. arXiv preprint arXiv:1806.08756 (2018)
6. Fu, C.Y., Shvets, M., Berg, A.C.: RetinaMask: learning to predict masks improves state-of-the-art single-shot detection for free. arXiv preprint arXiv:1901.03353 (2019)
7. He, K., Gkioxari, G., Dollár, P., Girshick, R.: Mask R-CNN. In: 2017 IEEE International Conference on Computer Vision (ICCV), pp. 2980–2988 (2017). https://doi.org/10.1109/ICCV.2017.322
8. Hinterstoisser, S., et al.: Multimodal templates for real-time detection of texture-less objects in heavily cluttered scenes. In: 2011 International Conference on Computer Vision. IEEE, November 2011. https://doi.org/10.1109/iccv.2011.6126326
9. Hodan, T., Haluza, P., Obdrzalek, S., Matas, J., Lourakis, M., Zabulis, X.: T-LESS: an RGB-d dataset for 6d pose estimation of texture-less objects. In: 2017 IEEE Winter Conference on Applications of Computer Vision (WACV). IEEE, March 2017. https://doi.org/10.1109/wacv.2017.103

10. Hodaň, T., et al.: BOP challenge 2020 on 6D object localization. In: Bartoli, A., Fusiello, A. (eds.) ECCV 2020. LNCS, vol. 12536, pp. 577–594. Springer, Cham (2020). https://doi.org/10.1007/978-3-030-66096-3_39
11. Hodaň, T., et al.: Photorealistic image synthesis for object instance detection. In: 2019 IEEE International Conference on Image Processing (ICIP), pp. 66–70 (2019). https://doi.org/10.1109/ICIP.2019.8803821
12. Jensen, A., Flood, P., Palm-Vlasak, L., Burton, W., Rullkoetter, P., Banks, S.: Joint track machine learning: an autonomous method for measuring 6dof tka kinematics from single-plane x-ray images. arXiv preprint arXiv:2205.00057 (2022)
13. Kaskman, R., Shugurov, I., Zakharov, S., Ilic, S.: 6 DoF pose estimation of textureless objects from multiple RGB frames. In: Bartoli, A., Fusiello, A. (eds.) ECCV 2020. LNCS, vol. 12536, pp. 612–630. Springer, Cham (2020). https://doi.org/10.1007/978-3-030-66096-3_41
14. Labbé, Y., Carpentier, J., Aubry, M., Sivic, J.: CosyPose: consistent multi-view multi-object 6D pose estimation. In: Vedaldi, A., Bischof, H., Brox, T., Frahm, J.-M. (eds.) ECCV 2020. LNCS, vol. 12362, pp. 574–591. Springer, Cham (2020). https://doi.org/10.1007/978-3-030-58520-4_34
15. Li, C., Bai, J., Hager, G.D.: A unified framework for multi-view multi-class object pose estimation. In: Ferrari, V., Hebert, M., Sminchisescu, C., Weiss, Y. (eds.) ECCV 2018. LNCS, vol. 11220, pp. 263–281. Springer, Cham (2018). https://doi.org/10.1007/978-3-030-01270-0_16
16. Li, Y., Wang, G., Ji, X., Xiang, Yu., Fox, D.: DeepIM: deep iterative matching for 6D pose estimation. Int. J. Comput. Vis. **128**(3), 657–678 (2019). https://doi.org/10.1007/s11263-019-01250-9
17. Lin, T.Y., Goyal, P., Girshick, R., He, K., Dollar, P.: Focal loss for dense object detection. In: 2017 IEEE International Conference on Computer Vision (ICCV). IEEE, October 2017. https://doi.org/10.1109/iccv.2017.324
18. Park, K., Patten, T., Vincze, M.: Pix2pose: Pixel-wise coordinate regression of objects for 6d pose estimation. In: 2019 IEEE/CVF International Conference on Computer Vision (ICCV). IEEE, October 2019. https://doi.org/10.1109/iccv.2019.00776
19. Peng, S., Liu, Y., Huang, Q., Zhou, X., Bao, H.: PVNet: pixel-wise voting network for 6dof pose estimation. In: 2019 IEEE/CVF Conference on Computer Vision and Pattern Recognition (CVPR). IEEE, June 2019. https://doi.org/10.1109/cvpr.2019.00469
20. Pitteri, G., Ramamonjisoa, M., Ilic, S., Lepetit, V.: On object symmetries and 6D pose estimation from images. In: 2019 International Conference on 3D Vision (3DV). IEEE, September 2019. https://doi.org/10.1109/3dv.2019.00073
21. Prewitt, J.M., et al.: Object enhancement and extraction. Pict. Process. Psychopictorics **10**(1), 15–19 (1970)
22. Ren, S., He, K., Girshick, R., Sun, J.: Faster R-CNN: towards real-time object detection with region proposal networks. IEEE Trans. Pattern Anal. Mach. Intell. **39**(6), 1137–1149 (2017). https://doi.org/10.1109/tpami.2016.2577031
23. Ronneberger, O., Fischer, P., Brox, T.: U-Net: convolutional networks for biomedical image segmentation. In: Navab, N., Hornegger, J., Wells, W.M., Frangi, A.F. (eds.) MICCAI 2015. LNCS, vol. 9351, pp. 234–241. Springer, Cham (2015). https://doi.org/10.1007/978-3-319-24574-4_28
24. Szeliski, R.: Computer Vision: Algorithms and Applications. Springer Science & Business Media, Berlin, Heidelberg (2010). https://doi.org/10.1007/978-1-84882-935-0

25. Tekin, B., Sinha, S.N., Fua, P.: Real-time seamless single shot 6d object pose prediction. In: 2018 IEEE/CVF Conference on Computer Vision and Pattern Recognition. IEEE, June 2018. https://doi.org/10.1109/cvpr.2018.00038
26. Tremblay, J., To, T., Sundaralingam, B., Xiang, Y., Fox, D., Birchfield, S.: Deep object pose estimation for semantic robotic grasping of household objects. In: 2nd Annual Conference on Robot Learning, CoRL 2018, Zürich, Switzerland, 29–31 October 2018, Proceedings. Proceedings of Machine Learning Research, vol. 87, pp. 306–316. PMLR (2018)
27. Virtanen, P., et al.: SciPy 1.0 contributors: SciPy 1.0: fundamental algorithms for scientific computing in python. Nat. Methods **17**, 261–272 (2020). https://doi.org/10.1038/s41592-019-0686-2
28. van der Walt, S., et al.: The scikit-image contributors: scikit-image: image processing in python. PeerJ **2**, e453 (2014). https://doi.org/10.7717/peerj.453
29. Wang, H., Sridhar, S., Huang, J., Valentin, J., Song, S., Guibas, L.J.: Normalized object coordinate space for category-level 6D object pose and size estimation. In: 2019 IEEE/CVF Conference on Computer Vision and Pattern Recognition (CVPR). IEEE, June 2019. https://doi.org/10.1109/cvpr.2019.00275
30. Xiang, Y., Schmidt, T., Narayanan, V., Fox, D.: PoseCNN: a convolutional neural network for 6d object pose estimation in cluttered scenes. In: Robotics: Science and Systems XIV. Robotics: Science and Systems Foundation, June 2018. https://doi.org/10.15607/rss.2018.xiv.019
31. Yang, J., Gao, Y., Li, D., Waslander, S.L.: ROBI: a multi-view dataset for reflective objects in robotic bin-picking. In: 2021 IEEE/RSJ International Conference on Intelligent Robots and Systems (IROS). IEEE, September 2021. https://doi.org/10.1109/iros51168.2021.9635871
32. Zakharov, S., Shugurov, I., Ilic, S.: DPOD: 6d pose object detector and refiner. In: 2019 IEEE/CVF International Conference on Computer Vision (ICCV). IEEE, October 2019. https://doi.org/10.1109/iccv.2019.00203
33. Zhou, X., Wang, D., Krähenbühl, P.: Objects as points. CoRR abs/1904.07850 (2019)
34. Zhou, Y., Barnes, C., Lu, J., Yang, J., Li, H.: On the continuity of rotation representations in neural networks. In: 2019 IEEE/CVF Conference on Computer Vision and Pattern Recognition (CVPR). IEEE, June 2019. https://doi.org/10.1109/cvpr.2019.00589

Trans6D: Transformer-Based 6D Object Pose Estimation and Refinement

Zhongqun Zhang[1]([✉]), Wei Chen[2], Linfang Zheng[1,3], Aleš Leonardis[1], and Hyung Jin Chang[1]

[1] University of Birmingham, Birmingham, UK
zxz064@student.bham.ac.uk
[2] Defense Innovation Institute, Boston, China
[3] SUSTech, Shenzhen, China

Abstract. Estimating 6D object pose from a monocular RGB image remains challenging due to factors such as texture-less and occlusion. Although convolution neural network (CNN)-based methods have made remarkable progress, they are not efficient in capturing global dependencies and often suffer from information loss due to downsampling operations. To extract robust feature representation, we propose a Transformer-based 6D object pose estimation approach (Trans6D). Specifically, we first build two transformer-based strong baselines and compare their performance: pure Transformers following the ViT (Trans6D-pure) and hybrid Transformers integrating CNNs with Transformers (Trans6D-hybrid). Furthermore, two novel modules have been proposed to make the Trans6D-pure more accurate and robust: (i) a patch-aware feature fusion module. It decreases the number of tokens without information loss via shifted windows, cross-attention, and token pooling operations, which is used to predict dense 2D-3D correspondence maps; (ii) a pure Transformer-based pose refinement module (Trans6D+) which refines the estimated poses iteratively. Extensive experiments show that the proposed approach achieves state-of-the-art performances on two datasets.

Keywords: 6D object pose estimation · Transformer

1 Introduction

In this paper, we are interested in estimating the 6D pose of objects from monocular RGB images. 6D object pose estimation has been gaining attention as it can be applied in many fields such as augmented reality, robotic manipulation, autonomous driving, etc. However, estimating the 6D pose from a monocular RGB image is still challenging, especially when the target object is under heavy occlusion or in changing illumination conditions.

With the explosive growth of deep learning, deep Convolutional Neural Networks (CNNs) have greatly improved monocular 6D object pose estimation [22], even at times surpassing RGB-D-based methods [1,22,31]. Recent works in this field can be roughly divided into two categories: i) approaches that use a CNN

L. Karlinsky et al. (Eds.): ECCV 2022 Workshops, LNCS 13808, pp. 112–128, 2023.
https://doi.org/10.1007/978-3-031-25085-9_7

to regress the 6D poses directly [5,8,33,36–38,40] and ii) indirectly [29,30]. A common drawback of the methods in the first category is their poor generalization ability due to the large search space of the rotation [15,40]. The second category overcomes this limitation by either utilizing CNNs to detect the 2D keypoints of the objects [30,34] or estimating the dense 2D-3D correspondence maps between the input image and the available 3D models [17,42]. Given the correspondences, one can recover the 6D pose parameters via Perspective-n-Point (PnP) algorithm [2]. Great success has been achieved with these approaches. However, these CNN-based methods are not efficient in capturing non-local spatial relationships. Therefore, their performance is limited in both accuracy and robustness. The codes will be publicly available on our website.

Even though Transformer [11] is originally designed for *Natural Language Processing* (NLP) tasks, recent works [4,9,14,21,26,41] show that both "Pure Transformer" and "CNN+Transformer" have the potential to become the universal models for computer vision tasks. The self-attention mechanism [41] makes them particularly effective in capturing the global dependencies. Therefore, we are interested in designing an algorithm that can leverage the advantage of Transformers in the 6D object pose estimation task.

In this work, we propose Trans6D, a simple yet effective framework that employs Transformers for 6D pose estimation. Firstly, we build two strong baseline frameworks (Trans6D-pure and Trans6D-hybrid) using Transformer. Among them, Trans6D-pure applies the ViT [12] directly, while Trans6D-hybrid uses ResNet-34 to learn local feature maps and then uses Transformer encoders to capture global dependencies. Since dividing the image into small patches will impact the accuracy of 6D object pose regression tasks, the Trans6D-Hybrid significantly outperforms the Trans6D-pure.

Secondly, we propose two novel modules to improve the performance of the Trans6D-pure: (i) a patch-aware feature fusion (PAFF) module is proposed to predict the 2D-3D correspondence map. We reshape the tokens from the last layer of Trans6D-pure into feature maps. Inspired by the pooling operation and stridden convolution in CNN, the PAFF module proposes to use token pooling and shifted windows to reduce the dimension of feature maps. To avoid information loss and alleviate the impact of image division, the PAFF module uses cross-attention to fuse the local feature of tokens in each window. (ii) The pure Transformer-based pose refinement module (Trans6D+) is introduced to use the input image and the initial pose estimation to learn an accurate transformation between the predicted object pose and the ground-truth pose.

Experimental results on two widely used benchmark datasets, LINEMOD [16] and Occlusion LINEMOD [3], demonstrate that Trans6D has state-of-the-art performances. The contributions of the paper are summarised:

- Two Transformer-based strong baselines are proposed and assessed for 6D object pose estimation, which achieve comparable performance with CNN-based frameworks.
- A patch-aware feature fusion (PAFF) module is designed to decrease the number of tokens without information loss, which is used to predict dense 2D-3D correspondence maps.

- A pure Transformer-based Refinement (Trans6D+) module is introduced to refine the estimated poses iteratively.
- For the first time, we show a simple but effective method based on Transformers achieves state-of-the-art performance, 96.9% on the LINEMOD dataset and 57.9% on the Occlusion LINEMOD dataset.

2   Related Work

2.1   6D Object Pose Estimation

Recent work in this field can be roughly divided into two categories: direct methods and correspondence-based approaches.

Direct methods use CNN-based networks to regress a 6D pose from a single RGB image directly. For Instance, PoseCNN [40] predicted 2D localization, depth information, and rotation from a CNN backbone. However, the direct methods usually exhibit poor generalization ability due to the large search space of the rotation and the lack of depth information. Therefore, Oberweger et al. [28] and BB8 [31] attempted to limit the rotation range; they discretize the pose space and seem the prediction of rotation as classification rather than regression. G2L-Net [6] harnesses the embedding vector features to regress the 6D pose. GDR [39] regress dense correspondences first and then use patch-PnP to learning 6D pose from the correspondences.

The correspondence-based methods are prevalent recently. These methods first built 2D-3D correspondence maps, then computed the 6D pose via PnP with the RANSAC algorithm. For example, Pixel2Pose [29] used an auto-encoder architecture to estimate the 3D coordinates per pixel to build dense correspondences, while PVNet [30], PVN3D [15], and PointPoseNet [7] adopted a voting net to select 2D keypoint or 3D keypoint respectively to build dense correspondences. Furthermore, HybridPose [34] suggested predicting hybrid correspondences (including keypoints, edge vectors, and symmetry) to enhance the robustness. DPOD [42] first estimateed multi-instance correspondences and then used a learning-based refiner to improve the accuracy. It is the first approach that unifies the correspondence-based rotation estimation and the direct regression-based translation estimation.

2.2   Vision Transformer

Transformer architecture was built for the NLP task, consisting of the multi-head self-attention mechanisms and feed-forward layer, to capture the long-term correlation between words. Recently, there is a growing interest in Transformer based computer-vision tasks. On the one hand, pure Transformer [12,20,23] is attracting more and more attention. ViT [12] applied pure Transformer directly to image classification tasks and attains excellent results compared to the state-of-the-art CNN-based method. TransReID [14] showed that a pure Transformer-based model can be used for the object ReID task. On the other hand, "CNN +

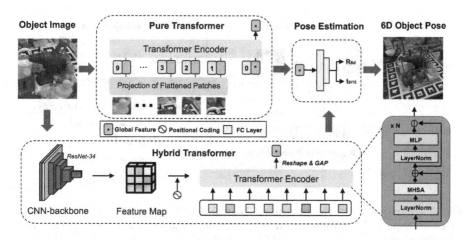

Fig. 1. An overview of the proposed Trans6D-pure and Trans6D-hybrid baselines. Given an input image of the target object, Trans6D-pure encodes the image as a sequence of patches and then models the global dependencies among each patch via ViT-like Transformer Layers. Trans6D-hybrid first extracts feature maps using a CNN-based backbone and flatten them to a sequence, then feeds the sequence into Transformer Layers. Output token marked with * is served as the global feature. The global feature is then used to regress 3D rotation and 3D translation.

Transformer" [4,43] also has better performance. DETR [4] extracted features from CNN-backbone, then viewed object detection as a direct set prediction problem that was suitable for Transformer structure. TransPose [41] predicted 2D heatmaps for human pose estimation by Transformer encoder and CNN, while METRO [26] tried to model non-local interactions among body joints and mesh vertices for human mesh reconstruction.

3 Methodology

In Fig. 1 and Fig. 2, we show an overview of our proposed framework that estimates the 6D object pose from a single RGB image. The input to Trans6D is an image of size 256×256 which contains only one object with a known class. The outputs of Trans6D-pure and Trans6D-hybrid are the predicted 3D translation and 3D rotation, while the outputs of Trans6D and Trans6D+ are 2D-3D correspondence maps of size 64×64. Given the correspondences, 6D pose is calculated by PnP and RANSAC algorithm. Our method consists of three modules: Two Transformer-based strong baselines, the Patch-Aware Feature Fusion (PAFF) Module, and the Pure Transformer-based Pose Refinement Module. In the following subsections, we describe each module in detail.

116   Z. Zhang et al.

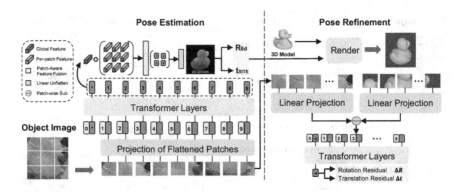

Fig. 2. An overview of the proposed Trans6D and Trans6D+. Trans6D is based on Trans6D-pure. The outputs of Trans6D-pure are reshaped as a feature map and feed it into patch-aware feature fusion (PAFF) module to downsample the tokens. Instead of directly regressing the 6D pose, Trans6D predicts the 2D-3D correspondence maps and compute the 6D pose by PnP algorithm. Trans6D+ renders the object at the estimated pose and learns to align the real image and the rendered image incrementally.

Following GDR-Net [39], the 6D pose is represented as a decoupled way, composed of a continuous 6-dimensional representation for rotation $\mathbf{R}_{6d}$ in $SO(3)$ and a scale-invariant representation for translation $\mathbf{t}_{\mathrm{SITE}}$.

3.1 Transformer-Based Baselines for 6D Object Pose Estimation

We build two Transformer-based strong baselines for 6D object pose estimation, which are based on pure Transformers following the ViT (Trans6D-pure) and hybrid Transformers integrating CNNs with Transformers (Trans6D-hybrid) separately.

Trans6D-Hybrid. Trans6D-hybrid consists of two components: a CNN backbone to extract low-level image feature maps; a Transformer Encoder to capture global dependencies between feature vectors, and each feature vector is distinguished by the position embedding, as shown in Fig. 1.

We employ a Convolutional Neural Network (CNN), ResNet34 [13], as the backbone for feature extraction. This backbone is pre-trained on ImageNet classification task [10], therefore Transformer can easily benefit from large-scale pre-trained CNNs. Given an initial image $x_{\mathrm{img}} \in \mathbb{R}^{3 \times H_0 \times W_0}$, the CNN-based backbone generates a feature map with lower-resolution $f \in \mathbb{R}^{C \times H \times W}$. Specifically, $C = 512$ and $H, W = \frac{H_0}{32}, \frac{W_0}{32}$.

The Transformer encoder is employed to model interaction among all the pixel-level features in the image. First, we transform the feature dimension to $f \in \mathbb{R}^{d \times H \times W}$ by a 1×1 convolution. Since the Transformer encoder expects a sequence as input, we then flattened the feature into a sequence $f \in \mathbb{R}^{L \times d}$, where $L = H \times W$, and this sequence is fed into the Transformer Encoder. As shown in Fig. 1, we follow the standard Transformer architecture as closely as possible,

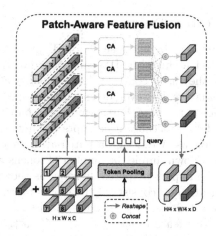

Fig. 3. Illustration of Patch-Aware Feature Fusion. We design a PAFF module which can not only downsample the number of tokens without information loss but also alleviate the impact by patch division.

which consists of a multi-head self-attention module and a fully connected feed-forward network (FFN).

Since the Transformer architecture is permutation-invariant, position embedding aims at giving an order to the sequence of the image feature map. Following [4], 2D Sine position embedding is used in Trans6D-hybrid. The output sequence of Transformer is reshaped to feature maps. We utilize a global average pooling operation to extract the global feature, which is fed into two FC layers to regress the 6D pose. The whole process can be formulated as:

$$\mathbf{R_{6d}}, \mathbf{t_{SITE}} = \text{FCLayers}(\text{GAP}(\text{Transformer}(\boldsymbol{f}))) \tag{1}$$

Trans6D-Pure. Given an image $x_{\text{img}} \in \mathbb{R}^{C \times H_0 \times W_0}$, Trans6D-pure first divides the images into non-overlapping $P \times P$ patches $\{x_n^i \mid i = 1, 2, \cdots, P\}$. These patches is then flattened into a sequence of 2D patches $\mathbf{x}_p \in \mathbb{R}^{N \times (P^2 \cdot C)}$ by a linear projection $\mathcal{F}$, P^2 is dimension of each feature vector and $N = HW/P^2$. An extra learnable embedding token ($*$) is added into the input sequences and this token is used to extract a global feature. Different from Trans6D-hybrid, position embeddings $z_{\text{pos}} \in \mathbb{R}^{(1+P^2) \times NC}$ in Tran6D-pure are learnable. Overall, The input feature matrix can be formulated as:

$$\mathcal{Z} = \left[x_*; \mathcal{F}\left(x_n^1\right); \mathcal{F}\left(x_n^2\right); \cdots; \mathcal{F}\left(x_n^P\right)\right] + z_{pos} \tag{2}$$

where [ ] represented concatenation operations. Then, we feed the feature into Transformer encoder layers. The whole process can be express as:

$$\mathbf{R_{6d}}, \mathbf{t_{SITE}} = \text{FCLayers}(\text{Transformer}(\mathcal{Z})[0, :]) \tag{3}$$

3.2 Patch-Aware Feature Fusion

Instead of directly regressing the 6D pose, Trans6D predicts 2D-3D correspondence maps. Standard ViT architecture [12] cannot be directly applied in a dense prediction task because they use a constant dimensionality of the hidden embeddings for all transformer layers. However, downsampling operations in CNNs (e.g. pooling) suffer from information loss. Furthermore, the patch division might also greatly impact the prediction because it corrupts the image structure. To solve the aforementioned problems, We design a patch-aware feature fusion (PAFF) module that can not only downsample the number of tokens without information loss but also alleviate the impact by patch division, as shown in Fig. 3.

The output of Trans6D-pure consists of the global token Z^* and a sequence of patch tokens $\{Z^i \mid i = 1, 2, \cdots, N\}$. As shown in Fig. 2, we first reshape the patch tokens $Z \in \mathbb{R}^{l \times c}$ as a feature map $f \in \mathbb{R}^{h \times w \times c}$ in the spatial dimension. After obtaining the re-organized feature map, we apply token pooling operation (1D convolution with 1D maxpooling) to downsample the feature map to $Z' \in \mathbb{R}^{\frac{h}{4} \times \frac{w}{4} \times c}$. Z' will serve as the learnable queries $\{q^i \mid i = 1, 2, \cdots, \frac{N}{4}\}$ in PAFF module. However, simply adding a token pooling will decrease the model's representation ability. Inspired by the strided convolution in CNN, we split the feature map into overlapping patches with a sliding window. As shown in Fig. 3, the patch tokens in each window are fed into PAFF module with the global token. Supposing each window contains $k \times k + 1$ patches and the sliding stride is s, the number of windows is

$$\frac{h}{4} \times \frac{w}{4} = \left\lfloor \frac{h-k}{s} + 1 \right\rfloor \times \left\lfloor \frac{w-k}{s} + 1 \right\rfloor. \tag{4}$$

The tokens of each sliding window server as query and key, and we compute their cross-attention (CA) [4] with the learnable queries. Thus the local information can be aggregated. The outputs of CA network is concatenated with each other, then they will be reshaped again to feature maps $f' \in \mathbb{R}^{\frac{h}{4} \times \frac{w}{4} \times D}$ and regress to 2D-3D correspondences map $M_{2D3D} \in \mathbb{R}^{\frac{h}{4} \times \frac{w}{4} \times 4}$ by Linear Unflatten.

3.3 Pure Transformer-Based Pose Refinement

In order to further improve the performance of Trans6D, we propose a pure Transformer-based pose refinement (Trans6D+) module. Inspired by DPOD [42], our refiner aims at regress the residual of rotation and translation with the loss function:

$$\mathcal{R}_{pose} = \frac{1}{M} \sum_j \min_{x \in \mathcal{M}_s} \left\| (Rx_j + t) - \left(\hat{R}_i x_k + \hat{t}_i \right) \right\| \tag{5}$$

where $\mathcal{M}_s$ represents the randomly selected 3D points from the object's 3D model. R and t is the ground truth of 6D pose, while $\hat{R}$ and $\hat{t}$ is the predicted rotation and translation.

In Fig. 2, we show the pipeline of our Trans6D+. Given the 3D model of an object and the predicted 6D pose parameters, we first render the object at the initial pose and crop the image around the object. The rendered image embeds the initial pose information and our idea is to make the Trans6D+ learning to align these two images incrementally. Trans6D+ contains two parallel Transformer branches, one receives the real image as input while the other extracts the feature from rendered image. Similar to Tran6D-pure, rendered image is also divided into the same number of patches with real image branch ($\mathbf{f}_c$), and then flattened them into a sequence of 2D patches $\mathbf{f}_r \in \mathbb{R}^{N \times (P^2 \cdot C)}$ by a linear projection. Then patches from two branches are subtracted and fed into the Transformer encoder layers to extract the global dependencies. An extra learnable embedding token $\mathbf{x}_*$ is added into the input sequences and this token is used to extract a global feature. Finally, the residual of rotation ΔR and translation Δt is regressed by the global feature. The whole process can be formulated as:

$$\Delta\mathbf{R}, \Delta\mathbf{t} = \text{FCLayers}(\text{Transformer}(\boldsymbol{f}_c - \boldsymbol{f}_r)[0, :]) \qquad (6)$$

3.4   Training

To train the transformer encoder, we apply loss functions on top of the transformer outputs, the network predicts a confidence value for each pixel to indicate whether it belongs to the object. The corresponding loss function is defined as

$$\begin{aligned}
\mathcal{L}_{Corr} = & \alpha \cdot \ell_1 \left(\sum_{j=1}^{n_c=3} \left(\bar{M}_{\text{vis}} \circ \left(\hat{M}_{\text{XYZ}_j} - \bar{M}_{\text{XYZ}_j} \right) \right) \right), \\
& + \beta \cdot \ell_1 \left(\hat{M}_{\text{vis}} - \bar{M}_{\text{vis}} \right)
\end{aligned} \qquad (7)$$

where $\bar{M}_{XYZ}$ represents the 3D coordinate of the available 3D model, $\bar{M}_{vis}$ represents the visible masks.

Once we get the confidence map and the correspondence map, the coordinates belong to the object can be obtained by setting a threshold for the confidence. However, the size of the object in an RGB image is different from the original image since we use a detector. To build the 2D-3D correspondences, we map the pixel from the coordinates map back to the RGB image.

4   Experiments

In this section, we first conduct ablation studies on the effectiveness of each module in Trans6D, and then evaluate Trans6D on prevalent benchmark datasets. The results show that our method achieves state-of-the-art performance.

4.1   Implementation Details

We implemented our framework using Pytorch and conducted all the experiments on an Intel i7-4930K CPU with one GTX 2080 Ti GPU. During training, we use Adam [27] for optimization. Also, we set the initial learning rate as 1e-4 and halve it every 50 epochs. The maximum epoch is 300. For object detection part, we fine-tune the YOLO-V3 [32] architecture which is pre-trained on the ImageNet [10] to locate the 2D object and fixed its size to 256 × 256.

4.2 Datasets

LINEMOD: is a widely used dataset for 6D object pose estimation. It consists of 13 objects, each containing about 1.2 k images with ground-truth poses for a single object. This dataset exhibits many challenges for pose estimation: textureless objects, cluttered scenes, and lighting condition variations. Following [25], we select 15% of the RGB images for training and 85% for testing. We also render 1000 images for each object as a supplement to the training set.

Occlusion LINEMOD: is a widely used dataset for 6D object pose estimation under severe occlusion. It consists of 8 objects, each containing about 1214 images with more occlusion are provided for testing. All Occlusion LINEMOD datasets are used for testing.

4.3 Evaluation Metrics

We evaluate Trans6D using average 3D distance of model points (ADD) metric [16].

ADD Metric. This metric computes the mean distance between two transformed object model using the estimated pose and the ground-truth pose. When the distance is less than 10% of the model's diameter, it is claimed that the estimated pose is correct. It can be computed by:

$$\frac{1}{|\mathcal{M}|} \sum_{x \in \mathcal{M}} \|(\mathbf{R} \cdot \mathbf{x} + \mathbf{T}) - (\widetilde{\mathbf{R}} \cdot \mathbf{x} + \widetilde{\mathbf{T}})\|, \tag{8}$$

where $|\mathcal{M}|$ is the number of points in the object model. x represents the point in object 3D model, $\mathbf{R}$ and $\mathbf{T}$ are the ground truth pose, and $\widetilde{\mathbf{R}}$ and $\widetilde{\mathbf{T}}$ are the estimated pose. For symmetric objects, we use the ADD-S metric [2], where the mean distance is computed based on the closest point distance. :

$$\frac{1}{|\mathcal{M}|} \sum_{x_1 \in \mathcal{M}} \min_{x_2 \in \mathcal{M}} \left\|(\mathbf{R} \cdot \mathbf{x}_1 + \mathbf{T}) - \left(\widetilde{\mathbf{R}} \cdot \mathbf{x}_2 + \widetilde{\mathbf{T}}\right)\right\|, \tag{9}$$

4.4 Ablation Studies

Compared to other methods [25], our proposed Trans6D has three novelties.

First, we build two Transformer-based baselines for 6D pose estimation: pure Transformers-based Trans6D-pure and Trans6D-hybrid combining CNNs with Transformers. As shown in Table 1, we compare Trans6D-hybrid (ResNet34

Table 1. Ablation studies of the effectiveness of two Transformer-based baselines on LINEMOD dataset. The metric we used to measure performance is ADD(-S) metric. "CNN" means CNNs-based method, "Trand6D-p" and "Trans6D-h" denote Trans6D-pure and Trans6D-hybrid, respectively.

Metric	ADD(-s)		
Method	CNN	Trans6D-p	Trans6D-h
Ape	11.43%	42.33%	**49.11%**
Benchvise	95.05%	94.18%	**96.46%**
Camera	75.49%	88.04%	**91.84%**
Can	89.57%	92.87%	**95.02%**
Cat	51.50%	77.73%	**81.51%**
Driller	**97.36%**	94.01%	96.79%
Duck	23.19%	54.34%	**55.99%**
Eggbox	**99.53%**	96.63%	98.75%
Glue	**94.21%**	89.14%	93.02%
Holepuncher	68.22%	87.26%	**89.49%**
Iron	93.77%	94.80%	**96.75%**
Lamp	97.02%	94.86%	**98.21%**
Phone	82.72%	90.15%	**92.96%**
Average	75.04%	84.34%	**87.73%**

Table 2. Ablation studies of the effectiveness of patch-aware feature fusion (PAFF) Module on Occlusion LINEMOD dataset. The metrics we used to measure performance are ADD(-S). "SW" means sliding windows operation, "TP" means token pooling operation and "CA" means cross-attention network.

Method	SW	TP	CA	ADD(-S)
EXP1	×	✓	×	40.9%
EXP2	✓	✓	×	43.1%
EXP3	×	✓	✓	45.4%
EXP4	✓	×	✓	**50.6%**

Table 3. Ablation studies of the effectiveness of pure Transformer-based refinement (Trans6D+) Module on LINEMOD dataset. The metrics we used to measure performance are ADD(-S). Compared approaches: PVNet [30], DPOD [42], DeepIm [24]

PVNet	+DPOD	+DeepIm	+Trans6D+
85.56%	92.83%	87.13%	**95.95%**
DPOD	+DPOD	+DPIM	+Trans6D+
82.98%	95.15%	88.6%	**95.86%**

Table 4. Quantitative evaluations on LINEMOD dataset. We use ADD metric to evaluate the methods. For symmetric objects *Egg Box* and *Glue*, we use the ADD-S metric. Note that, we summarize the pose estimation results reported in the original papers on LINEMOD dataset. Baseline approaches: BB8 [31], Pix2Pose [29], DPOD [42], PVNet [30], CDPN [25], Hybrid [34], GDRN [39].

Method	BB8	Pix2Pose	DPOD	PVNet	CDPN	Hybrid	GDRN	Trans6D	Trans6D+
Refinement	×	×	×(✓)	×	×	✓	×	×	✓
Ape	40.4%	58.1%	53.3% (87.7%)	43.6%	64.4%	63.1%	–	68.1%	**88.3%**
Bench Vise	91.0%	97.5%	95.3% (95.3%)	99.9%	97.8%	**99.9%**	–	99.5%	99.4%
Camera	55.7%	60.9%	90.4% (96.0%)	86.9%	91.7%	90.4%	–	93.7%	**97.8%**
Can	64.1%	84.4%	94.1% (**99.7%**)	95.5%	95.9%	98.5%	–	99.4%	99.1%
Cat	62.6%	65.0%	60.4% (**94.7%**)	79.3%	83.8%	89.4%	–	87.9%	93.2%
Driller	74.4%	76.3%	97.7% (98.8%)	96.4%	96.2%	98.5%	–	97.1%	**99.5%**
Duck	44.3%	43.8%	66.0% (86.3%)	52.6%	**66.8%**	65.0%	–	67.9%	**87.8%**
Egg Box	57.8%	96.8%	99.7% (99.9%)	99.2%	99.7%	**100.0%**	–	100%	**100%**
Glue	41.2%	79.4%	93.8% (96.8%)	95.7%	99.6%	98.8%	–	98.3%	**99.8%**
Hole Puncher	74.8%	52.8%	65.8% (86.9%)	81.9%	85.8%	89.7%	–	93.5%	**96.7%**
Iron	84.7%	83.4%	99.8% (**100%**)	98.9%	97.9%	**100.0%**	–	99.9%	99.9%
Lamp	76.5%	82.0%	88.1% (96.8%)	99.3%	97.9%	99.5%	–	99.5%	**99.7%**
Phone	54.0%	45.0%	74.2% (94.7%)	92.4%	90.8%	94.9%	–	98.7%	**99.5%**
Average	62.7%	72.4%	83.0% (95.2%)	86.3%	89.9 %	91.3%	93.7%	92.6%	**96.9%**

Table 5. Quantitative evaluations on Occlusion LINEMOD dataset. We use the ADD metric to evaluate the methods. For symmetric objects *Egg Box* and *Glue*, we use the ADD-S metric. Note that, we summarize the pose estimation results reported in the original papers on LINEMOD dataset. Baseline approaches: PoseCNN [40], Pix2Pose [29], DPOD [42], PVNet [30], Single-Stage [18], HybridPose [34], GDRN [39].

Method	PoseCNN	Pix2Pose	PVNet	Single-Stage	DPOD	Hybrid	GDRN	Trans6D	Trans6D+
Refinement	×	×	×	×	✓	✓	×	×	✓
Ape	9.6%	22.0%	15.8%	19.2%	–	20.9%	**46.8%**	31.2%	36.9%
Can	45.2%	44.7%	63.3%	65.1%	–	75.3%	90.8%	85.1%	**91.6%**
Cat	0.9%	22.7%	16.7%	18.9%	–	24.9%	40.5%	38.3%	**42.5%**
Driller	41.4%	44.7%	65.7%	69.0%	–	70.2%	**82.6%**	66.5%	70.8%
Duck	19.6%	15.0%	25.2%	25.3%	–	27.9%	**46.9%**	35.0%	41.1%
Egg Box	22.0%	25.2%	50.2%	52.0%	–	52.4%	54.2%	52.9%	**56.3%**
Glue	38.5%	32.4%	49.6%	51.4%	–	53.8%	**75.8%**	54.3%	62.0%
Hole Puncher	22.1%	49.5%	39.7%	45.6%	–	54.2%	60.1%	57.7%	**61.9%**
Average	24.9%	32.0%	40.8%	43.3%	47.3 %	47.5%	**62.2%**	52.6%	57.9%

+ Transformer) with CNN-based method (ResNet34 + CNN Head) using the same backbone, and also assess the performance of Trans6D-pure and Trans6D-hybrid. We observe that (i) using Transformer instead of CNN to estimate 6D parameters increased the accuracy from 75.04% to 87.73% when evaluated with ADD(-S) metric on LINEMOD dataset. (ii) The Trans6D-Hybrid (87.73%) significantly outperforms the Trans6D-pure (84.34%). The reason is that the pure

Table 6. Quantitative evaluations on YCB-Video Datasets. We use 2D-Proj, ADD AUC and ADD(-S) metrics. The threshold of the ADD(-S) metric is 2 cm. Note that, we summarize the pose estimation results reported in the original papers on LINEMOD dataset. State-of-the-art approaches: PoseCNN [40], DeepIM [24], Ameni et al. [35], PVNet [30], Singel-Stage [19], GDR-Net [39]

Methods	PoseCNN	DeepIM	PVNet	Single-Stage	GDR-Net	Ameni et al.	Trans6D	Trans6D+
2D-Proj.	–	–	47.4%	–	–	55.6%	53.2%	**62.4%**
ADD AUC	61.3%	81.9%	73.4%	–	84.4%	83.1%	82.5%	**85.9%**
ADD(-S)	21.3%	–	–	53.9%	60.1%	73.6%	67.7%	**75.2%**

Transformer-based method divides the image into small patches, which seriously impact the accuracy of regression tasks.

Second, we propose the patch-aware feature fusion (PAFF) module, which is used to predict dense 2D-3D correspondence maps. PAFF module decreases the number of tokens without information loss via shifted windows, cross-attention, and token pooling operations. We compare the impact of the three operations and show the results in Table 2. As it can be seen that using the token pooling operation only exhibits the worst performance since it will decrease the model's representation ability. When combining the cross-attention with either sliding window or token pooling operations, the method has better performance than combining the sliding window and token pooling. It is because that the cross-attention can aggregate the local information. Therefore, combining those three operations can avoid information loss and alleviate the impact of image division. Moreover, cross-attention play a decisive role.

Third, we propose a pure Transformer-based pose refinement module. In Table 3, we compare our Trans6D+ with DPOD and DeepIM using the same initial pose and number of iterations. Trans6D+ achieves almost 13% improvement on the PVNet while DPOD only has 7% improvement.

4.5 Comparison with State of the Arts

Object 6D Pose Estimation on LINEMOD: In Table 4, we compare our approach with state-of-the-art methods on LINEMOD Dataset. We use 15% of each object sequence to train and the rest of the sequence to test on LINEMOD dataset following other methods. The numbers in brackets are the results without post-refinement.

We use Trans6D-pure as our baseline. From Table 1 and Table 4, we can see that Trans6D outperforms the baseline by 8.3% in ADD metric. Trans6D also outperforms Pix2Pose by 20.2% predicts the 2D-3D correspondence by CNN encoder, while we use Transformer to regress such correspondence. Trans6D+ improves the Trans6D by 4.3% and achieves the state-of-the-art performance. Comparing to the second-best method GDR-Net [39] that using hybrid presentation to estimate the 6D pose, our method outperforms it by 3.2% in ADD accuracy. DPOD has two results, one is only using correspondence to estimate

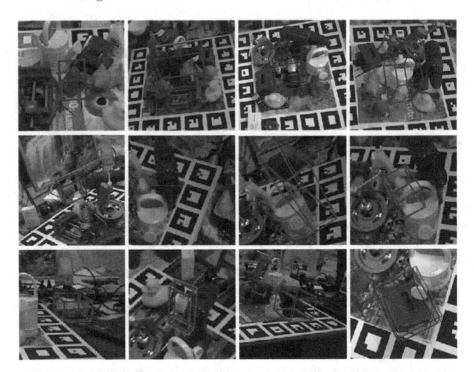

Fig. 4. Qualitative pose estimation results on LINEMOD and Occlusion LINEMOD dataset. Green 3D bounding boxes denote ground truth. Blue 3D bounding boxes represent our results. Our results match ground truth well. (Color figure online)

the 6D pose and Trans6D outperforms it by 9.6%, the other is a result after refinement and Trans6D+ outperforms it by 1.7%. In Fig. 4, we provide a visual comparison of predict pose versus ground truth pose.

Object 6D Pose Estimation on Occlusion LINEMOD: We use the model trained on the LINEMOD dataset for testing on the Occlusion LINEMOD dataset. Table 5 compares our method with other state-of-the-art methods [30,39] on Occlusion LINEMOD dataset in terms of ADD metric. From Table 5, we can see that Trans6D+ achieves a comparable accuracy (57.9%) and Trans6D achieves 52.6%. The improved performance demonstrates that the proposed method, enables Trans6D robust to partial occlusion.

Object 6D Pose Estimation on YCB-Video [40] Dataset: YCB-Video dataset contains 92 real video sequences for 21 YCB object instances. This dataset is challenging due to the image noise and occlusions. By following PoseCNN, we report the results on three metrics, 2D-Proj, ADD AUC and ADD(-S) metrics. From Table 6, we can see that Trans6D outperforms the baseline, PoseCNN, by 53.9% in ADD(-s) metric. Trans6D+ improves the Trans6D by 7.5% and achieves the state-of-the-art performance. Comparing to the second-best method, Ameni et al. [35] which also using pose refinement, our method outperforms it by 1.6% in ADD(-s) accuracy and 2.8% in AUC of ADD metric.

5 Conclusions

In this paper, we present a novel 6D object pose estimation framework build upon Transformers. We first construct two Transformer-based baselines and then compare their performance. One of the baselines uses pure Transformers (Trans6D-pure), and the other integrates CNNs with Transformers (Trans6D-hybrid). Then, we introduce two novel modules to improve the performance of the Trans6D-pure. The first is the patch-aware feature fusion (PAFF) module, which predicts the 2D-3D dense correspondence maps without information loss. The second is the pure Transformer-based pose refinement (Trans6D+) module, which iteratively refines the estimated pose. Our experiments demonstrate that the proposed method (Trans6D) achieves state-of-the-art performance in the LINEMOD and Occlusion LINEMOD datasets.

Furthermore, our method can be naturally extended to estimate the 6D object pose from the point cloud because the point cloud is sequence data and therefore suitable for Transformer. Despite the state-of-the-art performance, our method is not memory-friendly due to stacking a lot of self-attention modules. In future work, we plan to overcome the memory problem and extend Trans6D to more challenging scenes.

Acknowledgements. This work was supported by Institute of Information and communications Technology Planning and evaluation (IITP) grant funded by the Korea government (MSIT) (2021-0-00537, Visual common sense through self-supervised learning for restoration of invisible parts in images). ZQZ was supported by China Scholarship Council (CSC) Grant No. 202208060266. AL was supported in part by the EPSRC (grant number EP/S032487/1).

References

1. Billings, G., Johnson-Roberson, M.: SilhoNet: an RGB method for 6D object pose estimation. IEEE Robot. Autom. Lett. 4(4), 3727–3734 (2019). https://doi.org/10.1109/LRA.2019.2928776

2. Brachmann, E., Michel, F., Krull, A., Yang, M.Y., Gumhold, S., Rother, C.: Uncertainty-driven 6D pose estimation of objects and scenes from a single RGB image. In: 2016 IEEE Conference on Computer Vision and Pattern Recognition (CVPR), pp. 3364–3372 (2016). https://doi.org/10.1109/CVPR.2016.366

3. Brachmann, E., Krull, A., Michel, F., Gumhold, S., Shotton, J., Rother, C.: Learning 6D object pose estimation using 3D object coordinates. In: Fleet, D., Pajdla, T., Schiele, B., Tuytelaars, T. (eds.) ECCV 2014. LNCS, vol. 8690, pp. 536–551. Springer, Cham (2014). https://doi.org/10.1007/978-3-319-10605-2_35. https://www.microsoft.com/en-us/research/publication/learning-6d-object-pose-estimation-using-3d-object-coordinates/

4. Carion, N., Massa, F., Synnaeve, G., Usunier, N., Kirillov, A., Zagoruyko, S.: End-to-end object detection with transformers. In: Vedaldi, A., Bischof, H., Brox, T., Frahm, J.-M. (eds.) ECCV 2020. LNCS, vol. 12346, pp. 213–229. Springer, Cham (2020). https://doi.org/10.1007/978-3-030-58452-8_13

5. Chen, D., Li, J., Wang, Z., Xu, K.: Learning canonical shape space for category-level 6D object pose and size estimation. In: 2020 IEEE/CVF Conference on Computer Vision and Pattern Recognition (CVPR), pp. 11970–11979 (2020). https://doi.org/10.1109/CVPR42600.2020.01199

6. Chen, W., Jia, X., Chang, H.J., Duan, J., Leonardis, A.: G2L-Net: global to local network for real-time 6d pose estimation with embedding vector features. In: 2020 IEEE/CVF Conference on Computer Vision and Pattern Recognition (CVPR), pp. 4232–4241 (2020). https://doi.org/10.1109/CVPR42600.2020.00429

7. Chen, W., Duan, J., Basevi, H., Chang, H.J., Leonardis, A.: PonitPoseNet: point pose network for robust 6d object pose estimation. In: Proceedings of the IEEE Winter Conference on Applications of Computer Vision (WACV), March 2020

8. Chen, W., Jia, X., Chang, H.J., Duan, J., Shen, L., Leonardis, A.: FS-Net: fast shape-based network for category-level 6d object pose estimation with decoupled rotation mechanism (2021)

9. Cheng, Y., Lu, F.: Gaze estimation using transformer. arXiv preprint arXiv:2105.14424 (2021)

10. Deng, J., Dong, W., Socher, R., Li, L.J., Li, K., Fei-Fei, L.: ImageNet: a large-scale hierarchical image database. In: 2009 IEEE Conference on Computer Vision and Pattern Recognition, pp. 248–255 (2009). https://doi.org/10.1109/CVPR.2009.5206848

11. Devlin, J., Chang, M.W., Lee, K., Toutanova, K.: BERT: pre-training of deep bidirectional transformers for language understanding (2018)

12. Dosovitskiy, A., et al.: An image is worth 16×16 words: transformers for image recognition at scale. arXiv preprint arXiv:2010.11929 (2020)

13. He, K., Zhang, X., Ren, S., Sun, J.: Deep residual learning for image recognition. In: 2016 IEEE Conference on Computer Vision and Pattern Recognition (CVPR), pp. 770–778 (2016). https://doi.org/10.1109/CVPR.2016.90

14. He, S., Luo, H., Wang, P., Wang, F., Li, H., Jiang, W.: TransReID: transformer-based object re-identification. arXiv preprint arXiv:2102.04378 (2021)

15. He, Y., Sun, W., Huang, H., Liu, J., Fan, H., Sun, J.: PVN3D: a deep point-wise 3D keypoints voting network for 6dof pose estimation. In: 2020 IEEE/CVF Conference on Computer Vision and Pattern Recognition (CVPR), pp. 11629–11638 (2020). https://doi.org/10.1109/CVPR42600.2020.01165

16. Hinterstoisser, S., et al.: Model based training, detection and pose estimation of texture-less 3D objects in heavily cluttered scenes. In: Lee, K.M., Matsushita, Y., Rehg, J.M., Hu, Z. (eds.) ACCV 2012. LNCS, vol. 7724, pp. 548–562. Springer, Heidelberg (2013). https://doi.org/10.1007/978-3-642-37331-2_42

17. Hodaň, T., Baráth, D., Matas, J.: Epos: Estimating 6D pose of objects with symmetries. In: 2020 IEEE/CVF Conference on Computer Vision and Pattern Recognition (CVPR), pp. 11700–11709 (2020). https://doi.org/10.1109/CVPR42600.2020.01172

18. Hu, Y., Fua, P., Wang, W., Salzmann, M.: Single-stage 6D object pose estimation. In: 2020 IEEE/CVF Conference on Computer Vision and Pattern Recognition (CVPR), pp. 2927–2936 (2020). https://doi.org/10.1109/CVPR42600.2020.00300

19. Hu, Y., Fua, P., Wang, W., Salzmann, M.: Single-stage 6d object pose estimation. In: Proceedings of the IEEE/CVF Conference on Computer Vision and Pattern Recognition, pp. 2930–2939 (2020)

20. Hudson, D.A., Zitnick, C.L.: Generative adversarial transformers (2021)
21. Jiang, Y., Chang, S., Wang, Z.: TransGAN: two transformers can make one strong GAN. arXiv preprint arXiv:2102.07074 (2021)
22. Kehl, W., Manhardt, F., Tombari, F., Ilic, S., Navab, N.: SSD-6D: making RGB-based 3D detection and 6D pose estimation great again. In: 2017 IEEE International Conference on Computer Vision (ICCV), pp. 1530–1538 (2017). https://doi.org/10.1109/ICCV.2017.169
23. Kumar, M., Weissenborn, D., Kalchbrenner, N.: Colorization transformer (2021)
24. Li, Y., Wang, G., Ji, X., Xiang, Y., Fox, D.: Deepim: deep iterative matching for 6d pose estimation. In: European Conference on Computer Vision (ECCV) (2018)
25. Li, Z., Wang, G., Ji, X.: CDPN: coordinates-based disentangled pose network for real-time RGB-based 6-dof object pose estimation. In: 2019 IEEE/CVF International Conference on Computer Vision (ICCV), pp. 7677–7686 (2019). https://doi.org/10.1109/ICCV.2019.00777
26. Lin, K., Wang, L., Liu, Z.: End-to-end human pose and mesh reconstruction with transformers. In: Proceedings of the IEEE/CVF Conference on Computer Vision and Pattern Recognition, pp. 1954–1963 (2021)
27. Morrison, D., et al.: Cartman: the low-cost cartesian manipulator that won the amazon robotics challenge. In: 2018 IEEE International Conference on Robotics and Automation (ICRA), pp. 7757–7764. IEEE (2018)
28. Oberweger, M., Rad, M., Lepetit, V.: Making deep heatmaps robust to partial occlusions for 3D object pose estimation. In: Ferrari, V., Hebert, M., Sminchisescu, C., Weiss, Y. (eds.) ECCV 2018. LNCS, vol. 11219, pp. 125–141. Springer, Cham (2018). https://doi.org/10.1007/978-3-030-01267-0_8
29. Park, K., Patten, T., Vincze, M.: Pix2pose: pixel-wise coordinate regression of objects for 6d pose estimation. In: 2019 IEEE/CVF International Conference on Computer Vision (ICCV), pp. 7667–7676 (2019). https://doi.org/10.1109/ICCV.2019.00776
30. Peng, S., Zhou, X., Liu, Y., Lin, H., Huang, Q., Bao, H.: Pvnet: pixel-wise voting network for 6dof object pose estimation. IEEE Trans. Pattern Anal. Mach. Intell. 1 (2020). https://doi.org/10.1109/TPAMI.2020.3047388
31. Rad, M., Lepetit, V.: BB8: a scalable, accurate, robust to partial occlusion method for predicting the 3d poses of challenging objects without using depth. In: 2017 IEEE International Conference on Computer Vision (ICCV), pp. 3848–3856 (2017). https://doi.org/10.1109/ICCV.2017.413
32. Redmon, J., Farhadi, A.: Yolov3: an incremental improvement. arXiv preprint arXiv:1804.02767 (2018)
33. Shao, J., Jiang, Y., Wang, G., Li, Z., Ji, X.: PFRL: pose-free reinforcement learning for 6d pose estimation. In: 2020 IEEE/CVF Conference on Computer Vision and Pattern Recognition (CVPR), pp. 11451–11460 (2020). https://doi.org/10.1109/CVPR42600.2020.01147
34. Song, C., Song, J., Huang, Q.: Hybridpose: 6d object pose estimation under hybrid representations. In: 2020 IEEE/CVF Conference on Computer Vision and Pattern Recognition (CVPR), pp. 428–437 (2020). https://doi.org/10.1109/CVPR42600.2020.00051
35. Trabelsi, A., Chaabane, M., Blanchard, N., Beveridge, R.: A pose proposal and refinement network for better 6d object pose estimation. In: Proceedings of the IEEE/CVF Winter Conference on Applications of Computer Vision (WACV), pp. 2382–2391, January 2021

36. Wada, K., Sucar, E., James, S., Lenton, D., Davison, A.J.: MoreFusion: multi-object reasoning for 6d pose estimation from volumetric fusion. In: 2020 IEEE/CVF Conference on Computer Vision and Pattern Recognition (CVPR), pp. 14528–14537 (2020). https://doi.org/10.1109/CVPR42600.2020.01455

37. Wang, C., et al.: 6-pack: category-level 6d pose tracker with anchor-based keypoints. In: 2020 IEEE International Conference on Robotics and Automation (ICRA), pp. 10059–10066 (2020). https://doi.org/10.1109/ICRA40945.2020.9196679

38. Wang, C., et al.: Densefusion: 6d object pose estimation by iterative dense fusion. In: 2019 IEEE/CVF Conference on Computer Vision and Pattern Recognition (CVPR), pp. 3338–3347 (2019). https://doi.org/10.1109/CVPR.2019.00346

39. Wang, G., Manhardt, F., Tombari, F., Ji, X.: GDR-NET: geometry-guided direct regression network for monocular 6d object pose estimation. In: Proceedings of the IEEE/CVF Conference on Computer Vision and Pattern Recognition, pp. 16611–16621 (2021)

40. Xiang, Y., Schmidt, T., Narayanan, V., Fox, D.: PoseCNN: a convolutional neural network for 6d object pose estimation in cluttered scenes (2017)

41. Yang, S., Quan, Z., Nie, M., Yang, W.: Transpose: towards explainable human pose estimation by transformer (2020)

42. Zakharov, S., Shugurov, I., Ilic, S.: DPOD: 6d pose object detector and refiner. In: 2019 IEEE/CVF International Conference on Computer Vision (ICCV), pp. 1941–1950 (2019). https://doi.org/10.1109/ICCV.2019.00203

43. Zheng, S., et al.: Rethinking semantic segmentation from a sequence-to-sequence perspective with transformers (2020)

Learning to Estimate Multi-view Pose from Object Silhouettes

Yoni Kasten[1], True Price[2(✉)], David Geraghty[2], and Jan-Michael Frahm[2]

[1] NVIDIA Research, Santa Clara, USA
ykasten@nvidia.com
[2] Meta, Seattle, USA
{jtprice,dger,jmfrahm}@meta.com

Abstract. While Structure-from-Motion pipelines certainly have their success cases in the task of 3D object reconstruction from multiple images, they still fail on many common objects that lack distinctive texture or have complex appearance qualities. The central problem lies in 6DOF camera pose estimation for the source images: without the ability to obtain a good estimate of the epipolar geometries, all state-of-the-art methods will fail. Although alternative solutions exist for specific objects, general solutions have proved elusive. In this work, we revisit the notion that silhouette cues can provide reasonable constraints on multi-view pose configurations when texture and priors are unavailable. Specifically, we train a neural network to holistically predict camera poses and pose confidences for a given set of input silhouette images, with the hypothesis that the network will be able to learn cues for multi-view relationships in a data-driven way. We show that our network generalizes to unseen synthetic and real object instances under reasonable assumptions about the input pose distribution of the images, and that the estimates are suitable to initialize state-of-the-art 3D reconstruction methods.

1 Introduction

Three-dimensional object reconstruction, the process of converting imagery of an object into a representation of its geometry, is an increasingly mainstream component of augmented- and virtual-reality (AR/VR) research and applications, with much of this growth due to the increasing facility and scalability of capture technologies. In AR/VR entertainment, for example, commodity 3D scanning technology can efficiently generate photorealistic models for use in virtual worlds, reducing manual effort required by 3D artists. Likewise, many research applications now utilize realistic 3D models to drive synthetic data generation.

Historically, high-quality object capture methodologies have required a certain level of controlled capture, such as a fixed camera rig, or specific imaging equipment, such as a depth sensor [29,38]. Modernized pipelines driven

---

Y. Kasten—This work was completed while Yoni was an intern at Meta.

Supplementary Information The online version contains supplementary material available at https://doi.org/10.1007/978-3-031-25085-9_8.

L. Karlinsky et al. (Eds.): ECCV 2022 Workshops, LNCS 13808, pp. 129–147, 2023.
https://doi.org/10.1007/978-3-031-25085-9_8

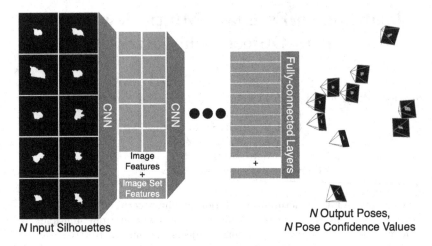

N Input Silhouettes

N Output Poses,
N Pose Confidence Values

Fig. 1. Our deep neural network takes as input a set of silhouette masks of an object observed from different viewpoints. After applying several permutation-equivariant layers that combine image-specific and image-set-generic features, the network outputs a 6DOF pose and a pose confidence for each input image.

by structure-from-motion (SfM) followed by multi-view stereo (MVS) and depthmap fusion [47,48] have increasingly democratized the process in recent years, enabling less-experienced users to run photogrammetry from a handheld camera, either with known temporal sequencing (*i.e.*, video capture) or capturing images as an unordered collection. These general-purpose pipelines also enable distributed collection, where photos from multiple users in different environments are leveraged to create a 3D model.

However, while casual 3D reconstruction is increasingly feasible, output reconstruction quality in these pipelines varies widely depending on the input imagery and target object, and there exist several categorical limitations, particularly for low-texture objects and objects having non-Lambertian surface reflectance properties. A number of approaches, for example the recent works of Yariv *et al.* [66] and Schmitt *et al.* [46], have pushed the envelope of dense surface estimation pipelines by jointly modeling object geometry, view-dependent lighting/reflectance effects, and – importantly – allowing for camera pose parameters to be refined as part of the optimization process, which is not generally possible with traditional MVS. These approaches have shown a remarkable improvement in completeness of the reconstructed object surface, as well as impressive quality for low-texture objects with complex appearance.

In this paper, we address a key remaining gap for state-of-the-art reconstruction methods: *camera pose initialization*, particularly when photometric methods fail. To this end, we introduce a neural-network-based alternative to SfM tackling the classical computer vision problem of multi-view pose from unknown object silhouettes [7,8,17,31]. Given a set of binary object masks obtained from multiple images of an object, the goal is to produce a camera pose for each image, relative to an arbitrary, unspecified object coordinate frame. The driving concept here is that, when constraints like point correspondences cannot be utilized, the object contour in the image still provides signal on the space of possible relative camera poses. For example, each foreground pixel in one image has a corresponding location in every other image, and thus the epipolar lines for those pixels must intersect the object silhouette in the other image. While previous work has wielded such principles using handcrafted features and/or controlled scenarios, our hypothesis is that a neural network should be able to naturally learn the joint space of camera viewpoints.

For the current work, we assume we are given a set of pre-extracted silhouette images with known camera intrinsic calibration, generally upright orientation, and medium-baseline camera motion. This image set is fed to an order-equivariant neural network (Fig. 1) that regresses a six-degrees-of-freedom (6DOF; *i.e.* rotation and translation) pose for each image simultaneously, as well as a confidence estimate that helps to identify images with higher levels of pose ambiguity. The network's task during training is to learn how to map the object contours into a common latent representation while also taking into account the global state of all contours together, and then to form a mapping from this representation into a final 6DOF pose.

For training, we render randomly posed silhouettes of CAD models and directly optimize the network's output to match the poses used for rendering. At test time, the network uses only the input silhouette images, without any knowledge about the 3D geometry of the observed object. Our method provides the following overall contributions:

- A deep-learning approach for silhouette-based multi-view 6DOF pose estimation for unknown objects. Previous works in this space have been very tailored to controlled settings or known objects [64], or have required carefully handcrafted features in a robust framework while only estimating 3DOF camera rotations [31].
- A neural network architecture leveraging DSS and DeepSets layers [33,68] to achieve unordered multi-view pose estimation. Such architectures have not been previously used for this task. In our case, the selection of the output global coordinate system is arbitrary, and there is not just one "correct" solution, in contrast to previous applications of permutation-equivariant layers. We thus introduce a new loss function that is agnostic to the output global coordinate system (Eqs. (2–7)). This formulation is crucial for making the training problem possible.
- A loss formulation that incorporates the von Mises-Fisher distribution to allow for pose confidence regression. We demonstrate that our network's confidence predictions reliably correspond to per-view pose accuracy results.
- Generalizability: While we train on only 15 object classes of CAD models, we show that the network generalizes to unseen object classes on a number of synthetic and real datasets, including datasets with imperfect masks.
- Putting silhouettes into practice: Considering the case of uncontrolled, unknown object capture, we demonstrate that silhouette-based reasoning offers a workable solution for low-texture objects where color-based reconstruction methods have inherent limitations. Examples are shown for a new

"Glass Figurines" dataset, where our method succeeds in several challenging cases where a state-of-the-art SfM pipeline [48] fails.

2   Related Work

Camera pose estimation for object reconstruction has a long history in the field of computer vision. For unknown objects and unordered images, possibly the most well-established approaches are photogrammetric methods like Structure-from-Motion (SfM) [48]. These methods are driven by 2D feature correspondence search, where distinct 2D image keypoints are detected, described, and matched between the input images. Assuming that such 2D image-to-image correspondences can be reliably found, additional geometric reasoning is used to begin recovering 6DOF image poses. In contrast to incremental SfM methods that build the final 3D reconstruction one image at a time, the method we propose is more in line with global SfM approaches [12,50] and recent holistic deep-learned approaches [37], where pose properties for all images are determined simultaneously. Typical global SfM methods rely on two-view pose estimates to initially solve for absolute image rotations, followed by a second stage to solve for absolute image positions. Recent work by Kasten *et al.* [25] has also suggested a one-step global approach by averaging essential matrices. While our neural network architecture does not leverage two-view relationships directly, it does employ a global representation of all images when deriving latent representations at different stages of the network.

It is also worth noting that many active-capture applications, for example object reconstruction pipelines that run on a smartphone [42,53], augment the camera pose estimates with inertial measurement unit readings available on the device, which provide a strong prior for the differential motion of the camera. In our work, we assume a different capture scenario, where the object of interest may be moved between different frames, or even where the collection of object images is derived from different locations at different times. Moreover, all SfM-type methods heavily depend on the reconstructibility of the object of interest. For objects with low texture or complex appearance, these methods often fail because the photometric assumptions underlying keypoint detection and description are violated.

Learning-Based Methods for 3D Object Reasoning. A litany of methods have been proposed for camera pose detection for known objects, especially in single-view contexts. Early methods leveraged deformable parts models for discrete viewpoint prediction [13,32,40]. Related work [16] achieved 6DOF pose estimation via view synthesis with brute-force evaluation of geometry priors. With the advent of deep learning, numerous approaches have advanced single-view object detection and pose estimation, including for discrete prediction, 3D bounding box estimation, direct pose regression, and direct 2D-to-3D point correspondence regression [2,5,27,41,43,54,62,63]. One recent extension to these works is HybridPose [51], which combines object pose regression with learned feature extraction and subsequent pose refinement. Also relevant to our work is SilhoNet [1], an object pose regressor that is trained to predict occlusion-aware and occlusion-agnostic object silhouette masks as an intermediate output. The silhouette is used as the primary cue for rotation, which allows the rotation regression submodule to train entirely on synthetic data.

Reconstruction-focused approaches have emphasized learning shape priors for object classes, especially for single-view geometry prediction. Choy *et al.* [4] trained a recurrent neural network for volumetric 3D reconstruction of multiple object classes. Beyond single-view shape estimation, this network is able to iteratively aggregate multiple images to refine the output, resulting in a coarse 3D model for instances of the trained-for classes. While this and related methods [6,10,35,61] penalize errors in 3D geometry, prior work leveraging deformable shapes [3,24] and subsequent works in differentiable projection and rendering have used object masks directly. Several methods [14,26,44,56,58,65] train single-view volumetric or mesh reconstruction models by reprojecting voxels into other views and optimizing the predicted voxel occupancy against the ground-truth object mask. Some such methods have also reported results on 2 to 5 input views [14,44,58].

Several learning-based reconstruction methods exist that estimate camera pose for canonical object frames [70] or between image pairs [20,55] with a silhouette-based loss. In the latter cases, a network is shown pair of images and jointly predicts (1) the relative pose between them and (2) a 3D geometry (a voxelization or a point cloud) for the object. The models are trained by reprojecting the predicted geometry into the first image and penalizing disagreements with the associated object mask. Each input image is independently processed, allowing for single-view applications of geometry and pose estimation at inference time.

A number of recent works learn a neural radiance field (NeRF) [36] while optimizing camera parameters [22,30,57,67]. These methods either require camera initialization, or can only handle roughly forward-facing scenes. Very recently, [34] used a generative adversarial training strategy without input camera poses in a general camera setup with a known camera distribution. For each scene, they train from scratch a (NeRF, discriminator) network pair by sampling camera poses according to the distribution and training the NeRF to fool the discriminator for whether a patch is fake (rendered) or real. This training process is heavy, on the order of hours, and must be done separately for each scene without any generalizability to new scenes. While the majority of the work focuses on color image processing, the authors do present a single proof-of-concept result taking in a large collection of silhouette images as input. In contrast to NeRF methods, our approach trains a neural network that holistically processes image sets in a single pass and generalizes to unseen objects and object classes.

Pose from Silhouettes of an Unknown Object. Methods for camera pose estimation from silhouettes date back more than two decades. Classical approaches utilize *epipolar mapping constraints*, where all epipolar silhouette lines mapped from one view must intersect the silhouette in another. For two views, epipolar constraints yield corresponding 2D object contour points with tangent epipolar lines. For multiple views, the constraints amount to finding a consistent visual hull for all images. Many early approaches optimized pose by identifying corresponding silhouette *frontier points*, either as single-take methods under controlled capture (*e.g.* a turntable or using mirrors) [7,18,19,59,60] or

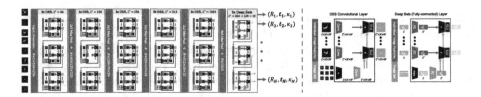

Fig. 2. Our permutation-equivariant network architecture. Starting from the N input silhouettes, five groups of three convolutional DSS layers are sequentially applied, interspersed by max-pooling operations. Then, three deep sets (fully-connected) layers are applied to finally obtain 10 output values per image, representing the image's rotation, translation, and pose confidence. Each convolution is followed by a batch normalization layer (not shown), and we use ELU activations throughout the network.

by optimizing a camera rig configuration under repeated observations [8,23,49]. Visual hull optimization [17] has also been proposed for controlled capture scenarios.

One similar work to ours is that of Littwin *et al.* [31], which aims to estimate the rotation distance between cameras using a handcrafted measure of silhouette contour similarity. While this two-view measure is quite noisy, the authors show that, given a sufficiently large source image set, an inlier set of relative rotation measurements can be determined via a robust fitting procedure. In contrast, our network considers all available images jointly when making its predictions, and it additionally can reason about camera translations and pose confidence.

Finally, Xiao *et al.* [64] trained a neural network to regress 6DOF pose for an novel object in an image under the assumption that the object geometry is known at inference time. Their approach first computes separate shape and appearance encodings and then feeds these to a pose regression sub-network. While the authors did not analyze the network's activations, it is quite possible that their network learns to encode object contours in the input image and compare these to possible projections of the 3D shape.

3   Method

We assume an input set of N images taken by N cameras capturing the same 3D object at different viewpoints. We further assume that the object silhouette masks are pre-extracted from the input images. In practice, this can be done either by classical or learning-based approaches [15] for 2D object segmentation. Our goal is to regress the camera poses solely from the silhouette masks.

To tackle this problem, we introduce a deep neural network architecture that learns to infer a set of 6DOF camera poses and pose confidences from silhouettes using a large training set of general 3D objects. To improve robustness against two-view ambiguities, our network considers all N input silhouettes jointly. While we directly optimize pose error during training, we observe that our network outputs poses that respect the silhouette constraints leveraged by earlier non-learning-based methods (see supplementary).

3.1   Network Architecture

Our network architecture (Fig. 2) is based on the recently introduced "deep sets of symmetric elements" (DSS) layers [33], which have shown to be effective across a variety of learning tasks involving inputs of unordered image sets. The input for each DSS layer is a set of N images with the same number of channels. Two learnable convolutional filters are then applied: a Siamese filter that is applied on each input image independently, and an aggregation module filter that is applied on the sum of all input images. The output of the second filter is then added to each output of the first filter, resulting in a new set of N images with a possibly different number of channels. Since summation is a permutation-invariant operation, it follows that a DSS layer is permutation-equivariant, meaning that applying a permutation to the N input images results in permuted outputs.

In our case, since the multi-view input silhouettes are unordered, we design our network to be permutation-equivariant. The inputs to the network are N, one-channel $\{0, 1\}$ silhouette binary masks, and the outputs are the corresponding N camera poses, each represented by 10 coordinates: 6 for the world-to-camera rotation using the $6D$ parameterization of [69], 3 for the world-to-camera translation, and another scalar to represent the confidence of the network in the estimated pose. We use a sequence of 5 DSS blocks, each consisting of 3 DSS layers with max-pooling operations between each block, followed by permutation-equivariant fully connected ("deep sets") layers as in [68].

3.2   Confidence-Based Loss Function

We use ground-truth camera poses (available at training time) for training the network. Let $(R_1, \mathbf{t}_1), \ldots, (R_N, \mathbf{t}_N)$ denote the output camera poses from the deep network and $(\bar{R}_1, \bar{\mathbf{t}}_1), \ldots, (\bar{R}_N, \bar{\mathbf{t}}_N)$ the respective ground-truth (GT) camera poses. For each input silhouette image, we predict a single pose confidence $\kappa \in \mathbb{R}$ corresponding to the scale parameter of a von Mises-Fisher (vMF) distribution [52].

First, it is worth taking a moment to discuss possible formulations for the network loss function and confidence prediction. On one hand, we could forego confidence estimation entirely and directly penalize the rotation and translation errors using, *e.g.*, an L2 penalty. We empirically found this approach to give similar accuracy to our formulation, with the caveat that the network no provides quality ratings for individual pose estimates. Alternatively, we could adopt a complete probability distribution like the Bingham distribution on rotation [11], which can properly model uncertainty directions in the tangent space of SO(3). In practice, however, we found that introducing more confidence parameters for rotation made the network more difficult to train. This may be caused in part by the fact that the full space of SO(3) is much larger than that of our assumed input viewpoints, which are generally upright and always object-centric. As such, we have chosen to model a single confidence parameter for rotation alone, and we show in our experiments that this approach is effective in separating good-quality pose estimates from those that are less certain.

To model 3DOF camera rotation and its confidence, we adopt a maximum-likelihood formulation where we predict a 2DOF probability distribution mean and scale for each axis of the local camera frame. We define three vMF distributions that share the same scale parameter κ, each with a probability density function of

$$f_i(\mathbf{x}; \mathbf{r}_i, \kappa) = C_3(\kappa)e^{\kappa \mathbf{r_i}^T \mathbf{x}}, \tag{1}$$

where $\mathbf{r}_i \in \mathbb{S}^2$ for $i \in \{1, 2, 3\}$ are the (unit) row vectors of the predicted rotation matrix for the image, and $C_3(\kappa) = \frac{\kappa}{2\pi(e^\kappa - e^{-\kappa})}$ forms a normalization factor.

We aim to predict distributions that explain the GT rotation axes with as high of a probability as possible. Denoting the GT row as $\bar{\mathbf{r}}_i \in \mathbb{S}^2$, the log-likelihood for this vector to be sampled from the corresponding distribution is

$$l_i(R_0) = \log\left(C_3(\kappa)\right) + \kappa \mathbf{r}_i^T (R_0 \bar{\mathbf{r}}_i), \tag{2}$$

where R_0 is a global rotation ambiguity of our solution relative to the GT cameras.

For log-likelihood l_i^j of camera j, we can compute R_0 as

$$R_0^* = \underset{R_0}{\operatorname{argmax}} \sum_{j=1}^{N} \sum_{i=1}^{3} l_i^j(R_0). \tag{3}$$

In the supplementary material, we show that this can simply done by weighted relative rotation averaging:

$$\tilde{\mathbf{q}}_0^* = \frac{1}{N} \sum_{j=1}^{N} \kappa_j \mathbf{q}_j^{-1} \bar{\mathbf{q}}_j, \qquad \mathbf{q}_0^* = \frac{\tilde{\mathbf{q}}_0^*}{\|\tilde{\mathbf{q}}_0^*\|}, \tag{4}$$

where $\mathbf{q_i}$, $\bar{\mathbf{q}}_i$, and $\mathbf{q}_0^*$ are the quaternions corresponding to R_i, $\bar{R}_i$, and R_0^*, respectively. The final loss function for the rotation and confidence outputs is

$$L_{R,\kappa} = \frac{1}{3N} \sum_{j=1}^{N} \sum_{i=1}^{3} -l_i^j(R_0^*). \tag{5}$$

For our predicted translation vectors, a camera-center loss is applied by considering global translation and scaling ambiguities. Denoting $\mathbf{c}_i = -R_0^{*T} R_i^T \mathbf{t}_i$ and $\bar{\mathbf{c}}_i = -\bar{R}_i^T \bar{\mathbf{t}}_i$ as predicted and GT camera centers, respectively, the camera-center loss is defined by

$$L_c = \frac{1}{N} \sum_{i=0}^{N} \left\| \frac{\mathbf{c}_i - \mathbf{c}}{s} - \frac{\bar{\mathbf{c}}_i - \bar{\mathbf{c}}}{\bar{s}} \right\|, \tag{6}$$

where the mean vectors $\mathbf{c} = \frac{1}{N} \sum_{j=1}^{N} \mathbf{c}_j, \bar{\mathbf{c}} = \frac{1}{N} \sum_{j=1}^{N} \bar{\mathbf{c}}_j$ account for the global translation ambiguity, and we divide by the mean distance between each camera center and the average center: $s = \frac{1}{N} \sum_{j=1}^{N} \|\mathbf{c} - \mathbf{c}_j\|, \bar{s} = \frac{1}{N} \sum_{j=1}^{N} \|\bar{\mathbf{c}} - \bar{\mathbf{c}}_j\|$. Our total loss function is:

$$L = \beta L_c + L_{R,\kappa} \tag{7}$$

with scalar weight β balancing the two loss parts. In our experiments, we use $\beta = 2$, which was chosen based on examining the validation set error. L_c values are in $[0, 1]$, and values of $L_{R,k}$ are typically around -10.

3.3 Training

For training our network, we use a collection of synthetic object models from multiple object categories. For each training iteration, we render N 256 × 256 px silhouettes with random camera poses around the given object. Each input set of N images is generated with azimuth and elevation sampled uniformly in the range $[-30°, 30°]$, while the camera roll from the scene vertical is sampled from a normal distribution with a standard deviation of 5°. This viewing range is selected to approximate a typical set of casually captured viewpoints of one side of an object; for example, the DTU dataset [21] used in our experiments has a similar range of viewing angles. Each camera is initially positioned to look at the object origin (defined as the median vertex), with a distance from the origin sampled uniformly within the range $[3.2r, 6r]$, where r is the object radius. The camera translations are then perturbed by an offset sampled from $\mathcal{N}(\mathbf{0}, (0.005r)I_{3\times3})$. The object itself is rotated randomly around its origin.

In all experiments, we use $N = 10$ input views for training. This number was reported by [31] to be a large-enough support set in the multi-view setting. We trained the network by minimizing Eq. (7) on a training split of 15 object categories. We used the ADAM optimizer [28] with a learning rate of 0.001. To improve initial training acceleration, we began with a batch size of 5 (*i.e.*, 5 groups of 10 random poses) with all images coming from the same object. To better maintain training acceleration in later epochs, we switched to a batch size of 1 after ∼60 epochs, and we trained overall for ∼250 epochs.

4   Experiments

4.1   Datasets

We evaluated our trained network on 3D objects from a validation split of object models from 3D Warehouse. We tested unseen objects and camera configurations from our 15 training classes, plus 5 unseen object classes.

In addition to manually created models, we further evaluated the network on a new dataset, RealScan, that consists of 30 high-resolution scans of a variety of real 3D objects ranging from stuffed animals to office supplies. We projected these scans with the same sampling described in Sect. 3.3 for 100 sets of 20 random views, with each set of views coming from either the front, back, top, sides, or bottom of the object. These high-polygon meshes, as well as their projected contours, are very different from the ones that are used for training the network.

We further applied our method to real images from (1) the DTU MVS dataset [21] and (2) a new "Glass Figurines" dataset containing objects that are difficult to reconstruct using traditional SfM methods. For DTU, we evaluated the 15 back-row cameras (available for scans with id number >80) whose cameras are far enough from the object such that most of the object is visible in the image. For the 8 scans, we used the input masks that were extracted manually by [39,66]. The Glass Figurines dataset consists of 11 objects with 10 images each, plus manually extracted objects masks and ground-truth camera poses computed using ArUco Tags [9,45]. We plan to publicly release the dataset.

Table 1. Camera pose accuracy and reprojection IOU for the 3D Warehouse dataset.

	Metrics	R [°]	t_s [ratio]	t_d [°]	R [IOU]	t [IOU]	t_s [IOU]	t_d [IOU]	$R+t$ [IOU]
Average per-class validation statistic over all 15 training classes									
Valid.	Mean (Med.)	6.40 (4.59)	0.03 (0.02)	2.36 (1.50)	0.84 (0.88)	0.70 (0.73)	0.92 (0.94)	0.70 (0.74)	0.46 (0.47)
	↑ 5 (Oracle)	5.17 (3.88)	0.03 (0.03)	2.17 (2.23)	0.86 (0.87)	0.71 (0.71)	0.93 (0.93)	0.72 (0.71)	0.49 (0.48)
	↓ 5 (Oracle)	7.63 (8.92)	0.04 (0.04)	2.55 (2.48)	0.82 (0.81)	0.68 (0.69)	0.91 (0.91)	0.69 (0.69)	0.42 (0.44)
Unseen test classes									
Bathtub	Mean (Med.)	6.61 (4.69)	0.03 (0.02)	2.07 (1.33)	0.92 (0.95)	0.86 (0.88)	0.95 (0.96)	0.87 (0.89)	0.67 (0.74)
	↑ 5 (Oracle)	5.33 (4.13)	0.03 (0.03)	1.90 (1.92)	0.94 (0.94)	0.87 (0.87)	0.95 (0.96)	0.88 (0.88)	0.70 (0.69)
	↓ 5 (Oracle)	7.89 (9.09)	0.03 (0.04)	2.23 (2.22)	0.91 (0.90)	0.85 (0.85)	0.94 (0.94)	0.87 (0.87)	0.64 (0.65)
Car	Mean (Med.)	6.12 (4.53)	0.03 (0.02)	2.27 (1.45)	0.92 (0.94)	0.84 (0.86)	0.94 (0.96)	0.85 (0.87)	0.62 (0.69)
	↑ 5 (Oracle)	4.92 (3.81)	0.03 (0.03)	1.99 (2.13)	0.93 (0.94)	0.85 (0.85)	0.95 (0.95)	0.86 (0.86)	0.66 (0.64)
	↓ 5 (Oracle)	7.32 (8.43)	0.03 (0.03)	2.55 (2.40)	0.90 (0.90)	0.83 (0.83)	0.94 (0.94)	0.84 (0.84)	0.58 (0.60)
Chair	Mean (Med.)	6.79 (4.96)	0.03 (0.02)	2.67 (1.70)	0.83 (0.87)	0.74 (0.78)	0.91 (0.94)	0.75 (0.79)	0.49 (0.51)
	↑ 5 (Oracle)	5.38 (4.14)	0.03 (0.03)	2.38 (2.52)	0.85 (0.87)	0.76 (0.75)	0.92 (0.92)	0.76 (0.76)	0.53 (0.51)
	↓ 5 (Oracle)	8.19 (9.43)	0.04 (0.04)	2.97 (2.83)	0.80 (0.78)	0.73 (0.73)	0.90 (0.91)	0.74 (0.74)	0.45 (0.47)
Lamp	Mean (Med.)	10.60 (7.26)	0.04 (0.03)	3.57 (2.24)	0.77 (0.83)	0.63 (0.68)	0.89 (0.93)	0.64 (0.69)	0.32 (0.27)
	↑ 5 (Oracle)	9.19 (6.57)	0.04 (0.04)	3.48 (3.51)	0.79 (0.80)	0.65 (0.65)	0.90 (0.90)	0.66 (0.66)	0.34 (0.33)
	↓ 5 (Oracle)	12.00 (14.62)	0.05 (0.05)	3.66 (3.63)	0.75 (0.73)	0.61 (0.62)	0.88 (0.88)	0.62 (0.63)	0.30 (0.31)
Mailbox	Mean (Med.)	11.15 (5.13)	0.06 (0.03)	3.98 (2.12)	0.82 (0.88)	0.73 (0.78)	0.93 (0.95)	0.74 (0.78)	0.36 (0.30)
	↑ 5 (Oracle)	8.98 (7.56)	0.05 (0.05)	3.49 (3.42)	0.84 (0.86)	0.74 (0.75)	0.94 (0.93)	0.75 (0.76)	0.38 (0.38)
	↓ 5 (Oracle)	13.32 (14.73)	0.07 (0.07)	4.47 (4.54)	0.81 (0.79)	0.72 (0.71)	0.93 (0.93)	0.73 (0.71)	0.33 (0.33)

4.2   Results

Camera pose accuracy results are presented for the 3D Warehouse dataset in Table 1 and for the RealScan dataset in Table 2. For both, we evaluate on 10 random views of the object per test instance. Due to space limitations, we only show a representative subset of the RealScan results, and for 3D Warehouse, we show the average per-class validation result across all 15 training classes, and for our 5 unseen testing classes. See our supplementary material for complete results. In each row, we show mean and median errors, plus a confidence-ordered breakdown of the mean error, for a variety of metrics. All "Top 5" and "Bottom 5" metrics are taken using our confidence ranking from highest to lowest; we also show "Oracle" rankings for these that consider the ordering of lowest rotation error to higher rotation error. (The oracle ordering is the same for all columns.) The oracle provides a lower bound on the Top-5 error and thus can be used to assess the effectiveness of our confidence predictions.

When our confidence output is near to or better than the oracle, this indicates that our network has learned a reasonable confidence for pose. We also report intersection-over-union (IOU), computed by rerendering the test object using our predicted poses after global alignment to the GT poses. An example IOU result is shown in Fig. 3, along with a visualization of the visual hull for our estimated poses.

In Tables 1 and 2, we report our mean rotation (R), translation-scale (t_s), and translation-direction (t_d) error. t_s is the absolute value of: one minus the magnitude ratio of the predicted and GT translation vectors. t_d is the angle between the predicted and GT translation vectors. Also in of Table 1, we isolate the different network outputs: the sixth column shows our rotation combined with GT translation, the next our translation with GT rotation, and so on.

Table 2. Pose accuracy for a representative subset of RealScan.

	Metrics	R [°]	t_s [ratio]	t_d [°]
Cheetah	Mean (Med.)	8.78 (6.43)	0.04 (0.03)	3.37 (2.57)
	↑ 5 (Oracle)	7.14 (5.50)	0.03 (0.03)	2.80 (3.13)
	↓ 5 (Oracle)	10.42 (12.06)	0.04 (0.04)	3.94 (3.62)
Chess knight	Mean (Med.)	10.29 (8.01)	0.04 (0.03)	4.84 (3.61)
	↑ 5 (Oracle)	8.12 (5.99)	0.04 (0.04)	4.97 (4.64)
	↓ 5 (Oracle)	12.47 (14.60)	0.05 (0.04)	4.70 (5.03)
Glasses	Mean (Med.)	9.15 (7.57)	0.05 (0.03)	4.39 (3.00)
	↑ 5 (Oracle)	7.45 (5.95)	0.04 (0.04)	3.91 (4.23)
	↓ 5 (Oracle)	10.85 (12.35)	0.05 (0.06)	4.87 (4.56)
Plastic cup	Mean (Med.)	15.33 (12.96)	0.05 (0.04)	6.45 (5.41)
	↑ 5 (Oracle)	12.84 (9.70)	0.05 (0.05)	6.50 (6.75)
	↓ 5 (Oracle)	17.82 (20.96)	0.05 (0.05)	6.41 (6.16)
Stapler	Mean (Med.)	8.88 (4.68)	0.03 (0.02)	2.77 (1.81)
	↑ 5 (Oracle)	7.44 (5.50)	0.03 (0.03)	2.43 (2.70)
	↓ 5 (Oracle)	10.31 (12.25)	0.03 (0.03)	3.10 (2.83)
Toy bunny	Mean (Med.)	8.29 (5.92)	0.03 (0.02)	2.92 (1.95)
	↑ 5 (Oracle)	6.06 (4.51)	0.03 (0.03)	2.78 (2.80)
	↓ 5 (Oracle)	10.52 (12.08)	0.03 (0.03)	3.06 (3.04)
Wooden spoon	Mean (Med.)	10.04 (8.18)	0.05 (0.03)	3.52 (2.57)
	↑ 5 (Oracle)	7.68 (5.76)	0.04 (0.04)	3.51 (3.65)
	↓ 5 (Oracle)	12.40 (14.32)	0.05 (0.05)	3.53 (3.40)

Table 3. Camera pose accuracy for the DTU dataset.

Id	R [°]		t_s [ratio]		t_d [°]	
	Mean	Median	Mean	Median	Mean	Median
83	4.69	4.11	0.02	0.02	3.63	3.73
97	16.44	16.73	0.11	0.08	9.54	8.72
105	4.35	4.16	0.01	0.01	2.57	2.37
106	6.20	5.08	0.02	0.02	1.78	1.84
110	4.59	3.43	0.02	0.01	0.68	0.68
114	3.13	3.12	0.02	0.02	0.78	0.66
118	7.17	6.34	0.03	0.02	5.55	4.89
122	9.76	8.74	0.03	0.03	6.23	5.43

Concerning the results themselves, we observe that we obtain consistent generalization from our training data to our unseen test classes and more realistic object scans. For many objects, rotation error is around 8° on average, with a substantially lower median error, and we observe a similar error distribution for validation and test instances (Fig. 4). We also observe generally low translation errors, and that our confidence ranking is consistently able to achieve rotation errors within a few degrees of the oracle. This ranking is also on par with the oracle in the IOU metrics. From the IOU metrics, we also see that our rotation estimates are generally high quality, achieving 80–90% IOU for nearly all test cases. Translation fares slightly worse, especially for the direction estimate, for which the IOU metric is very sensitive. We observe lower IOUs for both rotation and translation (rightmost column), which is expected since it reflects the full network output. See the supplementary for additional RealScan visualizations.

Qualitatively, our network understandably performs worse for objects with rotational symmetry, for example the 3D Warehouse lamps and the RealScan plastic cup. In the latter case, while the cup has a handle, this handle is not always visible in the input images, and so an unambiguous pose estimate cannot be determined. IOU is also a conservative metric for pose estimation, especially for thin structures like the RealScan eyeglasses and spoon, because even with a perfect rotation estimate, a small amount of translation error can cause the reprojection to shift considerably.

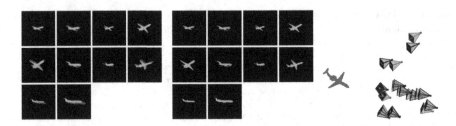

Fig. 3. Estimated poses for 10 silhouette masks of an airplane. Left: Object repro-jections by our cameras (red) versus the original input masks (green), ordered from greatest confidence (top left) to least (bottom right). Middle: Visual hull projection for our method (yellow) versus the original input masks (green), with the same ordering. Right: Predicted poses (blue frusta) relative to GT poses (red frusta), with the target object mesh in red. (Color figure online)

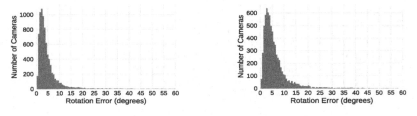

Fig. 4. Histogram of rotation errors for all test instances of airplanes (left, training class) and cars (right, unseen class).

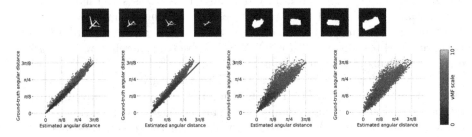

Fig. 5. Predicted versus ground-truth two-view angular distances for a validation class instance (airplane, left two columns) and an unseen class instance (car, right two columns), both from top-down viewpoints. Each point represents an image pair from 100 total images of the object, colored according to the image in the pair with the higher uncertainty (vMF angular spread at 95% of the vMF CDF). Top row: Example silhouettes. Second row: (1, 3) Result from running our network on 2000 samples of 10 images, showing the average error and lowest confidence per pair. (2, 4) Result from running our network once with all 100 images as input.

As for real-world datasets, results for the DTU and Glass Figurines datasets are presented in Tables 3 and 4, respectively. Different from the previous experi-ments, we provided our network with all 15 DTU images as input. We observe low rotation and translation errors for the majority of the objects in both datasets.

Comparison to Reconstruction Methods. For the Glass Figurines dataset, we also compare our method to GNeRF [34], a recent deep method that can optimize per-scene camera poses without accurate initialization. While GNeRF is built for color images, the authors also showed an example result on silhouettes. We evaluate both masked color images and silhouettes in Table 4. For this dataset, GNeRF performs much worse in pose estimation. This is understandable due to the limited size (10 images) and pose distribution of each image set. The NeRF is accordingly unable to generalize to novel viewpoints, especially for silhouettes where cross-view occupancy constraints must be leveraged. GNeRF also must train on a single image set at a time and takes hours to converge. In contrast, our network runs in a single pass without any additional training.

We also note in Table 4 whether COLMAP's SfM algorithm [48] could process the masked color images. When COLMAP succeeded, its poses tended to be very accurate (see supplementary). However, due to the lack of a consistent object appearance or background, COLMAP failed to reconstruct 5 of the 11 objects.

Pairwise Angular Distances. We conducted a small experiment to compare our method against the method of Littwin et al. [31], which is the only method we are aware of that can jointly estimate multi-view camera poses (albeit only relative rotations) for a collection of causally captured object silhouettes. We unfortunately were unable to obtain a copy of their implementation or data, and so we instead provide a qualitative comparison of Fig. 5 versus Fig. 2 in [31]. In the first and third graphs in Fig. 5, we have rendered 100 images of an object and from this sampled 2000 sets of 10 images. We plot the average estimated angular distance over all samples in which that pair appeared together, and we compare this to the ground-truth distance. Compared to [31], our estimates are much less noisy, and they match the ground truth with at least as much accuracy as [31] for a novel class.

We also show our confidence estimates in Fig. 5 and observe that they correlate to prediction accuracy and ground-truth distance, with nearer relative poses having higher confidence. Although confidence κ (Eq. (1)) is difficult to interpret directly in our loss formulation, we provide a rough sense of its scale by converting to an angular "spread" of the vMF distribution. Specifically, we consider the vMF CDF and, for a given value of κ, compute the angle $\arccos\left(\mathbf{r_i}^T\mathbf{x}\right)$ that covers 95% of the distribution over the surface of the sphere. Put more simply, a darker color in the plot indicates a tighter distribution and higher confidence.

Many Network Inputs. As evidenced by our DTU experiments, our network generalizes to more inputs than it was trained on. In second and fourth graphs in Fig. 5, we take this to the extreme and provide our network with all 100 views of the object. Surprisingly, our network easily handles this configuration, producing similar error distributions to our 10-image samples. Our confidence predictions also have a qualitatively higher sensitivity in this scenario.

Additional Results. We include a number of experiments in our supplementary, including complete results on the 3D Warehouse, RealScan, and Glass Figurines datasets; images of the RealScan IOU errors; and a visualization of the network satisfying epipolar constraints even for a failure case. We also include qualitative results on two real-world scenarios of a single object photographed in different environments: (1) a transparent swan sculpture with manually segmented masks, and (2) a chair with masks segmented via Mask R-CNN [15]. The latter case is a promising example of providing automatically extracted masks to our network. Finally, we include three proof-of-concept results of IDR [66] applied to our pose estimates for DTU. These results indicate that our approach has sufficient accuracy to initialize state-of-the-art reconstruction methods.

Table 4. Example images and mean camera pose accuracy for the Glass Figurines dataset. We compare our method to GNeRF [34] with color images and with silhouette inputs ([34]-S), and we note if SfM [48] succeeded or failed for the dataset. GNeRF failures are marked with dashes. See the supplementary for more information.

Object	SfM [48]	R [°] [34]	[34]-S	Ours	t_s [ratio] [34]	[34]-S	Ours	t_d [°] [34]	[34]-S	Ours
brown sq.	✓	18.71	–	3.74	0.04	–	0.03	7.12	–	2.59
dog (c.)	✓	14.00	18.43	6.92	0.13	0.12	0.04	3.02	2.19	4.88
dog (p.)	✗	17.56	–	8.14	0.05	–	0.02	9.36	–	7.32
dolphin	✗	24.70	–	5.10	0.03	–	0.03	7.52	–	3.01
flamingo	✓	21.05	–	6.13	0.10	–	0.05	2.97	–	2.02
flower	✗	–	17.27	3.94	–	0.06	0.02	–	16.79	3.33
frog	✓	21.66	15.48	3.59	0.02	0.07	0.02	2.40	11.41	0.91
parrot	✓	27.40	11.92	21.16	0.05	0.04	0.03	7.24	8.98	1.46
penguin	✗	20.04	–	9.09	0.03	–	0.04	3.38	–	1.97
rabbit	✗	25.32	20.75	5.38	0.02	0.05	0.02	6.38	1.96	3.21
snake	✓	29.85	24.59	11.29	0.05	0.05	0.02	7.85	4.24	1.90

5   Conclusion

The experimental results above support our hypothesis that neural networks can be trained to regress relative pose information, as well as pose confidences, for a given set of silhouette images of an unknown object. Our network model generalizes well to novel object classes and from the synthetic to the real domain. Although we train on a fixed number of 10 images, we observe that our network can capably regress poses for many more inputs at a time.

While the benefit of silhouette constraints for pose estimation has long been recognized, our work shows that silhouette cues on their own can effectively initialize pose estimates for state-of-the-art 3D reconstruction methods on untextured objects. Our work also suggests that permutation-equivariant processing may prove to be an invaluable tool in many-view object reconstruction pipelines, and that multi-view reasoning in neural networks (*e.g.*, aggregating features over all inputs in our pipeline) can yield more-robust estimates compared to two-view methods for 6DOF pose regression, particularly if confidence is also captured.

One limitation of our current work is its reliance on pre-segmented masks. Since our network takes masks as input, however, it could be integrated into an end-to-end pipeline that starts with an object segmentation network, applies a silhouette-based pose estimation, and then performs additional color-image-based pose refinement. Other future work includes leveraging symmetries and texture to resolve silhouette ambiguities when they arise. Also, while we show promising results on medium-baseline views, more work is needed to achieve full generalization w.r.t rotation, *e.g.*, in scenarios where object is viewed from opposite sides, or where the images have substantial relative roll.

References

1. Billings, G., Johnson-Roberson, M.: SilhoNet: an RGB method for 6d object pose estimation. IEEE Rob. Autom. Lett. 4(4), 3727–3734 (2019)
2. Cai, M., Reid, I.: Reconstruct locally, localize globally: a model free method for object pose estimation. In: Computer Vision and Pattern Recognition (CVPR), pp. 3153–3163 (2020)
3. Cashman, T.J., Fitzgibbon, A.W.: What shape are dolphins? Building 3D morphable models from 2D images. IEEE Trans. Pattern Anal. Mach. Intell. (PAMI) **35**(1), 232–244 (2012)
4. Choy, C.B., Xu, D., Gwak, J.Y., Chen, K., Savarese, S.: 3D-R2N2: a unified approach for single and multi-view 3D object reconstruction. In: Leibe, B., Matas, J., Sebe, N., Welling, M. (eds.) ECCV 2016. LNCS, vol. 9912, pp. 628–644. Springer, Cham (2016). https://doi.org/10.1007/978-3-319-46484-8_38
5. Do, T.T., Pham, T., Cai, M., Reid, I.: Real-time monocular object instance 6D pose estimation. In: British Machine Vision Conference (BMVC) (2019)
6. Fan, H., Su, H., Guibas, L.J.: A point set generation network for 3D object reconstruction from a single image. In: Computer Vision and Pattern Recognition (CVPR), pp. 605–613 (2017)
7. Forbes, K., Nicolls, F., de Jager, G., Voigt, A.: Shape-from-silhouette with two mirrors and an uncalibrated camera. In: Leonardis, A., Bischof, H., Pinz, A. (eds.) ECCV 2006. LNCS, vol. 3952, pp. 165–178. Springer, Heidelberg (2006). https://doi.org/10.1007/11744047_13
8. Forbes, K., Voigt, A., Bodika, N., et al.: Using silhouette consistency constraints to build 3D models. In: Pattern Recongnition Associaton of South Africa (PRASA), pp. 33–38 (2003)
9. Garrido-Jurado, S., Munoz-Salinas, R., Madrid-Cuevas, F.J., Medina-Carnicer, R.: Generation of fiducial marker dictionaries using mixed integer linear programming. Pattern Recogn. **51**, 481–491 (2016)
10. Girdhar, R., Fouhey, D.F., Rodriguez, M., Gupta, A.: Learning a predictable and generative vector representation for objects. In: Leibe, B., Matas, J., Sebe, N., Welling, M. (eds.) ECCV 2016. LNCS, vol. 9910, pp. 484–499. Springer, Cham (2016). https://doi.org/10.1007/978-3-319-46466-4_29
11. Glover, J., Popovic, S.: Bingham procrustean alignment for object detection in clutter. In: International Conference on Intelligent Robots and Systems, pp. 2158–2165. IEEE (2013)
12. Govindu, V.M.: Combining two-view constraints for motion estimation. In: Computer Vision and Pattern Recognition (CVPR), vol. 2, p. II. IEEE (2001)
13. Gu, C., Ren, X.: Discriminative mixture-of-templates for viewpoint classification. In: Daniilidis, K., Maragos, P., Paragios, N. (eds.) ECCV 2010. LNCS, vol. 6315, pp. 408–421. Springer, Heidelberg (2010). https://doi.org/10.1007/978-3-642-15555-0_30
14. Gwak, J., Choy, C.B., Chandraker, M., Garg, A., Savarese, S.: Weakly supervised 3D reconstruction with adversarial constraint. In: International Conference on 3D Vision (3DV), pp. 263–272. IEEE (2017)

15. He, K., Gkioxari, G., Dollár, P., Girshick, R.: Mask R-CNN. In: International Conference on Computer Vision (ICCV), pp. 2961–2969 (2017)
16. Hejrati, M., Ramanan, D.: Analysis by synthesis: 3D object recognition by object reconstruction. In: Computer Vision and Pattern Recognition (CVPR), pp. 2449–2456 (2014)
17. Hernández, C., Schmitt, F., Cipolla, R.: Silhouette coherence for camera calibration under circular motion. IEEE Trans. Pattern Anal. Mach. Intell. (PAMI) **29**(2), 343–349 (2007)
18. Huang, P.H., Lai, S.H.: Contour-based structure from reflection. In: Computer Vision and Pattern Recognition (CVPR), vol. 1, pp. 379–386. IEEE (2006)
19. Huang, P.H., Lai, S.H.: Silhouette-based camera calibration from sparse views under circular motion. In: Computer Vision and Pattern Recognition (CVPR), pp. 1–8. IEEE (2008)
20. Insafutdinov, E., Dosovitskiy, A.: Unsupervised learning of shape and pose with differentiable point clouds. In: Advances in Neural Information Processing Systems (NeurIPS), pp. 2802–2812 (2018)
21. Jensen, R., Dahl, A., Vogiatzis, G., Tola, E., Aanæs, H.: Large scale multi-view stereopsis evaluation. In: Computer Vision and Pattern Recognition (CVPR), pp. 406–413. IEEE (2014)
22. Jeong, Y., Ahn, S., Choy, C., Anandkumar, A., Cho, M., Park, J.: Self-calibrating neural radiance fields. In: Proceedings of the IEEE/CVF International Conference on Computer Vision, pp. 5846–5854 (2021)
23. Joshi, T., Ahuja, N., Ponce, J.: Structure and motion estimation from dynamic silhouettes under perspective projection. In: International Conference on Computer Vision (ICCV), pp. 290–295. IEEE (1995)
24. Kar, A., Tulsiani, S., Carreira, J., Malik, J.: Category-specific object reconstruction from a single image. In: Computer Vision and Pattern Recognition (CVPR), pp. 1966–1974 (2015)
25. Kasten, Y., Geifman, A., Galun, M., Basri, R.: Algebraic characterization of essential matrices and their averaging in multiview settings. In: Proceedings of the IEEE/CVF International Conference on Computer Vision (ICCV), October 2019
26. Kato, H., Ushiku, Y., Harada, T.: Neural 3D mesh renderer. In: Computer Vision and Pattern Recognition (CVPR), pp. 3907–3916 (2018)
27. Kehl, W., Manhardt, F., Tombari, F., Ilic, S., Navab, N.: SSD-6D making RGB-based 3D detection and 6d pose estimation great again. In: International Conference on Computer Vision (ICCV), pp. 1521–1529 (2017)
28. Kingma, D.P., Ba, J.: Adam: a method for stochastic optimization. arXiv preprint arXiv:1412.6980 (2014)
29. Levoy, M., et al.: The digital Michelangelo project: 3D scanning of large statues. In: Conference on Computer Graphics and Interactive Techniques, pp. 131–144 (2000)
30. Lin, C.H., Ma, W.C., Torralba, A., Lucey, S.: BARF: bundle-adjusting neural radiance fields. arXiv preprint arXiv:2104.06405 (2021)
31. Littwin, E., Averbuch-Elor, H., Cohen-Or, D.: Spherical embedding of inlier silhouette dissimilarities. In: Computer Vision and Pattern Recognition (CVPR), pp. 3855–3863 (2015)
32. López-Sastre, R.J., Tuytelaars, T., Savarese, S.: Deformable part models revisited: a performance evaluation for object category pose estimation. In: International Conference on Computer Vision (ICCV) Workshops, pp. 1052–1059. IEEE (2011)
33. Maron, H., Litany, O., Chechik, G., Fetaya, E.: On learning sets of symmetric elements. In: International Conference on Machine Learning (ICML) (2020)

34. Meng, Q., et al.: GNeRF: GAN-based neural radiance field without posed camera. In: Proceedings of the IEEE/CVF International Conference on Computer Vision (ICCV) (2021)

35. Mescheder, L., Oechsle, M., Niemeyer, M., Nowozin, S., Geiger, A.: Occupancy networks: learning 3D reconstruction in function space. In: Computer Vision and Pattern Recognition (CVPR), pp. 4460–4470 (2019)

36. Mildenhall, B., Srinivasan, P.P., Tancik, M., Barron, J.T., Ramamoorthi, R., Ng, R.: NeRF: representing scenes as neural radiance fields for view synthesis. In: Vedaldi, A., Bischof, H., Brox, T., Frahm, J.-M. (eds.) ECCV 2020. LNCS, vol. 12346, pp. 405–421. Springer, Cham (2020). https://doi.org/10.1007/978-3-030-58452-8_24

37. Moran, D., Koslowsky, H., Kasten, Y., Maron, H., Galun, M., Basri, R.: Deep permutation equivariant structure from motion. In: Proceedings of the IEEE/CVF International Conference on Computer Vision (ICCV), pp. 5976–5986, October 2021

38. Newcombe, R.A., et al.: KinectFusion: real-time dense surface mapping and tracking. In: International Symposium on Mixed and Augmented Reality (ISMAR), pp. 127–136. IEEE (2011)

39. Niemeyer, M., Mescheder, L., Oechsle, M., Geiger, A.: Differentiable volumetric rendering: learning implicit 3D representations without 3D supervision. In: Computer Vision and Pattern Recognition (CVPR) (2020)

40. Pepik, B., Stark, M., Gehler, P., Schiele, B.: Teaching 3D geometry to deformable part models. In: Computer Vision and Pattern Recognition (CVPR), pp. 3362–3369. IEEE (2012)

41. Poirson, P., Ammirato, P., Fu, C.Y., Liu, W., Kosecka, J., Berg, A.C.: Fast single shot detection and pose estimation. In: International Conference on 3D Vision (3DV), pp. 676–684. IEEE (2016)

42. Prisacariu, V.A., Kähler, O., Murray, D.W., Reid, I.D.: Simultaneous 3D tracking and reconstruction on a mobile phone. In: International Symposium on Mixed and Augmented Reality (ISMAR), pp. 89–98. IEEE (2013)

43. Rad, M., Lepetit, V.: BB8: a scalable, accurate, robust to partial occlusion method for predicting the 3D poses of challenging objects without using depth. In: International Conference on Computer Vision (ICCV), pp. 3828–3836 (2017)

44. Rezende, D.J., Eslami, S., Mohamed, S., Battaglia, P., Jaderberg, M., Heess, N.: Unsupervised learning of 3D structure from images. In: Advances in Neural Information Processing Systems (NeurIPS), vol. 29, pp. 4996–5004 (2016)

45. Romero-Ramirez, F.J., Muñoz-Salinas, R., Medina-Carnicer, R.: Speeded up detection of squared fiducial markers. Image Vis. Comput. **76**, 38–47 (2018)

46. Schmitt, C., Donne, S., Riegler, G., Koltun, V., Geiger, A.: On joint estimation of pose, geometry and svBRDF from a handheld scanner. In: Computer Vision and Pattern Recognition (CVPR), pp. 3493–3503 (2020)

47. Schönberger, J.L., Zheng, E., Frahm, J.-M., Pollefeys, M.: Pixelwise view selection for unstructured multi-view stereo. In: Leibe, B., Matas, J., Sebe, N., Welling, M. (eds.) ECCV 2016. LNCS, vol. 9907, pp. 501–518. Springer, Cham (2016). https://doi.org/10.1007/978-3-319-46487-9_31

48. Schönberger, J.L., Frahm, J.M.: Structure-from-motion revisited. In: Conference on Computer Vision and Pattern Recognition (CVPR) (2016)

49. Sinha, S.N., Pollefeys, M., McMillan, L.: Camera network calibration from dynamic silhouettes. In: Computer Vision and Pattern Recognition (CVPR), vol. 1, p. I. IEEE (2004)

146   Y. Kasten et al.

50. Sinha, S.N., Steedly, D., Szeliski, R.: A multi-stage linear approach to structure from motion. In: Kutulakos, K.N. (ed.) ECCV 2010. LNCS, vol. 6554, pp. 267–281. Springer, Heidelberg (2012). https://doi.org/10.1007/978-3-642-35740-4_21
51. Song, C., Song, J., Huang, Q.: HybridPose: 6D object pose estimation under hybrid representations. In: Computer Vision and Pattern Recognition (CVPR), pp. 431–440 (2020)
52. Sra, S.: Directional statistics in machine learning: a brief review. In: Applied Directional Statistics: Modern Methods and Case Studies, p. 225 (2018)
53. Tanskanen, P., Kolev, K., Meier, L., Camposeco, F., Saurer, O., Pollefeys, M.: Live metric 3D reconstruction on mobile phones. In: International Conference on Computer Vision (ICCV), pp. 65–72 (2013)
54. Tekin, B., Sinha, S.N., Fua, P.: Real-time seamless single shot 6d object pose prediction. In: Computer Vision and Pattern Recognition (CVPR), pp. 292–301 (2018)
55. Tulsiani, S., Efros, A.A., Malik, J.: Multi-view consistency as supervisory signal for learning shape and pose prediction. In: Computer Vision and Pattern Recognition (CVPR), pp. 2897–2905 (2018)
56. Tulsiani, S., Zhou, T., Efros, A.A., Malik, J.: Multi-view supervision for single-view reconstruction via differentiable ray consistency. In: Computer Vision and Pattern Recognition (CVPR), pp. 2626–2634 (2017)
57. Wang, Z., Wu, S., Xie, W., Chen, M., Prisacariu, V.A.: NeRF−: neural radiance fields without known camera parameters. arXiv preprint arXiv:2102.07064 (2021)
58. Wiles, O., Zisserman, A.: SilNet: single-and multi-view reconstruction by learning from silhouettes. In: British Machine Vision Conference (BMVC) (2017)
59. Wong, K.Y., Cipolla, R.: Structure and motion from silhouettes. In: International Conference on Computer Vision (ICCV), vol. 2, pp. 217–222. IEEE (2001)
60. Wong, K.Y., Cipolla, R.: Reconstruction of sculpture from its profiles with unknown camera positions. IEEE Trans. Image Process. 13(3), 381–389 (2004)
61. Wu, J., Wang, Y., Xue, T., Sun, X., Freeman, B., Tenenbaum, J.: MarrNet: 3D shape reconstruction via 2.5D sketches. In: Advances in Neural Information Processing Systems (NeurIPS), pp. 540–550 (2017)
62. Wu, J., et al.: Single image 3D interpreter network. In: Leibe, B., Matas, J., Sebe, N., Welling, M. (eds.) ECCV 2016. LNCS, vol. 9910, pp. 365–382. Springer, Cham (2016). https://doi.org/10.1007/978-3-319-46466-4_22
63. Xiang, Y., Schmidt, T., Narayanan, V., Fox, D.: PoseCNN: a convolutional neural network for 6D object pose estimation in cluttered scenes. arXiv preprint arXiv:1711.00199 (2017)
64. Xiao, Y., Qiu, X., Langlois, P.A., Aubry, M., Marlet, R.: Pose from shape: deep pose estimation for arbitrary 3D objects. In: British Machine Vision Conference (BMVC) (2019)
65. Yan, X., Yang, J., Yumer, E., Guo, Y., Lee, H.: Perspective transformer nets: Learning single-view 3D object reconstruction without 3D supervision. In: Advances in Neural Information Processing Systems (NeurIPS), pp. 1696–1704 (2016)
66. Yariv, L., et al.: Multiview neural surface reconstruction by disentangling geometry and appearance. In: Advances in Neural Information Processing Systems (NeurIPS), vol. 33 (2020)
67. Yen-Chen, L., Florence, P., Barron, J.T., Rodriguez, A., Isola, P., Lin, T.Y.: iNeRF: inverting neural radiance fields for pose estimation. In: IEEE/RSJ International Conference on Intelligent Robots and Systems (IROS) (2021)

68. Zaheer, M., Kottur, S., Ravanbakhsh, S., Poczos, B., Salakhutdinov, R.R., Smola, A.J.: Deep sets. In: Advances in neural information processing systems (NeurIPS), pp. 3391–3401 (2017)
69. Zhou, Y., Barnes, C., Lu, J., Yang, J., Li, H.: On the continuity of rotation representations in neural networks. In: Computer Vision and Pattern Recognition (CVPR) (2019)
70. Zhu, R., Kiani Galoogahi, H., Wang, C., Lucey, S.: Rethinking reprojection: closing the loop for pose-aware shape reconstruction from a single image. In: International Conference on Computer Vision (ICCV), pp. 57–65 (2017)

TransNet: Category-Level Transparent Object Pose Estimation

Huijie Zhang$^{(\boxtimes)}$ (iD), Anthony Opipari (iD), Xiaotong Chen (iD), Jiyue Zhu (iD), Zeren Yu (iD), and Odest Chadwicke Jenkins (iD)

University of Michigan, Ann Arbor, MI 48109, USA
{huijiezh,topipari,cxt,cormaczh,yuzeren,ocj}@umich.edu

Abstract. Transparent objects present multiple distinct challenges to visual perception systems. First, their lack of distinguishing visual features makes transparent objects harder to detect and localize than opaque objects. Even humans find certain transparent surfaces with little specular reflection or refraction, e.g. glass doors, difficult to perceive. A second challenge is that common depth sensors typically used for opaque object perception cannot obtain accurate depth measurements on transparent objects due to their unique reflective properties. Stemming from these challenges, we observe that transparent object instances within the same category (e.g. cups) look more similar to each other than to ordinary opaque objects of that same category. Given this observation, the present paper sets out to explore the possibility of category-level transparent object pose estimation rather than instance-level pose estimation. We propose *TransNet*, a two-stage pipeline that learns to estimate category-level transparent object pose using localized depth completion and surface normal estimation. TransNet is evaluated in terms of pose estimation accuracy on a recent, large-scale transparent object dataset and compared to a state-of-the-art category-level pose estimation approach. Results from this comparison demonstrate that TransNet achieves improved pose estimation accuracy on transparent objects and key findings from the included ablation studies suggest future directions for performance improvements. The project webpage is available at: https://progress.eecs.umich.edu/projects/transnet/.

Keywords: Transparent objects · Category-level object pose estimation · Depth completion · Surface normal estimation

1 Introduction

From glass doors and windows to kitchenware and all kinds of containers, transparent materials are prevalent throughout daily life. Thus, perceiving the pose (position and orientation) of transparent objects is a crucial capability for autonomous perception systems seeking to interact with their environment. However, transparent objects present unique perception challenges both in the RGB and depth domains. As shown in Fig. 2, for RGB, the color appearance

L. Karlinsky et al. (Eds.): ECCV 2022 Workshops, LNCS 13808, pp. 148–164, 2023.
https://doi.org/10.1007/978-3-031-25085-9_9

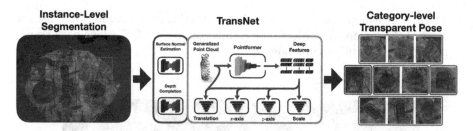

Instance-Level Segmentation   **TransNet**   **Category-level Transparent Pose**

Fig. 1. Overview of TransNet, a pipeline for category-level transparent object pose estimation. Given instance-level segmentation masks as input, TransNet estimates the 6 degrees of freedom pose and scale for each transparent object in the image. Internally, TransNet uses surface normal estimation, depth completion, and a transformer-based architecture for accurate pose estimation despite noisy sensor data.

of transparent objects is highly dependent on the background, viewing angle, material, lighting condition, etc. due to light reflection and refraction effects. For depth, common commercially available depth sensors record mostly invalid or inaccurate depth values within the region of transparency. Such visual challenges, especially missing detection in the depth domain, pose severe problems for autonomous object manipulation and obstacle avoidance tasks. This paper sets out to address these problems by studying how category-level transparent object pose estimation may be achieved using end-to-end learning.

Recent works have shown promising results on grasping transparent objects by completing the missing depth values followed by the use of a geometry-based grasp engine [9,12,29], or transfer learning from RGB-based grasping neural networks [36]. For more advanced manipulation tasks such as rigid body pick-and-place or liquid pouring, geometry-based estimations, such as symmetrical axes, edges [27] or object poses [26], are required to model the manipulation trajectories. Instance-level transparent object poses could be estimated from key-points on stereo RGB images [23,24] or directly from a single RGB-D image [38] with support plane assumptions. Recently emerged large-scale transparent object datasets [6,9,23,29,39] pave the way for addressing the problem using deep learning.

In this work, we aim to extend the frontier of 3D transparent object perception with three primary contributions.

- First, we explore the importance of depth completion and surface normal estimation in transparent object pose estimation. Results from these studies indicate the relative importance of each modality and their analysis suggests promising directions for follow-on studies.
- Second, we introduce *TransNet*, a category-level pose estimation pipeline for transparent objects as illustrated in Fig. 1. It utilizes surface normal estimation, depth completion, and a transformer-based architecture to estimate transparent objects' 6D poses and scales.

– Third, we demonstrate that TransNet outperforms a baseline that uses a state-of-the-art opaque object pose estimation approach [7] along with transparent object depth completion [9].

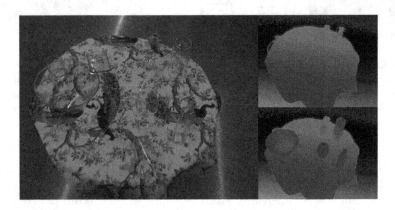

Fig. 2. Challenge for transparent object perception. Images are from Clearpose dataset [6]. The left is an RGB image. The top right is the raw depth image and the bottom right is the ground truth depth image.

2   Related Works

2.1   Transparent Object Visual Perception for Manipulation

Transparent objects need to be perceived before being manipulated. Lai *et al.* [18] and Khaing *et al.* [16] developed CNN models to detect transparent objects from RGB images. Xie *et al.* [37] proposed a deep segmentation model that achieved state-of-the-art segmentation accuracy. ClearGrasp [29] employed depth completion for use with pose estimation on robotic grasping tasks, where they trained three DeepLabv3+ [4] models to perform image segmentation, surface normal estimation, and boundary segmentation. Follow-on studies developed different approaches for depth completion, including implicit functions [47], NeRF features [12], combined point cloud and depth features [39], adversarial learning [30], multi-view geometry [1], and RGB image completion [9]. Without completing depth, Weng *et al.* [36] proposed a method to transfer the learned grasping policy from the RGB domain to the raw sensor depth domain. For instance-level pose estimation, Xu *et al.* [38] utilized segmentation, surface normal, and image coordinate UV-map as input to a network similar to [32] that can estimate 6 DOF object pose. Keypose [24] was proposed to estimate 2D keypoints and regress object poses from stereo images using triangulation. For other special sensors, Xu *et al.* [40] used light-field images to do segmentation using a graph-cut-based approach. Kalra *et al.* [15] trained Mask R-CNN [11] using polarization images as input to outperform the baseline that was trained on only RGB images by a large margin. Zhou *et al.* [44–46] employed light-field images to learn features for robotic grasping and object pose estimation. Along with the proposed methods, massive datasets, across different sensors and both synthetic and real-world domains, have been collected and made public for various related tasks [6,9,15,23,24,29,37,39,44,47]. Compared with these previous works, and to the best of our knowledge we propose the first category-level pose estimation approach for transparent objects. Notably, the proposed approach provides reliable 6D pose and scale estimates across instances with similar shapes.

2.2   Opaque Object Category-Level Pose Estimation

Category-level object pose estimation is aimed at estimating unseen objects' 6D pose within seen categories, together with their scales or canonical shape. To the best of our knowledge, there is not currently any category-level pose estimation works focusing on transparent objects, and the works mentioned below mostly consider opaque objects. They won't work well for transparency due to their dependence on accurate depth. Wang *et al.* [35] introduced the Normalized Object Coordinate Space (NOCS) for dense 3D correspondence learning, and used the Umeyama algorithm [33] to solve the object pose and scale. They also contributed both a synthetic and a real dataset used extensively by the following works for benchmarking. Later, Li *et al.* [19] extended the idea towards articulated objects. To simultaneously reconstruct the canonical point cloud and estimate the pose, Chen *et al.* [2] proposed a method based on canonical shape space (CASS). Tian *et al.* [31] learned category-specific shape priors from an autoencoder, and demonstrated its power for pose estimation and shape completion. 6D-ViT [48] and ACR-Pose [8] extended this idea by utilizing pyramid visual transformer (PVT) and generative adversarial network (GAN) [10] respectively. Structure-guided prior adaptation (SGPA) [3] utilized a transformer architecture for a dynamic shape prior adaptation. Other than learning a dense correspondence, FS-Net [5] regressed the pose parameters directly, and it proposed to learn two orthogonal axes for 3D orientation. Also, it contributed to an efficient data augmentation process for depth-only approaches. GPV-Pose [7] further improved FS-Net by adding a geometric consistency loss between 3D bounding boxes, reconstruction, and pose. Also with depth as the only input, category-level point pair feature (CPPF) [42] could reduce the sim-to-real gap by learning deep point pairs features. DualPoseNet [20] benefited from rotation-invariant embedding for category-level pose estimation. Differing from other works using segmentation networks to crop image patches as the first stage, CenterSnap [13] presented a single-stage approach for the prediction of 3D shape, 6D pose, and size.

Compared with opaque objects, we find the main challenge to perceive transparent objects is the poor quality of input depth. Thus, the proposed TransNet takes inspiration from the above category-level pose estimation works regarding feature embedding and architecture design. More specifically, TransNet leverages both Pointformer from PVT and the pose decoder from FS-Net and GPV-Pose. In the following section, the TransNet architecture is described, focusing on how to integrate the single-view depth completion module and utilize imperfect depth predictions to learn pose estimates of transparent objects.

3   TransNet

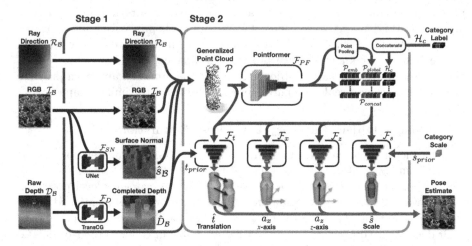

Fig. 3. Architecture for TransNet. TransNet is a two-stage deep neural network for category-level transparent object pose estimation. The first stage uses an object instance segmentation (from Mask R-CNN [11], which is not included in the diagram) to generate patches of RGB-D then used as input to a depth completion and a surface normal estimation network (RGB only). The second stage uses randomly sampled pixels within the objects' segmentation mask to generate a generalized point cloud formed as the per-pixel concatenation of ray direction, RGB, surface normal, and completed depth features. Pointformer [48], a transformer-based point cloud embedding architecture, transforms the generalized point cloud into high-dimensional features. A concatenation of embedding features, global features, and a one-hot category label (from Mask R-CNN) is provided for the pose estimation module. The pose estimation module is composed of four decoders, one each for translation, x-axis, z-axis, and scale regression respectively. Finally, the estimated object pose is recovered and returned as output.

Given an input RGB-D pair $(\mathcal{I}, \mathcal{D})$, our goal is to predict objects' 6D rigid body transformations $[\mathbf{R}|\mathbf{t}]$ and 3D scales $\mathbf{s}$ in the camera coordinate frame, where $\mathbf{R} \in SO(3), \mathbf{t} \in \mathbb{R}^3$ and $\mathbf{s} \in \mathbb{R}^3_+$. In this problem, inaccurate/invalid depth readings exist within the image region corresponding to transparent objects (represented as a binary mask $\mathcal{M}_t$). To approach the category-level pose estimation problem along with inaccurate depth input, we propose a novel two-stage deep neural network pipeline, called **TransNet**.

3.1   Architecture Overview

Following recent work in object pose estimation [5,7,34], we first apply a pretrained instance segmentation module (Mask R-CNN [11]) that has been finetuned on the pose estimation dataset to extract the objects' bounding box patches, masks, and category labels to separate the objects of interest from the entire image.

The first stage of TransNet takes the patches as input and attempts to correct the inaccurate depth posed by transparent objects. Depth completion (TransCG [9]) and surface normal estimation (U-Net [28]) are applied on RGB-D patches to obtain estimated depth-normal pairs. The estimated depth-normal pairs, together with RGB and ray direction patches, are concatenated to feature patches, followed by a random sampling strategy within the instance masks to generate generalized point cloud features.

In the second stage of TransNet, the generalized point cloud is processed through Pointformer [48], a transformer-based point cloud embedding module, to produce concatenated feature vectors. The pose is then separately estimated in four decoder modules for object translation, x-axis, z-axis, and scale respectively. The estimated rotation matrix can be recovered using the estimated two axes. Each component is discussed in more detail in the following sections.

3.2 Object Instance Segmentation

Similar to other categorical pose estimation work [7], we train a Mask R-CNN [11] model on the same dataset used for pose estimation to obtain the object's bounding box $\mathcal{B}$, mask $\mathcal{M}$ and category label $\mathcal{H}_c$. Patches of ray direction $\mathcal{R}_\mathcal{B}$, RGB $\mathcal{I}_\mathcal{B}$ and raw depth $\mathcal{D}_\mathcal{B}$ are extracted from the original data source following bounding box $\mathcal{B}$, before inputting to the first stage of TransNet.

3.3 Transparent Object Depth Completion

Due to light reflection and refraction on transparent material, the depth of transparent objects is very noisy. Therefore, depth completion is necessary to reduce the sensor noise. Given the raw RGB-D patch ($\mathcal{I}_\mathcal{B}$, $\mathcal{D}_\mathcal{B}$) pair and transparent mask $\mathcal{M}_t$ (a intersection of transparent objects' masks within bounding box $\mathcal{B}$), transparent object depth completion $\mathcal{F}_D$ is applied to obtain the completed depth of the transparent region $\{\hat{\mathcal{D}}_{(i,j)}|(i,j) \in \mathcal{M}_t\}$.

Inspired by one state-of-the-art depth completion method, TransCG [9], we incorporate a similar multi-scale depth completion architecture into TransNet.

$$\hat{\mathcal{D}}_\mathcal{B} = \mathcal{F}_D(\mathcal{I}_\mathcal{B}, \mathcal{D}_\mathcal{B}) \tag{1}$$

We use the same training loss as TransCG:

$$\mathcal{L} = \mathcal{L}_d + \lambda_{smooth}\mathcal{L}_s$$
$$\mathcal{L}_d = \frac{1}{N_p} \sum_{p \in \mathcal{M}_t \cap \mathcal{B}} \left\| \hat{\mathcal{D}}_p - \mathcal{D}_p^* \right\|^2$$
$$\mathcal{L}_s = \frac{1}{N_p} \sum_{p \in \mathcal{M}_t \cap \mathcal{B}} \left(1 - \cos \left\langle \mathcal{N}(\hat{\mathcal{D}}_p), \mathcal{N}(\mathcal{D}_p^*) \right\rangle \right) \tag{2}$$

where $\mathcal{D}^*$ is the ground truth depth image patch, $p \in \mathcal{M}_t \cap \mathcal{B}$ represents the transparent region in the patch, $\langle \cdot, \cdot \rangle$ denotes the dot product operator and $\mathcal{N}(\cdot)$ denotes the operator to calculate surface normal from depth. $\mathcal{L}_d$ is L_2 distance between estimated and ground truth depth within the transparency mask. $\mathcal{L}_s$ is the cosine similarity between surface normal calculated from estimated and ground truth depth. λ_{smooth} is the weight between the two losses.

3.4 Transparent Object Surface Normal Estimation

Surface normal estimation $\mathcal{F}_{SN}$ estimates surface normal $\mathcal{S}_\mathcal{B}$ from RGB image $\mathcal{I}_\mathcal{B}$. Although previous category-level pose estimation works [5,7] show that depth is enough to obtain opaque objects' pose, experiments in Sect. 4.3 demonstrate that surface normal is not a redundant input for transparent object pose estimation. Here, we slightly modify U-Net [28] to perform the surface normal estimation.

$$\hat{\mathcal{S}}_\mathcal{B} = \mathcal{F}_{SN}\left(\mathcal{I}_\mathcal{B}\right) \tag{3}$$

We use the cosine similarity loss:

$$\mathcal{L} = \frac{1}{N_p} \sum_{p \in \mathcal{B}} \left(1 - \cos\left\langle \hat{\mathcal{S}}_p, \mathcal{S}_p^* \right\rangle\right) \tag{4}$$

where $p \in \mathcal{B}$ means the loss is applied for all pixels in the bounding box $\mathcal{B}$.

3.5 Generalized Point Cloud

As input to the second stage, generalized point cloud $\mathcal{P} \in \mathbb{R}^{N \times d}$ is a stack of d-dimensional features from the first stage taken at N sample points, inspired from [38]. To be more specific, $d = 10$ in our work. Given the completed depth $\hat{\mathcal{D}}_\mathcal{B}$ and predicted surface normal $\hat{\mathcal{S}}_\mathcal{B}$ from Eq. (1), (3), together with RGB patch $\mathcal{I}_\mathcal{B}$ and ray direction patch $\mathcal{R}_\mathcal{B}$, a concatenated feature patch is given as $\left[\mathcal{I}_\mathcal{B}, \hat{\mathcal{D}}_\mathcal{B}, \hat{\mathcal{S}}_\mathcal{B}, \mathcal{R}_\mathcal{B}\right] \in \mathbb{R}^{H \times W \times 10}$. Here the ray direction $\mathcal{R}$ represents the direction from camera origin to each pixel in the camera frame. For each pixel (u, v):

$$p = \begin{bmatrix} u & v & 1 \end{bmatrix}^T$$
$$\mathcal{R} = \frac{K^{-1}p}{\|K^{-1}p\|^2} \tag{5}$$

where p is the homogeneous UV coordinate in the image plane and K is the camera intrinsic. The UV mapping itself is an important cue when estimating poses from patches [14], as it provides information about the relative position and size of the patches within the overall image. We use ray direction instead of UV mapping because it also contains camera intrinsic information.

We randomly sample N pixels within the transparent mask of the feature patch to obtain the generalized point cloud $\mathcal{P} \in \mathbb{R}^{N \times 10}$. A more detailed experiment in Sect. 4.3 explores the best choice of the generalized point cloud.

3.6   Transformer Feature Embedding

Given generalized point cloud $\mathcal{P}$, we apply an encoder and multi-head decoder strategy to get objects' poses and scales. We use Pointformer [48], a multi-stage transformer-based point cloud embedding method:

$$\mathcal{P}_{emb} = \mathcal{F}_{PF}\left(\mathcal{P}\right) \tag{6}$$

where $\mathcal{P}_{emb} \in \mathbb{R}^{N \times d_{emb}}$ is a high-dimensional feature embedding. During our experiments, we considered other common point cloud embedding methods such as 3D-GCN [21] demonstrating their power in many category-level pose estimation methods [5,7]. During feature aggregation for each point, they use the nearest neighbor algorithm to search nearby points within coordinate space, then calculate new features as a weighted sum of the features within surrounding points. Due to the noisy input $\hat{D}$ from Eq. (1), the nearest neighbor may become unreliable by producing noisy feature embeddings. On the other hand, Pointformer aggregates feature by a transformer-based method. The gradient back-propagates through the whole point cloud. More comparisons and discussions in Sect. 4.2 demonstrate that transformer-based embedding methods are more stable than nearest neighbor-based methods when both are trained on noisy depth data.

Then we use a Point Pooling layer (a multilayer perceptron (MLP) plus max-pooling) to extract the global feature $\mathcal{P}_{global}$, and concatenate it with local feature $\mathcal{P}_{emb}$ and the one-hot category $\mathcal{H}_c$ label from instance segmentation for the decoder:

$$\begin{aligned} \mathcal{P}_{global} &= \text{MaxPool}\left(\text{MLP}\left(\mathcal{P}_{emb}\right)\right) \\ \mathcal{P}_{concat} &= [\mathcal{P}_{emb}, \mathcal{P}_{global}, \mathcal{H}_c] \end{aligned} \tag{7}$$

3.7   Pose and Scale Estimation

After we extract the feature embeddings from multi-modal input, we apply four separate decoders for translation, x-axis, z-axis, and scale estimation.

Translation Residual Estimation. As demonstrated in [5], residual estimation achieves better performance than direct regression by learning the distribution of the residual between the prior and actual value. The translation decoder $\mathcal{F}_t$ learns a 3D translation residual from the object translation prior t_{prior} calculated as the average of predicted 3D coordinate over the sampled pixels in $\mathcal{P}$. To be more specific:

$$\begin{aligned} t_{prior} &= \frac{1}{N_p} \sum_{p \in N} K^{-1} \left[u_p \ v_p \ 1\right]^T \hat{\mathcal{D}}_p \\ \hat{t} &= t_{prior} + \mathcal{F}_t\left(\left[\mathcal{P}_{concat}, \mathcal{P}\right]\right) \end{aligned} \tag{8}$$

where K is the camera intrinsic and u_p, v_p are the 2D pixel coordinate for the selected pixel. We also use the L_1 loss between the ground truth and estimated position:

$$\mathcal{L}_t = \left|\hat{t} - t^*\right| \tag{9}$$

Pose Estimation. Similar to [5], rather than directly regress the rotation matrix R, it is more effective to decouple it into two orthogonal axes and estimate them separately. As shown in Fig. 3, we decouple R into the z-axis a_z (red axis) and x-axis a_x (green axis). Following the strategy of confidence learning in [7], the network learns confidence values to deal with the problem that the regressed two axes are not orthogonal:

$$[\hat{a}_i, c_i] = \mathcal{F}_i\left(\mathcal{P}_{concat}\right), \ i \in \{x, z\}$$
$$\theta_z = \frac{c_x}{c_x + c_z}\left(\theta - \frac{\pi}{2}\right)$$
$$\theta_x = \frac{c_z}{c_x + c_z}\left(\theta - \frac{\pi}{2}\right) \tag{10}$$

where c_x, c_z denote the confidence for the learned axes. θ represents the angle between a_x and a_z. θ_x, θ_z are obtained by solving an optimization problem and then used to rotate the a_x and a_z within their common plane. More details can be found in [7]. For the training loss, first, we use L_1 loss and cosine similarity loss for axis estimation:

$$\mathcal{L}_{r_i} = |\hat{a}_i - a_i^*| + 1 - \langle \hat{a}_i, a_i^* \rangle, \ i \in \{x, z\} \tag{11}$$

Then to constrain the perpendicular relationship between two axes, we add the angular loss:

$$\mathcal{L}_a = \langle \hat{a}_x, \hat{a}_z \rangle \tag{12}$$

To learn the axis confidence, we add the confidence loss, which is the L_1 distance between estimated confidence and exponential L_2 distance between the ground truth and estimated axis:

$$\mathcal{L}_{con_i} = |c_i - \exp\left(\alpha \|\hat{a}_i - a_i^*\|_2\right)|, \ i \in \{x, z\} \tag{13}$$

where α is a constant to scale the distance.

Thus the overall loss for the second stage is:

$$\mathcal{L} = \lambda_s\mathcal{L}_s + \lambda_t\mathcal{L}_t + \lambda_{r_x}\mathcal{L}_{r_x} + \lambda_{r_z}\mathcal{L}_{r_z}$$
$$+ \lambda_{r_a}\mathcal{L}_a + \lambda_{con_x}\mathcal{L}_{con_x} + \lambda_{con_z}\mathcal{L}_{con_z} \tag{14}$$

To deal with object symmetry, we apply specific treatments for different symmetry types. For axial symmetric objects (those that remain the same shape when rotating around one axis), we ignore the loss for the x-axis, i.e., $\mathcal{L}_{con_x}, \mathcal{L}_{r_x}$. For planar symmetric objects (those that remain the same shape when mirrored about one or more planes), we generate all candidate x-axis rotations. For example, for an object symmetric about the $x - z$ plane and $y - z$ plane, rotating the x-axis about the z-axis by π radians will not affect the object's shape. The new x-axis is denoted as a_{x_π} and the loss for the x-axis is defined as the minimum loss of both candidates:

$$\mathcal{L}_x = \min\left(\mathcal{L}_x(a_x), \mathcal{L}_x(a_{x_\pi})\right) \tag{15}$$

Scale Residual Estimation. Similar to the translation decoder, we define the scale prior s_{prior} as the average of scales of all object 3D CAD models within each category. Then the scale of a given instance is calculated as follows:

$$\hat{s} = s_{prior} + \mathcal{F}_s \left(\mathcal{P}_{concat} \right) \tag{16}$$

The loss function is defined as the L_1 loss between the ground truth scale and estimated scale:

$$\mathcal{L}_s = |\hat{s} - s^*| \tag{17}$$

4   Experiments

Dataset. We evaluated TransNet and baseline models on the Clearpose Dataset [6] for categorical transparent object pose estimation. The Clearpose Dataset contains over 350K real-world labeled RGB-D frames in 51 scenes, 9 sets, and around 5M instance annotations covering 63 household objects. We selected 47 objects and categorize them into 6 categories, *bottle, bowl, container, tableware, water cup, wine cup*. We used all the scenes in set2, set4, set5, and set6 for training and scenes in set3 and set7 for validation and testing. The division guaranteed that there were some unseen objects for testing within each category. Overall, we used 190K images for training and 6K for testing. For training depth completion and surface normal estimation, we used the same dataset split.

Implementation Details. Our model was trained in several stages. For all the experiments in this paper, we were using the ground truth instance segmentation as input, which could also be obtained by Mask R-CNN [11]. The image patches were generated from object bounding boxes and re-scaled to a fixed shape of 256×256 pixels. For TransCG, we used AdamW optimizer [25] for training with $\lambda_{smooth} = 0.001$ and the overall learning rate is 0.001 to train the model till converge. For U-Net, we used the Adam optimizer [17] with a learning rate of $1e^{-4}$ to train the model until convergence. For both surface normal estimation and depth completion, the batch size was set to 24 images. The surface normal estimation and depth completion model were frozen during the training of the second stage.

For the second stage, the training hyperparameters for Pointformer followed those used in [48]. We used data augmentation for RGB features and instance mask for sampling generalized point cloud. A batch size of 18 was used. To balance sampling distribution across categories, 3 instance samples were selected randomly for each of 6 categories. We followed GPV-Pose [7] on training hyperparameters. The learning rate for all loss terms were kept the same during training, $\{\lambda_{r_x}, \lambda_{r_z}, \lambda_{r_a}, \lambda_t, \lambda_s, \lambda_{con_x}, \lambda_{con_z}\} = \{8, 8, 4, 8, 8, 1, 1\} \times 0.0001$. We used the Ranger optimizer [22,41,43] and used a linear warm-up for the first 1000 iterations, then used a cosine annealing method at the 0.72 anneal point. All the experiments for pose estimation were trained on a 16G RTX3080 GPU for 30 epochs with 6000 iterations each. All the categories were trained on the same model, instead of one model per category.

Evaluation Metrics. For category-level pose estimation, we followed [5,7] using 3D intersection over union (IoU) between the ground truth and estimated 3D bounding box (we used the estimated scale and pose to draw an estimated 3D bounding box) at 25%, 50% and 75% thresholds. Additionally, we used 5°2 cm, 5°5 cm, 10°5 cm, 10°10 cm as metrics. The numbers in the metrics represent the percentage of the estimations with errors under such degree and distance. For Sect. 4.4, we also used separated translation and rotation metrics: 2 cm, 5 cm, 10 cm, 5°, 10° that calculate percentage with respect to one factor.

For depth completion evaluation, we calculated the root of mean squared error (RMSE), absolute relative error (REL) and mean absolute error (MAE), and used $\delta_{1.05}$, $\delta_{1.10}$, $\delta_{1.25}$ as metrics, while δ_n was calculated as:

$$\delta_n = \frac{1}{N_p} \sum_p \mathbf{I}\left(\max\left(\frac{\hat{\mathcal{D}}_p}{\mathcal{D}_p^*}, \frac{\mathcal{D}_p^*}{\hat{\mathcal{D}}_p} \right) < n \right) \tag{18}$$

where $\mathbf{I}(\cdot)$ represents the indicator function. $\hat{\mathcal{D}}_p$ and $\mathcal{D}_p^*$ mean estimated and ground truth depth for each pixel p.

For surface normal estimation, we calculated RMSE and MAE errors and used 11.25°, 22.5°, and 30° as thresholds. Here 11.25° represents the percentage of estimates with an angular distance less than 11.25° from ground truth surface normal.

4.1 Comparison with Baseline

Table 1. Comparison with the baseline on the Clearpose Dataset.

Method	$3D_{25}\uparrow$	$3D_{50}\uparrow$	$3D_{75}\uparrow$	5°2 cm↑	5°5 cm↑	10°5 cm↑	10°10 cm↑
GPV-Pose	**93.7**	58.3	10.5	0.4	1.5	7.4	9.1
TransNet	90.3	**67.4**	**22.1**	**2.4**	**7.5**	**23.6**	**27.6**

We chose one state-of-the-art categorical opaque object pose estimation model (GPV-Pose [7]) as a baseline, which was trained with estimated depth from TransCG [9] for a fair comparison. From Table 1, TransNet outperformed the baseline in most of the metrics on the Clearpose dataset. $3D_{25}$ is very easy to learn, so there is no huge difference between them. For the rest of the metrics, TransNet achieved around 2× the percentage on $3D_{50}$, 3× on 10°5 cm, 10°10 cm and 5× on 5°5 cm, 5°2 cm over the baseline. Qualitative results are shown in Fig. 4 for TransNet.

4.2 Embedding Method Analysis

In Table 2, we compared the embedding method between 3D-GCN [21] and Pointformer [48] on TransNet. Modalities for generalized point cloud were depth,

Fig. 4. Qualitative results of category-level pose estimates from TransNet. The left column is the original RGB image within our test set and the right column is the pose estimation results. The white bounding box is the ground truth and the colored one is the estimation result. Different colors represent different categories. For axial symmetric objects, because we only care about the scale and z-axis, we use the ground truth x-axis and estimated z-axis to calculate the estimated x-axis, for better visualization. In the figure, there is a pitcher without either ground truth or estimated bounding box because it is not within any of the defined categories, so we ignore it for both training and testing.

RGB and ray direction (without surface normal) for all the trials. The only differences between them were depth type and embedding methods. With ground truth input, 3D-GCN and Pointformer achieved similar results. For some metrics, *i.e.* 5°5 cm, 3D-GCN was even better. But when the ground truth depth was changed to estimated depth (modeling the change from opaque to transparent setting), Pointformer retained much more accuracy than 3D-GCN. Here is our explanation. Like many point cloud embedding methods, 3D-GCN propagates information between nearest neighbors. It is a very efficient method given a point cloud with low noise. But given the completed depth, high noise makes it unstable to pass data among neighbors. While for Pointformer, information is passed through the whole point cloud, no matter how large the noise is. Therefore, given depth information with large uncertainty, the transformer-based embedding method might be more powerful than embedding methods using nearest neighbors.

Table 2. Comparison between different embedding methods

Depth type	Embedding	$3D_{25}$↑	$3D_{50}$↑	$3D_{75}$↑	5°2 cm↑	5°5 cm↑	10°5 cm↑	10°10 cm↑
Ground truth	3D-GCN	**90.0**	**84.1**	43.0	21.4	**48.0**	**61.8**	**64.7**
	Pointformer	**90.0**	81.8	**56.5**	**24.1**	39.3	59.0	60.7
Estimation	3D-GCN	**88.8**	59.8	10.4	0.9	3.4	12.3	15.4
	Pointformer	88.5	**62.2**	**17.6**	**1.6**	**5.0**	**17.4**	**20.9**

4.3 Ablation Study of Generalized Point Cloud

We explored different combinations of feature inputs for the generalized point cloud to find the one most suitable for TransNet. Results are shown in Table 3. For trials 1 and 2, we compared the effect of adding estimated surface normal to the generalized point cloud. All the metrics demonstrated that the inclusion of surface normal does improve the resulting pose estimation accuracy.

Table 3. Ablation study for a different combination of the generalized point cloud. For both trials, we also use RGB as an input feature for the generalized point cloud.

Trial	Depth	Normal	Ray-direction	$3D_{25}$↑	$3D_{50}$↑	$3D_{75}$↑	5°2 cm↑	5°5 cm↑	10°5 cm↑	10°10 cm↑
1	✓		✓	88.5	62.2	17.6	1.6	5.0	17.4	20.9
2	✓	✓	✓	**90.3**	**67.4**	**22.1**	**2.4**	**7.5**	**23.6**	**27.6**

4.4 Depth and Surface Normal Exploration on TransNet

We explored the combination of depth and surface normal with different accuracy. Results in Table 4 and Table 5 show performance for TransCG and U-Net separately. "GT" and "EST" in Table 6 represent ground truth and estimated input for depth and surface normal respectively. From the comparison

Table 4. Accuracy for depth completion on Clearpose dataset. All the metrics are calculated within the transparent mask.

Metric	RMSE↓	REL↓	MAE↓	$\delta_{1.05}$↑	$\delta_{1.10}$↑	$\delta_{1.25}$↑
Value	0.055	0.044	0.041	68.93	89.40	98.93

Table 5. Accuracy for surface normal estimation on Clearpose dataset.

Metric	RMSE↓	MAE↓	11.25°↑	22.5°↑	30°↑
Value	0.1915	0.1334	56.75	88.45	96.64

Table 6. Evaluation for depth and surface normal accuracy on TransNet.

Trial	Depth	Normal	$3D_{25}$↑	$3D_{50}$↑	$3D_{75}$↑	5°2 cm↑	5°5 cm↑	10°5 cm↑	10°10 cm↑	5°↑	10°↑	2 cm↑	5 cm↑	10 cm↑
1	GT	GT	95.1	87.7	66.7	31.8	48.4	66.5	66.7	47.3	66.3	63.3	97.9	99.9
2	GT	EST	90.9	82.1	56.3	23.4	36.5	58.0	59.6	37.3	59.6	53.6	97.2	99.9
3	EST	GT	94.0	83.8	34.3	8.1	29.9	47.8	60.3	37.3	61.8	22.2	77.1	97.4
4	EST	EST	90.3	67.4	22.1	2.4	7.5	23.6	27.6	8.8	28.1	16.6	77.4	96.8

of results among trials 1–3, accurate depth is more essential than surface normal for category-level transparent object pose estimation. For instance, as the ground truth depth changes to the estimated depth from trial 1 to trial 3, 5°2 cm decreases by 23.7. Compared with surface normal estimation, 5°2 cm only decreases by 8.4 between trial 1 and trial 2. More specifically, from decoupled rotation and translation metrics, we can see that 2 cm decreases by 41.1 between trial 1 and trial 3 compared to 9.7 between trial 1 and trial 2, meaning that depth accuracy is more important for translation estimation. Focusing on 2 cm, 5 cm, 10 cm between trial 1 and trial 4, the first metric decreases by 46.7 but the latter two lose much less (20.5 for 5 cm and 3.1 for 10 cm). This can be explained by the result of depth completion accuracy shown in Table 4 (MAE = 0.041 m, between 2 cm and 5 cm). From the comparison of trial 1–4 on metrics 5° and 10°, we can see that either accurate surface normal or accurate depth can support good performance in rotation metrics (for either trial 2 or trial 3, 5° decreases by 10.0 and 10° decreased by around 7). Once we use the estimation version of both, 5° decreases by 38.5 and 10° decreases by 38.2.

5 Conclusions

In this paper, we proposed *TransNet*, a two-stage pipeline for category-level transparent object pose estimation. *TransNet* outperformed a baseline by taking advantage of both state-of-the-art depth completion and opaque object category pose estimation. Ablation studies about multi-modal input and feature embedding modules were performed to guide deeper explorations. In the future, we plan to explore how category information can be used earlier in the network for better accuracy, improve depth completion potentially using additional consistency losses, and extend the model to be category-level across both transparent and opaque instances.

References

1. Chang, J., et al.: Ghostpose*: multi-view pose estimation of transparent objects for robot hand grasping. In: 2021 IEEE/RSJ International Conference on Intelligent Robots and Systems (IROS), pp. 5749–5755. IEEE (2021)
2. Chen, D., Li, J., Wang, Z., Xu, K.: Learning canonical shape space for category-level 6d object pose and size estimation. In: Proceedings of the IEEE/CVF Conference on Computer Vision and Pattern Recognition, pp. 11973–11982 (2020)
3. Chen, K., Dou, Q.: SGPA: structure-guided prior adaptation for category-level 6D object pose estimation. In: Proceedings of the IEEE/CVF International Conference on Computer Vision, pp. 2773–2782 (2021)
4. Chen, L.C., Zhu, Y., Papandreou, G., Schroff, F., Adam, H.: Encoder-decoder with atrous separable convolution for semantic image segmentation. In: Proceedings of the European Conference on Computer Vision (ECCV), pp. 801–818 (2018)
5. Chen, W., Jia, X., Chang, H.J., Duan, J., Shen, L., Leonardis, A.: FS-NET: fast shape-based network for category-level 6d object pose estimation with decoupled rotation mechanism. In: Proceedings of the IEEE/CVF Conference on Computer Vision and Pattern Recognition, pp. 1581–1590 (2021)
6. Chen, X., Zhang, H., Yu, Z., Opipari, A., Jenkins, O.C.: Clearpose: large-scale transparent object dataset and benchmark. arXiv preprint arXiv:2203.03890 (2022)
7. Di, Y., et al.: GPV-pose: category-level object pose estimation via geometry-guided point-wise voting. arXiv preprint arXiv:2203.07918 (2022)
8. Fan, Z., et al.: ACR-pose: adversarial canonical representation reconstruction network for category level 6d object pose estimation. arXiv preprint arXiv:2111.10524 (2021)
9. Fang, H., Fang, H.S., Xu, S., Lu, C.: TRANSCG: a large-scale real-world dataset for transparent object depth completion and grasping. arXiv preprint arXiv:2202.08471 (2022)
10. Goodfellow, I., et al.: Generative adversarial nets. Adv. Neural Inf. Process. Syst. **27** (2014)
11. He, K., Gkioxari, G., Dollár, P., Girshick, R.: Mask R-CNN. In: Proceedings of the IEEE International Conference on Computer Vision, pp. 2961–2969 (2017)
12. Ichnowski, J., Avigal, Y., Kerr, J., Goldberg, K.: DEX-NERF: using a neural radiance field to grasp transparent objects. arXiv preprint arXiv:2110.14217 (2021)
13. Irshad, M.Z., Kollar, T., Laskey, M., Stone, K., Kira, Z.: Centersnap: single-shot multi-object 3d shape reconstruction and categorical 6d pose and size estimation. arXiv preprint arXiv:2203.01929 (2022)
14. Jiang, X., Li, D., Chen, H., Zheng, Y., Zhao, R., Wu, L.: UNI6D: a unified cnn framework without projection breakdown for 6d pose estimation. In: Proceedings of the IEEE/CVF Conference on Computer Vision and Pattern Recognition, pp. 11174–11184 (2022)
15. Kalra, A., Taamazyan, V., Rao, S.K., Venkataraman, K., Raskar, R., Kadambi, A.: Deep polarization cues for transparent object segmentation. In: Proceedings of the IEEE/CVF Conference on Computer Vision and Pattern Recognition, pp. 8602–8611 (2020)

16. Khaing, M.P., Masayuki, M.: Transparent object detection using convolutional neural network. In: Zin, T.T., Lin, J.C.-W. (eds.) ICBDL 2018. AISC, vol. 744, pp. 86–93. Springer, Singapore (2019). https://doi.org/10.1007/978-981-13-0869-7_10

17. Kingma, D.P., Ba, J.: Adam: a method for stochastic optimization. arXiv preprint arXiv:1412.6980 (2014)

18. Lai, P.J., Fuh, C.S.: Transparent object detection using regions with convolutional neural network. In: IPPR Conference on Computer Vision, Graphics, and Image Processing, vol. 2 (2015)

19. Li, X., Wang, H., Yi, L., Guibas, L.J., Abbott, A.L., Song, S.: Category-level articulated object pose estimation. In: Proceedings of the IEEE/CVF Conference on Computer Vision and Pattern Recognition, pp. 3706–3715 (2020)

20. Lin, J., Wei, Z., Li, Z., Xu, S., Jia, K., Li, Y.: DUALPOSENET: category-level 6D object pose and size estimation using dual pose network with refined learning of pose consistency. In: Proceedings of the IEEE/CVF International Conference on Computer Vision, pp. 3560–3569 (2021)

21. Lin, Z.H., Huang, S.Y., Wang, Y.C.F.: Convolution in the cloud: learning deformable kernels in 3D graph convolution networks for point cloud analysis. In: Proceedings of the IEEE/CVF Conference on Computer Vision and Pattern Recognition (CVPR) (2020)

22. Liu, L., et al.: On the variance of the adaptive learning rate and beyond. arXiv preprint arXiv:1908.03265 (2019)

23. Liu, X., Iwase, S., Kitani, K.M.: STEREOBJ-1M: large-scale stereo image dataset for 6d object pose estimation. In: Proceedings of the IEEE/CVF International Conference on Computer Vision, pp. 10870–10879 (2021)

24. Liu, X., Jonschkowski, R., Angelova, A., Konolige, K.: KeyPose: multi-view 3D labeling and keypoint estimation for transparent objects. In: Proceedings of the IEEE/CVF Conference on Computer Vision and Pattern Recognition, pp. 11602–11610 (2020)

25. Loshchilov, I., Hutter, F.: Decoupled weight decay regularization. arXiv preprint arXiv:1711.05101 (2017)

26. Lysenkov, I., Eruhimov, V., Bradski, G.: Recognition and pose estimation of rigid transparent objects with a kinect sensor. Robotics **273**(273–280), 2 (2013)

27. Phillips, C.J., Lecce, M., Daniilidis, K.: Seeing glassware: from edge detection to pose estimation and shape recovery. In: Robotics: Science and Systems, vol. 3, p. 3 (2016)

28. Ronneberger, O., Fischer, P., Brox, T.: U-Net: convolutional networks for biomedical image segmentation. In: Navab, N., Hornegger, J., Wells, W.M., Frangi, A.F. (eds.) MICCAI 2015. LNCS, vol. 9351, pp. 234–241. Springer, Cham (2015). https://doi.org/10.1007/978-3-319-24574-4_28

29. Sajjan, S., et al.: Clear Grasp: 3D shape estimation of transparent objects for manipulation. In: 2020 IEEE International Conference on Robotics and Automation (ICRA), pp. 3634–3642. IEEE (2020)

30. Tang, Y., Chen, J., Yang, Z., Lin, Z., Li, Q., Liu, W.: Depthgrasp: depth completion of transparent objects using self-attentive adversarial network with spectral residual for grasping. In: 2021 IEEE/RSJ International Conference on Intelligent Robots and Systems (IROS), pp. 5710–5716. IEEE (2021)

31. Tian, M., Ang, M.H., Lee, G.H.: Shape prior deformation for categorical 6D object pose and size estimation. In: Vedaldi, A., Bischof, H., Brox, T., Frahm, J.-M. (eds.) ECCV 2020. LNCS, vol. 12366, pp. 530–546. Springer, Cham (2020). https://doi.org/10.1007/978-3-030-58589-1_32

32. Tian, M., Pan, L., Ang, M.H., Lee, G.H.: Robust 6d object pose estimation by learning rgb-d features. In: 2020 IEEE International Conference on Robotics and Automation (ICRA), pp. 6218–6224. IEEE (2020)
33. Umeyama, S.: Least-squares estimation of transformation parameters between two point patterns. IEEE Trans. Pattern Anal. Mach. Intell. **13**(04), 376–380 (1991)
34. Wang, C., et al.: Densefusion: 6d object pose estimation by iterative dense fusion. In: Proceedings of the IEEE/CVF Conference on Computer Vision and Pattern Recognition, pp. 3343–3352 (2019)
35. Wang, H., Sridhar, S., Huang, J., Valentin, J., Song, S., Guibas, L.J.: Normalized object coordinate space for category-level 6d object pose and size estimation. In: Proceedings of the IEEE/CVF Conference on Computer Vision and Pattern Recognition, pp. 2642–2651 (2019)
36. Weng, T., Pallankize, A., Tang, Y., Kroemer, O., Held, D.: Multi-modal transfer learning for grasping transparent and specular objects. IEEE Rob. Autom. Lett. **5**(3), 3791–3798 (2020)
37. Xie, E., Wang, W., Wang, W., Ding, M., Shen, C., Luo, P.: Segmenting transparent objects in the wild. In: Vedaldi, A., Bischof, H., Brox, T., Frahm, J.-M. (eds.) ECCV 2020. LNCS, vol. 12358, pp. 696–711. Springer, Cham (2020). https://doi.org/10.1007/978-3-030-58601-0_41
38. Xu, C., Chen, J., Yao, M., Zhou, J., Zhang, L., Liu, Y.: 6DoF pose estimation of transparent object from a single RGB-D image. Sensors **20**(23), 6790 (2020)
39. Xu, H., Wang, Y.R., Eppel, S., Aspuru-Guzik, A., Shkurti, F., Garg, A.: Seeing glass: joint point cloud and depth completion for transparent objects. arXiv preprint arXiv:2110.00087 (2021)
40. Xu, Y., Nagahara, H., Shimada, A., Taniguchi, R.I.: Transcut: transparent object segmentation from a light-field image. In: Proceedings of the IEEE International Conference on Computer Vision, pp. 3442–3450 (2015)
41. Yong, H., Huang, J., Hua, X., Zhang, L.: Gradient centralization: a new optimization technique for deep neural networks. In: Vedaldi, A., Bischof, H., Brox, T., Frahm, J.-M. (eds.) ECCV 2020. LNCS, vol. 12346, pp. 635–652. Springer, Cham (2020). https://doi.org/10.1007/978-3-030-58452-8_37
42. You, Y., Shi, R., Wang, W., Lu, C.: CPPF: towards robust category-level 9D pose estimation in the wild. arXiv preprint arXiv:2203.03089 (2022)
43. Zhang, M., Lucas, J., Ba, J., Hinton, G.E.: Lookahead optimizer: k steps forward, 1 step back. Adv. Neural Inf. Process. Syst. **32** (2019)
44. Zhou, Z., Chen, X., Jenkins, O.C.: Lit: Light-field inference of transparency for refractive object localization. IEEE Rob. Autom. Lett. **5**(3), 4548–4555 (2020)
45. Zhou, Z., Pan, T., Wu, S., Chang, H., Jenkins, O.C.: Glassloc: plenoptic grasp pose detection in transparent clutter. In: 2019 IEEE/RSJ International Conference on Intelligent Robots and Systems (IROS), pp. 4776–4783. IEEE (2019)
46. Zhou, Z., Sui, Z., Jenkins, O.C.: Plenoptic monte carlo object localization for robot grasping under layered translucency. In: 2018 IEEE/RSJ International Conference on Intelligent Robots and Systems (IROS), pp. 1–8. IEEE (2018)
47. Zhu, L., et al.: Rgb-d local implicit function for depth completion of transparent objects. In: Proceedings of the IEEE/CVF Conference on Computer Vision and Pattern Recognition, pp. 4649–4658 (2021)
48. Zou, L., Huang, Z., Gu, N., Wang, G.: 6d-vit: category-level 6d object pose estimation via transformer-based instance representation learning. arXiv preprint arXiv:2110.04792 (2021)

W34 - Drawings and Abstract Imagery: Representation and Analysis

W34 - Drawings and Abstract Imagery: Representation and Analysis

The DIRA workshop aims to bring together researchers from both industry and academia who are working in diverse areas of research in representation, analysis and applications of abstract images, such as illustrations, drawings, technical diagrams, charts, and plots. The goal of the workshop is to explore and highlight technical challenges, insights and solutions that exist in representation, information extraction, semantic understanding, content-based retrieval, question-answering and various other topics pertaining to representation and analysis of abstract images, which has traditionally lagged behind progress in their natural image counterparts.

October 2022

Diane Oyen
Kushal Kafle
Michal Kucer
Pradyumna Reddy
Cory Scott

Fuse and Attend: Generalized Embedding Learning for Art and Sketches

Ujjal Kr Dutta$^{(\boxtimes)}$ [iD]

Myntra, Bengaluru, India
ukdacad@gmail.com

Abstract. While deep Embedding Learning approaches have witnessed widespread success in multiple computer vision tasks, the state-of-the-art methods for representing natural images need not necessarily perform well on images from other domains, such as paintings, cartoons, and sketch. This is because of the huge shift in the distribution of data from across these domains, as compared to natural images. Domains like sketch often contain sparse informative pixels. However, recognizing objects in such domains is crucial, given multiple relevant applications leveraging such data, for instance, sketch to image retrieval. Thus, achieving an Embedding Learning model that could perform well across multiple domains is not only challenging, but plays a pivotal role in computer vision. To this end, in this paper, we propose a novel Embedding Learning approach with the goal of generalizing across different domains. During training, given a query image from a domain, we employ gated fusion and attention to generate a positive example, which carries a broad notion of the semantics of the query object category (from across multiple domains). By virtue of Contrastive Learning, we pull the embeddings of the query and positive, in order to learn a representation which is robust across domains. At the same time, to teach the model to be discriminative against examples from different semantic categories (across domains), we also maintain a pool of negative embeddings (from different categories). We show the prowess of our method using the DomainBed framework, on the popular PACS (Photo, Art painting, Cartoon, and Sketch) dataset.

1 Introduction

Embedding Learning plays a crucial role in the success of a plethora of computer vision applications, such as object recognition, clustering, content-based retrieval, information extraction, question-answering, semantic understanding, to name a few. It essentially aims at learning a vector/feature representation (*embedding*) of the raw data in question. The goal is to learn a discriminative embedding space, where similar examples are grouped together, while pushing away the dissimilar ones. The notion of similarity varies with an application, and usually pertains to the semantics of the object contained in an image.

While Embedding Learning has seen remarkable success in natural images, their success in image based applications from other domains like paintings, cartoons, and sketch, is yet to reach the full potential. This is mainly because of

© The Author(s), under exclusive license to Springer Nature Switzerland AG 2023
L. Karlinsky et al. (Eds.): ECCV 2022 Workshops, LNCS 13808, pp. 167–183, 2023.
https://doi.org/10.1007/978-3-031-25085-9_10

the fact that the pixel level representation of such domains is very different from natural images. While the latter may benefit more from properties like color, texture, and shading, the former may benefit more from spatial and shape information of the objects. At the same time, they may have sparse pixel information.

In this paper, we try to address this question: "*Can we learn a single, common embedding which is good across multiple domains (be it natural photos, art paintings, cartoons, or sketch)*?" To this end, we pose this problem as a Domain Generalization (DG) approach, wherein, the idea is to leverage a number of labeled domains $\mathcal{D}_1, \cdots, \mathcal{D}_{tr}$ to train a model, which could perform well on any unseen domain $\mathcal{D}_{tr+1}$. Following are the motivations for formulating the problem as DG: i) One may have labeled data from across a few domains, and it would be expected that a model trained on those domains should be able to perform well on data from any unseen domain, ii) If the domains are related (say, containing similar semantics), then it could be beneficial to leverage the common information present in them, to arrive at a better representation.

For example, let us assume that we have labeled data from 3 domains: natural photos, art paintings, and cartoons, with a common set of semantic categories among them. If we could collectively represent the *global information* for a semantic category, say, dog, and force the embedding of a dog image from any domain to lie close to that *global information*, then we could perhaps make our embedding learning model robust even for an unseen domain, such as, sketch. By global information, we refer to those attributes, which are essential to identify a semantic category (eg, tail, nose, ears of a dog category). To achieve this, we make use of gated-fusion and attention mechanism to form a *positive example* which could capture such *global information*. Then, we make use of a Contrastive Learning loss to align the embeddings of a query and the positive corresponding to the semantic class of the query. At the same time, to ensure that examples of different semantic categories are separated far apart, we also make use of a pool of *negative examples*. We show the merits of our method for the classification task, using the DomainBed framework, on the popular PACS (Photo, Art painting, Cartoon, and Sketch) dataset.

2   Proposed Method

Our proposed method is illustrated in Fig. 1. During training, given a query image, we maintain a pool of positive (same semantic class as query, but different domain) and negative examples (different semantic class, irrespective of domain). Given an encoder (eg, ResNet) with some initial model weights, the embedding/feature of a positive from a domain is computed while being oblivious to the feature of the query. Thus, we call it as an *irresponsible representative*.

It is good if we can update this positive feature (into a *responsible representative*) so as to take into account the feature of the query as well, as both of them belong to the same semantic category, albeit from different domains (hence, these features must share some common semantic attribute as well, for instance, a natural image and a sketch image of a dog should contain nose, ear, tail, etc.).

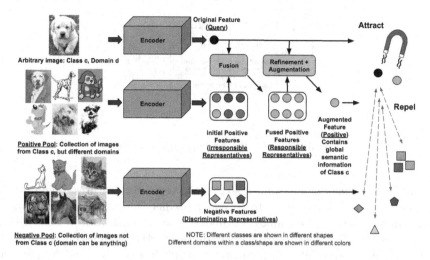

Fig. 1. An illustration of the proposed approach. The figure is best viewed in color. (Color figure online)

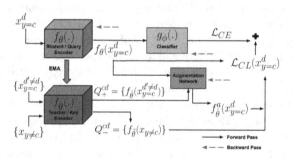

Fig. 2. Architecture of our method.

We propose a learnable (based on the end loss) gated-fusion mechanism to address this. Specifically, the *gated-fusion learns what percentage of original domain information of an irresponsible positive to keep, and what percentage of new information from the domain of the query to be learnt, so as to update and obtain a responsible positive.*

Now that we have a set of *responsible representatives* for a query feature, we can use them to attend the query feature, and use the corresponding attention weights to fuse them together with a linear combination to obtain a final augmented positive feature. Intuitively, this augmented positive may be interpreted as containing global semantic information of the category/class of the query, towards which the query embedding should be pulled closer, to make our encoder robust to domains. This is done with Contrastive Learning. We also push the embeddings for the negative examples away from the query embedding.

Hence, the focus of our work is the contrastive learning based training of a feature Encoder $f_\theta(.)$, that is discriminative enough, of semantically dissimilar

$f_\theta(x^d_{y=c})$

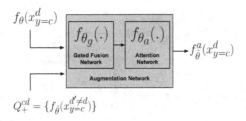

$Q^{cd}_+ = \{f^a_{\tilde\theta}(x^{d'\neq d}_{y=c})\}$

Fig. 3. Blocks of the Augmentation Network.

content, while being domain robust, so that the generated features could directly be utilized by a classifier $g_\phi(.)$ to correctly predict the class label of an input image. In our method, we employ a Student-Teacher framework to learn such a feature Encoder. The overall architecture of our method is illustrated in Fig. 2.

Let us denote an arbitrary example/raw image from class c, and domain d as $x^d_{y=c}$, and its corresponding embedding/feature vector obtained using our (Student/Query) Encoder $f_\theta(.)$, as $f_\theta(x^d_{y=c})$ (Fig. 2). Here, $y = c$ denotes the semantic class label for the example, and θ denotes the learnable parameters of the Query Encoder $f_\theta(.)$. Our objective is to learn θ in such a way that the embeddings $f_\theta(x^d_{y=c})$ and $f_\theta(x^{d'}_{y=c})$ for a pair $(x^d_{y=c}, x^{d'}_{y=c})$ of semantically similar examples (i.e., $y = c$) from two different domains d and d' are grouped together, with the end goal of correctly predicting their class labels with a classifier $g_\phi(.)$, with parameters ϕ. In other words, $f_\theta(.)$ should be *domain robust* (or generalizable).

We maintain a copy of $f_\theta(.)$, denoted as $f_{\tilde\theta}(.)$, which we call as the (Teacher/Key) Encoder. Here, $\tilde\theta$ is obtained as an *Exponential Moving Average* (EMA) of θ, i.e., we first initialize $\tilde\theta$ as $\tilde\theta = \theta$, and then iteratively update it as: $\tilde\theta = \mu\tilde\theta + (1 - \mu)\theta$, $\mu > 0$. In order to make $f_\theta(.)$ *domain robust*, the focus of our method is to leverage $f_{\tilde\theta}(.)$ to obtain an augmented feature (positive) $f^a_{\tilde\theta}(x^d_{y=c})$ corresponding to $f_\theta(x^d_{y=c})$, and pull the embeddings $f_\theta(x^d_{y=c})$ and $f^a_{\tilde\theta}(x^d_{y=c})$ closer to each other.

We obtain $f^a_{\tilde\theta}(x^d_{y=c})$ in such a way that it could capture a broad notion of the semantics of the class $y = c$, to which $x^d_{y=c}$ belongs. To do so, we maintain a pool/set $Q^{cd}_+ = \bigcup_{d'\neq d} Q(x^{d'}_{y=c})$, such that $Q(x^{d'}_{y=c}) = \{f_{\tilde\theta}(x^{d'\neq d}_{y=c})\}$ is a set of embeddings of examples from the same class as $x^d_{y=c}$, but from domain $d' \neq d$. The detailed blocks of the Augmentation Network from Fig. 2 are shown in the Fig. 3. The Augmentation Network takes as input the query embedding $f_\theta(x^d_{y=c})$ and the embeddings of Q^{cd}_+, to produce $f^a_{\tilde\theta}(x^d_{y=c})$. It consists of two major components: i) gated fusion, and ii) attention network, which we discuss next.

2.1   Gated Fusion for Representative Refinement

Q^{cd}_+ is treated as a set of *positive representatives* which take the responsibility of refining their own initial representations (via the Key Encoder) $\{f_{\tilde\theta}(x^{d'\neq d}_{y=c})\}$

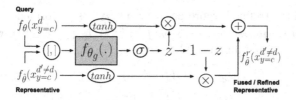

Fig. 4. The Gated Fusion Network takes one (irresponsible) positive representative $f_{\tilde{\theta}}(x_{y=c}^{d'\neq d})$ at a time from Q_+^{cd}, and produces a fused/refined (responsible) positive representative.

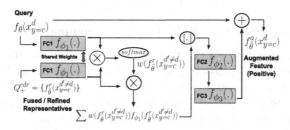

Fig. 5. The Attention Network takes the fused/refined positive representatives to compute attention weights of the query embedding against them, and forms an augmented feature/positive, corresponding to the query.

into $Q_+^{cdr} = \{f_{\tilde{\theta}}^r(x_{y=c}^{d'\neq d})\}$ (we consolidate/abuse the notation by using a single set of same class examples from other domains, to denote the union of sets of examples). This refinement is done by taking into account the representation $f_\theta(x_{y=c}^d)$, using a gated fusion mechanism (Fig. 4). By considering one representative $f_{\tilde{\theta}}(x_{y=c}^{d'\neq d})$ from Q_+^{cd} at a time, the refined/fused representative $f_{\tilde{\theta}}^r(x_{y=c}^{d'\neq d})$ is obtained as:

$$f_{\tilde{\theta}}^r(x_{y=c}^{d'\neq d}) = z \odot tanh(f_\theta(x_{y=c}^d)) + (1-z) \odot tanh(f_{\tilde{\theta}}(x_{y=c}^{d'\neq d})) \tag{1}$$

Here, z is a learnable gate vector obtained as: $z = \sigma(f_{\theta_g}([f_\theta(x_{y=c}^d), f_{\tilde{\theta}}(x_{y=c}^{d'\neq d})]))$, such that θ_g represents parameters of the gated fusion network. Also, $\odot$, $\sigma(.)$, $tanh(.)$ and $[,]$ are the element-wise product, sigmoid, tanh and concatenation operations respectively.

2.2 Attention-Based Query Embedding Refinement and Augmentation for Positive Generation

The refined features from Q_+^{cdr} are then attended by $f_\theta(x_{y=c}^d)$, one representative at a time (Fig. 5), to obtain attention weights: $\{w(f_{\tilde{\theta}}^r(x_{y=c}^{d'\neq d})) = softmax(f_{\phi_1}(f_\theta(x_{y=c}^d))^\top f_{\phi_1}(f_{\tilde{\theta}}^r(x_{y=c}^{d'\neq d})))\}$. Here, $softmax(.)$ is used to normalize the attention weights across the representatives. Using these weights, a linear combination of the (refined) representatives in Q_+^{cdr} is performed to obtain a single *positive representative* $p_{y=c} = \sum w(f_{\hat{\theta}}^r(x_{y=c}^{d'\neq d}))f_{\phi_1}(f_{\hat{\theta}}^r(x_{y=c}^{d'\neq d}))$, which is then used to obtain the final augmented feature as:

$$f_{\hat{\theta}}^a(x_{y=c}^d) = relu(f_{\theta}(x_{y=c}^d)$$
$$+ f_{\phi_3}(relu(f_{\phi_2}([f_{\phi_1}(f_{\theta}(x_{y=c}^d)), p_{y=c}])))). \tag{2}$$

$f_{\hat{\theta}}^a(x_{y=c}^d)$ now captures the accumulated semantic information of the class $y = c$ from across all the domains, and serves as a positive for the query embedding. Here, $\theta_a = \{\phi_1, \phi_2, \phi_3\}$ represents parameters of the attention network.

2.3   Student-Teacher Based Contrastive Learning

The Student-Teacher framework arises in our method due to the fact that the Key Encoder serves as a *Teacher* to provide a *positive* embedding (as a *target*/guidance), with respect to which the query/*anchor* embedding $f_{\theta}(x_{y=c}^d)$ obtained by the (Student/Query) Encoder needs to be pulled closer. The notions of (anchor/query, positive/key) are quite popular in the *embedding learning* literature, where, in order to group similar examples, triplets of examples in the form of (anchor, positive, negative) are sampled. Usually, the anchor and positive are semantically similar (and thus, need to be pulled closer), while the negative is dissimilar to the other two (and thus, needs to be pushed away).

Now, although we have obtained an augmented feature to serve as the positive for the query encoding $f_{\theta}(x_{y=c}^d)$, we also need to push away embeddings of dissimilar examples, in order to learn a robust, yet semantically discriminative embedding. For this purpose, corresponding to each $x_{y=c}^d$, we also maintain a pool/set of negatives denoted as $Q_-^{cd} = \bigcup_{\forall d':d'\neq d,d'=d}\{\bigcup_{c'\neq c} Q(x_{y=c'}^{d'})\} = \{f_{\hat{\theta}}(x_{y\neq c})\}$ (abusing the notation to concisely represent the entire pool of negatives as a single set). Essentially, this pool contains embeddings of example images from other classes, across all the domains (including the same domain as $x_{y=c}^d$). It should be carefully noted that the embeddings in the pools Q_+^{cd} and Q_-^{cd} are obtained using the Key Encoder $f_{\hat{\theta}}(.)$.

Keeping in mind the computational aspects, we maintain a dynamically updated queue $Q^{c'd'} \approx Q(x_{y=c'}^{d'})$ of fixed size *queue_sz*, for each class c', from across all domains d'. While training, for each considered example $x_{y=c}^d$, the pools Q_+^{cd} and Q_-^{cd} are obtained using union from the collection of queues $Q^{c'd'}$. During the network updates, the examples present in a mini-batch are used to replace old examples from the collection $Q^{c'd'}$.

Now, for a given $x_{y=c}^d$, in order to pull its embedding closer to the augmented feature $f_{\hat{\theta}}^a(x_{y=c}^d)$ while moving away the negatives from Q_-^{cd}, we compute an adapted Normalized Temperature-scaled cross entropy (NT-Xent) loss [3] as follows:

$$\mathcal{L}_{CL}(x_{y=c}^d)$$
$$= -\log \frac{\exp(f_{\theta}(x_{y=c}^d)^\top f_{\hat{\theta}}^a(x_{y=c}^d)/\tau)}{\sum_{f_{\hat{\theta}}(x_{y\neq c})\in Q_-^{cd}} \exp(f_{\theta}(x_{y=c}^d)^\top f_{\hat{\theta}}(x_{y\neq c})/\tau)}. \tag{3}$$

Algorithm 1. Pseudocode of RCERM

```
1   def enQ_deQ(queue,data_emb):# enqueue+dequeue
2       queue=torch.cat((queue, data_emb), 0)
3       if queue.size(0) > queue_sz:
4           queue = queue[-queue_sz:]
5
6   ''' In algorithms.py '''
7   class RCERM(Algorithm):
8       def __init__(self):# initialize networks
9       # featurizer, classifier, f_gated, f_attn
10          key_encoder=copy.deepcopy(featurizer)
11          # optim.Adam(featurizer, classifier, f_gated, f_attn)
12      def update(self,minibatch):
13      # minibatch: domainbed styled minibatch
14          loss_cl,all_x,all_y=0,None,None
15          for id_c in range(nclass):
16              for id_d in range(ndomains):
17                  #data_tensor: minibatch(id_c,id_d)
18                  q = featurizer(data_tensor)
19                  all_x = torch.cat((all_x, q), 0)
20                  all_y = torch.cat((all_y, labels), 0)
21                  pos_Q,neg_Q=get_posneg_queues(id_c,id_d)
22                  # Using f_gated+f_attn obtain:
23                  k=self.get_augmented_batch(q,pos_Q)
24                  # .detach() k and l2 normalize q,k
25                  loss_cl += loss_func(q, k, neg_Q)
26                  data_emb=key_encoder(data_tensor)
27                  #.detach()+l2 normalize data_emb
28                  enQ_deQ(queues[id_c][id_d],data_emb)
29          all_pred=classifier(all_x)
30          loss_ce=F.cross_entropy(all_pred, all_y)
31          loss = loss_ce+lambda*loss_cl
32          optim.zero_grad()
33          loss.backward()
34          optim.step()
35          for thtild, tht in (key_encoder.params(),featurizer.params()):
36              thtild=mu*thtild + (1 - mu)*tht
```

Here, $\tau > 0$ denotes the temperature parameter. $\mathcal{L}_{CL}(x^d_{y=c})$ in (3) is summed over all the examples from across all classes c and all domains d to obtain the following aggregated *contrastive loss*:

$$\mathcal{L}_{CL} = \sum_c \sum_d \sum_{x^d_{y=c}} \mathcal{L}_{CL}(x^d_{y=c}). \tag{4}$$

Now, assuming that we have a classifier $g_\phi(.)$ that predicts the class label of an example $x^d_{y=c}$, using the embedding obtained by the domain robust Query Encoder $f_\theta(.)$, the *Empirical Risk Minimization* (ERM) can be approximated by minimizing the following loss:

$$\mathcal{L}_{CE} = \sum_d \sum_c l_{CE}(g_\phi(f_\theta(x^d_{y=c})), y = c). \tag{5}$$

Here, $l_{CE}(.)$ is the standard cross-entropy loss for classification. Then, the total loss for our method can be expressed as:

$$\mathcal{L}_{total} = \mathcal{L}_{CE} + \lambda \mathcal{L}_{CL}. \tag{6}$$

Here, $\lambda > 0$ is a hyperparameter. Thus, the overall optimization problem of our method can be expressed as:

$$\min_{\theta, \phi, \theta_g, \theta_a} \mathcal{L}_{total}. \tag{7}$$

Here, $\theta, \phi, \theta_g, \theta_a$ respectively denote the parameters of the (Query) Encoder $f_\theta(.)$, classifier $g_\phi(.)$, gated-fusion network $f_{\theta_g}(.)$ and the attention network $f_{\theta_a}(.)$. It should be noted that as the Key Encoder $f_{\hat{\theta}}(.)$ is obtained as an EMA of the Query Encoder, we do not backpropagate gradients through it. In Algorithm 1, we provide a pseudo-code, roughly outlining the integration of our method into the PyTorch-based DomainBed framework. We call our method as the Refined Contrastive ERM (RCERM), owing to its refinement based feature augmentation, for positive generation in Contrastive learning.

3   Related Work and Experiments

Dataset: To evaluate the prowess of our proposed method, we make use of the **Photos, Art, Cartoons, and Sketches (PACS) dataset**. It consists of images from 4 different domains: i) Photos (1,670 images), ii) Art Paintings (2,048 images), iii) Cartoons (2,344 images) and iv) Sketches (3,929 images). There are seven semantic categories shared across the domains, namely, 'dog', 'elephant', 'giraffe', 'guitar', 'horse', 'house', 'person'. It is a widely popular dataset to test the robustness of Domain Generalization (DG) models. In particular, *in this work, we are specifically interested in figuring out how our proposed method performs in the domains containing drawings and abstract imagery, such as, Art, Cartoons, and Sketches domains*. The images contained in these domains are significantly different from that of the Photos domain containing natural scene images.

Baseline/Related Methods: Contrary to the traditional supervised learning assumption of the training and test data belonging to an identical distribution, Domain Generalization (DG) methods assume that training data is divided into a number of different, but semantically related domains, with an underlying causal mechanism, and test data could be from a different distribution. To generalize well in unseen data from a different distribution, and account for real-world situations, they try to learn invariance criteria among these domains.

Following are some of the related DG methods that have been used as baselines in this paper:

1. ERM [17]: Naive approach of minimizing consolidated domain errors.
2. IRM [1]: Attempts at learning correlations that are invariant across domains.
3. GroupDRO [14]: Minimizes the worst-case training loss over the domains.
4. Mixup [18]: Pairs of examples from random domains along with their labels are interpolated to perform ERM.
5. MLDG [8]: MAML based meta-learning for DG.

Table 1. Performance of SOTA methods on PACS using the training-domain validation model selection criterion

Method	A	C	P	S	Avg
ERM	84.7 ± 0.4	80.8 ± 0.6	97.2 ± 0.3	79.3 ± 1.0	85.5
IRM	84.8 ± 1.3	76.4 ± 1.1	96.7 ± 0.6	76.1 ± 1.0	83.5
GroupDRO	83.5 ± 0.9	79.1 ± 0.6	96.7 ± 0.3	78.3 ± 2.0	84.4
Mixup	86.1 ± 0.5	78.9 ± 0.8	97.6 ± 0.1	75.8 ± 1.8	84.6
MLDG	85.5 ± 1.4	80.1 ± 1.7	97.4 ± 0.3	76.6 ± 1.1	84.9
CORAL	88.3 ± 0.2	80.0 ± 0.5	97.5 ± 0.3	78.8 ± 1.3	86.2
MMD	86.1 ± 1.4	79.4 ± 0.9	96.6 ± 0.2	76.5 ± 0.5	84.6
DANN	86.4 ± 0.8	77.4 ± 0.8	97.3 ± 0.4	73.5 ± 2.3	83.6
CDANN	84.6 ± 1.8	75.5 ± 0.9	96.8 ± 0.3	73.5 ± 0.6	82.6
MTL	87.5 ± 0.8	77.1 ± 0.5	96.4 ± 0.8	77.3 ± 1.8	84.6
SagNet	87.4 ± 1.0	80.7 ± 0.6	97.1 ± 0.1	80.0 ± 0.4	86.3
ARM	86.8 ± 0.6	76.8 ± 0.5	97.4 ± 0.3	79.3 ± 1.2	85.1
VREx	86.0 ± 1.6	79.1 ± 0.6	96.9 ± 0.5	77.7 ± 1.7	84.9
RSC	85.4 ± 0.8	79.7 ± 1.8	97.6 ± 0.3	78.2 ± 1.2	85.2
AND-mask	85.3 ± 1.4	79.2 ± 2.0	96.9 ± 0.4	76.2 ± 1.4	84.4
SAND-mask	85.8 ± 1.7	79.2 ± 0.8	96.3 ± 0.2	76.9 ± 2.0	84.6
Fishr	88.4 ± 0.2	78.7 ± 0.7	97.0 ± 0.1	77.8 ± 2.0	85.5

6. CORAL [16]: Aligning second-order statistics of a pair of distributions.
7. MMD [9]: MMD alignment of a pair of distributions.
8. DANN [4]: Adversarial approach to learn features to be domain agnostic.
9. CDANN [10]: Variant of DANN conditioned on class labels.
10. MTL [2]: Mean embedding of a domain is used to train a classifier.
11. SagNet [11]: Preserves the image content while randomizing the style.
12. ARM [19]: Meta-learning based adaptation of test time batches.
13. VREx [6]: Variance penalty based IRM approximation.
14. RSC [7]: Iteratively discarding challenging features to improve generalizability.
15. AND-mask [12]: Hessian matching based DG approach.
16. SAND-mask [15]: An enhanced Gradient Masking strategy is employed to perform DG.
17. Fishr [13]: Matching the variances of domain-level gradients.

Table 2. Performance of SOTA methods on PACS using the leave-one-domain-out cross-validation model selection criterion

Method	A	C	P	S	Avg.
ERM	83.2 ± 1.3	76.8 ± 1.7	97.2 ± 0.3	74.8 ± 1.3	83.0
IRM	81.7 ± 2.4	77.0 ± 1.3	96.3 ± 0.2	71.1 ± 2.2	81.5
GroupDRO	84.4 ± 0.7	77.3 ± 0.8	96.8 ± 0.8	75.6 ± 1.4	83.5
Mixup	85.2 ± 1.9	77.0 ± 1.7	96.8 ± 0.8	73.9 ± 1.6	83.2
MLDG	81.4 ± 3.6	77.9 ± 2.3	96.2 ± 0.3	76.1 ± 2.1	82.9
CORAL	80.5 ± 2.8	74.5 ± 0.4	96.8 ± 0.3	78.6 ± 1.4	82.6
MMD	84.9 ± 1.7	75.1 ± 2.0	96.1 ± 0.9	76.5 ± 1.5	83.2
DANN	84.3 ± 2.8	72.4 ± 2.8	96.5 ± 0.8	70.8 ± 1.3	81.0
CDANN	78.3 ± 2.8	73.8 ± 1.6	96.4 ± 0.5	66.8 ± 5.5	78.8
MTL	85.6 ± 1.5	78.9 ± 0.6	97.1 ± 0.3	73.1 ± 2.7	83.7
SagNet	81.1 ± 1.9	75.4 ± 1.3	95.7 ± 0.9	77.2 ± 0.6	82.3
ARM	85.9 ± 0.3	73.3 ± 1.9	95.6 ± 0.4	72.1 ± 2.4	81.7
VREx	81.6 ± 4.0	74.1 ± 0.3	96.9 ± 0.4	72.8 ± 2.1	81.3
RSC	83.7 ± 1.7	82.9 ± 1.1	95.6 ± 0.7	68.1 ± 1.5	82.6

Table 3. Performance of SOTA methods on PACS using the test-domain validation model selection criterion

Method	A	C	P	S	Avg.
ERM	86.5 ± 1.0	81.3 ± 0.6	96.2 ± 0.3	82.7 ± 1.1	86.7
IRM	84.2 ± 0.9	79.7 ± 1.5	95.9 ± 0.4	78.3 ± 2.1	84.5
GroupDRO	87.5 ± 0.5	82.9 ± 0.6	97.1 ± 0.3	81.1 ± 1.2	87.1
Mixup	87.5 ± 0.4	81.6 ± 0.7	97.4 ± 0.2	80.8 ± 0.9	86.8
MLDG	87.0 ± 1.2	82.5 ± 0.9	96.7 ± 0.3	81.2 ± 0.6	86.8
CORAL	86.6 ± 0.8	81.8 ± 0.9	97.1 ± 0.5	82.7 ± 0.6	87.1
MMD	88.1 ± 0.8	82.6 ± 0.7	97.1 ± 0.5	81.2 ± 1.2	87.2
DANN	87.0 ± 0.4	80.3 ± 0.6	96.8 ± 0.3	76.9 ± 1.1	85.2
CDANN	87.7 ± 0.6	80.7 ± 1.2	97.3 ± 0.4	77.6 ± 1.5	85.8
MTL	87.0 ± 0.2	82.7 ± 0.8	96.5 ± 0.7	80.5 ± 0.8	86.7
SagNet	87.4 ± 0.5	81.2 ± 1.2	96.3 ± 0.8	80.7 ± 1.1	86.4
ARM	85.0 ± 1.2	81.4 ± 0.2	95.9 ± 0.3	80.9 ± 0.5	85.8
VREx	87.8 ± 1.2	81.8 ± 0.7	97.4 ± 0.2	82.1 ± 0.7	87.2
RSC	86.0 ± 0.7	81.8 ± 0.9	96.8 ± 0.7	80.4 ± 0.5	86.2
AND-mask	86.4 ± 1.1	80.8 ± 0.9	97.1 ± 0.2	81.3 ± 1.1	86.4
SAND-mask	86.1 ± 0.6	80.3 ± 1.0	97.1 ± 0.3	80.0 ± 1.3	85.9
Fishr	87.9 ± 0.6	80.8 ± 0.5	97.9 ± 0.4	81.1 ± 0.8	86.9

Evaluation Protocol: The presence of multiple domains, the numerous available choices to construct a fair evaluation protocol, due to the multiple ways of forming the training data from across multiple domains makes model selection in DG a non-trivial problem. Inconsistencies in evaluation protocol, network architectures, etc, also makes a fair comparison among the plethora of available approaches difficult. To address this, the recently proposed framework DomainBed [5] was introduced, that ensures an uniform evaluation protocol to inspect and compare DG approaches, by running a large number of hyperparameter and model combinations (called as a *sweep*) automatically.

It also introduces 3 model selection criteria:

1. **Training-domain validation:** The idea is to partition each training domain into a big split and a small split. Sizes of respective big and small splits would be the same for all training domains. The union of the big splits is used to train a model configuration, and the one performing the best in the union of the small splits is chosen as the best model for evaluating on the test data.

2. **Leave-one-out validation:** As the name indicates, it requires to train a model on all train domains except one, and evaluate on the left out domain. This process is repeated by leaving out one domain at a time, and then choosing the final model with the best average performance.

3. **Test-domain validation:** A model is trained on the union of the big splits of the train domains (similar to Training-domain validation), and the final epoch performance is evaluated on a small split of the test data itself. The third criterion though not suited for real-world applications, is often chosen only for evaluating methods.

Performance of State-Of-The-Art (SOTA) approaches on PACS: Following the standard sweep of DomainBed, in Table 1, Table 2 and Table 3, we respectively report the performance of the SOTA DG methods on PACS dataset (in terms of classification accuracy, meaning that a higher value is better). The domains Art painting, Cartoons, Photos (natural images), and Sketches are shown as columns A, C, P and S respectively, along with the average performance of a method across these as column Avg. As also claimed by the DomainBed paper, we observed that *no single method outperforms the classical ERM by more than one point, thus making ERM a reasonable baseline for DG*. Very recently, the Fishr method [13] has been proposed which performs competitive to ERM, on average. The most important observation is the fact that none of the methods performs the best across all the domains, showcasing the difficult nature of the dataset, and the problem of generalization in particular.

3.1   Comparison of Our Method RCERM Against the SOTA ERM and Fishr Methods:

Due to the SOTA performances of the ERM and the Fishr methods, we now compare our proposed method against them. In Table 4, we report the comparison of our method against the ERM method (best method for a column, within a selection criterion, is shown in bold). *For the first two model selection criteria (train domain validation and leave-one-out) our proposed method outperforms the ERM method on all the domains, except on Cartoons*. While the performance gain on natural scene Photos domain is not large, we found that *on Arts and Sketches, our method performs better than ERM by a large margin (upto 2% point). In the sketch domain, our method performs better by 3.7% point over ERM using the leave-one-out criterion*. Our method also performs better on Cartoons than ERM, when feedback from the test set is provided. However, in all other cases, ERM in itself performs quite competitive, in the first place.

Table 4. Comparison of our proposed method against the state-of-the-art ERM method.

Model Selection train-domain validation					
Method	A	C	P	S	Avg.
ERM	84.7 ± 0.4	$\mathbf{80.8} \pm 0.6$	97.2 ± 0.3	79.3 ± 1.0	85.5
RCERM	$\mathbf{86.6} \pm 1.5$	78.1 ± 0.4	$\mathbf{97.3} \pm 0.1$	$\mathbf{81.4} \pm 0.6$	**85.9**

Model Selection leave-one-domain out					
Method	A	C	P	S	Avg
ERM	83.2 ± 1.3	$\mathbf{76.8} \pm 1.7$	97.2 ± 0.3	74.8 ± 1.3	**83.0**
RCERM	$\mathbf{84.9} \pm 1.7$	70.6 ± 0.1	$\mathbf{97.6} \pm 0.8$	$\mathbf{78.5} \pm 0.2$	82.9

Model Selection test-domain validation set					
Method	A	C	P	S	Avg
ERM	$\mathbf{86.5} \pm 1.0$	81.3 ± 0.6	96.2 ± 0.3	$\mathbf{82.7} \pm 1.1$	**86.7**
RCERM	85.0 ± 0.7	$\mathbf{83.2} \pm 1.4$	$\mathbf{96.6} \pm 0.2$	80.8 ± 2.0	86.4

We also compare our method with the most recently proposed Fishr approach in Table 5, and observe that *our method outperforms Fishr on the Sketch domain by 3.6% points using train-domain validation criterion, and the Cartoons domain by 2.4% points using test-domain validation criterion* (NOTE: Results of Fishr on leave-one-out have not been evaluated by the authors). In fact, *on average, we outperform Fishr using the train-validation criterion by 0.4% points*, which despite being a *low looking value*, is actually quite significant in the DG problem for the PACS dataset.

Table 5. Comparison of our proposed method against the state-of-the-art Fishr method.

Model Selection train-domain validation					
Method	A	C	P	S	Avg
Fishr	**88.4**$\pm$ 0.2	**78.7**$\pm$ 0.7	97$\pm$ 0.1	77.8 $\pm$ 2.0	85.5
RCERM	86.6 $\pm$ 1.5	78.1 $\pm$ 0.4	**97.3** $\pm$ 0.1	**81.4** $\pm$ 0.6	**85.9**
Model Selection test-domain validation set (oracle)					
Method	A	C	P	S	Avg
Fishr	**87.9**$\pm$ 0.6	80.8$\pm$ 0.5	**97.9**$\pm$ 0.4	**81.1**$\pm$ 0.8	**86.9**
RCERM	85.0 $\pm$ 0.7	**83.2** $\pm$ 1.4	96.6 $\pm$ 0.2	80.8 $\pm$ 2.0	86.4

3.2 Ablation/Analysis of Our Method:

While DomainBed frees the user of the hyperparameter searches automatically, we perform an ablation analysis of our method by removing the gated-fusion component of our method, and call it as RCERM with No Gated fusion (RCERMNG). The results are reported in Table 6. We found that *RCERM by virtue of its gated fusion component indeed performs better than RCERMNG without the gated fusion.* This is because the positives update their representation by taking into account a query. This helps in contributing to the alignment of the embeddings of semantically similar objects from across domains.

Table 6. Ablation studies showing role of the gated-fusion component

Model Selection train-domain validation					
Method	A	C	P	S	Avg.
RCERMNG	86.5 $\pm$ 0.8	**80.4** $\pm$ 0.2	96.3 $\pm$ 0.4	77.1 $\pm$ 1.4	85.1
RCERM	**86.6** $\pm$ 1.5	78.1 $\pm$ 0.4	**97.3** $\pm$ 0.1	**81.4** $\pm$ 0.6	**85.9**
Model Selection leave-one-domain out					
Method	A	C	P	S	Avg
RCERMNG	82.9 $\pm$ 3.4	**75.4** $\pm$ 1.5	96.5 $\pm$ 1.3	76$\pm$ 0.7	82.7
RCERM	**84.9** $\pm$ 1.7	70.6 $\pm$ 0.1	**97.6** $\pm$ 0.8	**78.5** $\pm$ 0.2	**82.9**

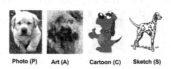

Photo (P)   Art (A)   Cartoon (C)   Sketch (S)

Fig. 6. Illustration of PACS type images.

Fig. 7. Illustration of a few Cartoon images with abnormal semantics.

3.3   Key Takeaways:

1. The DG methods when compared in a fair, uniform setting using the DomainBed framework, perform more or less similar to the classical ERM method on the PACS dataset. The recently proposed Fishr method performs competitive to ERM.
2. On the PACS dataset, for all methods in general, the classification performance is the best for the Photos domain containing natural scene images (~97% using training-domain validation), because, the underlying backbone (eg, ResNet50) has been pretrained on the ImageNet dataset which contains natural scenes. The models perform relatively well on the Art paintings domain (~84−88% using training-domain validation), as the distribution of art images is relatively closer to that of natural images (in terms of texture, shades, etc.). The performance is poorer on both Sketches and Cartoons (~73−80% using training-domain validation), which have greater distribution shift compared to that of natural images, and where attributes like shape play a prominent role than texture, color etc. Figure 6 illustrates PACS type of images.
3. *Despite the challenging nature of the Art and Sketches domains, our proposed method outperforms/performs competitive against SOTA methods like ERM and Fishr*. This validates the merit of our work, to be a suitable alternative to address images with drawings and abstract imagery, while being robust across different domains. The promise of our method lies on the fact that **we are able to learn a single, common embedding which performs well on domains like art and sketches**.
4. Regarding the Cartoons domain, by inspecting certain images qualitatively, we suspect the very nature of the images to be the hurdle in obtaining a better performance. This is because of the fact that Cartoon images very often violate the semantics of objects as observed in real life, by virtue of abnormal attributes (Fig. 7). For instance, a house having eyes, an animal having extremely large/small eyes relative to the size of the entire head of the animal, or, even unusually large head. We believe, that a separate study could be dedicated to such cartoon images.

Table 7. Ranking of all the SOTA methods (including our RCERM) using two model selection criteria, on the Art (A) and Sketch (S) domains. Avg (A, S) denotes the average ranks across these two domains. A lower value indicates a better rank.

Training-domain validation				Leave-one-domain-out cross-validation			
Method	A	S	Avg (A, S)				
SagNet	4	2	3				
CORAL	2	5	3.5	Method	A	S	**Avg (A, S)**
RCERM	**6**	**1**	**3.5**	**RCERM**	**5**	**2**	**3.5**
ARM	5	3	4	MMD	4	4	4
Fishr	1	8	4.5	Mixup	3	8	5.5
MTL	3	10	6.5	MTL	2	9	5.5
VREx	10	9	9.5	GroupDRO	6	6	6
ERM	16	4	10	ARM	1	11	6
RSC	13	7	10	CORAL	14	1	7.5
MMD	8	13	10.5	ERM	9	7	8
SAND-mask	11	11	11	SagNet	13	3	8
GroupDRO	18	6	12	MLDG	12	5	8.5
MLDG	12	12	12	DANN	7	13	10
DANN	7	17	12	VREx	11	10	10.5
Mixup	9	16	12.5	IRM	10	12	11
AND-mask	14	14	14	RSC	8	14	11
IRM	15	15	15	CDANN	15	15	15
CDANN	17	18	17.5				

5. Based on the results from Table 1, Table 2 and Table 3, in Table 7 we additionally report the rankings of the performances of all the SOTA methods, including our RCERM, using two model selection criteria, on the Art (A) and Sketch (S) domains. It can be concluded that **across all DG methods, for Art and Sketches, RCERM obtains the second best average rank using train-domain validation, and best average rank using leave-one-out criterion.**

4   Conclusion

In this paper, we propose a novel Embedding Learning approach that seeks to generalize well across domains such as Drawings and Abstract images. During training, for a given query image, we obtain an augmented positive example for Contrastive Learning by leveraging gated fusion and attention. At the same time, to make the model discriminative, we push away examples from different semantic categories (across domains). We showcase the prowess of our method using the DomainBed framework, on the popular PACS (Photo, Art painting, Cartoon, and Sketch) dataset.

Acknowledgment. I would like to thank Professor Haris, MBZUAI, for the insightful conversations.

References

1. Arjovsky, M., Bottou, L., Gulrajani, I., Lopez-Paz, D.: Invariant risk minimization. arXiv preprint arXiv:1907.02893 (2019)
2. Blanchard, G., Lee, G., Scott, C.: Generalizing from several related classification tasks to a new unlabeled sample. Adv. Neural Inf. Process. Syst. **24** (2011)
3. Chen, T., Kornblith, S., Norouzi, M., Hinton, G.: A simple framework for contrastive learning of visual representations. In: Proceedings of International Conference on Machine Learning (ICML), pp. 1597–1607. PMLR (2020)
4. Ganin, Y., Ustinova, E., Ajakan, H., Germain, P., Larochelle, H., Laviolette, F., Marchand, M., Lempitsky, V.: Domain-adversarial training of neural networks. J. Mach. Learn. Res. **17**(1), 2030–2030 (2016)
5. Gulrajani, I., Lopez-Paz, D.: In search of lost domain generalization. In: Proceedings of International Conference on Learning Representations (ICLR) (2020)
6. Krueger, D., Caballero, E., Jacobsen, J.H., Zhang, A., Binas, J., Zhang, D., Le Priol, R., Courville, A.: Out-of-distribution generalization via risk extrapolation (rex). In: International Conference on Machine Learning. pp. 5815–5826. PMLR (2021)
7. Krueger, D., et al.: Out-of-distribution generalization via risk extrapolation (rex). In: International Conference on Machine Learning, pp. 5815–5826. PMLR (2021)
8. Li, D., Yang, Y., Song, Y.Z., Hospedales, T.: Learning to generalize: meta-learning for domain generalization. In: Proceedings of the AAAI Conference on Artificial Intelligence, vol. 32 (2018)
9. Li, H., Pan, S.J., Wang, S., Kot, A.C.: Domain generalization with adversarial feature learning. In: Proceedings of the IEEE Conference on Computer Vision and Pattern Recognition, pp. 5400–5409 (2018)
10. Li, Y., Tian, X., Gong, M., Liu, Y., Liu, T., Zhang, K., Tao, D.: Deep domain generalization via conditional invariant adversarial networks. In: Proceedings of the European Conference on Computer Vision (ECCV), pp. 624–639 (2018)
11. Nam, H., Lee, H., Park, J., Yoon, W., Yoo, D.: Reducing domain gap via style-agnostic networks, vol. 2, no. 7, p. 8. arXiv preprint arXiv:1910.11645 (2019)
12. Parascandolo, G., Neitz, A., Orvieto, A., Gresele, L., Schölkopf, B.: Learning explanations that are hard to vary. In: International Conference on Learning Representations (2020)
13. Rame, A., Dancette, C., Cord, M.: Fishr: invariant gradient variances for out-of-distribution generalization. In: International Conference on Machine Learning, pp. 18347–18377. PMLR (2022)
14. Sagawa, S., Koh, P.W., Hashimoto, T.B., Liang, P.: Distributionally robust neural networks for group shifts: on the importance of regularization for worst-case generalization. arXiv preprint arXiv:1911.08731 (2019)
15. Shahtalebi, S., Gagnon-Audet, J.C., Laleh, T., Faramarzi, M., Ahuja, K., Rish, I.: Sand-mask: an enhanced gradient masking strategy for the discovery of invariances in domain generalization. arXiv preprint arXiv:2106.02266 (2021)

16. Sun, B., Saenko, K.: Deep CORAL: correlation alignment for deep domain adaptation. In: Hua, G., Jégou, H. (eds.) ECCV 2016. LNCS, vol. 9915, pp. 443–450. Springer, Cham (2016). https://doi.org/10.1007/978-3-319-49409-8_35
17. Vapnik, V.N.: An overview of statistical learning theory. IEEE Trans. Neural Netw. **10**(5), 988–999 (1999)
18. Xu, M., et al.: Adversarial domain adaptation with domain mixup. In: Proceedings of the AAAI Conference on Artificial Intelligence, vol. 34, pp. 6502–6509 (2020)
19. Zhang, M., Marklund, H., Gupta, A., Levine, S., Finn, C.: Adaptive risk minimization: a meta-learning approach for tackling group shift, vol. 8. arXiv preprint arXiv:2007.02931 (2020)

3D Shape Reconstruction from Free-Hand Sketches

Jiayun Wang[1]([✉])[iD], Jierui Lin[1], Qian Yu[1,2], Runtao Liu[1], Yubei Chen[1], and Stella X. Yu[1][iD]

[1] UC Berkeley/ICSI, Berkeley, CA, USA
{peterwg,jerrylin0928,runtao_liu,yubeic,stellayu}@berkeley.edu
[2] Beihang University, Beijing, China
qianyu@buaa.edu.cn

Abstract. Sketches are arguably the most abstract 2D representations of real-world objects. Although a sketch usually has geometrical distortion and lacks visual cues, humans can effortlessly envision a 3D object from it. This suggests that sketches encode the information necessary for reconstructing 3D shapes. Despite great progress achieved in 3D reconstruction from distortion-free line drawings, such as CAD and edge maps, little effort has been made to reconstruct 3D shapes from free-hand sketches. We study this task and aim to enhance the power of sketches in 3D-related applications such as interactive design and VR/AR games.

Unlike previous works, which mostly study distortion-free line drawings, our 3D shape reconstruction is based on free-hand sketches. A major challenge for free-hand sketch 3D reconstruction comes from the insufficient training data and free-hand sketch diversity, e.g. individualized sketching styles. We thus propose data generation and standardization mechanisms. Instead of distortion-free line drawings, synthesized sketches are adopted as input training data. Additionally, we propose a sketch standardization module to handle different sketch distortions and styles. Extensive experiments demonstrate the effectiveness of our model and its strong generalizability to various free-hand sketches. Our code is available.

Keywords: Sketch · Interactive design · Shape reconstruction · Data insufficiency · 3D reconstruction

1 Introduction

Human free-hand sketches are the most abstract 2D representations for 3D visual perception. Although a sketch may consist of only a few colorless strokes and exhibit various deformation and abstractions, humans can effortlessly envision the corresponding real-world 3D object from it. It is of interest to develop a computer vision model that can replicate this ability. Although sketches and 3D representations have drawn great interest from researchers in recent years, these two modalities have been studied relatively independently. We explore the

© The Author(s), under exclusive license to Springer Nature Switzerland AG 2023
L. Karlinsky et al. (Eds.): ECCV 2022 Workshops, LNCS 13808, pp. 184–202, 2023.
https://doi.org/10.1007/978-3-031-25085-9_11

plausibility of bridging the gap between sketches and 3D, and build a computer vision model to recover 3D shapes from sketches. Such a model will unleash many applications, like interactive CAD design and VR/AR games.

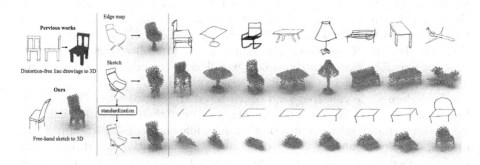

Fig. 1. Left: We study 3D reconstruction from a single-view free-hand sketch, differing from previous works [7,30,49] which use multi-view distortion-free line-drawings as training data. **Center**: While previous works [7,49] employ distortion-free line drawings (e.g. edge-maps) as proxies for sketches, our model trained on synthesized sketches can generalize better to free-hand sketches. Further, the proposed sketch standardization module makes the method generalizes well to free-hand sketches by standardizing different sketching styles and distortion levels. **Right**: Our model unleashes many practical applications such as real-time 3D modeling with sketches. A demo is here.

With the development of new devices and sensors, sketches and 3D shapes, as representations of real-world objects beyond natural images, become increasingly important. The popularity of touch-screen devices makes sketching not a privilege to professionals anymore and increasingly popular. Researchers have applied sketch in tasks like image retrieval [12,26,45,51,58,62] and image synthesis [12,13,29,40,52,63] to leverage its power in expression. Furthermore, as depth sensors, such as structured light device, LiDAR, and TOF cameras, become more ubiquitous, 3D data become an emerging modality in computer vision. 3D reconstruction, the process of capturing the shape and appearance of real objects, is an essential topic in 3D computer vision. 3D reconstruction from multi-view images has been studied for many years [1,4,11,38]. Recent works [8,10,43] have further explored 3D reconstruction from a single image.

Despite these trends and progress, there are limited works connecting 3D and sketches. We argue that sketches are abstract 2D representations of 3D perception, and it is of great significance to study sketches in a 3D-aware perspective and build connections between two modalities. Researchers have explored the potential of *distortion-free* line drawings (e.g. edge maps) for 3D modeling [27,28,57]. These works are based on *distortion-free* line drawings and generalize poorly to free-hand sketches (Fig. 1L). Furthermore, the role of line drawings in such works is to provide geometrical information for the subsequent 3D modeling. Some other works [7,30] employ neural networks to reconstruct 3D shapes directly from line drawings. However, their decent reconstructions come with two major limitations: **a)** they use distortion-free line drawings as training data, which makes such models hard to generalize to free-hand sketches; **b)** they usually require inputs depicting the object from multi-views to achieve satisfactory outcomes. Therefore, such methods cannot reconstruct the 3D shape from a single-view free-hand sketch well, as we show later in the experiment section. Other works such as [19,49] tackle 3D retrieval instead of 3D shape reconstruction from sketches. Retrieved shapes come from the pre-collected gallery set and may not resembles novel sketches well. Overall, reconstructing a 3D shape from a single *free-hand* sketch is still left not well explored.

We explore single-view free-hand sketch-based 3D reconstruction (Fig. 1**C**). A free-hand sketch is defined as a line drawing created without any additional tool. As an abstract and concise representation, it is different from distortion-free line drawings (e.g. edge maps) since it commonly has some spatial distortions, but it can still reflect the essential geometric shape. 3D reconstruction from sketch is challenging due to the following reasons: **a)** Data insufficiency. Paired sketch-3D datasets are rare although there exist several large-scale sketch datasets and 3D shape datasets, respectively. Furthermore, collecting sketch-3D pairs can be very time-consuming and expensive than collecting sketch-image pairs, as each 3D shape could be sketched from various viewing angles. **b)** Misalignment between two representations. A sketch depicts an object from a certain view while a 3D shape can be viewed from multiple angles due to the encoded depth information. **c)** Due to the nature of hand drawing, a sketch is usually geometrically imprecise with a individual style compared to the real object. Thus a sketch can only provide suggestive shape and structural information. In contrast, a 3D shape is faithful to its corresponding real-world object with no geometric deformation.

To address these challenges, we propose a single-view sketch-to-3D shape reconstruction framework. Specifically, it takes a sketch from an *arbitrary* angle as input and reconstructs a 3D point cloud. Our model cascades a sketch standardization module U and a reconstruction module G. U handles various sketching styles/distortions and transfers inputs to standardized sketches while G takes a standardized sketch to reconstruct the 3D shape (point cloud) *regardless of the object category*. The key novelty lies in the mechanisms we propose to tackle the data insufficiency issue. Specifically, we first train an photo-to-sketch model on unpaired large-scale datasets. Based on the model, sketch-3D pairs can be automatically generated from 2D renderings of 3D shapes. Together with the standardization module U which unifies input sketch styles, the synthesized sketches provide sufficient information to train the reconstruction model G. We conduct extensive experiments on a composed sketch-3D dataset, spanning 13 classes, where sketches are synthesized and 3D objects come from the ShapeNet dataset [3]. Furthermore, we collect an evaluation set, which consists of 390 real sketch-3D pairs. Results demonstrate that our model can reconstruct 3D shapes with certain geometric details from real sketches under different styles, stroke line-widths, and object categories. Our model also enables practical applications such as real-time 3D modeling with sketches (Fig. 1**R**).

To summarize our contributions: **a)** We are one of the pioneers to study the plausibility of reconstructing 3D shapes from single-view free-hand sketches. **b)** We propose a novel framework for this task and explore various design choices. **c)** To handle data insufficiency, we propose to train on synthetic sketches. Moreover, sketch standardization is introduced to make the model generalize to free-hand sketches better. It is a general method for zero-shot domain translation, and we show applications on zero-shot image translation tasks.

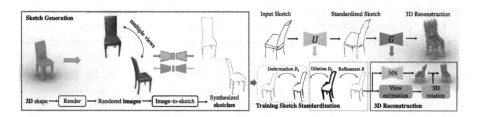

Fig. 2. Model overview. The model consists of three major components: sketch generation, sketch standardization, and 3D reconstruction. To generate synthesized sketches, we first render 2D images for a 3D shape from multiple viewpoints. We then employ an image-to-sketch translation model to generate sketches of corresponding views. The standardization module standardize sketches with different styles and distortions. *Deformation D_2* is only used in training for augmentation such that the model is robust to geometric distortions of sketches. For inference, sketches are dilated (D_1) and refined (R) so their style matches that of training sketches. For 3D reconstruction, a view estimation module is adopted to align the output's view and the ground-truth 3D shape. (Color figure online)

2   Related Works

3D Reconstruction from Images. While SfM [38] and SLAM [11] achieve success in handling multi-view 3D reconstructions in various real-world scenarios, their reconstructions can be limited by insufficient input viewpoints and 3D scanning data. Deep-learning-based methods have been proposed to further improve reconstructions by completing 3D shapes with occluded or hollowed-out areas [4,22,59]. In general, recovering the 3D shape from a single-view image is an ill-posed problem. Attempts to tackle the problem include 3D shape reconstructions from silhouettes [8], shading [43], and texture [54]. However, these methods need strong presumptions and expertise in natural images [65], limiting their usage in real-world scenarios. Generative adversarial networks (GANs) [14] and variational autoencoders (VAEs) [24] have achieved success in image synthesis and enabled [55] 3D shape reconstruction from a single-view image. Fan *et al.* [10] further adopt point clouds as 3D representation, enabling models to reconstruct certain geometric details from an image. They may not directly work on sketches as many visual cues are missing.

3D reconstruction networks are designed differently depending on the output 3D representation. 3D voxel reconstruction networks [18,48,56] benefit from many image processing networks as convolutions are appropriate for voxels. They are usually constrained to low resolution due to the computational overhead. Mesh reconstruction networks [25,53] are able to directly learn from meshes, where they suffer from topology issues and heavy computation [39]. We adopt point cloud representation as it can capture certain 3D geometric details with low computational overhead. Reconstructing 3D point clouds from images has been shown to benefit from well-designed network architectures [10,32], latent embedding matching [31], additional image supervision [35], etc.

Sketch-Based 3D Retrievals/Reconstructions. Free-hand sketches are used for 3D shape retrieval [19,49] given their power in expression. However, retrieval methods are significantly constrained by the gallery dataset. Precise sketching is also studied in the computer graphics community for 3D shape modeling or procedural modeling [20,27,28]. These works are designed for professionals and require additional information for shape modeling, e.g., surface-normal, procedural model parameters. Delanoy *et al.* [7] first employ neural networks to learn 3D voxels from line-drawings. While it achieves impressive performance, this model has several limitations: **a)** The model uses distortion-free edge map as training data. While working on some sketches with small distortions, it cannot generalize to general free-hand sketches. **b)** The model requires multiple inputs from different viewpoints for a satisfactory result. These limitations prevent the model from generalizing to real free-hand sketches. Recent works also explore reconstructing 3D models from sketches with direct shape optimization [17], shape contours [16], differential renderer [64], and unsupervised learning [50]. Compared to existing works, the proposed method in this work reconstructs the 3D point cloud based on a single-view free-hand sketch. Our model may make 3D reconstruction and its applications more accessible to the general public.

Fig. 3. Synthesized sketches are visually more similar to free-hand sketches than edge maps as they contain distortions and emphasize perceptually significant contours. After standardization, the free-hand sketches share a uniform style similar to training data.

3   3D Reconstruction from Sketches

The proposed framework has three modules (Fig. 2). To deal with data insufficiency, we first synthesize sketches as the training set. The module U transfers an input sketch to a standardized sketch. Then, the module G takes the standardized sketch to reconstruct a 3D shape (point clouds). We also present details of a new sketch-3D dataset, which is collected for evaluating the proposed model.

3.1   Synthetic Sketch Generation

To the best of our knowledge, there exists no paired sketch-3D dataset. While it is possible to resort to edge maps [7], edge maps are different from sketches (as shown in the 3rd and 4th rows of Fig. 3). We show that the reconstruction model trained on edge maps cannot generalize well to real free-hand sketches in Sect. 4.4. Thus it is crucial to find an efficient and reliable way to synthesize sketches for 3D shapes. Inspired by [29], we employ a generative model to synthesize sketches from rendered images of 3D shapes. Figure 2L depicts the procedure. Specifically, we first render m images for each 3D shape, where each image corresponds to a particular view of a 3D shape. We then adopt the model introduced in [29] to synthesize gray-scale sketches images, denoted as $\{S_i | S_i \in \mathbb{R}^{W \times H}\}$, as our training data. W, H refer to the width and height of a sketch image.

3.2   Sketch Standardization

Sketches usually have strong individual styles and geometric distortions. Due to the gap between the free-hand sketches and the synthesized sketches, directly using the synthesized sketches as training data would not lead to a robust model. The main issues are that the synthesized sketches have a uniform style and they do not contain enough geometric distortions. Rather, the synthesized sketches can be treated as an intermediate representation if we can find a way to project a free-hand sketch to the synthesized sketch domain. We propose a zero-shot domain translation technique, the sketch standardization module, to achieve this domain adaption goal without using the free-hand sketches as the training data. The training of the sketch standardization module only involves synthesized sketches. The general idea is to project a distorted synthesized sketch to the original synthesized sketch. The training consists of two parts: **a)** since the free-hand sketches usually have geometric distortions, we apply predefined distortion augmentation to the input synthesized sketches first. **b)** A geometrically distorted synthesized sketch still has a different style and line style compared to the free-hand sketches. Thus, the first stage of the standardization is to apply a dilation operation. The dilation operation would project distorted synthesized sketches and the free-hand sketches to the same domain. Then, a refinement network follows to project the dilated sketch back to the synthesized sketch domain.

In summary, as in Fig. 2, the standardization module U first applies a dilation operator D_2 to the input sketch, which is followed by a refinement operator R to transfer to the standardized synthesized-sketch style (or training-sketch style) $\widetilde{S}_i$, i.e. $U = R \circ D_2$. R is implemented as an image translation network. During training, a synthesized sketch $\widetilde{S}_i$ is first augmented by the deformation operator D_1 to mimic the drawing distortion, and then U aims to project it back to $\widetilde{S}_i$. Please note that D_1 would not be used during the testing. We illustrate the standardization process in Fig. 2**R** with more details in the following.

Deformation. When training U, each synthesized sketch is deformed with moving least squares [46] for random, local and rigid distortion. Specifically, we randomly sample a set of control points on sketch strokes and denote them as p, and denote the corresponding deformed point set as q. Following moving least squares, we solve for the best affine transformation $l_v(x)$ such that: $\min \sum_i w_i |l_v(p_i) - q_i|^2$, where p_i and q_i are row vectors and weights $w_i = \frac{1}{|p_i - v|^{2\alpha}}$. Affine transformation can be written as $l_v(p_i) = p_i M + T$. We add constraint $M^T M = I$ to make the deformation is rigid to avoid too much distortion. Details can be found in [46].

Style Translation. Adapting to unknown input free-hand sketch style during inference can be considered as a zero-shot domain translation problem, which is

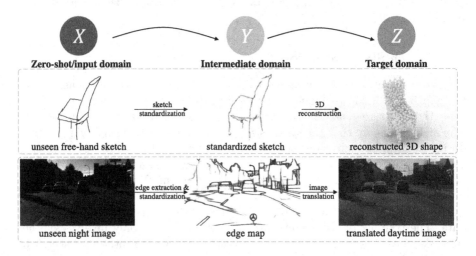

Fig. 4. Sketch standardization can be considered as a general zero-shot domain translation method. Given a sample from a zero-shot (input) domain X, we first translate it to a universal intermediate domain Y, and finally to the target domain Z. **1st example** (2nd row): the input domain is an unseen free-hand sketch. With sketch standardization, it is translated to an intermediate domain: standardized sketch, which shares similar style as synthesized sketch for training. With 3D reconstruction, the standardized sketch can be translated to the target domain: 3D point clouds. **2nd example** (last row): the input domain is an unseen nighttime image. With edge extraction, it gets translated to an intermediate domain: edge map. With the image-to-image translation model, the standardized edge map can be translated to the target domain: daytime image.

challenging. Inspired by [61], we first dilate the augmented training sketch strokes with 4 pixels and then use image-to-image translation network Pix2Pix [21] to translate the dilated sketches to the un-distorted synthesized sketches. During inference, we also dilate the free-hand sketches and apply the trained Pix2Pix model such that the style of an input free-hand sketch could be adapted to the synthesized sketch style during training. The dilation step can be considered as introducing uncertainty for the style adaption. Further, we show in Sect. 4.5 that the proposed style standardization module could be used as a general zero-shot domain translation technique, which generalizes to more applications such as sketch classification and zero-shot image-to-image translation.

A More General Message: Zero-Shot Domain Translation. We illustrate in Fig. 4 a more general message of the standardization module: it can be considered as a general method for zero-shot domain translation. Consider the following problem: we would like to build a model to transfer domain X to domain Z but we do not have any training data from domain X. We propose a general idea to solve this problem is to build an intermediate domain Y as a bridge such that: **a)** we can translate data from domain X to domain Y and **2)** we can further translate data from domain Y to domain Z. We give two examples in the caption of Fig. 4 and provided experimental results in Sect. 4.5.

3.3 Sketch-Based 3D Reconstruction

Our 3D reconstruction network G (pipeline in Fig. 2R) consists of several components. Given a standardized sketch $\widetilde{S}_i$, the view estimation module first estimates its viewpoint. $\widetilde{S}_i$ is then fed to the sketch-to-3D module to generate a point cloud $P_{i,pre}$, whose pose aligns with the sketch viewpoint. A 3D rotation corresponding to the viewpoint is then applied to $P_{i,pre}$ to output the canonically-posed point cloud P_i. The objective of G is to minimize distances between reconstructed point cloud P_i and the ground-truth point cloud $P_{i,gt}$.

View Estimation Module. The view estimation module g_1 aims to determine the 3D pose from an input sketch $\widetilde{S}$. Similar to the input transformation module of the PointNet [42], g_1 estimates a 3D rotation matrix A from a sketch $\widetilde{S}$, i.e., $A = g_1(\widetilde{S})$. A regularization loss $L_{\text{orth}} = \|I - AA^T\|_F^2$ is applied to ensure A is a rotation (orthogonal) matrix. The rotation matrix A rotates a point cloud from the viewpoint pose to a canonical pose, which matches the ground truth.

3D Reconstruction Module. The reconstruction network g_2 learns to reconstruct a 3D point cloud P_{pre} from a sketch $\widetilde{S}$, i.e., $P_{pre} = g_2(\widetilde{S})$. P_{pre} is further transformed by the corresponding rotation matrix A to P so that P aligns with the ground-truth 3D point cloud P_{gt}'s canonical pose. Overall, we have $P = g_1(\widetilde{S}) \cdot g_2(\widetilde{S})$. To train G, we penalize the distance between an output point cloud P and the ground-truth point cloud P_{gt}. We employ the Chamfer distance (CD) between two point clouds $P, P_{gt} \subset \mathbb{R}^3$:

$$d_{CD}(P\|P_{gt}) = \sum_{\mathbf{p} \in P} \min_{\mathbf{q} \in P_{gt}} \|\mathbf{p} - \mathbf{q}\|_2^2 + \sum_{\mathbf{q} \in P_{gt}} \min_{\mathbf{p} \in P} \|\mathbf{p} - \mathbf{q}\|_2^2 \qquad (1)$$

The final loss of the entire network is $L = \sum_i d_{CD} \left(G \circ U(S_i) \| P_{i,gt}\right) + \lambda L_{\text{orth}} = \sum_i d_{CD} \left(A_i \cdot P_{i,pre} \| P_{i,gt}\right) + \lambda L_{\text{orth}} = \sum_i d_{CD} \left(g1(\widetilde{S}_i) \cdot g_2(\widetilde{S}_i) \| P_{i,gt}\right) + \lambda L_{\text{orth}}$ where λ is the weight of the orthogonal regularization loss and $\widetilde{S}_i = R \circ D_2 \circ D_1(S_i)$ is the standardized sketch from S_i. We employ CD rather than EMD (Sect. 4.2) to penalize the difference between the reconstruction and the ground-truth point clouds because CD emphasizes the geometric outline of point clouds and leads to reconstructions with better geometric details. EMD, however, emphasizes the point cloud distribution and may not preserve the geometric details well at locations with low point density.

4   Experimental Results

We first present the datasets, training and evaluation details, followed by qualitative and quantitative results. Then, we provide comparisons with some state-of-the-art methods. We also conduct ablation studies to understand each module.

4.1   3D Sketching Dataset

To evaluate the performance of our method, we collected a real-world evaluation set containing paired sketch-3D data. Specifically, we randomly choose ten 3D shapes from each of the 13 categories of the ShapeNet dataset [3]. Then we randomly render 3 images from different viewpoints for each 3D shape. Totally, there are 130 different 3D shapes and 390 rendered images. We recruited 11 volunteers to draw the sketches for the rendered images. Final sketches are reviewed for quality control. We present several examples in Fig. 3.

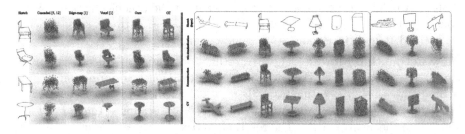

Fig. 5. Left: Performance on free-hand sketches with different design choices. The design pool includes the model with a cascaded two-stage structure (2nd column), the model trained on edge maps (3rd column), the model whose 3D output is represented by voxel (4th column), and the proposed model (5th column). Overall, the proposed method achieves better performance and keeps more fine-grained details, e.g., the legs of chairs. **Right**: 3D reconstructions on our newly-collected free-hand sketch evaluation dataset. *Outlined in green:* Examples of some good reconstruction results. Our model reconstructs 3D shapes with fine geometric fidelity of multiple categories *unconditionally*. *Outlined in red:* Examples of failure cases. Our model may not handle detailed structures well (e.g., *watercraft*), recognize the wrong category (e.g., *display* as a *lamp*) due to the ambiguity of the sketch, as well as not able to generate 3D shape from very abstract sketches where few geometric information is available (e.g., *rifle*). (Color figure online)

4.2   Training Details and Evaluation Metrics

Training. The proposed model is trained on a subset of ShapeNet [3] dataset, following settings of [56]. The dataset consists of 43,783 3D shapes spanning 13 categories, including car, chair, table, etc. For each category, we randomly select 80% 3D shapes for training and the rest for evaluation. As mentioned in Sect. 3.1, corresponding sketches of rendered images from 24 viewpoints of each 3D shape of ShapeNet are synthesized with our synthetic sketch generation module.

Evaluation. To evaluate our method's 3D reconstruction performance on freehand sketches, we use our proposed sketch-3D datasets (Sect. 4.1). To evaluate the generalizability of our model, we also evaluate on three additional free-hand sketch datasets, including the Sketchy dataset [45], the TU-Berlin dataset [9], and the QuickDraw dataset [15]. For these additional datasets, only sketches from categories that overlap with the ShapeNet dataset are considered.

Following the previous works [10,31,60], we adopt two evaluation metrics to measure the similarity between the reconstructed 3D point cloud P and the ground-truth point cloud P_{gt}. The first one is the Chamfer Distance (Eq. 1), and another one is the Earth Mover's Distance (EMD): $d_{EMD}(P, P_{gt}) = \min_{\phi:P \mapsto P_{gt}} \sum_{x \in P} \|x - \phi(x)\|$, where P, P_{gt} has the same size $|P| = |P_{gt}|$ and $\phi : P \mapsto P_{gt}$ is a bijection. CD and EMD evaluate the similarity between two point clouds from two different perspectives (more details can be found in [10]).

4.3   Implementation Details

Sketch Generation. We utilize an off-the-shelf sketch-image translation model [29] to synthesize sketches for training. Given the appropriate quality of the generated sketches on the ShapeNet dataset (with some samples depicted in Fig. 3), we directly use the model without any fine-tuning.

Table 1. Ours outperforms baselines for 3D reconstruction. [7] uses edge-maps rather than sketches as input. [56] uses voxels rather than point clouds as output. [37] represents the 3D shapewith multi-view depth maps. "cas." refers to the two-stage cascaded training following [10,29]. CD and EMD measure distances between reconstructions and ground-truths from different perspectives (see text for details). The lower, the better.

Error	Chamfer Distance (×10⁻⁴)							Earth Mover's Distance (×10⁻²)						
	Points	Edge [7]	Voxel [56]	Cas.	[37]	Retrieval [49]	Ours	Points	Edge [7]	Voxel [56]	Cas.	[37]	Retrieval [49]	Ours
Airplane	11.4	7.8	35.1	71.7	8.0	11.2	**6.1**	8.5	7.3	10.8	12.7	8.5	11.9	**6.5**
Bench	29.2	16.7	202.8	414.1	16.8	14.5	**13.0**	11.1	8.7	22.0	25.8	10.0	8.6	**7.8**
Cabinet	61.7	50.4	59.1	354.5	51.5	45.3	**39.2**	17.6	17.8	17.0	29.6	18.4	17.2	**16.0**
Car	20.8	13.3	173.2	114.2	14.1	14.2	**10.4**	**8.9**	20.0	25.2	20.0	21.6	21.2	18.0
Chair	41.8	36.4	108.6	237.1	36.1	33.0	**26.9**	15.1	15.6	19.4	22.8	16.1	15.3	**13.0**
Display	68.6	48.3	**33.1**	340.2	49.3	38.2	37.7	15.5	15.1	**13.1**	27.9	16.4	14.6	14.4
Lamp	63.3	59.4	107.0	214.0	60.2	63.5	**46.3**	21.3	22.6	21.2	24.9	22.3	22.6	**20.4**
Speaker	88.2	79.7	203.2	406.4	81.2	72.3	**62.1**	19.4	19.2	23.8	28.0	21.8	20.0	**17.9**
Rifle	17.0	12.1	170.1	15.4	12.3	14.2	**10.1**	11.2	13.8	23.7	15.4	15.2	17.6	12.4
Sofa	32.8	20.9	141.2	482.4	22.3	20.3	**16.3**	11.1	8.5	18.6	25.4	9.1	8.6	**7.7**
Table	55.2	49.4	134.7	469.5	50.5	49.1	**40.7**	19.1	17.7	18.5	26.5	18.2	18.2	**17.3**
Telephone	30.7	27.3	26.9	259.8	27.1	27.4	**21.3**	13.4	13.6	15.1	27.2	15.1	15.3	**12.3**
Watercraft	32.9	26.0	129.1	53.8	26.0	27.3	**20.3**	12.5	11.1	23.1	17.8	12.2	12.7	**10.6**
Avg.	42.6	34.4	117.2	264.1	35.0	33.1	**26.9**	14.2	14.7	19.3	23.4	15.8	15.7	**13.4**
Free-hand sketch	87.1	89.0	162.5	334.2	91.8	89.2	**86.1**	18.6	16.4	22.9	26.1	17.0	16.8	**16.0**

Data Augmentation. During training, to improve the model's generalizability and robustness, we perform data augmentation for synthetic sketches before feeding them to the standardization module. Specifically, we apply image spatial translation (up to ± 10 pixels) and rotation (up to $\pm 10°$) on each input sketch.

Sketch Standardization. Each input sketch S_i is first randomly deformed with moving least squares [46] both globally and locally (D_1), and then binarized and dilated five times iteratively (D_2) to obtain a rough sketch S_r. The rough sketch S_r is then used to train a Pix2Pix model [21], R, to reconstruct the input sketch S_i. The network is trained for 100 epochs with an initial learning rate of 2e-4. Adam optimizer [23] is used for the parameter optimization. During evaluation, random deformation D_1 is discarded.

3D Reconstruction. The 3D reconstruction network is based on [10]'s framework with hourglass network architecture [36]. We compare several different network architectures (simple encoder-decoder architecture, two-prediction-branch architecture, etc.) and find that hourglass network architecture gives the best performance. This may be due to its ability to extract key points from images [2,36]. We train the network for 260 epochs with an initial learning rate of 3e-5. The weight λ of the orthogonal loss is 1e-3. To enhance the performance on every category, all categories of 3D shapes are trained together. The class-aware mini-batch sampling [47] is adopted to ensure a balanced category-wise distribution for each mini-batch. We choose Adam optimizer [23] for the parameter optimization. 3D point clouds are visualized with the rendering tool from [33].

4.4 Results and Comparisons

We first present our model's 3D shape reconstruction performance, along with the comparisons with various baseline methods. Then we present the results on sketches from different viewpoints and of different categories, as well the results on other free-hand sketch datasets. Note that unless specifically mentioned, all evaluations are on the free-hand sketches rather than synthesized sketches.

Baseline Methods. Our 3D reconstruction network is a one-stage model where the input sketch is treated as an image, and point clouds represent the output 3D shape. As conducting the first work for single-view sketch-based 3D reconstruction, we explore different design options adopted by previous works on distortion-free line drawings and/or 3D reconstruction, including architectures, representation of sketches and 3D shapes. We compare with different variants to demonstrate the effectiveness of each choice of our model.

1) Model Design: End-to-End vs. Two-Stage. Although the task of reconstructing 3D shapes from free-hand sketches is new, sketch-to-image synthesize and 3D shape reconstruction from images have been studied before [10,29,56]. Is a straight combination of the two models, instead of an end-to-end model, enough to perform well for the task? To compare these two architectures' performance, We implement a cascaded model by composing a sketch-to-image model [66] and an image-to-3D model [10] to reconstruct 3D shapes.

2) Sketch: Point-Based vs. Image-Based. Considering a sketch is relatively sparse in pixel space and consists of colorless strokes, we can employ 2D point clouds to represent a sketch. Specifically, 512 points are randomly sampled from strokes of each binarized sketch, and we use a point-to-point network architecture (adapted from PointNet [42]) to reconstruct 3D shapes from the 2D point clouds.

3) Sketch: Using Edge Maps as Proxy. We compare with a previous work [7]. Our proposed model uses synthetic sketch for training. However, an alternative option is using edge maps as a proxy of the free-hand sketch. As edge maps can be generated automatically (we use the Canny edge detector in implementation), the comparison helps us understand if our proposed synthesizing method is necessary.

4) 3D Shape: Voxel vs. Point Cloud. We compare with a previous work [56]. In this variant, we follow their settings and represent a 3D shape with voxels. As the voxel representation is adopted from the previous method, the comparison helps to understand if representing 3D shapes with point clouds has benefits.

5) 3D Shape: Depth Map vs. Point Cloud. In this variant, we exactly follow a previous work [37] and represent the 3D shape with multi-view depth maps.

Comparison and Results. Table 1 and Fig. 5L present quantitative and qualitative results of our method and different design variants. Specifically for quantitative comparisons (Table 1), we report 3D shape reconstruction performance on both synthesized (evaluation set) and free-hand sketches. This is due to that the collected free-hand sketch dataset is relatively small and together they provide a more comprehensive evaluation. We have the following observations: **a)** Representing sketches as images outperforms representing them as 2D point clouds (points vs. ours). **b)** The model trained on synthesized sketches performs better on real free-hand sketches than the model trained on edge maps (89.0 vs. 86.1 on CD, 16.4 vs. 16.0 on EMD). Training with edge maps could reconstruct okay overall coarse shape. However, the unsatisfactory performance on geometric details reveals such methods are hard to generalize to free-hand sketches with distortions. It also shows the necessity of the proposed sketch generation and standardization modules. **c)** For model design, the end-to-end model outperforms the two-stage model by a large margin (cas. vs. ours). **d)** For 3D shape representation, while the voxel representation can reconstruct the general shape well, the fine-grained details are frequently missing due to its low resolution ($32 \times 32 \times 32$). Thus, point clouds outperform voxels. The proposed method also outperforms a previous work that uses depth maps as 3D shape representation [37]. Note that the resolution of voxels can hardly improve much due to the complexity and computational overhead. However, we show that increasing the number of points improves the reconstruction quality (details in supplementary).

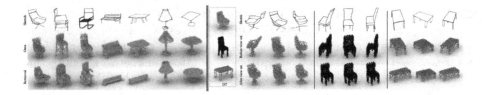

Fig. 6. Left: Ours (2nd row) versus nearest-neighbor retrieval results (last row) of given sketches. Our model generalizes to unseen 3D shapes better and has higher geometry fidelity. **Right**: 3D reconstructions of sketches from different viewpoints. Before the view estimation module, the reconstructed 3D shape aligns with the input sketch's viewpoint. The module transforms the pose of the output 3D shape to align with canonical pose, i.e. the pose of the ground-truth 3D shape.

Fig. 7. Left: Our approach trained on ShapeNet can be directly applied to other unseen sketch datasets [9,15,45] and generalize well. Our model is able to reconstruct 3D shapes from sketches with different styles and line-widths, and even low-resolution data. **Right**: Standardized sketches converted from different individual styles (by different volunteers). For each rendered image of a 3D object, we show free-hand sketches from two volunteers and the standardized sketches from these free-hand sketches. Contents are preserved after the standardization process, and standardized sketches share the style similar to the synthesized ones.

Retrieval Results. We compare with nearest-neighbor retrievals, following methods and settings of [49] (Fig. 6L). We could generalize to unseen 3D shapes and reconstructs with higher geometry fidelity (e.g., stand of the lamp).

Reconstruction with Different Categories and Views. Figure 5R shows 3D reconstruction results with sketches from different object categories. Our model reconstructs 3D shapes of multiple categories *unconditionally*. There are some failure cases that the model may not handle well.

Figure 6R depicts reconstructions with sketches from different views. Our model can reconstruct 3D shapes from different views even if certain parts are occluded (e.g. legs of the table). Slight variations in details exist for different views.

Evaluation on Other Free-Hand Sketch Datasets. We also evaluate on three other free-hand sketch datasets [9,15,45]. Our model can reconstruct 3D shapes from sketches with different styles, line-widths, and levels of distortions even at low resolution (Fig. 7L).

Table 2. Ablation studies of standardization and view estimation module. CD is enlarged by 10^4, and EMD by 10^2. **(a)** 3D shape reconstruction errors of ablation studies of standardization and view estimation module. Having both standardization and view estimation module gives the highest performance. The lower, the better. **(b)** Reconstruction error with different components of the standardization module: deformation and style translation. Having both parts gives the highest performance. **(c)** The sketch standardization module improves cross-dataset sketch classification accuracy. A ResNet-50 model is trained on TU-Berlin [9] and evaluated Sketchy dataset [45]. The sketch standardization module gives 3% points gain.

(a)					(b)				(c)	
error	no standard.	no view est.	ours		deform.	trans.	CD	EMD	(%)	acc.
CD	92.6	86.8	**86.1**		✗	✗	92.6	18.2	w/o std.	75.1
EMD	18.2	16.2	**16.0**		✗	✓	87.2	16.3	w/ std.	**78.1**
					✓	✗	90.1	17.4		
					✓	✓	**86.1**	**16.0**		

4.5 Sketch Standardization Module

Visualization. The standardization module can be considered as a domain translation module designed for sketches. We show the standardized sketches of these free-hand sketches and compare them to the synthesized ones Fig. 7R. With the standardization module, sketches share a style similar to synthesized sketches which are used as training data. Thus, standardization diminishes the domain gap of sketches with various styles and enhances the generalizability.

Ablation Studies of the Entire Module. The sketch standardization module is introduced to handle various drawing styles of humans. We thus verify this module's effectiveness on real sketches, both quantitatively (Table 2a) and qualitatively (Fig. 5R). As shown in Table 2a, the reconstruction performance has a significant drop when removing the standardization module. Its effect is also proved in visualizations. In Fig. 5R, we can observe that our full model equipping with the standardization module can produce 3D shapes with higher quality, being more similar to GT shapes, e.g., the airplane and the lamp.

Ablation Studies of Different Components. The standardization module consists of two components: sketch deformation and style translation. We study each module's performance and report in Table 2b. We observe that the style transformation part improves the reconstruction performance better compared with the deformation part, while having both parts gives the highest performance.

Additional Applications. We show the effectiveness of the proposed sketch standardization with two more applications. The first applications is on cross-dataset sketch classification. We identify the common 98 categories of TU-Berlin sketch dataset [9] and Sketchy dataset [45]. Then we train on TU-Berlin and evaluate on Sketchy. As reported in Table **2c**, adding additional sketch standardization module, the classification accuracy improves 3% points.

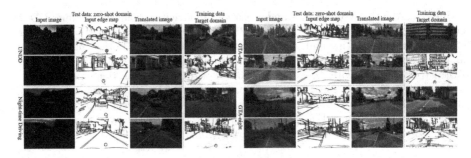

Fig. 8. Zero-shot domain translation results. We aim to translate zero-shot images (**Left**) to the training data domain (**Right**). Specifically, we evaluate the proposed zero-shot domain translation performance on three new datasets: UNDD [34], Night-Time Driving [6] and GTA [44]. The novel domains of night-time and simulated images can be translated to the target domain of daytime and real-world images by leveraging the synthetic edge map domain as a bridge. The target domain is CityScapes dataset [5]. We extract corresponding edge maps and train an image-to-image translation model [21] to translate edge maps to the corresponding RGB images. 1st, 3rd, 5th, 7th rows of column 1 depict some sample training RGB images and 2nd, 4th, 6th, 8th rows of column 1 depict the corresponding edge maps respectively.

The second application corresponds to the second example depicted in Fig. 4. The target domain is CityScapes dataset [5], where the training data comes from. We extract corresponding edge maps with a deep learning approach [41] and train an image-to-image translation model [21] to translate edge maps to the corresponding RGB images. We evaluate the zero-shot domain translation performance on three new datasets: UNDD [34] (night images), Night-Time Driving [6] (night images) and GTA [44] (synthetic images; screenshots taken from simulated environment). The novel domains of night and simulated images can be translated to the target domain of daytime and real-world images. We visualize the results in Fig. 8.

4.6 View Estimation Module

Removing the view estimation module leads to a performance drop of CD and EMD (Table 2a). For qualitative results (Fig. 6**R**), without the 3D rotation, the reconstructed 3D shape has the pose aligned with the input sketch. With the 3D rotation, the 3D shape is aligned to the ground truth's canonical pose.

5 Summary

We study 3D shape reconstruction from a single-view free-hand sketch. The major novelty is that we use synthesized sketches as training data and introduce a sketch standardization module, in order to tackle the data insufficiency and sketch style variation issues. Extensive experimental results shows that the proposed method is able to successfully reconstruct 3D shapes from single-view free-hand sketches *unconditioned* on viewpoints and categories. The work may unleash more potentials of the sketch in applications such as sketch-based 3D design/games, making them more accessible to the general public.

References

1. Alexiadis, D.S., et al.: An integrated platform for live 3D human reconstruction and motion capturing. IEEE Trans. Circuits Syst. Video Technol. **27**(4), 798–813 (2016)
2. Cao, Z., Simon, T., Wei, S.E., Sheikh, Y.: Realtime multi-person 2D pose estimation using part affinity fields. In: Proceedings of the IEEE Conference on Computer Vision and Pattern Recognition (2017)
3. Chang, A.X., et al.: ShapeNet: an information-rich 3D model repository. arXiv preprint arXiv:1512.03012 (2015)
4. Choy, C.B., Xu, D., Gwak, J.Y., Chen, K., Savarese, S.: 3D-R2N2: a unified approach for single and multi-view 3D object reconstruction. In: Leibe, B., Matas, J., Sebe, N., Welling, M. (eds.) ECCV 2016. LNCS, vol. 9912, pp. 628–644. Springer, Cham (2016). https://doi.org/10.1007/978-3-319-46484-8_38
5. Cordts, M., et al.: The cityscapes dataset for semantic urban scene understanding. In: Proceedings of the IEEE Conference on Computer Vision and Pattern Recognition (CVPR) (2016)
6. Dai, D., Van Gool, L.: Dark model adaptation: Semantic image segmentation from daytime to nighttime. In: 2018 21st International Conference on Intelligent Transportation Systems (ITSC), pp. 3819–3824. IEEE (2018)
7. Delanoy, J., Aubry, M., Isola, P., Efros, A.A., Bousseau, A.: 3D sketching using multi-view deep volumetric prediction. Proc. ACM Comput. Graph. Interact. Tech. **1**(1), 1–22 (2018)
8. Dibra, E., Jain, H., Oztireli, C., Ziegler, R., Gross, M.: Human shape from silhouettes using generative HKS descriptors and cross-modal neural networks. In: Proceedings of the IEEE Conference on Computer Vision and Pattern Recognition (2017)
9. Eitz, M., Hays, J., Alexa, M.: How do humans sketch objects? ACM Trans. Graph. **31**(4), 44:1–44:10 (2012)
10. Fan, H., Su, H., Guibas, L.J.: A point set generation network for 3D object reconstruction from a single image. In: Proceedings of the IEEE Conference on Computer Vision and Pattern Recognition (2017)
11. Fuentes-Pacheco, J., Ruiz-Ascencio, J., Rendón-Mancha, J.M.: Visual simultaneous localization and mapping: a survey. Artif. Intell. Rev. **43**(1), 55–81 (2015)
12. Gao, X., Wang, N., Tao, D., Li, X.: Face sketch-photo synthesis and retrieval using sparse representation. IEEE Trans. Circuits Syst. Video Technol. **22**(8), 1213–1226 (2012)
13. Ghosh, A., et al.: Interactive sketch & fill: multiclass sketch-to-image translation. In: Proceedings of the IEEE International Conference on Computer Vision (2019)
14. Goodfellow, I., et al.: Generative adversarial nets. In: Advances in Neural Information Processing Systems (2014)
15. Google: The quick, draw! Dataset (2017). quickdraw.withgoogle.com/data
16. Guillard, B., Remelli, E., Yvernay, P., Fua, P.: Sketch2Mesh: reconstructing and editing 3D shapes from sketches. arXiv preprint arXiv:2104.00482 (2021)

17. Han, Z., Ma, B., Liu, Y.S., Zwicker, M.: Reconstructing 3D shapes from multiple sketches using direct shape optimization. IEEE Trans. Image Process. **29**, 8721–8734 (2020)
18. Häne, C., Tulsiani, S., Malik, J.: Hierarchical surface prediction for 3D object reconstruction. In: 2017 International Conference on 3D Vision. IEEE (2017)
19. He, X., Zhou, Y., Zhou, Z., Bai, S., Bai, X.: Triplet-center loss for multi-view 3D object retrieval. In: Proceedings of the IEEE Conference on Computer Vision and Pattern Recognition (2018)
20. Huang, H., Kalogerakis, E., Yumer, E., Mech, R.: Shape synthesis from sketches via procedural models and convolutional networks. IEEE Trans. Visual Comput. Graphics **23**(8), 2003–2013 (2016)
21. Isola, P., Zhu, J.Y., Zhou, T., Efros, A.A.: Image-to-image translation with conditional adversarial networks. In: Proceedings of the IEEE Conference on Computer Vision and Pattern Recognition (2017)
22. Kar, A., Häne, C., Malik, J.: Learning a multi-view stereo machine. In: Advances in Neural Information Processing Systems (2017)
23. Kingma, D.P., Ba, J.: Adam: a method for stochastic optimization. arXiv preprint arXiv:1412.6980 (2014)
24. Kingma, D.P., Welling, M.: Auto-encoding variational Bayes. arXiv preprint arXiv:1312.6114 (2013)
25. Kolotouros, N., Pavlakos, G., Daniilidis, K.: Convolutional mesh regression for single-image human shape reconstruction. In: Proceedings of the IEEE Conference on Computer Vision and Pattern Recognition (2019)
26. Lei, J., Song, Y., Peng, B., Ma, Z., Shao, L., Song, Y.Z.: Semi-heterogeneous three-way joint embedding network for sketch-based image retrieval. IEEE Trans. Circuits Syst. Video Technol. **30**(9), 3226–3237 (2019)
27. Li, C., Pan, H., Liu, Y., Tong, X., Sheffer, A., Wang, W.: BendSketch: modeling freeform surfaces through 2D sketching. ACM Trans. Graph. **36**(4), 1–14 (2017)
28. Li, C., Pan, H., Liu, Y., Tong, X., Sheffer, A., Wang, W.: Robust flow-guided neural prediction for sketch-based freeform surface modeling. ACM Trans. Graph. (TOG) **37**(6), 1–12 (2018)
29. Liu, R., Yu, Q., Yu, S.: An unpaired sketch-to-photo translation model. arXiv preprint arXiv:1909.08313 (2019)
30. Lun, Z., Gadelha, M., Kalogerakis, E., Maji, S., Wang, R.: 3D shape reconstruction from sketches via multi-view convolutional networks. In: 2017 International Conference on 3D Vision (3DV), pp. 67–77. IEEE (2017)
31. Mandikal, P., Navaneet, K., Agarwal, M., Babu, R.V.: 3D-LMNet: latent embedding matching for accurate and diverse 3D point cloud reconstruction from a single image. arXiv preprint arXiv:1807.07796 (2018)
32. Mandikal, P., Radhakrishnan, V.B.: Dense 3D point cloud reconstruction using a deep pyramid network. In: 2019 IEEE Winter Conference on Applications of Computer Vision. IEEE (2019)
33. Mo, K., Guerrero, P., Yi, L., Su, H., Wonka, P., Mitra, N., Guibas, L.J.: StructureNet: hierarchical graph networks for 3D shape generation. ACM Trans. Graph. **38**(6) (2019)
34. Nag, S., Adak, S., Das, S.: What's there in the dark. In: 2019 IEEE International Conference on Image Processing (ICIP), pp. 2996–3000. IEEE (2019)
35. Navaneet, K., Mandikal, P., Agarwal, M., Babu, R.V.: CapNet: continuous approximation projection for 3D point cloud reconstruction using 2D supervision. In: Proceedings of the AAAI Conference on Artificial Intelligence, vol. 33, pp. 8819–8826 (2019)

36. Newell, A., Yang, K., Deng, J.: Stacked hourglass networks for human pose estimation. In: Leibe, B., Matas, J., Sebe, N., Welling, M. (eds.) ECCV 2016. LNCS, vol. 9912, pp. 483–499. Springer, Cham (2016). https://doi.org/10.1007/978-3-319-46484-8_29

37. Nozawa, N., Shum, H.P., Ho, E.S., Morishima, S.: Single sketch image based 3D car shape reconstruction with deep learning and lazy learning. In: VISIGRAPP (1: GRAPP), pp. 179–190 (2020)

38. Özyeşil, O., Voroninski, V., Basri, R., Singer, A.: A survey of structure from motion*. Acta Numer. **26**, 305–364 (2017)

39. Pan, J., Han, X., Chen, W., Tang, J., Jia, K.: Deep mesh reconstruction from single RGB images via topology modification networks. In: Proceedings of the IEEE International Conference on Computer Vision (2019)

40. Peng, C., Gao, X., Wang, N., Li, J.: Superpixel-based face sketch-photo synthesis. IEEE Trans. Circuits Syst. Video Technol. **27**(2), 288–299 (2015)

41. Poma, X.S., Riba, E., Sappa, A.: Dense extreme inception network: towards a robust CNN model for edge detection. In: Proceedings of the IEEE/CVF Winter Conference on Applications of Computer Vision, pp. 1923–1932 (2020)

42. Qi, C.R., Su, H., Mo, K., Guibas, L.J.: PointNet: deep learning on point sets for 3D classification and segmentation. In: Proceedings of the IEEE Conference on Computer Vision and Pattern Recognition (2017)

43. Richter, S.R., Roth, S.: Discriminative shape from shading in uncalibrated illumination. In: Proceedings of the IEEE Conference on Computer Vision and Pattern Recognition (2015)

44. Richter, S.R., Vineet, V., Roth, S., Koltun, V.: Playing for data: ground truth from computer games. In: Leibe, B., Matas, J., Sebe, N., Welling, M. (eds.) ECCV 2016. LNCS, vol. 9906, pp. 102–118. Springer, Cham (2016). https://doi.org/10.1007/978-3-319-46475-6_7

45. Sangkloy, P., Burnell, N., Ham, C., Hays, J.: The sketchy database: learning to retrieve badly drawn bunnies. ACM Trans. Graph. **35**(4), 1–12 (2016)

46. Schaefer, S., McPhail, T., Warren, J.: Image deformation using moving least squares. In: ACM SIGGRAPH 2006 Papers, pp. 533–540 (2006)

47. Shen, L., Lin, Z., Huang, Q.: Relay backpropagation for effective learning of deep convolutional neural networks. In: Leibe, B., Matas, J., Sebe, N., Welling, M. (eds.) ECCV 2016. LNCS, vol. 9911, pp. 467–482. Springer, Cham (2016). https://doi.org/10.1007/978-3-319-46478-7_29

48. Tatarchenko, M., Dosovitskiy, A., Brox, T.: Octree generating networks: efficient convolutional architectures for high-resolution 3D outputs. In: Proceedings of the IEEE International Conference on Computer Vision (2017)

49. Wang, F., Kang, L., Li, Y.: Sketch-based 3d shape retrieval using convolutional neural networks. In: Proceedings of the IEEE Conference on Computer Vision and Pattern Recognition (2015)

50. Wang, L., Qian, C., Wang, J., Fang, Y.: Unsupervised learning of 3D model reconstruction from hand-drawn sketches. In: Proceedings of the 26th ACM International Conference on Multimedia, pp. 1820–1828 (2018)

51. Wang, L., Qian, X., Zhang, X., Hou, X.: Sketch-based image retrieval with multi-clustering re-ranking. IEEE Trans. Circuits Syst. Video Technol. **30**(12), 4929–4943 (2019)

52. Wang, N., Gao, X., Sun, L., Li, J.: Anchored neighborhood index for face sketch synthesis. IEEE Trans. Circuits Syst. Video Technol. **28**(9), 2154–2163 (2017)

53. Wang, N., Zhang, Y., Li, Z., Fu, Y., Liu, W., Jiang, Y.-G.: Pixel2Mesh: generating 3D mesh models from single RGB images. In: Ferrari, V., Hebert, M., Sminchisescu, C., Weiss, Y. (eds.) ECCV 2018. LNCS, vol. 11215, pp. 55–71. Springer, Cham (2018). https://doi.org/10.1007/978-3-030-01252-6_4

54. Witkin, A.P.: Recovering surface shape and orientation from texture. Artif. Intell. **17**(1–3), 17–45 (1981)

55. Wu, J., Zhang, C., Xue, T., Freeman, B., Tenenbaum, J.: Learning a probabilistic latent space of object shapes via 3D generative-adversarial modeling. In: Advances in Neural Information Processing Systems (2016)

56. Xie, H., Yao, H., Sun, X., Zhou, S., Zhang, S.: Pix2Vox: context-aware 3D reconstruction from single and multi-view images. In: Proceedings of the IEEE International Conference on Computer Vision (2019)

57. Xu, B., Chang, W., Sheffer, A., Bousseau, A., McCrae, J., Singh, K.: True2Form: 3D curve networks from 2D sketches via selective regularization. Trans. Graph. **33**(4) (2014). 2601097.2601128

58. Xu, P., et al.: Fine-grained instance-level sketch-based video retrieval. IEEE Trans. Circuits Syst. Video Technol. **31**(5), 1995–2007 (2020)

59. Yang, B., Rosa, S., Markham, A., Trigoni, N., Wen, H.: Dense 3D object reconstruction from a single depth view. IEEE Trans. Pattern Anal. Mach. Intell. **41**(12), 2820–2834 (2018)

60. Yang, G., Huang, X., Hao, Z., Liu, M.Y., Belongie, S., Hariharan, B.: PointFlow: 3D point cloud generation with continuous normalizing flows. In: Proceedings of the IEEE International Conference on Computer Vision (2019)

61. Yang, S., Wang, Z., Liu, J., Guo, Z.: Deep plastic surgery: robust and controllable image editing with human-drawn sketches. arXiv preprint arXiv:2001.02890 (2020)

62. Yu, Q., Liu, F., Song, Y.Z., Xiang, T., Hospedales, T.M., Loy, C.C.: Sketch me that shoe. In: Proceedings of the IEEE Conference on Computer Vision and Pattern Recognition (2016)

63. Zhang, S., Gao, X., Wang, N., Li, J.: Face sketch synthesis from a single photosketch pair. IEEE Trans. Circuits Syst. Video Technol. **27**(2), 275–287 (2015)

64. Zhang, S.H., Guo, Y.C., Gu, Q.W.: Sketch2Model: view-aware 3D modeling from single free-hand sketches. In: Proceedings of the IEEE/CVF Conference on Computer Vision and Pattern Recognition, pp. 6012–6021 (2021)

65. Zhang, Y., Liu, Z., Liu, T., Peng, B., Li, X.: RealPoint3D: an efficient generation network for 3D object reconstruction from a single image. IEEE Access **7**, 57539–57549 (2019)

66. Zhu, J.Y., Park, T., Isola, P., Efros, A.A.: Unpaired image-to-image translation using cycle-consistent adversarial networks. In: Proceedings of the IEEE International Conference on Computer Vision (2017)

Abstract Images Have Different Levels of Retrievability Per Reverse Image Search Engine

Shawn M. Jones[(✉)] and Diane Oyen

Los Alamos National Laboratory, Los Alamos, NM 87545, USA
{smjones,doyen}@lanl.gov

Abstract. Much computer vision research has focused on natural images, but technical documents typically consist of abstract images, such as charts, drawings, diagrams, and schematics. How well do general web search engines discover abstract images? Recent advancements in computer vision and machine learning have led to the rise of reverse image search engines. Where conventional search engines accept a text query and return a set of document results, including images, a reverse image search accepts an image as a query and returns a set of images as results. This paper evaluates how well common reverse image search engines discover abstract images. We conducted an experiment leveraging images from Wikimedia Commons, a website known to be well indexed by Baidu, Bing, Google, and Yandex. We measure how difficult an image is to find again (retrievability), what percentage of images returned are relevant (precision), and the average number of results a visitor must review before finding the submitted image (mean reciprocal rank). When trying to discover the same image again among similar images, Yandex performs best. When searching for pages containing a specific image, Google and Yandex outperform the others when discovering photographs with precision scores ranging from 0.8191 to 0.8297, respectively. In both of these cases, Google and Yandex perform better with natural images than with abstract ones achieving a difference in retrievability as high as 54% between images in these categories. These results affect anyone applying common web search engines to search for technical documents that use abstract images.

Keywords: Image similarity · Image retrieval

1 Introduction

Technical documents, such as manuals and research papers, contain visualizations in the form of abstract images such as charts, diagrams, schematics, and illustrations. Yet, technical information is often searched and retrieved by text alone [12,17,40]. Meanwhile, computer vision has significantly advanced the ability to search and retrieve images [28,52], including sketch-based retrieval

L. Karlinsky et al. (Eds.): ECCV 2022 Workshops, LNCS 13808, pp. 203–222, 2023.
https://doi.org/10.1007/978-3-031-25085-9_12

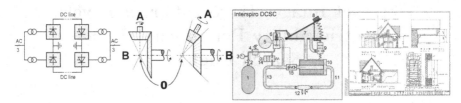

from left: Bipolar Hi Voltage DC, Bevel Gear, diving rebreather, carriage house

(a) schematic

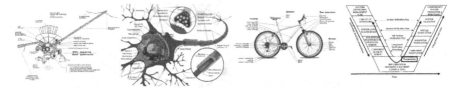

from left: Galileo spacecraft, neuron, bicycle, systems engineering V diagram

(b) diagram

from left: manatee, downtown Toronto 1890, James Garfield,
World Trade Center ground zero

(c) photo

from left: the Snake River, sunrise over Lake Michigan, New York City,
photographing a model

Fig. 1. Example images from Wikipedia Commons of our image categories. The categories of *schematic* and *diagram* represent abstract images while *photo* and *photograph* represent natural ones.

(a) Baidu   (b) Bing

(c) Google   (d) Yandex

Fig. 2. Baidu, Bing, Google, and Yandex support similar outputs if a visitor uploads an image file. A visitor can choose to view images similar to the input, or view the pages with the image that they uploaded.

[54,57,59,75]. Yet there are few computer vision research papers focused on querying and retrieving abstract, technical drawings [38,51,69], and so it is unclear whether the computer vision approaches that are successful with general images are similarly successful with technical drawings. Rather than evaluating computer vision algorithms directly, we used common search engines to evaluate whether these black-box systems (presumably incorporating state-of-the-art image retrieval algorithms) work as well in retrieving diagrams as they do with retrieving photographs. Our goal is to highlight the gap between the success of searching general images on the web and the capability to search images containing technical information on the web. We achieve this goal through the development of a method to quantify the retrievability of schematics (Fig. 1a) and diagrams (Fig. 1b) – abstract images – versus photos (Fig. 1c) and photographs (Fig. 1d) – natural images – using common search engines. In this work, we provide an answer to the following research question:

When using the reverse image search capability of general web search engines, are natural images more easily discovered than abstract ones?

Our experiment uses the **reverse image search** capability of Baidu, Bing, Google, and Yandex. Rather than submitting textual queries and receiving images as **search engine results (SERs)**, with reverse image search, we upload image files. As seen in Fig. 2, we uploaded an image of the Snake River to all four search engines. From here, a visitor can choose to explore images that are **similar-to** their uploaded image, or they can also view the list of **pages-with** the uploaded image, including images that are highly visually similar.

We used Wikimedia Commons as a source of images to upload. Prior work has focused on the presence of *Wikipedia* content in Google SERs [44] as well as those from DuckDuckGo and Bing [66]. These efforts established that Wikipedia articles have high retrievability for certain types of queries. Search engines discover images through their source documents. Thus we assume that the images in Wikipedia documents are equally retrievable.

We experimented with these four search engines. We acquired 200 abstract images from Wikimedia Commons with the search terms *diagram* and *schematic* and 199 natural images with the search terms *photo* and *photograph*. We submitted these images to the reverse image search engines at Baidu, Bing, Google, and Yandex. From there, we evaluated how often the search engine returned the same image in the results, as established by perceptual hash [36,49,77]. We evaluated the SERs in terms of precision, mean reciprocal rank [67], and retrievability [7]. Yandex performed best in all cases. When searching for pages containing a specific image, Google and Yandex outperform the others when discovering photographs with precision scores ranging from 0.8191 to 0.8297, respectively. In both of these cases, Google and Yandex perform better with natural images than with abstract ones achieving a difference in retrievability as high as 54% between images in these categories. These results indicate a clear difference in capability between natural images and abstract images, which will affect a technical document reader's ability to complete the task of discovering similar images or other documents containing an image of interest.

2   Background

Our study primarily focuses on using image files as queries, which we refer to as **query images**. Using image files as queries makes our work different from many other studies evaluating public search engines. Similar constructs exist within the SERs at the search engines in this study. As noted above, these four search engines support **similar-to** and **pages-with** SERs. The screenshots in Fig. 2 demonstrate how a visitor is presented with summaries of these options before they choose the type of SER they would like to explore further. Each SER contains an **image URL** that identifies the image indexed by the search engine. Each SER also provides a **page URL** identifying the page where a visitor can find that image in use. If the image from the image URL is no longer available, each SER also contains a **thumbnail URL** that represents a smaller version of the image. The search engine hosts these thumbnails to speed up the presentation of results.

2.1 Metrics to Evaluate Search Engines

$$Precision@k = \begin{cases} \frac{|\text{relevant SERs} \cap \text{retrieved SERs}|}{k} & \text{if } k < |\text{retrieved SERs}| \\ \frac{|\text{relevant SERs}|}{|\text{retrieved SERs}|} & \text{otherwise} \end{cases} \quad (1)$$

We apply different metrics to measure the performance of the different Reverse Image Search engines. Equation 1 demonstrates how to calculate **precision** at a cutoff k, also written $p@k$. This measure helps us understand the performance of the search engine by answering the question "What is the percentage of images in the SERs that are relevant if we stop at k results?" If the most relevant results exist at the beginning of a set of SERs, then precision decreases as the value of k increases.

$$r(d) = \sum_{q \in Q} o_q \cdot f(k_{dq}, c) \quad (2)$$

Second, we want to understand if the search engine returned the query image within a given set of results. Equation 2 shows how to calculate Azzopardi's retrievability [7]. Retrievability helps us answer the question, "Given a document, was it retrieved within the cutoff c?" Retrievability leverages a cutoff c, similar to $p@k$'s k, to indicate how many results the user reviews before stopping. Its cost function $f(k_{dq}, c)$ returns 1 if document d is found within cutoff c for query q, and 0 otherwise. As we increase c, we increase the chance of the search engine returning a relevant document; hence retrievability increases.

$$MRR = \frac{1}{Q} \sum_{i=1}^{|Q|} \frac{1}{rank_i} \quad (3)$$

Finally, we measure a visitor's effort before finding their query image in the results. Equation 3 demonstrates how to calculate **mean reciprocal rank** (**MRR**). MRR helps us answer the question "How many results, on average across all queries, must a visitor review before finding a relevant one?" For example, if the first result is relevant for an individual query, its reciprocal rank is $\frac{1}{1}$. Its reciprocal rank is $\frac{1}{2}$ if the second is relevant. If the third is relevant, it is $\frac{1}{3}$. MRR is the mean of these reciprocal ranks across all queries. An MRR of 1.0 indicates that the first result was relevant for all queries. An MRR of 0 indicates that no queries returned a relevant result.

2.2   Similarity Approaches to Establish Relevance

Each of these metrics requires that we identify which SERs are relevant. In our case, "relevant" means that it is the same image we uploaded. To determine if an image is relevant, we apply a perceptual hash [77] to each image. This perceptual hash provides a non-unique hash describing the content of the image. We compute the distances between two hashes to determine if two images are the same. If they score below some threshold, then we consider them the same. Perceptual hashes are intended to be a stand-in for human evaluation of the similarity of two images. Many perceptual hash algorithms exist. For this work, we apply ImageHash's [10] implementation of pHash [36] and GoFigure's [50] implementation of VisHash [49].

ImageHash's pHash, hereafter referred to as **pHash**, was designed to compare photographs. It applies Discrete Cosine Transforms (DCT). A user compares pHashes by computing the Hamming distance between them. As established by Krawetz [36] we use a distance of 5-bits to indicate that two images are likely the same. pHash [36,77] demonstrates improvements over simpler algorithms such as the average hash algorithm (**aHash**), which computes a hash based on the mean of color values in each pixel. Fei et al. [23], Chamoso [13], Arefkhani [3] and Vega [64] have demonstrated that pHash is more accurate than aHash in a variety of applications. These studies also evaluated the Difference hashing [37] (**dHash**) algorithm and found that it performs almost as well as pHash.

VisHash [49] was designed as an extension to dHash with a focus on providing a similarity metric for diagrams and technical drawings. VisHash relies more heavily on the shapes in the image, whereas pHash relies on frequencies. A user compares VisHashes by computing a normalized L2 distance between them. As established by Oyen et al. [49], we use a distance of ≤ 0.3 to determine if a query image is the same as the one in its SER. VisHash has fewer collisions than pHash and fewer false positive matches, especially for diagrams and technical drawings.

3   Related Work

Precision and MRR [18,42] are well-known metrics applied in evaluating search engines, especially with the established TREC dataset [68]. Retrievability is the newer metric. Azzopardi conducted several studies [5–7,14,70–74] to support the effectiveness of the retrievability metric with different search engines and corpora. Retrievability has been used to evaluate simulated user queries [63], assess the impact of OCR errors on information retrieval [60], examine retrievability in web archives [32,56], and expose issues with retrievability of patents and legal documents [8]. These studies give us confidence in this metric's capability. As far as we know, we are the first to apply it to image SERs.

In a 2013 manual evaluation, Nieuwenhuysen [47] found Google's image search to have higher precision than competitor Tineye[1]. They repeated their

---

[1] https://tineye.com/.

manual study in 2018 [48], and included Yandex but still discovered that Google performed best. In 2015, Kelly [35] conducted a similar study and found both Google and Tineye performed poorly. In 2020, Bitirim et al. [9] manually evaluated Google's reverse image search capability using 25 query images and declared its precision poor (52%) across multiple categories of items. Our study differs because we use 380 query images and use perceptive hashes to evaluate similarity, thus avoiding human disagreements about relevance. We also include both types of SER results.

d'Andréa and Mintz [20,21] noted that reverse image searches could help researchers study events for which they do not speak the language but wish to find related sources. Similar studies were conducted on images about repealing Ireland's 8th Amendment [19], Dutch politics [34], biological research [41], disinformation/misinformation [1,4,27], dermatologic diagnoses [31,53,58], and the spread of publicly available images [61]. Our work differs because it evaluates the search engines themselves for retrievability rather than as a tool to address another research question.

Horváth [30] applied reverse image search to construct a dataset useful for training classifiers. Similarly, Guinness et al. [29] performed reverse image searches with Google and scraped captions for the visually impaired. Chutel and Sakare [15,16] summarized existing reverse image search techniques. Gaillard and Egyed-Zsigmond [24,25], Araujo et al. [2], Mawoneke et al. [43], Diyasa et al. [22], Veres et al. [65], Gandhi et al. [26] evaluated their own reverse image search algorithms. Rather than evaluating our own system, we evaluate how well the state of the art publicly-available reverse image search engines function for discovering the same image again. We are more interested in whether the type of image matters to the results.

Perceptual hashes have been applied to verify images converted for digital preservation [62], to detect plagiarism [45], to evaluate machine learning results [33], and to uncover disinformation [76]. Alternatives to our chosen pHash and VisHash methods include work by Ruchay et al. [55], Monga and Evans [46], Cao et al. [11], and Lei et al. [39]. We chose ImageHash's pHash for this preliminary study due to its prevalence in other literature and its robust implementation as part of ImageHash. We selected GoFigure's VisHash as an alternative for comparison because it was specifically designed for drawings and diagrams.

4   Methods

To conduct this experiment, we needed a set of query images that we could reasonably be assured were indexed by search engines, so we turned to Wikimedia Commons. We wanted images from the categories of *diagram* and *schematic* (abstract imagery) as well as *photo* and *photograph* (natural images), so we submitted these search terms to the Wikimedia Commons API in January 2022 and recorded the resultant images. Figure 1 shows example Wikimedia Commons images from each of these queries. Those returned from the search terms *diagram* or *schematic* reflect abstract imagery. Those returned from the search terms *photo* and *photograph* represent natural images. We took the first 100 images returned from each category. One image failed to download. Two images overlap between the *schematic* and *diagram* categories. Seventeen images overlap between the *photo* and *photograph* categories. Of the 400 Wikimedia Commons images returned by the API, we were left with 380 unique images to use as queries for the reverse image search engines.

Because the images have many different sizes and search engines have file size limitations for uploaded images, we asked the Wikimedia API to generate images that were 640px wide, scaling their height to preserve their aspect ratio.

Table 1. The number of queries for which we acquired *similar-to* results

	Baidu	Bing	Google	Yandex	Category total
Diagram	82	83	79	97	100
Schematic	74	87	82	100	100
Photo	80	78	72	93	100
Photograph	77	85	79	94	99
Total	313	333	310	384	399 (380 unique)

Table 2. The number of queries for which we acquired *pages-with* results

	Baidu	Bing	Google	Yandex	Category total
Diagram	64	79	97	94	100
Schematic	54	57	87	95	100
Photo	38	44	71	94	100
Photograph	38	50	76	94	99
Total	194	230	331	377	399 (380 unique)

We developed scraping programs to submit images and extract both pages-with and similar-to SERs from Baidu, Bing, Google, and Yandex. To provide the best evaluation, we removed any context from the image files that might impart additional metadata to the search engine. To avoid the possibility that the search engines might apply the filename to its analysis, we changed the file name of each image to `upload_file.ext` where `ext` reflected the appropriate extension for the content type. We opted to use each search engine's file upload feature rather than providing the URLs directly from Wikimedia to ensure that the URL did not influence the results. We ran the scraping programs in June 2022 to successfully acquire the number of results shown in Tables 1 and 2.

We submitted the same 380 query images to each search engine's scraping program and captured the pages-with and similar-to results. We recorded the SER URL, page URL, image URL, thumbnail URL, and position of a maximum of 100 SERs from each search engine. In some cases, we did not have 100 SERs. Because downloading image URLs resulted in some images that were no longer available, we downloaded all 154,191 thumbnail images because they were more reliable. Thumbnail sizes differ by search engine, ranging in size from 61px wide with Google's page-with results to 480px tall with Yandex's similar-to results. All of our metrics require a relevance determination for each result. Due to the scale, rather than applying human judgment, we considered an SER relevant if the distance between its thumbnail image and the query image fell under the VisHash or pHash thresholds mentioned earlier.

We encountered issues downloading 8,268/154,191 (5.362%) thumbnail images. VisHash encountered issues processing 170/154,191 (0.110%) thumbnail images. Because these results could not be analyzed, we treated them as if they were not relevant for metrics purposes.

5 Results and Discussion

Tables 1 and 2 list the number of queries that returned results. These numbers are close to the total number of images per category (right column), giving us confidence in our scrapers' ability. In the rest of this section, we break down the results for the SERs with results.

Precision helps us understand how many relevant images were found within the cutoff k. Figure 3 features the average precision scores for these queries with VisHash. Figure 4 shows the same for pHash. Green lines represent natural images in the categories of *photo* and *photograph*. Blue lines represent abstract images in the categories of *diagram* and *schematic*. We keep these colors consistent across all figures in this paper. We compute precision for each value of the cutoff k within the first 10 SERs. A score of 1.0 is ideal.

We see similar patterns with both image similarity measures (pHash and VisHash). For detecting if the same image exists in the result, recall that we selected a VisHash threshold ≤ 0.3 as recommended by Oyen et al. [49] and a pHash threshold of ≤ 5 bits as recommended by Krawetz [36]. For similar-to results, we see that Baidu and Bing start with higher average precisions for *diagram* than other categories, with VisHash showing higher scores. Google's precision is lower than 0.15 regardless of category and more severe with pHash than VisHash. Yandex starts strong with the highest precision. For pages-with results, the natural image categories score the highest precision across all search engines except for Bing. Google's precision scores are much higher with pages-with than with similar-to and are competitive with Yandex. For *photograph* at $k = 1$ and page-with results, Google scores a precision of 0.8158, which is much better than the 0.52 discovered by Bitirim et al. [9] in 2020. Yandex performs slightly better than Google with *photograph* at $k = 1$ and pages-with results, scoring a precision of 0.8280. For Google, the difference in precision at $k = 1$

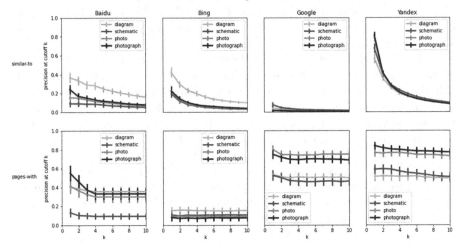

Fig. 3. Average precision at cutoff k between the values of 1 and 10 for Wikimedia Commons query images, using a VisHash threshold of 0.3. Error bars represent standard error at that value of k.

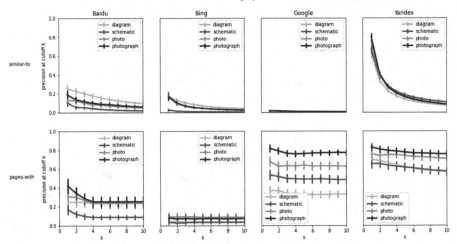

Fig. 4. Average precision at cutoff k between the values of 1 and 10 for Wikimedia Commons query images, using a pHash threshold of 5 bits. Error bars represent standard error at that value of k.

Mean Retrievability Across Image Search Engines
(relevance established by VisHash distance ≤ 0.3)

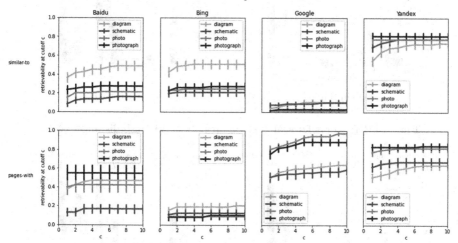

Fig. 5. Mean retrievability at cutoff c between the values of 1 and 10 for the Wikimedia Commons query images, using a VisHash threshold of 0.3. Error bars represent standard error at that value of c.

Mean Retrievability Across Image Search Engines
(relevance established by pHash distance ≤ 5 bits)

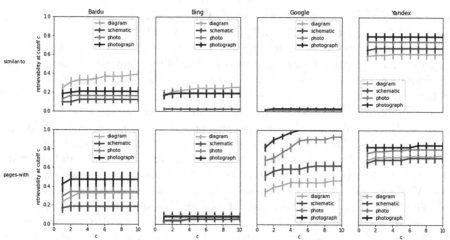

Fig. 6. Mean retrievability at cutoff c between the values of 1 and 10 for the Wikimedia Commons query images, using a pHash threshold of 5 bits. Error bars represent standard error at that value of c.

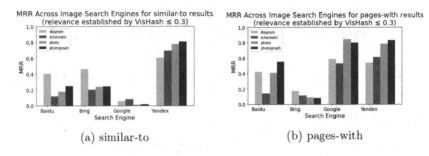

(a) similar-to   (b) pages-with

Fig. 7. MRR scores for search engine results at VisHash distance ≤0.3

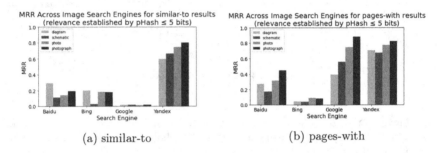

(a) similar-to   (b) pages-with

Fig. 8. MRR scores for search engine results at pHash distance ≤5 bits

and page-with results is greatest at 0.4491 between abstract images and natural images with *diagram* at 0.3667 and *photograph* at 0.8158.

Retrievability helps us determine if our query image was found within the cutoff c. Figure 5 demonstrates the mean retrievability of images within the first ten results, as compared by VisHash. Figure 6 does the same for pHash. A score of 1.0 is ideal.

For similar-to results, we see the highest mean retrievability scores with *diagram* for Baidu and Bing. Google performs poorly here, regardless of the image similarity method. Yandex performs best but better for natural images compared to abstract ones. At $c = 1$, Yandex experiences a maximum of 27% difference in retrievability between *photograph* (0.8085) and *diagram* (0.5417) when measured with VisHash. This difference goes down to 19% when measured with pHash.

When comparing the retrievability of pages-with results, for the most part, we see that all search engines tend to favor natural images over abstract ones. Bing has the poorest retrievability (less than 0.21) regardless of the image similarity method, so it is difficult to determine if it truly favors images in the category of *diagram*. Yandex has high retrievability for natural images at $c = 1$, but Google's retrievability passes Yandex by $c = 10$. Google's and Yandex's retrievability is higher than the others for abstract images. At $c = 1$, Yandex has a max 32% difference in retrievability between *photograph* (0.8297) and *diagram* (0.5054) for pages-with results when measured with VisHash. At $c=1$, Google has a max 48% difference in retrievability between *photograph* (0.8157) and *diagram* (0.3402) for pages-with results when measured with pHash. This number reaches 54% by $c = 10$.

Our final measure, MRR, establishes how many results a visitor must review before finding the query image. A score of 1.0 is ideal because it indicates that the query image was found as the first result every time. Figure 7 shows a series of bar charts demonstrating the MRR scores established with VisHash. Figure 8 shows the same with pHash. We represent each search engine with four bars, each bar representing an image category's performance. Again, green represents natural image categories while blue represents abstract images.

Yandex is the clear winner regardless of image similarity measure for similar-to results. Yandex consistently provides an MRR score of around 0.8 for natural images. Bing and Baidu are competitive for second place, and Google does not perform well for this result type. For pages-with results, Google performs competitively and scores slightly higher than Yandex in some cases. Baidu comes in a distant third, and Bing does not appear to be competitive.

Search engines crawl the web and add each page (and its images) to its **search index**. Maybe some of these search engines have not indexed the query image, which may explain the disparity in results. Unfortunately, it is difficult to assess the size of a search engine's index. Assuming that each search engine encountered these images via Wikipedia, we can search for the term "Wikipedia" in each search engine and discover how many pages the search engine has indexed with that term. Each page represents an opportunity to index an image from Wikimedia Commons. Using this method, the search engines, in order of result count, are Google (12.5 billion), Bing (339 million), Baidu (89.8 million), and Yandex (5 million). Assuming these numbers are an accurate estimate, then different index sizes do not explain the disparity in our results because Yandex, with the smallest number of indexed Wikipedia pages, still scores best by every measure.

Through our black-box analysis, by all measures, Google and Yandex perform best when returning pages-with image results, and they score best with natural images. Yandex performs best with similar-to image results and scores best with natural images. Google's results for similar-to may indicate that it is not trying to return the query image at all, instead favoring diversity over similarity.

6   Future Work

Here we report the preliminary results of a much larger experiment. We aim to build a stable dataset of Wikimedia Commons images that all four search engines have indexed; thus, we intend to repeat this experiment with more images. Once we have this list of images, we intend to conduct a follow-on study where we transform the images (e.g., cropping, resizing, rotating, color changes) to see how well the search engines perform despite each transformation.

7   Conclusion

Technical documents typically contain abstract images, but often users employ text queries to discover technical information. Computer vision has advanced such that we no longer need to only rely on text queries to retrieve images and find associated content. We can leverage the reverse image search capability of many general web search engines. Reverse image search allows one to upload an image as a query and review other images as search engine results. Few computer vision papers focus on retrieving abstract imagery; thus, we are uncertain if computer vision approaches that are successful with natural images perform as well with abstract ones. We leveraged the reverse image search capabilities of Baidu, Bing, Google, and Yandex to evaluate these black-box systems to answer the research question *When using the reverse image search capability of general web search engines, are natural images more easily discovered than abstract ones?*

We experimented with 200 abstract images and 199 natural images from Wikimedia Commons. Using scrapers, we submitted these to the reverse image search engines of Baidu, Bing, Google, and Yandex to produce 154,191 results. Each search engine has two types of results: pages-with and similar-to. Pages-with results help the visitor discover which pages contain the same image as the one uploaded. Similar-to results help the visitor discover images that are like the one uploaded. Applying perceptual hashes to compare image queries to their SER images, we discovered that Yandex performs best in all cases through precision-at-k, retrievability, and mean reciprocal rank. Yandex also performs better for natural images than abstract ones. When considering pages-with results, Google, with a precision of 0.8191, and Yandex, with a precision of 0.8297, are competitive when reviewing the first result against its uploaded image. Both favor natural images in their performance, achieving a retrievability difference as high as 54% between natural and abstract imagery at ten results.

Many reasons exist for users to conduct reverse image searches with abstract imagery. They may wish to protect intellectual property, build datasets, provide evidence for legal cases, establish scholarly evidence, or justify funding through image reuse in the community. That abstract images are at a disadvantage hurts users leveraging search engines for these use cases and provides opportunities for computer vision and information retrieval researchers.

References

1. Aprin, F., Chounta, I.A., Hoppe, H.U.: "See the image in different contexts": using reverse image search to support the identification of fake news in Instagram-like social media. In: Intelligent Tutoring Systems, pp. 264–275. Springer, Cham (2022). https://doi.org/10.1007/978-3-031-09680-8_25
2. Araujo, F.H., et al.: Reverse image search for scientific data within and beyond the visible spectrum. Exp. Syst. Appl. **109**, 35–48 (2018). https://doi.org/10.1016/j.eswa.2018.05.015
3. Arefkhani, M., Soryani, M.: Malware clustering using image processing hashes. In: Proceedings of the 9th Iranian Conference on Machine Vision and Image Processing (MVIP), pp. 214–218 (2015). https://doi.org/10.1109/IranianMVIP.2015.7397539

4. Askinadze, A.: Fake war crime image detection by reverse image search. In: Datenbanksysteme für Business, Technologie und Web (BTW 2017) - Workshopband, pp. 345–354. Gesellschaft für Informatik e.V., Bonn (2017). https://dl.gi.de/handle/20.500.12116/930

5. Azzopardi, L.: Theory of retrieval: the retrievability of information. In: Proceedings of the 2015 International Conference on The Theory of Information Retrieval, ICTIR 2015, pp. 3–6. Association for Computing Machinery, Northampton (2015). https://doi.org/10.1145/2808194.2809444

6. Azzopardi, L., English, R., Wilkie, C., Maxwell, D.: Page retrievability calculator. Adv. Inf. Retrieval, 737–741 (2014). https://doi.org/10.1007/978-3-319-06028-6_85

7. Azzopardi, L., Vinay, V.: Retrievability: an evaluation measure for higher order information access tasks. In: Proceeding of the 17th ACM Conference on Information and Knowledge Mining, p. 561. ACM Press, Napa Valley (2008). https://doi.org/10.1145/1458082.1458157

8. Bashir, S., Rauber, A.: Improving retrievability of patents with cluster-based pseudo-relevance feedback documents selection. In: Proceedings of the 18th ACM Conference on Information and Knowledge Management, Hong Kong, China, pp. 1863–1866 (2009). https://doi.org/10.1145/1645953.1646250

9. Bitirim, Y., Bitirim, S., Celik Ertugrul, D., Toygar, O.: An evaluation of reverse image search performance of google. In: 2020 IEEE 44th Annual Computers, Software, and Applications Conference (COMPSAC), pp. 1368–1372 (2020). https://doi.org/10.1109/COMPSAC48688.2020.00-65

10. Buchner, J.: Johannesbuchner/imagehash (2021). https://github.com/JohannesBuchner/imagehash

11. Cao, Y., Qi, H., Kato, J., Li, K.: Hash ranking with weighted asymmetric distance for image search. IEEE Trans. Comput. Imaging $3(4)$, 1008–1019 (2017). https://doi.org/10.1109/TCI.2017.2736980

12. Caragea, C., et al.: CiteSeer^x: A Scholarly Big Dataset. In: de Rijke, M., et al. (eds.) ECIR 2014. LNCS, vol. 8416, pp. 311–322. Springer, Cham (2014). https://doi.org/10.1007/978-3-319-06028-6_26

13. Chamoso, P., Rivas, A., Martín-Limorti, J.J., Rodríguez, S.: A hash based image matching algorithm for social networks. In: De la Prieta, F., et al. (eds.) PAAMS 2017. AISC, vol. 619, pp. 183–190. Springer, Cham (2018). https://doi.org/10.1007/978-3-319-61578-3_18

14. Chen, R.C., Azzopardi, L., Scholer, F.: An empirical analysis of pruning techniques: performance, retrievability and bias. In: Proceedings of the 2017 ACM on Conference on Information and Knowledge Management, Singapore, Singapore, pp. 2023–2026 (2017). https://doi.org/10.1145/3132847.3133151

15. Chutel, P.M., Sakhare, A.: Evaluation of compact composite descriptor based reverse image search. In: Proceedings of the 2014 International Conference on Communication and Signal Processing, Melmaruvathur, India, pp. 1430–1434 (2014). https://doi.org/10.1109/ICCSP.2014.6950085

16. Chutel, P.M., Sakhare, A.: Reverse image search engine using compact composite descriptor. Int. J. Adv. Res. Comput. Sci. Manag. Stud. $2(1)$ (2014). https://www.ijarcsms.com/docs/paper/volume2/issue1/V2I1-0106.pdf

17. Clark, C., Divvala, S.: PDFFigures 2.0: mining figures from research papers. In: IEEE/ACM Joint Conference on Digital Libraries (JCDL), pp. 143–152 (2016). https://ieeexplore.ieee.org/abstract/document/7559577

18. Croft, W.B., Metzler, D., Strohman, T.: Information Retrieval in Practice. Pearson Education, Boston (2015). https://ciir.cs.umass.edu/irbook/

19. Curran, A.: Ordinary and extraordinary images: making visible the operations of stock photography in posters against the repeal of the 8th amendment. Feminist Encounters **6**(1) (2022). https://doi.org/10.20897/femenc/11746

20. d'Andrea, C., Mintz, A.: Studying 'Live' cross-platform circulation of images with a computer vision API: an experiment based on a sports media event. In: The 19th Annual Conference of the Association of Internet Researchers, Montréal, Canada (2018). https://doi.org/10.5210/spir.v2018i0.10477

21. d'Andrea, C., Mintz, A.: Studying the live cross-platform circulation of images with computer vision API: an experiment based on a sports media event. Int. J. Commun. **13**(0) (2019). https://ijoc.org/index.php/ijoc/article/view/10423

22. Diyasa, I.G.S.M., Alhajir, A.D., Hakim, A.M., Rohman, M.F.: Reverse image search analysis based on pre-trained convolutional neural network model. In: Proceedings of the 6th Information Technology International Seminar (ITIS), Surabaya, Indonesia, pp. 1–6 (2020). https://doi.org/10.1109/ITIS50118.2020.9321037

23. Fei, M., Li, J., Liu, H.: Visual tracking based on improved foreground detection and perceptual hashing. Neurocomputing **152**, 413–428 (2015). https://doi.org/10.1016/j.neucom.2014.09.060

24. Gaillard, M., Egyed-Zsigmond, E.: Large scale reverse image search. XXXVème Congrès INFORSID, p. 127 (2017). https://inforsid.fr/actes/2017/INFORSID_2017_paper_34.pdf

25. Gaillard, M., Egyed-Zsigmond, E., Granitzer, M.: CNN features for Reverse Image Search. Document numérique **21**(1–2), 63–90 (2018). https://www.cairn.info/revue-document-numerique-2018-1-page-63.htm

26. Gandhi, V., Vaidya, J., Rana, N., Jariwala, D.: Reverse image search using discrete wavelet transform, local histogram and canny edge detector. Int. J. Eng. Res. Technol. **7**(6) (2018). https://www.ijert.org/reverse-image-search-using-discrete-wavelet-transform-local-histogram-and-canny-edge-detector

27. Ganti, D.: A novel method for detecting misinformation in videos, utilizing reverse image search, semantic analysis, and sentiment comparison of metadata. In: SSRN (2022). https://ssrn.com/abstract=4128499

28. Gordo, A., Almazan, J., Revaud, J., Larlus, D.: End-to-end learning of deep visual representations for image retrieval. Int. J. Comput. Vis. **124**(2), 237–254 (2017). https://doi.org/10.1007/s11263-017-1016-8

29. Guinness, D., Cutrell, E., Morris, M.R.: Caption crawler: enabling reusable alternative text descriptions using reverse image search. In: Proceedings of the 2018 CHI Conference on Human Factors in Computing Systems, Montreal, QC, Canada, pp. 1–11 (2018). https://doi.org/10.1145/3173574.3174092

30. Horváth, A.: Object recognition based on Google's reverse image search and image similarity. In: Proceedings of the Seventh International Conference on Graphic and Image Processing (ICGIP 2015), vol. 9817, pp. 162–166. International Society for Optics and Photonics, SPIE (2015). https://doi.org/10.1117/12.2228505

31. Jia, J.L., Wang, J.Y., Mills, D.E., Shen, A., Sarin, K.Y.: Fitzpatrick phototype disparities in identification of cutaneous malignancies by google reverse image. J. Am. Acad. Dermatol. **84**(5), 1415–1417 (2021). https://doi.org/10.1016/j.jaad.2020.05.005

32. Jones, S.M.: Improving collection understanding for web archives with storytelling: shining light into dark and stormy archives. Ph.D. thesis, Old Dominion University (2021). https://doi.org/10.25777/zts6-v512

33. Jones, S.M., Weigle, M.C., Klein, M., Nelson, M.L.: Automatically selecting striking images for social cards. In: Proceedings of the 13th ACM Web Science Conference, pp. 36–45 (2021). https://doi.org/10.1145/3447535.3462505
34. Kateřina, Z.: Propaganda on social media: the case of geert wilders. Master's thesis, Charles University (2018). https://hdl.handle.net/20.500.11956/99767
35. Kelly, E.: Reverse image lookup of a small academic library digital collection. Codex J. Louisiana Chap. ACRL **3**(2) (2015). https://journal.acrlla.org/index.php/codex/article/view/101
36. Krawetz, N.: Looks like it (2011). https://hackerfactor.com/blog/index.php%3F/archives/432-Looks-Like-It.html
37. Krawetz, N.: Kind of like that (2013). https://www.hackerfactor.com/blog/index.php?/archives/529-Kind-of-Like-That.html
38. Kucer, M., Oyen, D., Castorena, J., Wu, J.: DeepPatent: large scale patent drawing recognition and retrieval. In: Proceedings of the IEEE/CVF Winter Conference on Applications of Computer Vision (WACV), pp. 2309–2318 (2022). https://openaccess.thecvf.com/content/WACV2022/html/Kucer_DeepPatent_Large_Scale_Patent_Drawing_Recognition_and_Retrieval_WACV_2022_paper.html
39. Lei, Y., Wang, Y., Huang, J.: Robust image hash in Radon transform domain for authentication. Sig. Process. Image Commun. **26**(6), 280–288 (2011). https://doi.org/10.1016/j.image.2011.04.007
40. Li, S., Hu, J., Cui, Y., Hu, J.: DeepPatent: patent classification with convolutional neural networks and word embedding. Scientometrics **117**(2), 721–744 (2018). https://doi.org/10.1007/s11192-018-2905-5
41. Mamrosh, J.L., Moore, D.D.: Using google reverse image search to decipher biological images. Current Protoc. Mol. Biol. **111**(1), 19.13.1–19.13.4 (2015). https://doi.org/10.1002/0471142727.mb1913s111
42. Manning, C.D., Raghavan, P., Schütze, H.: Introduction to Information Retrieval. Cambridge University Press (2008). https://nlp.stanford.edu/IR-book/html/htmledition/irbook.html
43. Mawoneke, K.F., Luo, X., Shi, Y., Kita, K.: Reverse image search for the fashion industry using convolutional neural networks. In: 2020 IEEE 5th International Conference on Signal and Image Processing (ICSIP), pp. 483–489 (2020). https://doi.org/10.1109/ICSIP49896.2020.9339350
44. McMahon, C., Johnson, I., Hecht, B.: The substantial interdependence of wikipedia and google: a case study on the relationship between peer production communities and information technologies. In: Proceedings of the Eleventh International AAAI Conference on Web and Social Media (ICWSM 2017), Montréal, Québec, Canada, p. 10 (2017). https://www.aaai.org/ocs/index.php/ICWSM/ICWSM17/paper/viewPaper/15623
45. Meuschke, N., Gondek, C., Seebacher, D., Breitinger, C., Keim, D., Gipp, B.: An adaptive image-based plagiarism detection approach. In: Proceedings of the 18th ACM/IEEE on Joint Conference on Digital Libraries, JCDL 2018, pp. 131–140. Association for Computing Machinery, Fort Worth (2018). https://doi.org/10.1145/3197026.3197042
46. Monga, V., Evans, B.: Perceptual image hashing via feature points: performance evaluation and tradeoffs. IEEE Trans. Image Process. **15**(11), 3452–3465 (2006). https://doi.org/10.1109/TIP.2006.881948
47. Nieuwenhuysen, P.: Search by image through the WWW: an additional tool for information retrieval. In: The International Conference on Asia-Pacific Library and Information Education and Practices A-LIEP, p. 38 (2013)

48. Nieuwenhuysen, P.: Finding copies of an image: a comparison of reverse image search systems on the WWW. In: Proceedings of 14th International Conference on Webometrics, Informetrics and Scientometrics, Macau, China, pp. 97–106 (2018). https://doi.org/10.22032/dbt.39355

49. Oyen, D., Kucer, M., Wohlberg, B.: VisHash: visual similarity preserving image hashing for diagram retrieval. In: Applications of Machine Learning 2021, vol. 11843, pp. 50–66. International Society for Optics and Photonics, SPIE (2021). https://doi.org/10.1117/12.2594720

50. Oyen, D., Wohlberg, B., Kucer, M.: GoFigure-LANL/VisHash (2021). https://github.com/GoFigure-LANL/VisHash

51. Piroi, F., Lupu, M., Hanbury, A., Zenz, V.: CLEF-IP 2011: retrieval in the intellectual property domain. In: Conference and Labs of the Evaluation Forum (2011). https://ceur-ws.org/Vol-1177/CLEF2011wn-CLEF-IP-PiroiEt2011.pdf

52. Radenović, F., Tolias, G., Chum, O.: Fine-tuning CNN image retrieval with no human annotation. IEEE Trans. Pattern Anal. Mach. Intell. **41**(7), 1655–1668 (2018). https://doi.org/10.1109/TPAMI.2018.2846566

53. Ransohoff, J.D., Li, S., Sarin, K.Y.: Assessment of accuracy of patient-initiated differential diagnosis generation by google reverse image searching. JAMA Dermatol. **152**(10), 1164–1166 (2016). https://doi.org/10.1001/jamadermatol.2016.2096

54. Ribeiro, L.S.F., Bui, T., Collomosse, J., Ponti, M.: Sketchformer: transformer-based representation for sketched structure. In: The IEEE/CVF Conference on Computer Vision and Pattern Recognition (CVPR), June 2020. https://openaccess.thecvf.com/content_CVPR_2020/html/Ribeiro_Sketchformer_Transformer-Based_Representation_for_Sketched_Structure_CVPR_2020_paper.html

55. Ruchay, A., Kober, V., Yavtushenko, E.: Fast perceptual image hash based on cascade algorithm. In: Applications of Digital Image Processing XL, vol. 10396, pp. 424–430. International Society for Optics and Photonics, SPIE (2017). https://doi.org/10.1117/12.2272716

56. Samar, T., Traub, M.C., van Ossenbruggen, J., Hardman, L., de Vries, A.P.: Quantifying retrieval bias in Web archive search. Int. J. Dig. Libr. **19**(1), 57–75 (2018). https://doi.org/10.1007/s00799-017-0215-9

57. Sangkloy, P., Burnell, N., Ham, C., Hays, J.: The sketchy database: learning to retrieve badly drawn bunnies. ACM Trans. Graph. (Proc. SIGGRAPH) (2016). https://doi.org/10.1145/2897824.2925954

58. Sharifzadeh, A., Smith, G.P.: Inaccuracy of Google reverse image search in complex dermatology cases. J. Am. Acad. Dermatol. **84**(1), 202–203 (2021). https://doi.org/10.1016/j.jaad.2020.04.107

59. Song, J., Song, Y.Z., Xiang, T., Hospedales, T.M.: Fine-grained image retrieval: the text/sketch input dilemma. In: BMVC, vol. 2, p. 7 (2017). https://doi.org/10.5244/C.31.45

60. van Strien, D., Beelen, K., Ardanuy, M.C., Hosseini, K., McGillivray, B., Colavizza, G.: Assessing the impact of OCR quality on downstream NLP tasks. In: Proceedings of the 12th International Conference on Agents and Artificial Intelligence (2020). https://doi.org/10.5220/0009169004840496

61. Thompson, S., Reilly, M.: "A picture is worth a thousand words": reverse image lookup and digital library assessment. J. Assoc. Inf. Sci. Technol. **68**(9), 2264–2266 (2017). https://doi.org/10.1002/asi.23847

62. Tikhonov, A.: Preservation of digital images: question of fixity. Heritage **2**(2), 1160–1165 (2019). https://doi.org/10.3390/heritage2020075

63. Traub, M.C., Samar, T., van Ossenbruggen, J., He, J., de Vries, A., Hardman, L.: Querylog-based assessment of retrievability bias in a large newspaper corpus. In: Proceedings of the 16th ACM/IEEE-CS on Joint Conference on Digital Libraries, JCDL 2016, pp. 7–16. Association for Computing Machinery, New York (2016). https://doi.org/10.1145/2910896.2910907

64. Vega, F., Medina, J., Mendoza, D., Saquicela, V., Espinoza, M.: A robust video identification framework using perceptual image hashing. In: 2017 XLIII Latin American Computer Conference (CLEI), pp. 1–10 (2017). https://doi.org/10.1109/CLEI.2017.8226396

65. Veres, O., Rusyn, B., Sachenko, A., Rishnyak, I.: Choosing the method of finding similar images in the reverse search system. In: COLINS, pp. 99–107 (2018). https://ceur-ws.org/Vol-2136/10000099.pdf

66. Vincent, N., Hecht, B.: A deeper investigation of the importance of Wikipedia links to search engine results. Proc. ACM Hum. Comput. Interact. 5(CSCW1), 1–15 (2021). https://doi.org/10.1145/3449078

67. Voorhees, E.M.: The TREC-8 question answering track report. In: Proceedings of the 8th Text Retrieval Conference (TREC-8) (1999). https://nvlpubs.nist.gov/nistpubs/Legacy/SP/nistspecialpublication500-246.pdf

68. Voorhees, E.M., Harman, D.K.: TREC: Experiment and Evaluation in Information Retrieval. MIT Press (2005)

69. Vrochidis, S., Moumtzidou, A., Kompatsiaris, I.: Concept-based patent image retrieval. World Patent Inf. 34(4), 292–303 (2012). https://doi.org/10.1016/j.wpi.2012.07.002

70. Wilkie, C., Azzopardi, L.: Relating retrievability, performance and length. In: Proceedings of the 36th International ACM SIGIR Conference on Research and Development in Information Retrieval, Dublin, Ireland, pp. 937–940 (2013). https://doi.org/10.1145/2484028.2484145

71. Wilkie, C., Azzopardi, L.: Efficiently estimating retrievability bias. Adv. Inf. Retrieval, 720–726 (2014). https://doi.org/10.1007/978-3-319-06028-6_82

72. Wilkie, C., Azzopardi, L.: A retrievability analysis: exploring the relationship between retrieval bias and retrieval performance. In: Proceedings of the 23rd ACM International Conference on Conference on Information and Knowledge Management, pp. 81–90. Association for Computing Machinery, Shanghai (2014). https://doi.org/10.1145/2661829.2661948

73. Wilkie, C., Azzopardi, L.: Retrievability and retrieval bias: a comparison of inequality measures. Adv. Inf. Retrieval, 209–214 (2015). https://doi.org/10.1007/978-3-319-16354-3_22

74. Wilkie, C., Azzopardi, L.: Algorithmic bias: do good systems make relevant documents more retrievable? In: Proceedings of the 2017 ACM Conference on Information and Knowledge Management, Singapore, Singapore, pp. 2375–2378 (2017). https://doi.org/10.1145/3132847.3133135

75. Xu, P., Hospedales, T.M., Yin, Q., Song, Y.Z., Xiang, T., Wang, L.: Deep learning for free-hand sketch: a survey. IEEE Trans. Pattern Anal. Mach. Intell. (2022). https://doi.org/10.1109/TPAMI.2022.3148853

76. Zannettou, S., Caulfield, T., Bradlyn, B., De Cristofaro, E., Stringhini, G., Blackburn, J.: Characterizing the use of images in state-sponsored information warfare operations by Russian Trolls on Twitter. In: Proceedings of the International AAAI Conference on Web and Social Media, vol. 14, no. (1), pp. 774–785 (2020). https://ojs.aaai.org/index.php/ICWSM/article/view/7342

77. Zauner, C.: Implementation and benchmarking of perceptual image hash functions. Master's thesis, Upper Austria University of Applied Sciences (2010). https://www.phash.org/docs/pubs/thesis_zauner.pdf

W35 - Sign Language Understanding

Sign languages are spatial-temporal languages and constitute a key form of communication for Deaf communities. Recent progress in fine-grained gesture and action classification, machine translation and image captioning, point to the possibility of automatic sign language understanding becoming a reality. The Sign Language Understanding Workshop is a full-day event which brings together two workshops: firstly, Open Challenges in Continuous Sign Language Recognition (morning session), and secondly, Sign Language Recognition, Translation & Production (afternoon session - virtual).

The Open Challenges in Continuous Sign Language Recognition workshop will focus on advances and new challenges on the topic of SLR. To advance and motivate the research on the field, this workshop will have an associated challenge on fine-grain sign spotting for continuous SLR. The aim of this workshop is to put a spotlight on the strengths and limitations of the existing approaches, and define the future directions of the field.

The Sign Language Recognition, Translation & Production workshop will bring together computer vision researchers, sign language linguists and members of the Deaf community. The workshop will consist of invited talks and also a challenge with three tracks: individual sign recognition; English sentence to sign sequence alignment; and sign spotting. The focus of this workshop is to engage with members of the Deaf community, broaden participation in sign language research, and cultivate collaborations.

October 2022

Liliane Momeni
Gul Varol
Hannah Bull
Prajwal K. R.
Neil Fox
Ben Saunders
Necat Cihan Camgöz
Richard Bowden
Andrew Zisserman
Bencie Woll
Sergio Escalera
Jose L. Alba-Castro
Thomas B. Moeslund
Julio C. S. Jacques Junior
Manuel Vazquez-Enriquez

ECCV 2022 Sign Spotting Challenge: Dataset, Design and Results

Manuel Vázquez Enríquez[1]([✉])[ID], José L. Alba Castro[1][ID],
Laura Docio Fernandez[1][ID], Julio C. S. Jacques Junior[2][ID],
and Sergio Escalera[2,3][ID]

[1] University of Vigo, Vigo, Spain
{mvazquez,jalba,ldocio}@gts.uvigo.es
[2] Computer Vision Center, Bellaterra, Spain
jjacques@cvc.uab.cat
[3] University of Barcelona, Barcelona, Spain
sergio@maia.ub.es

Abstract. The ECCV 2022 Sign Spotting Challenge focused on the problem of fine-grain sign spotting for continuous sign language recognition. We have released and made publicly available a new dataset of Spanish sign language of around 10 h of video data in the health domain performed by 7 deaf people and 3 interpreters. The added value of this dataset over existing ones is the frame-level precise annotation of 100 signs with their corresponding glosses and variants made by sign language experts. This paper summarizes the design and results of the challenge, which attracted 79 participants, contextualizing the problem and defining the dataset, protocols and baseline models, as well as discussing top-winning solutions and future directions on the topic.

Keywords: Sign spotting · Sign language · 3DCNN · ST-GCN

1 Introduction

In recent years, the scientific community has accelerated research in automatic Sign Language Recognition (SLR), supported by the latest advances in deep learning models with flexible spatial-temporal representation capacities [27]. It is well known that these methods need to be fed with huge amounts of data and this is one of the main drawbacks for a faster progress of automatic SLR. Sign languages are purely visual languages lacking a fully accepted writing system. Transcribing the visual information of continuous signing to glosses or any of the few alternative codes (e.g., HamNoSys [26]) can only be made by few specialists and it is a extremely time-consuming task.

SLR can be roughly classified into Isolated (ISLR) and Continuous (CSLR) sign language recognition. ISLR is the most studied scenario by the scientific community and for which more annotated datasets are available, because of the relative easy annotation protocols of a discrete and predefined set of signs. CSLR, however, is much more complex due to three main factors. First, the co-articulation between signs drastically increases the variability of the sign realization with respect to isolated signs and reflects the signing style of each person.

L. Karlinsky et al. (Eds.): ECCV 2022 Workshops, LNCS 13808, pp. 225–242, 2023.
https://doi.org/10.1007/978-3-031-25085-9_13

Second, the speed is much higher than in the examples of isolated signs. Another point is the interplay of non-manual components: torso, head, eyebrows, eyes, mouth, lips, and even tongue [9]. The third factor is the already mentioned shortage of experts capable of annotating signs in sentences.

The ultimate goal consist in Sign Language Translation (SLT) from a Sign Language (SL) to a spoken language. Therefore, available corpora composed of a visual sign language and the corresponding transcription in glosses are scarce. Recent works [1,4,36] are leveraging captioned and signed broadcast media to develop direct SLT without passing explicitly through any intermediate transcribed representation. Nevertheless, in those restricted-domain cases where the transcription to glosses is available, either decoupling SLR to glosses and translation from glosses to spoken language, or using the glosses as a training guidance for the end-to-end translation process yields better performance [5].

Sign Spotting is a special case of CSLR where the specific grammar of each SL is not taken into account but only delimiting the localization of a particular sign. It allows the development of a wide number of applications such as indexing SL content, enabling efficient search and "intelligent fast-forward" to topics of interest, helping to linguistic studies or even learning sign language. Sign spotting can also be reliably used for collecting co-articulated samples of the query sign to improve an ISLR model without the need of extensive expert annotations [20].

2   Related Work

Early works on sign/gesture localization in video were supported by time-series pattern spotting techniques like Dynamic Time Warping [35], Global Alignment Kernel [25], Hidden Markov Models [2], Conditional Random Fields [39] or Hierarchical Sequential interval Pattern Trees [24] to build models over hand-crafted features usually directed by linguistic knowledge. Most of these approaches were tested over ad-hoc datasets of isolated signs in lab conditions or continuous sign language with limited variability of signers or annotated signs. The statistical approach used in speech recognition [18] was adapted to SL and tested on the "new" RWTH-Phoenix-Weather database [12], which set a milestone in CSLR.

With the release of the RWTH-Phoenix-Weather database of German continuous SL and the advent of deep learning, CSLR started to attract the attention of many research groups. Cui et al. [10] proposed the first CSLR system completely trained with deep networks, based on weakly supervision and end-to-end training. They used a convolutional neural network (CNN) with temporal convolution and pooling for spatio-temporal representation learning, and a RNN model with a long short-term memory (LSTM) module to learn the mapping of feature sequences to sequences of glosses. The trend shifted to learn short-term spatial-temporal features of signs directly from the RGB(+D) sequences or modalities derived from it (e.g., optical flow or skeletal data), mostly pushed by the success of 3D Convolutional Networks and the two-stream Inflated 3D ConvNet (I3D model) used in action classification [7]. Spatio-temporal Graph Convolutional Networks (ST-GCNs) [38], inherited from the action recognition field, also helped to stimulate research on ISLR and CSLR, using skeleton data from RGB body sequences [6,8] to learn spatio-temporal patterns.

Sign spotting has benefited from the advances in spatial-temporal representation and sequence learning from weakly labeled signs. Sign spotting has a specific domain mismatch problem between the query (usually isolated signs) and target (co-articulated signs). Jiang et al. [16] designed a one-shot sign spotting approach using I3D for learning the spatial-temporal representation to address the temporal scale discrepancies between the query and the target videos, building multiple queries from a single video using different frame-level strides. They proposed a transformer-based network that exploits self attention and mutual attention between the query and target features to learn the self and mutual temporal dependency. Their results on the BSLCORPUS [30] (continuous) using SignBank [11] (isolated) as queries outperformed previous approaches. The recently released BSL-1K [1] dataset, with weakly-aligned subtitles from broadcast footage and a vocabulary of 1000 signs automatically located in 1000 h hours of video allows training larger CSLR models and testing sign spotting techniques. In [34], the authors proposed an approach to one/few shot sign spotting through three supervision cues: mouthing, subtitled words and isolated sign dictionary. They used a unified learning framework using the principles of Noise Contrastive Estimation and Multiple Instance Learning [22], which allows the learning of representations from weakly-aligned subtitles while exploiting sparse labels from mouthings [1] and explicitly accounts for sign variations. The approach is based on I3D and was evaluated for sign spotting on BSL-1K using the isolated signs dataset collected by the same research group, BSLDICT [23].

3   Challenge Design

The ECCV 2022 Sign Spotting Challenge[1] was aimed to attract attention on the strengths and limitations of the existing approaches, and help to define the future directions of the field. It was divided into two competition tracks:

- **MSSL (multiple shot supervised learning).** MSSL is a classical machine learning track where signs to be spotted are the same in training, validation and test sets. The three sets contain samples of signs cropped from the continuous stream of Spanish sign language, meaning that all of them have co-articulation influence. The training set contains the begin-end timestamps (in milisecs) annotated by a deaf person and a SL-interpreter (with an homogeneous criteria) of multiple instances for each of the query signs. Participants needed to spot those signs in a set of test videos.
- **OSLWL (one shot learning and weak labels).** OSLWL is a realistic variation of a one-shot learning adapted to the sign language specific problem, where it is relatively easy to obtain a couple of examples of a sign using just a sign language dictionary, but it is much more difficult to find co-articulated

---

[1] Challenge - https://chalearnlap.cvc.uab.cat/challenge/49/description/.

versions of that specific sign. When subtitles are available, as in broadcast-based datasets, the typical approach consists in using the text to predict a likely interval where the sign might be performed. In this track, we simulated that case by providing a set of queries (20 isolated signs performed by a signer of the dataset and 20 by an external one) and a set of 4 sec video intervals around each and every co-articulated instance of the queries. Intervals with no instances of the queries were provided to simulate the typical case where the subtitle or translated text shows a word that the signer selected not to perform or even a subtitle error. Participants needed to spot the exact location of the sign instances in the provided video intervals. In this track, only one sign needs to be located for each video.

The participants were free to join any of these challenge tracks. Each track was composed of two phases, i.e., development and test phase. At the development phase, public train data was released and participants needed to submit their predictions with respect to a validation set. At the test phase, participants needed to submit their results with respect to the test data. Participants were ranked, at the end of the challenge, using the test data. Note that this competition involved the submission of results (and not code). Therefore, participants were required to share their codes and trained models after the end of the challenge so that the organizers could reproduce their results in a "code verification stage". At the end of the challenge, top ranked methods that passed the code verification stage (discussed in Sect. 4) were considered as valid submissions.

3.1   The Dataset

LSE_eSaude_UVIGO, released for the ECCV 2022 Sign Spotting Challenge, is a dataset of Spanish Sign Language (LSE: "Lengua de Signos Española") in the health domain (around 10 h of video data), signed by 10 people (7 deaf and 3 interpreters) partially annotated with the exact location of 100 signs. This dataset has been collected under different conditions than the typical Continuous Sign Language datasets, which are mainly based on subtitled broadcast and real-time translation. In our case, the signs are performed to explain health contents by translating printed cards, so reliance on signers is large due to the richer expressivity and naturalness. The dataset was acquired in studio conditions with blue chroma-key, no shadow effects and uniform illumination, at 25 FPS and a resolution of 1080×720. The added value of the dataset is the rich and rigorous hand-made annotations. Experts interpreters and deaf people were in charge of annotating the location of the selected signs. Additional details about the dataset, data split and annotation protocol can be found in the dataset webpage[2].

The signers in the test set can be the same or different to the training and validation set. Signers are men, women, right and left-handed. The amount of samples per sign was not uniform, as some signs are common terms but others are related to a specific health topic. The total number of hand annotations

---

[2] Dataset - https://chalearnlap.cvc.uab.cat/dataset/42/description/.

are 4822 in MSSL track and 1921 in OSLWL track. The duration of the co-articulated signs is not a Gaussian variable, starting from 120 ms (3 frames) up to 2 s (50 frames), with a mean duration of 520 ms (13 frames).

3.2 Evaluation Protocol

The evaluation protocol is the same for both tracks, based on the F1 score. Matching score per sign instance is evaluated as the Intersection over Unit (IoU) of the ground-truth interval and the predicted interval. In order to allow relaxed locations, the IoU threshold is swept from $t = 0.2$ to 0.8 using 0.05 as step size.

Given a specific video file n and query sign i to spot, the submitted solutions are evaluated as follows. First, let us define $Ret_i(k, n)$ as the kth interval in video file n retrieved as a prediction of sign i, and $Rel_i(n)$ an interval annotated for an instance of sign i. Then, $IoU_i(k, n)$ is obtained as:

$$IoU_i(k, n) = \frac{Ret_i(k, n) \cap Rel_i(n)}{Ret_i(k, n) \cup Rel_i(n)}. \tag{1}$$

A True Positive (TP) occurs if $IoU_i(k, n) \geq t$. Note that IoU is calculated only between intervals with prediction and ground-truth matching. Moreover, no overlap is allowed among Ret_i intervals. Then, for all $Ret_i(k, n)$ reported from each and every video file n containing L_n ground-truth instances, we compute $TP(t) = \sum_n \sum_i \sum_k (IoU_i(k, n) \geq t)$, $FP(t) = (\sum_n \sum_i \sum_k 1) - TP(t)$, and $FN(t) = (\sum_n L_n) - TP(t)$ for each IoU threshold t. Then, the accumulated over all t values as $TP = \sum_t TP(t)$, $FP = \sum_t FP(t)$, and $FN = \sum_t FN(t)$.

Precision (P) and Recall (R) averages the amounts from different thresholds as $P = TP/(TP + FP)$ and $R = TP/(TP + FN)$. Finally, the F1 score is obtained using in Eq. 2.

$$F1 = 2 * (P * R)/(P + R). \tag{2}$$

3.3 The Baseline

The baseline model is similar for both tracks, with a common pipeline and two branches that take into account the requirements of MSSL and OSLWL tracks. First, a person detection model [28] locates the position of the signer and a 512×512 ROI is extracted from the original footage. Then, the Mediapipe Holistic keypoint detector [13] extracts $11 + 21 \times 2 = 53$ coordinate points each 40ms. The core of the model is a multi-scale ST-GCN, MSG3D [21], trained with upper body and hands skeleton [37]. Two models are independently trained and averaged, one with coordinates as input features (joints), and the other with distances between connected coordinates (bones). No explicit RGB or motion information is used for spatial-temporal modeling. The model was pre-trained on the AUTSL [31] dataset, fine-tuned using 3 signers of MSSL train set and validated over the other 2. Ground-truth annotations of the 60 class-signs were used with 2 different context windows: 400 ms to train a short-context model and 1000 ms to train a long-context model. Small random shifts around the ground-truth interval in the training window allowed data augmentation.

At inference stage, the short- and long-context models are applied in a sequential decision pipeline. The short-context model is applied first with a stride of one video-frame (40 ms). It yields one class decision per frame if the wining class surpasses a threshold of 0.75. As this output is noisy, a non-linear filter designed as a morphological operator eliminates potential short-time false positives and false negatives by building convex-like sign outputs. The isolated candidate signs are then fed to the long-context model in charge of eliminating false positive candidates with a stricter threshold of 0.8. Figure 1 shows the main blocks of the pipeline for the MSSL (upper branch) and OSLWL (lower branch) tracks.

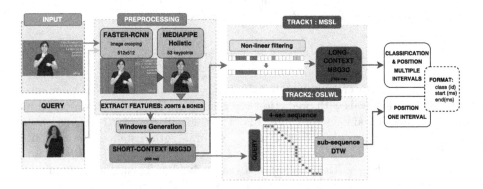

Fig. 1. Pipeline of the baseline solution for MSSL and OSLWL tracks.

The solution for OSLWL track had no training at all. The inference pipeline shares all the blocks with MSSL up to the short-context model. At this point, the baseline extracts an internal embedding vector of 384D just before the last FC layer. For every 4 s interval a specific sign can appear at any place (or not appear at all), so a sub-sequence Dynamic Time Warping (DTW) is applied with cosine distance between the specific query of the searched sign and the 4 s 384D-sequence. A fixed threshold of 0.2 was set for discarding low-valued matching scores. So, the result of the sub-sequence DTW is the time-interval of the spotted sign only if the score surpasses the threshold.

4   Challenge Results and Winning Methods

The challenge ran from 21 April 2022 to 24 June 2022 through Codalab[3], a powerful open source framework for running competitions. It attracted a total of 79 participants, 37 in MSSL track and 42 in OSLWL track, suggesting where the research community is paying more attention, given the two challenge tracks.

---

[3] Codalab - https://codalab.lisn.upsaclay.fr.

4.1   The Leaderboard

The leaderboard at the test phase of both tracks, for the submissions that passed the code verification and improved the baseline scores, are shown in Table 1, ranked by average F1 using multiple IOU thresholds from 0.2 to 0.8 with a stride 0.05. As it can be observed, the top-3 winning methods improved the baseline scores on both tracks by a large margin.

Table 1. Leaderboard of MSSL and OSLWL tracks at the test phase.

MSSL			OSLWL		
Rank	Participant	Avg F1	Rank	Participant	Avg F1
1	ryanwong	0.606554	1	th	0.595802
2	th	0.566752	2	Mikedddd	0.559295
3	Random_guess	0.564260	3	ryanwong	0.514309
4	Baseline	0.300123	4	Baseline	0.395083

Table 2 shows the scores for different thresholds where the performance on the relaxed to strict localization requirements can be observed. It is interesting to highlight three points: i) the top-3 winning methods surpassed the baseline regardless the IoU threshold; ii) the reduction of F1 when requiring strict spatial spotting; iii) the *th* team would win both tracks under the stricter IoU threshold.

Table 2. IoU score for different threshold values (0.2, 0.5, 0.8). Bold values highlight the overall winning method. In italic, when swapped leadeboard position.

MSSL				OSLWL			
Participant	1@20	F1@50	F1@80	Participant	F1@20	F1@50	F1@80
ryanwong	**0.744**	**0.677**	0.280	th	0.784	**0.647**	**0.269**
th	0.660	0.626	**0.282**	Mikedddd	**0.809**	0.594	0.160
Random_guess	0.715	0.652	0.204	ryanwong	0.744	0.552	*0.164*
Baseline	0.465	0.339	0.056	Baseline	0.621	0.437	0.080

Table 3 shows general information about the top-3 winning methods. As can be seen, common strategies employed by top-winning solutions are the use of pre-trained models (most of them for feature extraction, as detailed in the next section), some face, hand, body detection, alignment or segmentation strategy, combined with pose estimation and/or spatio-temporal information modeling.

Table 3. General information about the top-3 winning approaches.

Track	MSSL			OSLWL		
Rank	1	2	3	1	2	3
Participant	ryanwong	th	Random_guess	th	Mikedddd	ryanwong
Pre-trained models	✓	✓	✓	✓	✓	✓
External data	✗	✗	✗	✗	✗	✗
Any kind of depth information (e.g., 3D pose estimation)	✗	✗	✓	✗	✓	✗
Use of validation set as part of the training data (at test stage)	✓	✓	✓	✗	✓	✗
Handcrafted features	✗	✗	✗	✗	✗	✗
Face / hand / body detection, alignment or segmentation	✓	✓	✗	✓	✓	✓
Use of any pose estimation method	✗	✓	✓	✓	✓	✗
Spatio-temporal feature extraction	✓	✓	✓	✓	✓	✓
Explicitly classify any attribute (e.g., gender/handedness)	✗	✗	✗	✗	✗	✗
Bias mitigation technique (e.g. rebalancing training data)	✓	✗	✓	✗	✗	✗

Next, we briefly introduce the top winning methods based on the information provided by the authors. For a detailed information, we refer the reader to the associated fact sheets, available for download in the challenge webpage(See footnote 4).

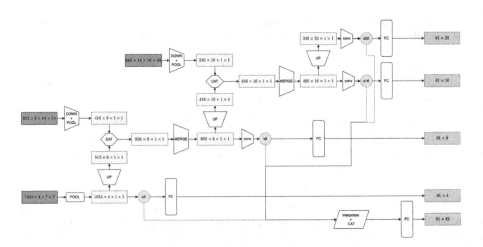

Fig. 2. Top-1, MSSL track (*ryanwong* team): proposed pipeline.

4.2 Top Winning Approaches: MSSL Track

Top-1: *ryanwong* **team.** The *ryanwong* team proposed to modify and use an I3D [7] model from [1], pretrained on WLASL [19] dataset. Originally, the I3D model outputs a single sign prediction (i.e., for a given region in time), given a sign video sequence of 32 frames, which limits the network to coarse temporal predictions. Thus, they proposed hierarchical I3D model, which can predict signs at frame level. Instead of taking the output at the final layer with spatial temporal global average pooling and applying a FC layer for class predictions, they take the output before global average pooling and additional feature outputs before the 3D max pool layers in the I3D model, obtaining 3 feature outputs each with a higher temporal resolution. For a given sequence length of 32 frames of dimensions 224×224, the base I3D model outputs the following features $1024 \times 4 \times 7 \times 7$, $832 \times 8 \times 14 \times 14$ and , $480 \times 16 \times 28 \times 28$, with a temporal resolution of 4, 8 and 16, respectively. The proposed network, illustrated in Fig. 2, uses these inputs to output coarse-to-fine temporal predictions ranging from 4, 8, 16 and 32 temporally aligned predictions. That is, making 1 prediction every 8, 4, 2 and 1 frame(s), respectively. Cross Entropy loss is used to predict the sign at each time segment for the coarse-to-fine predictions. A trade off between precision and recall is obtained with different random sampling probabilities were instead of selecting only frame regions around only known sign classes, they randomly select frame regions from other areas. The final predictions are based on temporally interpolating the softmax of the logit features for each of the predictions to the original sequence length (32) and averaging the 5 output results to obtain the probabilities for each class prediction at frame level.

Top-2: *th* **team.** The *th* team proposed a two-stage framework, which consists of feature extraction and temporal sign action localization, illustrated in Fig. 3a.

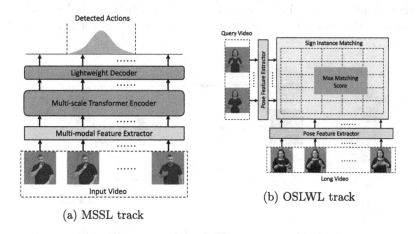

(a) MSSL track

(b) OSLWL track

Fig. 3. a) Top-2 and b) Top-1 winning solutions on MSSL and OSLWL track, respectively, both from *th* team.

First, multiple modalities (RGB, optical flow [33] and pose) are used to extract robust spatio-temporal feature representation. Then, a Transformer backbone is adopted to identify actions in time and recognize their categories. As pre-processing, MMDetection [8] is employed to detect the signer spatial location, followed by MMPose [6] (from OpenMMLab project) for extracting body and hand poses, used to crop the upper-body patch of the signer. Finally, three types of data are generated, one for each modality. Different backbones are used for feature extraction, given their complementary representations. RGB modality is fed into the BSL-1k [1] pre-trained I3D [7]. Flow modality is processed by the AUTSL [31] pre-trained I3D [7]. For the pose modality, pre-trained GCN [3,14] is used to extract body and hand cues. The extracted features are concatenated for the next localization stage, where Transformer is used to perform localization similar to [40]. Specifically, it combines a multiscale feature representation with local self-attention, using a light-weighted decoder to classify every moment in time and to estimate the corresponding action boundaries. During training, focal loss for sign action classification and generalized IoU loss for distance regression are adopted. At the inference stage, the output of every time step is taken, organized as the triplet sign action confidence score, onset and offset of the action. These candidates are further processed via NMS to remove highly overlapping instances, which leads to the final localization outputs.

Top-3: *Random_guess* **team.** The *Random_guess* team proposed a two-stage pipeline for sign spotting. The aim of the first stage is to condense sign language relevant information from multiple domain experts into a compact representation for sign spotting, while the goal of the second stage is to spot signs from longer video with a more powerful temporal module. Their workflow diagram is presented in Fig. 4. At the inference stage, the processed video clips (cropped video, masked video, optical flow, and 3D skeleton) are fed into four trained expert modules. The extracted features of each clip are concatenated into a vector to represent sign information for spotting. The temporal module takes vectors as input and extracts contextual information. It is composed of a two-

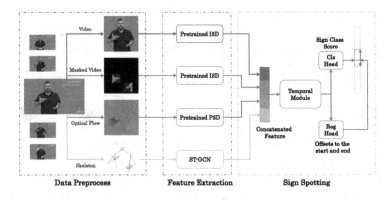

Fig. 4. Top-3, MSSL track (*Random_guess* team): workflow diagram.

layer 1D convolutional and a two-layer BiLSTM for local and contextual temporal information extraction, respectively. Their outputs are integrated through a convolutional layer. The integrated features are fed into the classification head and the regression head, separately, to predict the corresponding class and the offsets to its start-end. Then, a vote-based method is applied to provide better localization results. During training, they first train the backbone for feature extraction. Then, the whole model is fine tuned for sign spotting.

4.3 Top Winning Approaches: OSLWL Track

Top-1: *th* team. The *th* team presented a two-stage framework, consisting of feature extraction and sign instance matching, illustrated in Fig. 3b. First, they extract frame-level feature representation from pose modality. Then, they build a similarity graph to perform frame-wise matching between the query isolated sign and long video. MMDetection [8] is used to detect the signer spatial location, and MMPose for extracting body and hand poses [6]. The compact pose is used to indicate the gesture state of a certain frame, and for trimming the effective time span of the query sign. Pre-trained GCN [3,14] is used to extract body and hand representations in a frame-wise manner. Given the extracted feature of the query and long video, an ad-hoc similarity graph is built to perform sign instance matching. Authors suggest incorporating non-manual cues (e.g., facial expressions) as a possible way to further boost the performance.

Top-2: *Mikedddd* team. The *Mikedddd* team proposed a multi-modal framework for extracting sign language features from RGB images using I3D-MLP [7], 2D-pose [32] and 3D-pose [29] based on SL-GCN [15]. They introduced a novel sign spotting loss function by combining the triplet loss and cross-entropy loss to obtain more discriminative feature representations and training each model separately. At inference stage, the three models are employed to extract visual features before multi-modality features fusion for the sign spotting task. The pipeline of the proposed framework is shown in Fig. 5. More precisely, to achieve the goal of the challenge they proposed the following steps: 1) first, they observed there is an obvious domain gap between isolated signs and continuous signs. Therefore, they trimmed the isolated sign videos by removing the video frames before hands up and after hands down. Then, they fed the gallery sign videos into the model to generate feature representations. As a result, 20 feature representations are obtained in total, each of which represents an isolated sign; 2) For

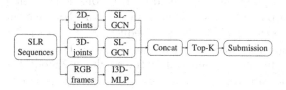

Fig. 5. *Mikedddd* team (top-2, OSLWL track): the inference pipeline.

each video from the query set they use a sliding window to crop video clips from the beginning to the end. Then, they input all clips into the proposed model to extract feature representations; 3) All feature representations from this video are used to calculate the Cosine distance with respect to the corresponding feature representation of the sign provided in the gallery set. The clip that has the maximal similarity with the gallery sign representation is regarded as their retrieval result; 4) They propose a $top - K$ transferring technique to address the domain gap between the gallery set and the query set. After the second step, they obtain several retrieved results for each isolated sign, which are sorted given the computed distance to find the most similar K clips. Each feature representation from the gallery set will be updated by the average of the feature representations of the $top - K$ retrieved clips. Iteratively, the updated feature representation for a certain sign will be used to retrieve signs from continuous sign videos again.

Top-3: *ryanwong* **team.** The *ryanwong* team presented a method for sign spotting using existing I3D [7] models pretrained on sign language datasets. First, they show how these models can be used for identifying important frames from isolated sign dictionaries. The goal at this stage is to discard the most common frames where the signer is in their resting pose or those frames with irrelevant information. To this end, feature vectors are obtained for each frame using I3D. Then, the Cosine similarity between each of the frame features across all of the isolated sign videos is computed, which creates a cosine similarity matrix that is used to determine how common the sign frame is within the sign video dictionary. The "key" frames are kept for each isolated sign sequence based on the obtained similarity matrix and predefined thresholds. Next, for each isolated query video they randomly select 8 frames (sorted by indices) of the important "key" frames previously identified. This query sequence is used as input into an I3D model for feature extraction, where a feature vector q is obtained. Similarly, the co-articulated sign video is used as input into the I3D model for feature extraction, and a feature vector k is obtained. The cosine similarity between q and k is calculated to obtain the similarity matrix s. This process is repeated with 64 different combinations of query sequences and 64 different co-articulated sign video (with random data augmentation and random frame selection), obtaining 4096 similarity scores. The mean of all similarity scores at each time step is calculated to obtain the final similarity score s_f. Finally, they compute the "normalized similarity score" s_n by making the assumption that there exists at least 1 occurrence of the isolated sign in the co-articulated sequence by dividing s_f by the maximum value in the sequence. Any indices in s_n greater than a predefined threshold (0.9) is considered a match between the isolated sign video and the continuous sign segment. For each time index with a matched spotting they include the 8 subsequent frames as spotting matches and combine matching spottings if they are in range of 10 frames of each other. The final results are obtained by ensembling the results of 2 models [1] pretrained on WLASL [19] and MSASL [17], and taking the average between the normalized cosine similarities.

4.4 Performance on Marginal Distributions of the Test Set

In this section, we analyze the performance of the top-winning methods on marginal distributions of the test set. Table 4 shows model performance in MSSL track on signers seen during training (SD) and new signers (SI). As expected, top-winning methods performed worse on the signer independent (i.e., SI) scenario, with a performance improvement greater than 10% *w.r.t* the signer-dependent scenario. Note that all methods had models pre-trained with external sign language data. However, they used the challenge training data for fine-tuning. The *baseline* also performed worse on the SI scenario but at a smaller ratio if compared with its SD F1 score. This may be explained by the fact that it was tuned with skeleton data only. It is also worth noting that ranking would stay almost the same if looking just to SI performance, with a swap between the 2nd and 3rd position, not surprising as both had already really close F1 values.

Table 4. Signer dependency of the systems of MSSL track at test phase.

MSSL: SD → p01, p05, SI → p03, p08								
Rank	Participant	SD/SI	TP	FP	FN	Pr	Re	F1
1	ryanwong	SD	5803	2439	2985	0.704	0.660	0.682
		SI	3813	2102	4949	0.645	0.435	0.520
2	th	SD	5607	2973	3181	0.653	0.638	0.646
		SI	3448	2376	5314	0.592	0.394	0.473
3	Random_guess	SD	4982	2298	3806	0.684	0.567	0.620
		SI	3406	1495	5356	0.695	0.389	0.499
4	baseline	SD	2763	5076	6025	0.352	0.314	0.332
		SI	1638	2301	7124	0.416	0.187	0.258

Note that this results account for the 13th IoU threshold (i.e., 0.8) tested.

Table 5. Per signer performance of MSSL track at test phase.

MSSL: Signers → p05(interpreter), p01, p03, p08 (deaf)								
Rank	Participant	Signer	TP	FP	FN	Pr	Re	F1
1	ryanwong	p05	4823	1573	1781	0.754	0.730	0.742
		p01	980	866	1204	0.531	0.449	0.486
		p03	3348	1865	4478	0.642	0.427	0.514
		p08	465	237	471	0.662	0.497	0.568
2	th	p05	4635	1852	1969	0.714	0.702	0.708
		p01	972	1121	1212	0.464	0.445	0.454
		p03	3067	2224	4759	0.580	0.392	0.468
		p08	381	152	555	0.715	0.407	0.518
3	Random_guess	p05	4143	1473	2461	0.738	0.627	0.678
		p01	839	825	1345	0.504	0.384	0.436
		p03	2953	1324	4873	0.690	0.377	0.488
		p08	453	171	483	0.725	0.484	0.581
4	Baseline	p05	2389	3864	4215	0.382	0.362	0.372
		p01	374	1212	1810	0.236	0.171	0.198
		p03	1354	2104	6472	0.391	0.173	0.240
		p08	284	197	652	0.590	0.303	0.401

Table 5 shows the results marginalized per signer. Signer *p05* is a sign language interpreter, the remainder in MSSL test set are native deaf signers. Participants methods obtained a +10% larger F1 on the interpreter videos, *w.r.t* deaf signers. A possible explanation is that interpreters display less naturalness, hence variability, while signing. This hypothesis should be contrasted further, as *p05* is also the signer with more training data. The F1 ranking among signers holds among the three competing methods (*p05* > *p08* > *p03* > *p01*), so there is a clear dependency on signing style that makes it difficult to separate from the dependency of amount of training data (*p03* and *p08* not seen during training). We leave this important question open to further testing because many of the current larger datasets are based on broadcast video with interpreters' signing.

Table 6 shows per signer performance, precision (Pr), recall (Re) and F1 in OSLWL track. Note that OSLWL does not have explicit training signs/signers, but 2 of the analyzed methods, 2nd,*Mikedddd* and 4th, *Baseline*, used MSSL data in the model that extracts feature vectors from the query and MSSL videos. *Mikedddd* also used the OSLWL validation data. We can make some observations from Table 6. First, performance on interpreters (*p05*, *p09*) is better than on deaf signers if the model used training data containing them, which is not observed on methods ranked 1st and 3rd, where deaf signer *p04* obtained better or similar performance. So, as in MSSL, performance is quite signer dependent. Second, there's not a better performance on signer *p03* who is the one that signs half of the queries. This observation shows that, even though the scenario and signer is the same, the domain change between isolated signs and co-articulated signs plays a more important role.

In short, the top-winning methods show a consistent performance on marginal distributions of the test set. Per signer evaluation shows a remarkable performance difference that can not be fully blamed to the amount of training data available for each signer but, probably, to factors related to signing style. It seems that interpreters' style is easier to learn that deafs' style, but there's not enough evidence to support this hypothesis, which deserves further investigation.

Table 6. Per signer performance of OSLWL track at test phase.

OSLWL: Signers → p05, p09 (interpreters), p01, p03, p04, p08 (deaf)									
Rank	Participant	Pr	Re	F1	Signer	F1	Re	Pr	Participant	Rank
1	th	0.588	0.596	0.592	p05	0.584	0.617	0.554	Mikedddd	2
		0.645	0.645	0.645	p09	0.651	0.680	0.624		
		0.507	0.497	0.502	p01	0.460	0.477	0.444		
		0.453	0.453	0.453	p03	0.462	0.510	0.421		
		0.702	0.728	0.715	p04	0.582	0.612	0.555		
		0.632	0.438	0.517	p08	0.361	0.361	0.361		
3	ryanwong	0.488	0.521	0.504	p05	0.422	0.372	0.487	Baseline	4
		0.562	0.576	0.569	p09	0.485	0.448	0.529		
		0.415	0.484	0.447	p01	0.264	0.239	0.294		
		0.474	0.568	0.517	p03	0.277	0.226	0.357		
		0.529	0.553	0.541	p04	0.423	0.393	0.457		
		0.405	0.467	0.434	p08	0.278	0.225	0.365		

5   Conclusions

This paper summarized the ECCV 2022 Sign Spotting Challenge. We analysed and discussed the challenge design, top winning solutions and results. The new released dataset allowed training and testing sign spotting methods under challenging conditions with deaf and interpreter signers. Although having around 10 h of video data, rich and rigorous hand-made annotations, the dataset is still limited by its small number of participants which have an impact on generalization. To address this problem, future research directions should move on the development of novel large-scale and public datasets and on the research and development of methods that are both fair and accurate, where people from different gender, age, demographics, sign style, among others, are considered.

Top winning solutions of this challenge shared similar pipeline blocks regarding spatial-temporal representation using pre-trained models on larger sign language datasets from different languages, both from 3DCNN and ST-GCN deep learning models. That is, they benefited from state-of-the-art approaches for feature extraction, modeling and/or fusion. Main differences are based on how the domain adaptation problem is handled and the use of the target dataset to train/fine-tune the models or their meta-parameters. The post-challenge experiments showed that signer style can affect the quality of the sign spotting performance, which can be affected by the amount of train data and data distribution previously mentioned. In this line, future work should consider paying more attention to explainability/interpretability, which are key to understand what part or components of the model are more relevant to solve a particular problem, or to explain possible sources of bias or misclassification.

Acknowledgments. This work has been supported by the Spanish projects PID2019-105093GB-I00 and RTI2018-101372-B-I00, by ICREA under the ICREA Academia programme and by the Xunta de Galicia and ERDF through the Consolidated Strategic Group AtlanTTic (2019–2022). Manuel Vázquez is funded by the Spanish Ministry of Science and Innovation through the predoc grant PRE2019-088146.

References

1. Albanie, Samuel, et al.: BSL-1K: scaling up co-articulated sign language recognition using mouthing cues. In: Vedaldi, Andrea, Bischof, Horst, Brox, Thomas, Frahm, Jan-Michael. (eds.) ECCV 2020. LNCS, vol. 12356, pp. 35–53. Springer, Cham (2020). https://doi.org/10.1007/978-3-030-58621-8_3
2. Alon, J., Athitsos, V., Yuan, Q., Sclaroff, S.: A unified framework for gesture recognition and spatiotemporal gesture segmentation. IEEE Trans. Pattern Anal. Mach. Intell. **31**(9), 1685–1699 (2009)
3. Cai, Y., et al.: Exploiting spatial-temporal relationships for 3d pose estimation via graph convolutional networks. In: International Conference on Computer Vision (ICCV), pp. 2272–2281 (2019)
4. Camgoz, N.C., Hadfield, S., Koller, O., Ney, H., Bowden, R.: Neural sign language translation. In: Proceedings of the IEEE conference on Computer Vision and Pattern Recognition (CVPR) (2018)

5. Camgoz, N.C., Koller, O., Hadfield, S., Bowden, R.: Sign language transformers: joint end-to-end sign language recognition and translation. In: Conference on Computer Vision and Pattern Recognition (CVPR), pp. 10023–10033 (2020)
6. Cao, Z., Hidalgo, G., Simon, T., Wei, S., Sheikh, Y.: Openpose: realtime multiperson 2d pose estimation using part affinity fields. IEEE Trans. Pattern Anal. Mach. Intell. **43**(01), 172–186 (2021)
7. Carreira, J., Zisserman, A.: Quo vadis, action recognition? a new model and the kinetics dataset. In: Conference on Computer Vision and Pattern Recognition (CVPR) (2017)
8. Chen, K., et al.: Mmdetection: Open mmlab detection toolbox and benchmark. CoRR abs/1906.07155 (2019)
9. Cooper, H., Holt, B., Bowden, R.: Sign language recognition. In: Moeslund, T., Hilton, A., Krüger, V., Sigal, L. (eds.) Visual Analysis of Humans, Springer, London, pp. 539–562 (2011). https://doi.org/10.1007/978-0-85729-997-0_27
10. Cui, R., Liu, H., Zhang, C.: Recurrent convolutional neural networks for continuous sign language recognition by staged optimization. In: Conference on Computer Vision and Pattern Recognition (CVPR), pp. 1610–1618 (2017)
11. Fenlon, J.B., et al.: Bsl signbank: a lexical database and dictionary of British sign language 1st edn (2014)
12. Forster, J., Schmidt, C., Koller, O., Bellgardt, M., Ney, H.: Extensions of the sign language recognition and translation corpus RWTH-PHOENIX-weather. In: Proceedings of the Ninth International Conference on Language Resources and Evaluation (LREC 2014), pp. 1911–1916 (2014)
13. Grishchenko, I., Bazarevsky, V.: Mediapipe holistic - simultaneous face, hand and pose prediction, on device. https://ai.googleblog.com/2020/12/mediapipe-holisticsimultaneous-face.html (2022). Accessed 18 Jul 2022
14. Hu, H., Zhao, W., Zhou, W., Wang, Y., Li, H.: SignBERT: pre-training of hand-model-aware representation for sign language recognition. In: International Conference on Computer Vision (ICCV), pp. 11087–11096 (2021)
15. Jiang, S., Sun, B., Wang, L., Bai, Y., Li, K., Fu, Y.: Skeleton aware multi-modal sign language recognition. In: Conference on Computer Vision and Pattern Recognition Workshops (CVPRW), pp. 3408–3418 (2021)
16. Jiang, T., Camgoz, N.C., Bowden, R.: Looking for the signs: identifying isolated sign instances in continuous video footage. In: IEEE International Conference on Automatic Face and Gesture Recognition (FG 2021), pp. 1–8 (2021)
17. Joze, H.R.V., Koller, O.: MS-ASL: a large-scale data set and benchmark for understanding American sign language. CoRR abs/1812.01053 (2018)
18. Koller, O., Forster, J., Ney, H.: Continuous sign language recognition: towards large vocabulary statistical recognition systems handling multiple signers. Comput. Vis. Image Understand. **141**, 108–125 (2015)
19. Li, D., Rodriguez, C., Yu, X., Li, H.: Word-level deep sign language recognition from video: a new large-scale dataset and methods comparison. In: Winter Conference on Applications of Computer Vision (WACV) (2020)
20. Li, D., Yu, X., Xu, C., Petersson, L., Li, H.: Transferring cross-domain knowledge for video sign language recognition. In: Conference on Computer Vision and Pattern Recognition (CVPR), pp. 6204–6213 (2020)
21. Liu, Z., Zhang, H., Chen, Z., Wang, Z., Ouyang, W.: Disentangling and unifying graph convolutions for skeleton-based action recognition. In: Conference on Computer Vision and Pattern Recognition (CVPR), pp. 140–149 (2020)

22. Miech, A., Alayrac, J.B., Smaira, L., Laptev, I., Sivic, J., Zisserman, A.: End-to-End Learning of Visual Representations from Uncurated Instructional Videos. In: Conference on Computer Vision and Pattern Recognition (CVPR) (2020)
23. Momeni, L., Varol, G., Albanie, S., Afouras, T., Zisserman, A.: Watch, read and lookup: learning to spot signs from multiple supervisors. In: ACCV (2020)
24. Ong, E.J., Koller, O., Pugeault, N., Bowden, R.: Sign spotting using hierarchical sequential patterns with temporal intervals. In: 2014 IEEE Conference on Computer Vision and Pattern Recognition, pp. 1931–1938 (2014)
25. Pfister, T., Charles, J., Zisserman, A.: Domain-adaptive discriminative one-shot learning of gestures. In: European Conference on Computer Vision (ECCV), vol. 8694, pp. 814–829 (2014)
26. Prillwitz, S.: HamNoSys Version 2.0. Hamburg Notation System for Sign Languages: An Introductory Guide. Intern. Arb. z. Gebärdensprache u. Kommunik, Signum Press, Dresden (1989)
27. Rastgoo, R., Kiani, K., Escalera, S.: Sign Language recognition: a deep Survey. Expert Syst. Appl. **164**, 113794 (2021)
28. Ren, S., He, K., Girshick, R., Sun, J.: Faster r-cnn: towards real-time object detection with region proposal networks. In: Cortes, C., Lawrence, N., Lee, D., Sugiyama, M., Garnett, R. (eds.) Advances in Neural Information Processing Systems, vol. 28, pp. 91–99. Curran Associates, Inc. (2015)
29. Rong, Y., Shiratori, T., Joo, H.: Frankmocap: A monocular 3d whole-body pose estimation system via regression and integration. In: International Conference on Computer Vision Workshops (ICCVW), pp. 1749–1759 (2021)
30. Schembri, A.C., Fenlon, J.B., Rentelis, R., Reynolds, S., Cormier, K.: Building the British sign language corpus. Lang. Documentation Conserv. **7**, 136–154 (2013)
31. Sincan, O.M., Keles, H.Y.: AUTSL: a large scale multi-modal Turkish sign language dataset and baseline methods. IEEE Access **8**, 181340–181355 (2020)
32. Sun, K., Xiao, B., Liu, D., Wang, J.: Deep high-resolution representation learning for human pose estimation. In: Conference on Computer Vision and Pattern Recognition (CVPR), pp. 5686–5696 (2019)
33. Sánchez Pérez, J., Meinhardt-Llopis, E., Facciolo, G.: TV-L1 optical flow estimation. Image Process. Line **3**, 137–150 (2013)
34. Varol, G., Momeni, L., Albanie, S., Afouras, T., Zisserman, A.: Scaling up sign spotting through sign language dictionaries. Int. J. Comput. Vis. 1–24 (2022). https://doi.org/10.1007/s11263-022-01589-6
35. Viitaniemi, V., Jantunen, T., Savolainen, L., Karppa, M., Laaksonen, J.: S-pot - a benchmark in spotting signs within continuous signing. In: International Conference on Language Resources and Evaluation (LREC), pp. 1892–1897 (2014)
36. Voskou, A., Panousis, K.P., Kosmopoulos, D., Metaxas, D.N., Chatzis, S.: Stochastic transformer networks with linear competing units: application to end-to-end SL translation. In: International Conference on Computer Vision, ICCV 2021, Montreal, QC, Canada, 10–17 October 2021, pp. 11926–11935 (2021)
37. Vázquez-Enríquez, M., Alba-Castro, J.L., Docío-Fernández, L., Rodríguez-Banga, E.: Isolated sign language recognition with multi-scale spatial-temporal graph convolutional networks. In: Conference on Computer Vision and Pattern Recognition (CVPR) Workshops (2021)
38. Yan, S., Xiong, Y., Lin, D.: Spatial temporal graph convolutional networks for skeleton-based action recognition. In: Proceedings of the Thirty-Second AAAI Conference on Artificial Intelligence, pp. 7444–7452 (2018)

Hierarchical I3D for Sign Spotting

Ryan Wong[(✉)], Necati Cihan Camgöz, and Richard Bowden

University of Surrey, Guildford, UK
{r.wong,n.camgoz,r.bowden}@surrey.ac.uk

Abstract. Most of the vision-based sign language research to date has focused on Isolated Sign Language Recognition (ISLR), where the objective is to predict a single sign class given a short video clip. Although there has been significant progress in ISLR, its real-life applications are limited. In this paper, we focus on the challenging task of Sign Spotting instead, where the goal is to simultaneously identify and localise signs in continuous co-articulated sign videos. To address the limitations of current ISLR-based models, we propose a hierarchical sign spotting approach which learns coarse-to-fine spatio-temporal sign features to take advantage of representations at various temporal levels and provide more precise sign localisation. Specifically, we develop Hierarchical Sign I3D model (HS-I3D) which consists of a hierarchical network head that is attached to the existing spatio-temporal I3D model to exploit features at different layers of the network. We evaluate HS-I3D on the ChaLearn 2022 Sign Spotting Challenge - MSSL track and achieve a state-of-the-art 0.607 F1 score, which was the top-1 winning solution of the competition.

Keywords: Sign language recognition · Sign spotting

1 Introduction

Sign Languages are visual languages that incorporate the motion of the hands, facial expression and body movement [3]. They are the primary form of communication amongst deaf communities with most countries having their own sign languages with different dialects across different regions. Although there are many commonalities across sign languages in terms of linguistics and grammatical rules, each has a very different lexicon [24].

There has been increasing interest in computational sign language research. A popular research topic has been Isolated Sign Language Recognition (ISLR), where the goal is to identify which single sign is present in a short isolated sign video clip [15,17]. Although there are still challenges to solve, such as signer-independent recognition [25], the real-life applications of ISLR are limited.

In this work we focus on the closely related field of Sign Spotting, where the objective is to identify and localise instances of signs within a co-articulated continuous sign video. Sign spotting is beneficial for several real-life applications, such as Sign Content Retrieval, where spotting models are used to search through large unlabelled corpora to locate instances of signs.

© The Author(s), under exclusive license to Springer Nature Switzerland AG 2023
L. Karlinsky et al. (Eds.): ECCV 2022 Workshops, LNCS 13808, pp. 243–255, 2023.
https://doi.org/10.1007/978-3-031-25085-9_14

Current sign spotting approaches can be categorized under two groups. The first is dictionary based sign spotting approaches where given an isolated sign, the objective is to identify and locate co-articulate instances of that sign in a continuous sign video. This usually involves one-shot/few shot learning where minimal annotated examples are available [26].

The second group of sign spotting approaches align closer to ISLR, but instead of a sign video containing an isolated sign, the video segment is usually longer with one or multiple instances of co-articulated signs that needs to be identified from a set vocabulary. This involves multiple shot supervised learning where there are multiple examples of a set vocabulary within a larger corpus of continuous sign videos.

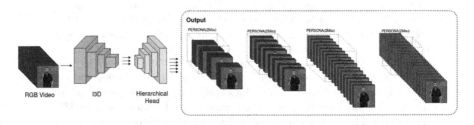

Fig. 1. Concept of the Hierarchical Sign I3D model which takes an input video sequence and predicts the localisation of signs at various temporal resolutions

In this work we build up on the latter group of sign spotting. To address the limitations of the previous approaches, we propose a novel hierarchical spatio-temporal network architecture, named a Hierarchical Sign I3D model (HS-I3D), and identify coarse-to-fine temporal locations of signs in continuous sign videos as shown in Fig. 1.

HS-I3D comprises of a backbone and a head. Although our approach can be used with any spatio-temporal backbone, we've chosen I3D due to its success in related sign tasks [26], which enables the use of other pretrained SLR models. The additional hierarchical spatio-temporal network head has the ability to predict sign labels at the frame level for better estimation of the boundaries between signs.

The main contributions of this work can be summarised as:

1. We introduce a novel hierarchical spatio-temporal network head which can be attached to existing spatio-temporal sign models to learn the coarse-to-fine temporal locations of signs.
2. We demonstrate the importance of incorporating random sampling techniques during training and show the impact and trade-off it has between precision and recall.
3. Our architecture achieves state-of-the-art results on the 2022 ChaLearn Sign Spotting Challenge in the multiple shot supervised learning (MSSL) track.

2   Related Work

2.1   Sign Language Recognition

Over the last few decades, significant progress has been made towards Sign Language Recognition. Traditional feature engineered approaches, such as hand shape and motion modeling techniques [2,6,9,11] have been replaced by data driven, machine learning approaches. These data driven approaches require large annotated datasets therefore many ISLR datasets have been created, including but not limited to Turkish Sign Language (TID) [25], American Sign Language (ASL) [15,17], Chinese Sign Language [30] and British Sign Language (BSL) [1,7].

Most of the current ISLR approaches use either raw RGB videos or pose-based input. The pose based input utilizes human pose estimators such as Open-Pose [4] and MediaPipe [19], which distill the signer to a set of keypoints. By using keypoints, irrelevant appearance information is discarded, such as the background and a person's visual appearance. Various models have been developed to allow keypoint input, Pose→Sign [1] makes use of a 2D ResNet architecture [10] and found that the keypoint inputs underperformed compared to RGB based approaches. More recently, Graph Convolutional Networks (GCNs) using human keypoints have achieved comparable results to RGB models [13].

RGB based approaches, which use the raw frames as input, have been extended to spatio-temporal architectures, building on existing action recognition models, such as 3DCNNs [21] and more recently the I3D model [5]. Such architectures achieve strong classification performance on ISLR datasets [13,15,17].

Like in other areas of computer vision, transfer learning has been shown to be effective in improving results of sign language recognition [1], which is especially important for transferring domain knowledge across different sign languages. Motivated by this, we use a pretrained I3D model as the backbone model for our proposed Hierarchical Sign I3D model which allows us to leverage models pretrained on larger scale ISLR dataset.

2.2   Sign Spotting

While ISLR aims to identify an isolated sign in a given sequence, Sign Spotting requires identification of both the start and end of a sign instance from a set vocabulary within a continuous sign video. Early methods utilized hand crafted features, such as thresholding-based approaches using Conditional Random Fields, to distinguish the difference between signs in the vocabulary and non-sign patterns [28]. Another technique detected skin-coloured regions in frames and utilized temporal alignment techniques, such as Dynamic Time Warping [27]. Sequential Interval Patterns were also proposed, which used hierarchical trees to learn a strong classifier to spot signs [20].

One-shot approaches, which use sign dictionaries have recently been explored, where given a set vocabulary from a dictionary, the objective is to locate these signs within a continuous sign video. Sign-Lookup makes use of a transformer-based network which utilises cross-attention to spot signs from a query dictionary to the target continuous video [14].

Most ISLR models are trained on datasets that are collected specifically for the task, where the signs are slower and not co-articulated. This makes transfer learning from such models to tasks that focus on co-articulated sign videos, such as sign spotting, difficult. However, there have been recent work [18] which focus on transferring knowledge from models trained on isolated sign language recognition datasets to find signs in continuous co-articulated sign videos.

In this work we focus on multiple shot supervised learning using the LSE eSaude UVIGO dataset [8], where we have multiple instances of the precise locations of signs from a vocabulary to be spotted in a co-articulated sign video. This allows transfer learning to address the domain adaptation problem from models trained on ISLR datasets to the sign spotting dataset.

3   Approach

The HS-I3D model consists of a spatio-temporal backbone based on the Inception I3D model with a hierarchical network head to locate sign instances from a set vocabulary. We first briefly give an overview of the feature extraction method used to extract spatio-temporal representations at different resolutions. Then we introduce the hierarchical network head to make predictions at different temporal resolutions. Finally we describe the learning objectives for the localisation of signs.

3.1   I3D Feature Extraction Layers

The I3D model is a general video classification architecture where given a sequence of frames, one class is predicted. This does not naturally allow for the localisation of signs or multiple predictions within a sequence.

To address the objective of localisation, we use the I3D model to extract features at different spatio-temporal resolutions. Instead of taking the output at the final layer, after spatio-temporal global average pooling and applying a fully connected layer for classification, we take the output before global average pooling and additional intermediary features obtained from the outputs before the spatio-temporal max pooling layers in the I3D model. We therefore obtain three spatio-temporal features, each with different temporal resolutions, where features extracted from earlier layers have larger temporal dimensions.

For the rest of this work we assume that the input to the I3D model contains 32 consecutive frames with a resolution of 224×224. For a given input sequence, the dimensions of the features extracted from the I3D model are as follows $1024 \times 4 \times 7 \times 7$, $832 \times 8 \times 14 \times 14$ and $480 \times 16 \times 28 \times 28$, with a temporal dimension of 4, 8 and 16, and feature channel dimensions of 1024, 832 and 480, respectively.

3.2   Hierarchical Network Head

The features extracted from the I3D model are used as input to a hierarchical network head to predict signs at different temporal resolutions. The hierarchical network head follows a U-Net [23] design, but instead of increasing the spatial resolution we increase the temporal resolution, making class prediction outputs at different temporal levels by creating skip connections between the different feature layers.

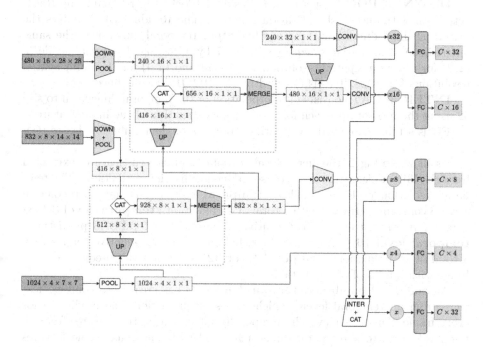

Fig. 2. Flow of the introduced hierarchical network head when given a sequence of 32 frames with dimensions 224 × 224. The three inputs are taken from outputs of various stages of the I3D model. The output consists of five temporal segment predictions of various segment lengths

An overview describing the flow of the hierarchical network architecture is given in Fig. 2, where we define the following network blocks as follows:

POOL consists of 3D Global Average Pooling which reduces the spatial dimensions to 1 × 1 while keeping the original temporal and feature dimensions.

UP consists of 3D transpose convolution which doubles the input temporal dimension and halves the feature dimension. The temporal dimension is doubled using a kernel size of $(2, 1, 1)$ and a stride of $(2, 1, 1)$.

CAT represents concatenating the feature outputs, which is used as a form of skip connection between the different feature layers.

MERGE consists of a 3D convolution layer followed by batch normalisation [12] and swish activation [22] which is repeated twice. The first 3D convolution has a kernel size of $(1,1,1)$ and the same number of output channels as the input channels. The second 3D convolution has a kernel size of $(1,1,1)$ with output channel dimensions being double the input of the lowest feature channel dimension before the respective **CAT** block.

CONV consists of a 3D convolution with a kernel size of $(1,1,1)$ keeping the same output dimension as the input dimension.

DOWN + POOL consists of a Residual Block [10], replacing the ReLU activation with the swish activation function. The Residual Block halves the feature channel dimension, keeping the output temporal dimensions the same but reducing the spatial dimensions to 7×7 by adjusting the spatial stride to $(2,2)$ for the input spatial resolution of 14×14 and $(4,4)$ for the input spatial resolution of 28×28. This is followed by the **POOL** block defined above.

INTER + CAT consists of interpolation of the temporal dimension to size 32 using the nearest approach followed by concatenation of the input features.

FC is a fully connected layer with output size of the number of classes (C).

As shown in Fig. 2, the hierarchical network head uses the features extracted from the I3D model as input to coarse-to-fine localisation predictions. The temporal prediction levels are as follows: **temporal level 4** $(x4)$ makes 1 prediction every 8 consecutive frames with a total of 4 predictions; **temporal level 8** $(x8)$ makes 1 prediction every 4 consecutive frames with a total of 8 predictions; **temporal level 16** $(x16)$ makes 1 prediction every 2 consecutive frames with a total of 16 predictions; **temporal level 32** $(x32)$ makes 1 prediction every frame with a total of 32 predictions.

An additional **combined temporal level** (x) is created to combine features from all temporal levels, which makes 32 prediction, one prediction for every frame. This is achieved by temporally interpolating the features from all temporal levels to a temporal dimension of 32 and concatenates the features before a fully connected layer, which produces the interpolated prediction. We therefore obtain a hierarchy of 5 sign localisation predictions at different temporal resolutions.

3.3   Learning Objectives

Cross entropy loss is used as the learning objective, where the target is set to the sign class label, if it exists within the prediction window, else the target is set to an additional *"background"* class. To obtain the final prediction, softmaxed logits from all five levels are temporally interpolated to the original sequence length. We then take their average for each frame to get sign class probabilities. A greedy search is applied which selects the highest probability class as the sign label for each frame. Consecutive frames of the same sign label, excluding the background class, are then selected as the temporal window for the identified sign.

4 Experiments

In this section, we evaluate the proposed HS-I3D model on the challenging LSE_eSaude_ UVIGO dataset. We first introduce the data set and the sign spotting objective. Then we give an overview of the evaluation metric used to measure the sign spotting performance of the model. Next we demonstrate qualitative results of the proposed approach to give reader more insight. Finally, we perform ablation studies, demonstrating the importance of the hierarchical components of our network and the impact of random data sampling on the precision and recall.

4.1 LSE_eSaude_UVIGO Dataset

LSE_eSaude_UVIGO [8] is a Spanish Sign Language (LSE: Lengua de Signos Española) dataset in the health domain with around 10 h of continuous sign videos. The dataset contains 10 signers which include seven deaf and three interpreters, captured in studio conditions with a constant blue background. It is partially annotated with the exact location of 100 signs, namely beginning and end timestamps, which were annotated by interpreters and deaf signers.

For our work the experiments are performed using the ChaLearn 2022 MSSL (multiple shot supervised learning) track protocol. MSSL is the classical machine learning track where sign classes are the same in the training, validation and test sets (i.e. no out of vocabulary samples).

We use the official dataset splits, namely; MSSL_Train_Set, which is the training dataset containing around 2.5 h of videos with annotations of 60 signs performed by five people; MSSL_Val_Set, which is the validation dataset containing around 1.5 h of videos with annotations of the same 60 signs performed by four people; and the MSSL_Test_Set, which is the test set with around 1.5 h of videos with annotations of the same 60 signs performed by four people.

4.2 Evaluation Metric

For evaluation we use the matching score per sign instance metric, which is based on the intersection over union (IoU) of the ground-truth and predicted intervals. The IoU results are thresholded using the values that are spread between 0.20 to 0.80 with 0.05 steps (i.e. $\{0.20, 0.25, ..., 0.75, 0.80\}$), yielding a set of 13 true positive, false positives and false negatives scores. These are then summed and used to calculate the final F1-score.

4.3 Random Sampling Probabilities

SLR approaches conventionally select consecutive frames around a known sign vocabulary during training. But since the objective is sign spotting with an F1-score evaluation metric, it requires a balance between precision and recall. Some sign videos may have signs outside of vocabulary that are visually similar to the signs in the vocabulary, which will impact the precision if predicted as false positives. We address this issue by introducing random sampling probabilities (rsp), where instead of selecting frame regions around only known sign classes, we randomly select consecutive frame regions from other areas in the continuous sign video based on the rsp. A higher rsp indicates higher probability of selecting consecutive frames from random locations in the video.

4.4 Implementation Details

We use the Inception I3D model [5] as our spatio-temporal backbone, which was pretrained on the large scale BSL1K dataset and then on the WLASL dataset [1]. The ReLU activation functions are replaced by the Swish activation function [22], as it has been shown to improve results for SLR [13].

During training, HS-I3D is trained in an end-to-end manner. The input to the model is 32 consecutive frames of size 224×224 with random data augmentations, such as random cropping, rotation, horizontal flipping, colour jitter and gray scaling. Mixup [29] is also applied with an α value of 1.0. Since the LSE_eSaude_UVIGO dataset has large class imbalances, we re-balance the class distributions and under-sample majority classes and over-sample minority classes.

The HS-I3D model is trained for 200 epochs with a batch size of 8 using an Adam optimizer [16] with an initial learning rate of 3×10^{-4} and cosine annealing decay until the model converges. The final predictions of the model are based on the average probability output of the different temporal levels, which are temporally interpolated to a temporal length of 32, to match the frame input size.

Four fold cross validation is used, where each fold is separated by signer ID using the MSSL_Train_Set, always keeping signer number five in the training set due to the large number of annotations for their samples. This training processes is repeated three times with different random sampling probabilities. The rsp values used during training is 0.0, 0.1 and 0.5.

An additional six fold cross validation set is used consisting of a mixture of training and validation dataset and follows the same process as the original four fold cross validation approach.

Therefore an ensemble based on the mean probability outputs of 30 models $(4 \times 3 + 6 \times 3)$ is used for the final predictions, where we apply a temporal stride of one over the test videos, taking the average probabilities between overlapping time segments.

4.5 Results

The results of the HS-I3D model are compared to the baseline approach in Table 1. Our model was able to surpass the baseline F1-score by 0.307, which is a significant improvement. HS-I3D using the averaged output from all temporal levels has a 0.022 performance improvement over just using the main temporal level x output.

Table 1. Performance of the ensemble HS-I3D models on the LSE_eSaude_UVIGO test set

Model	F1-score
Baseline	0.300
HS-I3D ($[x]$)	0.585
HS-I3D (avg$[x,x4,x8,x16,x32]$)	**0.607**

4.6 Ablation Studies

In this section, we analyze the impact of the hierarchical network head at different levels. Then, we demonstrate the importance and benefits of different random sampling probabilities. We report results with models trained only on the training dataset (MSSL_Train_Set) and display results on the validation dataset (MSSL_Val_Set).

Temporal Levels. In Table 2, we demonstrate the impact of the different temporal levels on the precision, recall and overall F1-score. We find that the most coarse temporal level $x4$ which makes a prediction every 8 frames, has the lowest F1-score, while the temporal level $x8$ has the highest F1-score. Averaging the prediction probabilities shows small improvements to the F1-score with the highest precision score compared to the individual temporal levels.

Table 2. Impact of the different temporal levels from the HS-I3D model on the LSE_eSaude_UVIGO validation set

Temporal Level	Precision	Recall	F1-score
$x4$	0.628	0.496	0.554
$x8$	0.631	**0.543**	0.583
$x16$	0.643	0.531	0.581
$x32$	0.634	0.513	0.567
x	0.636	0.498	0.558
Average $(x4, x8, x16, x32, x)$	**0.655**	0.531	**0.586**

In Fig. 3 we see the prediction probabilities between the different temporal levels, where in the top diagram we tend to see the temporal level $x4$ probabilities starts to identify the sign sightly earlier and ends slightly later compared to more fine-grained temporal levels. This has a negative impact when the same signs are in close proximity as shown in the bottom figure for the sign CERTIFICADO, where there is not such an easy distinction for temporal level $x4$ compared to other temporal levels.

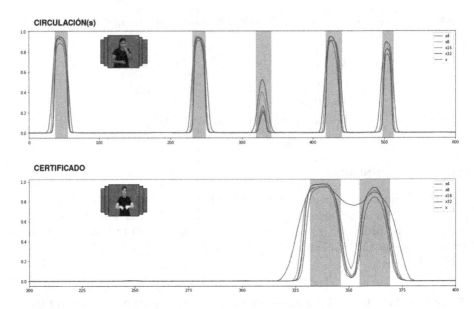

Fig. 3. Comparison of prediction probabilities (y-axis) between different temporal levels over the frames (x-axis) in a video sequence. Graphs shows probabilities for the sign CIRCULACIÓN(s) (top) and CERTIFICADO (bottom) over a continuous sign video with the ground truth interval highlighted by the red region. (Color figure online)

Random Sampling Probability. In Table 3, we demonstrate the impact of the sampling on the precision, recall and overall F1-score. We note a strong correlation between the precision and random sampling probability, where a rsp of 0.5 has improved precision compared to a model trained with 0.0 rsp with a 0.216 increase in precision. While higher rsp increases the precision, there is a trade off in the recall, where we find a significant decrease of 0.117 in recall.

Table 3. Impact of random sampling probabilities on the LSE_eSaude_UVIGO validation set

Random sampling prob.	Precision	Recall	F1-score
0.0	0.439	**0.648**	0.523
0.1	0.621	0.553	0.584
0.5	**0.655**	0.531	0.586
Ensemble (0.0,0.1,0.5)	0.634	0.576	**0.604**

In Fig. 4 we see the prediction probabilities for the sign HÍGADO over time from a continuous sign video. We find that a model trained with an rsp of 0.0 is able to recall more instances of the sign but at the cost of producing more false positives than a model trained with an rsp of 0.5.

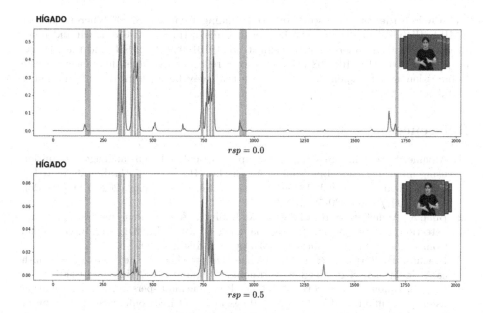

Fig. 4. Comparison of probabilities (y-axis) of spotting HÍGADO (blue line) over frames (x-axis) within a continuous sign video with the ground truth interval (red regions). The top graph represents the probabilities obtained with models trained with $rsp = 0.0$ while the bottom graph represents the probabilities obtained with models trained with $rsp = 0.5$ (Color figure online)

We find improved results, i.e. better balance between the precision and recall, with an overall higher F1-score, by ensembling 3 different rsp models. The ensembling is done by averaging the prediction probabilities.

5   Conclusion

In this paper, we propose the novel Hierarchical Sign I3D (HS-I3D) model, which takes advantage of intermediary layers of an I3D model for coarse-to-fine temporal sign representations. We develop a hierarchical network head, which combines spatio-temporal features for spotting signs at various temporal resolutions.

Our approach is able to make use of existing I3D models pretrained on large scale SLR datasets. We demonstrate the effectiveness of our model and show the importance of random data sampling while noting the trade-off between precision and recall. Our model achieve state-of-the-art performance of 0.607 F1-score on the ChaLearn 2022 MSSL track for the LSE_eSaude_UVIGO dataset winning the Multiple Shot Supervised Learning sign spotting challenge.

Future work involves exploration of better ensemble techniques with the different rsp models and temporal windows instead of the simple averaging prediction probabilities. Additionally, keypoint based models, such as Pose→Sign or GCNs, can be adapted to work with our hierarchical network head as an alternative input modality for sign spotting.

Acknowledgments. This work received funding from the SNSF Sinergia project 'SMILE II' (CRSII5 193686), the European Union's Horizon2020 research and innovation programme under grant agreement no. 101016982 'EASIER' and the EPSRC project 'ExTOL' (EP/R03298X/1). This work reflects only the authors view and the Commission is not responsible for any use that may be made of the information it contains.

References

1. Albanie, S., et al.: BSL-1K: scaling up co-articulated sign language recognition using mouthing cues. In: Vedaldi, A., Bischof, H., Brox, T., Frahm, J.-M. (eds.) ECCV 2020. LNCS, vol. 12356, pp. 35–53. Springer, Cham (2020). https://doi.org/10.1007/978-3-030-58621-8_3

2. Bilal, S., Akmeliawati, R., El Salami, M.J., Shafie, A.A.: Vision-based hand posture detection and recognition for sign language-a study. In: 2011 4th International Conference on Mechatronics (ICOM), pp. 1–6. IEEE (2011)

3. Braem, P.B., Sutton-Spence, R.: The hands are the head of the Mouth. The Mouth as Articulator in Sign Languages. Hamburg: Signum Press, Hamburg (2001)

4. Cao, Z., Simon, T., Wei, S.E., Sheikh, Y.: Realtime multi-person 2d pose estimation using part affinity fields. In: Proceedings of the IEEE Conference on Computer Vision and Pattern Recognition, pp. 7291–7299 (2017)

5. Carreira, J., Zisserman, A.: Quo vadis, action recognition? a new model and the kinetics dataset. In: proceedings of the IEEE Conference on Computer Vision and Pattern Recognition, pp. 6299–6308 (2017)

6. Cooper, H., Holt, B., Bowden, R.: Sign language recognition. In: Moeslund, T., Hilton, A., Krüger, V., Sigal, L. (eds.) Visual Analysis of Humans, pp. 539–562. Springer, London (2011). https://doi.org/10.1007/978-0-85729-997-0_27

7. Cormier, K., et al.: From corpus to lexical database to online dictionary: issues in annotation of the bsl corpus and the development of bsl signbank. In: 5th Workshop on the Representation of Sign Languages: Interactions between Corpus and Lexicon [workshop part of 8th International Conference on Language Resources and Evaluation, Turkey, Istanbul LREC 2012. ELRA, Paris, pp. 7–12 (2012)

8. Enríquez, M.V., Alba-Castro, J.L., Docio-Fernandez, L., Junior, J.C.S.J., Escalera, S.: Eccv 2022 sign spotting challenge: dataset, design and results. In: European Conference on Computer Vision Workshops (ECCVW) (2022)

9. Fillbrandt, H., Akyol, S., Kraiss, K.F.: Extraction of 3d hand shape and posture from image sequences for sign language recognition. In: 2003 IEEE International SOI Conference. Proceedings (Cat. No. 03CH37443), pp. 181–186. IEEE (2003)

10. He, K., Zhang, X., Ren, S., Sun, J.: Deep residual learning for image recognition. In: Proceedings of the IEEE Conference on Computer Vision and Pattern Recognition, pp. 770–778 (2016)

11. Holden, E.-J., Owens, R.: Visual sign language recognition. In: Klette, R., Gimel'farb, G., Huang, T. (eds.) Multi-Image Analysis. LNCS, vol. 2032, pp. 270–287. Springer, Heidelberg (2001). https://doi.org/10.1007/3-540-45134-X_20

12. Ioffe, S., Szegedy, C.: Batch normalization: accelerating deep network training by reducing internal covariate shift. In: International Conference on Machine Learning, pp. 448–456. PMLR (2015)

13. Jiang, S., Sun, B., Wang, L., Bai, Y., Li, K., Fu, Y.: Skeleton aware multi-modal sign language recognition. In: Proceedings of the IEEE/CVF Conference on Computer Vision and Pattern Recognition, pp. 3413–3423 (2021)

14. Jiang, T., Camgöz, N.C., Bowden, R.: Looking for the signs: identifying isolated sign instances in continuous video footage. In: 2021 16th IEEE International Conference on Automatic Face and Gesture Recognition (FG 2021), pp. 1–8. IEEE (2021)

15. Joze, H.R.V., Koller, O.: Ms-asl: a large-scale data set and benchmark for understanding American sign language. arXiv preprint arXiv:1812.01053 (2018)

16. Kingma, D.P., Ba, J.: Adam: a method for stochastic optimization. arXiv preprint arXiv:1412.6980 (2014)

17. Li, D., Rodriguez, C., Yu, X., Li, H.: Word-level deep sign language recognition from video: a new large-scale dataset and methods comparison. In: Proceedings of the IEEE/CVF Winter Conference on Applications of Computer Vision, pp. 1459–1469 (2020)

18. Li, D., Yu, X., Xu, C., Petersson, L., Li, H.: Transferring cross-domain knowledge for video sign language recognition. In: Proceedings of the IEEE/CVF Conference on Computer Vision and Pattern Recognition, pp. 6205–6214 (2020)

19. Lugaresi, C., et al.: Mediapipe: a framework for building perception pipelines. arXiv preprint arXiv:1906.08172 (2019)

20. Ong, E.J., Koller, O., Pugeault, N., Bowden, R.: Sign spotting using hierarchical sequential patterns with temporal intervals. In: Proceedings of the IEEE Conference on Computer Vision and Pattern Recognition, pp. 1923–1930 (2014)

21. Qiu, Z., Yao, T., Mei, T.: Learning spatio-temporal representation with pseudo-3d residual networks. In: proceedings of the IEEE International Conference on Computer Vision, pp. 5533–5541 (2017)

22. Ramachandran, P., Zoph, B., Le, Q.V.: Searching for activation functions. arXiv preprint arXiv:1710.05941 (2017)

23. Ronneberger, O., Fischer, P., Brox, T.: U-net: convolutional networks for biomedical image segmentation. In: Navab, N., Hornegger, J., Wells, W.M., Frangi, A.F. (eds.) MICCAI 2015. LNCS, vol. 9351, pp. 234–241. Springer, Cham (2015). https://doi.org/10.1007/978-3-319-24574-4_28

24. Sandler, W., Lillo-Martin, D.: Sign Language and Linguistic Universals. Cambridge University Press, Cambridge (2006)

25. Sincan, O.M., Keles, H.Y.: Autsl: a large scale multi-modal Turkish sign language dataset and baseline methods. IEEE Access 8, 181340–181355 (2020)

26. Varol, G., Momeni, L., Albanie, S., Afouras, T., Zisserman, A.: Scaling up sign spotting through sign language dictionaries. Int. J. Comput. Vis. 130(6), 1416–1439 (2022)

27. Viitaniemi, V., Jantunen, T., Savolainen, L., Karppa, M., Laaksonen, J.: S-pot-a benchmark in spotting signs within continuous signing. In: LREC Proceedings of European Language Resources Association (LREC) (2014)

28. Yang, H.D., Sclaroff, S., Lee, S.W.: Sign language spotting with a threshold model based on conditional random fields. IEEE Trans. Pattern Anal. Mach. Intell. 31(7), 1264–1277 (2008)

29. Zhang, H., Cisse, M., Dauphin, Y.N., Lopez-Paz, D.: mixup: beyond empirical risk minimization. arXiv preprint arXiv:1710.09412 (2017)

30. Zhang, J., Zhou, W., Xie, C., Pu, J., Li, H.: Chinese sign language recognition with adaptive hmm. In: 2016 IEEE International Conference on Multimedia and Expo (ICME), pp. 1–6. IEEE (2016)

Multi-modal Sign Language Spotting by Multi/One-Shot Learning

Landong Liu, Wengang Zhou$^{(\boxtimes)}$, Weichao Zhao, Hezhen Hu, and Houqiang Li$^{(\boxtimes)}$

CAS Key Laboratory of GIPAS, EEIS Department, University of Science and Technology of China (USTC), Hefei, China
{china07,saruka,alexhu}@mail.ustc.edu.cn, {zhwg,lihq}@ustc.edu.cn

Abstract. The sign spotting task aims to identify *whether* and *where* an isolated sign of interest exists in a continuous sign language video. Recently, it has received substantial attention since it is a promising tool to annotate large-scale sign language data. Previous methods utilized multiple sources of available supervision information to localize the sign actions under the RGB domain. However, these methods overlook the complementary nature of different modalities, *i.e.,* RGB, optical flow, and pose, which are beneficial to the sign spotting task. To this end, we propose a framework to merge multiple modalities for multiple-shot supervised learning. Furthermore, we explore the sign spotting task with the one-shot setting, which needs fewer annotations and has broader applications. To evaluate our approach, we participated in the Sign Spotting Challenge, organized by ECCV 2022. The competition contains two tracks, *i.e.,* multiple-shot supervised learning (MSSL for track 1) and one-shot learning with weak labels (OSLWL for track 2). In track 1, our method achieves around 0.566 F1-score and is ranked 2nd. In track 2, we are ranked the 1st, with a 0.6 F1-score. These results demonstrate the effectiveness of our proposed method. We hope our solution will provide some insight for future research in the community.

Keywords: Sign language spotting · Action localization · One-shot learning

1 Introduction

Sign spotting is a task that aims to determine the onset and offset of a specific sign category in a continuous sign language video. The results of sign spotting can be used in downstream tasks such as sign language translation [5] and recognition [23]. What's more, it is significantly useful for potential applications to automatically annotate datasets, which is a gloss-free [15] method. For instance, datasets like BSL-1K [1] utilize the sign spotting technique to substitute people for sign language datasets' annotation. This technique is expected to significantly lower costs and speed up the labeling processes.

L. Karlinsky et al. (Eds.): ECCV 2022 Workshops, LNCS 13808, pp. 256–270, 2023.
https://doi.org/10.1007/978-3-031-25085-9_15

However, current sign spotting algorithms still suffer from the following limitations. (1) It is very difficult to utilize data-driven algorithms due to the lack of data. (2) Different signers exhibit variations (fast or slow, left-handed or right-handed), and there are similar sign language words. (3) Sign actions exhibit co-articulated characteristics in the continuous video.

Traditional sign spotting methods mainly deploy handcrafted features like SURF, SIFT, and HOG to reduce the reliance on big data. However, these methods achieve relatively poor performance. After that, researchers study either with dynamic time warping and skin color histograms [32] or with hierarchical sequential patterns [25]. These approaches consider a fully-supervised setting with a single source of supervision. Recently, some methods have conducted a thorough analysis of sophisticated sign spotting algorithms like contrastive learning and Transformers [3,31] with auxiliary information like mouth shapes [1] and subtitles [3]. However, these methods generally overlook the importance of multi-modal features, *i.e.,* RGB, optical flow, and pose, which is beneficial to the sign spotting task. To explore those different modalities, we fully exploit them in different settings of sign spotting.

Inspired by the success of action localization and one-shot learning, we borrow the mature algorithm of action localization for multiple-shot supervised learning (MSSL) (track 1) [11]. Specifically, we merge RGB, optical flow, and pose features, which exhibit complementary effects. Then features are fed into the multi-scale Transformer Encoder. With the powerful Transformer backbone, we can get more powerful hidden state features. Finally, we utilize a lightweight 1D-CNN decoder to generate whether an action exists and where the onset and offset of this action are. For track 2 [11], *i.e.,* one-shot learning with a weak label, we propose a new graph-based one-shot learning method. Specifically, we first extract pose features and then build a graph based on the cosine similarity between them. After that, we use the longest path searching algorithm to match query and inference videos.

The rest of this paper is organized as follows. In Sect. 2, we provide a review on related work, including sign spotting, action localization, and video copy detection. In Sect. 3, we demonstrate the top-winning method in OSLWL and the 2nd-place strategy on MSSL. In Sect. 4, we show some ablation studies and additional experiments. In Sect. 5, we conclude the paper with a discussion and make some discussions for future research.

2   Related Work

In this section, we will briefly review the related topics, including sign language spotting, action localization, and video copy detection.

Sign Spotting. The objective of sign spotting is to find temporal boundaries between signs in continuous sign language. Sign spotting has been proposed to ease the workload of human labeling costs. Albanie *et al.* have pioneered the use of sign spotting to automatically classify datasets on datasets like BSL-1K [1]. Additionally, the phenomenon of co-articulate between sign language motions,

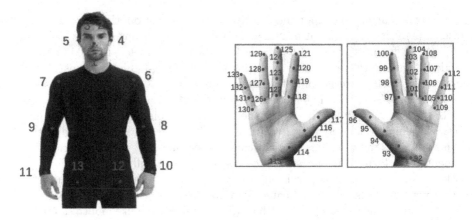

Fig. 1. The annotations of key points obtained by MMPose [9].

which will confound the model, increase the difficulty of sign spotting. To address this issue, some researchers advocate using auxiliary cues like mouth shape and video subtitles [1, 24, 28] to assist in training.

Action Localization. The methods of action localization can be divided into two categories, *i.e.*, two-stage methods and one-stage methods. The two-stage methods start by generating a lot of proposals, classifying them, returning the onset and offset of the action, and then using post-processing techniques like non-maximal suppression (NMS) to get the outcome. Detecting action boundaries [12, 17, 19, 22, 36], graph representation [2, 34], and Transformers [7, 30, 33] are examples of representative works. Most one-stage methods incorporate the anchor concept and are based on the temporal detection method. For instance, Lin et al. [18] localize action via a detection scheme [21]. We use the action localization method Actionformer [35] for MSSL the competition's track one.

Video Copy Detection. Video copy detection detects whether and where a long video contains the short video variant. Early works use the Hough voting algorithm [10] and network flow algorithm [29] on cosine similarity graph to detect target video. However, they use hand-crafted features, which limits their performance. In [14], researchers resort to CNN features to further improve the performance. This task is similar to the one-shot setting of sign spotting. However, the latter is more challenging since it involves temporal variants between the isolated sign and its continuous counterpart.

3   Methods

In this section, we will first introduce the feature extraction for RGB, optical flow, and pose, respectively. Then, we present our solutions to multiple-shot supervised learning (MSSL) and one-shot learning with weak labels (OSLWL) tasks separately.

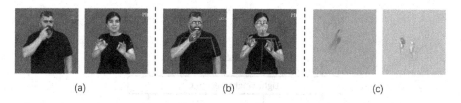

Fig. 2. Two examples of our utilized modality visualization results are presented. (a) RGB information. (b) Skeleton information generated by MMPose. (c) Optical flow extracted by TV-L1 flow.

3.1 Feature Extraction

We utilize multiple modalities to extract robust spatio-temporal feature representations depicting the continuous sign video.

Preprocessing. Since the signer only occupies a relatively small part of the original video frame, we utilize MMDetection [8] to first detect the signer's spatial location, which is followed by MMPose [9] Fig. 1 to extract the body and hand poses. Then, the detected pose is utilized to crop the upper-body patch of the signer. Besides, the flow is calculated by the TV-L1 algorithm. Finally, three types of data are obtained, *i.e.,* RGB, flow and pose modalities (Fig. 2).

RGB. Given the cropped RGB video V^r, we utilize the I3D backbone [6] pretrained on the BSL-1k [1] to extract representations $\mathbf{F}^r$. Specifically, we extract the clip level feature with the receptive field and stride as 8 and 2.

Optical Flow. Since the features extracted from RGB images are easily affected by the appearance of signers, we also extract the motion information from the flow modality. First, we use the TV-L1 algorithm to extract optical flow, and then we utilize the I3D backbone [6] pre-trained on the AUTSL [27] to extract flow features $\mathbf{F}^f$ with the same receptive field and stride setting.

Pose. To provide sufficient information on sign language for downstream tasks, we also extract the body and hand cues with pre-trained GCN [4,13]. The pose sequence $\mathbf{J}^p$ include three parts, *i.e.,* right hand J^{right}, left hand J^{left} and body J^{body}. The pose sequence is fed into the GCN backbone and converted into frame-level features $\mathbf{F}^{out}$, which is computed as follows,

$$\mathbf{F}^{out} = \mathbf{D}^{-\frac{1}{2}}(\mathbf{A} + \mathbf{I})\mathbf{D}^{-\frac{1}{2}}\mathbf{J}^p\mathbf{W}, \tag{1}$$

where adjacent matrix $\mathbf{A}$ represents intra-body connections and an identity matrix $\mathbf{I}$ represents self-connections, $\mathbf{D}$ presents the diagonal degree of $(\mathbf{A} + \mathbf{I})$, and $\mathbf{W}$ is a trainable weight matrix of the convolution. Then, we utilize an average pooling operation with the same receptive field and stride setting to generate pose features $\mathbf{F}^p$.

3.2 MSSL Framework

Given the representations of multiple modalities, we first utilize the multi-scale Transformer encoder to map these features into a multiscale feature represen-

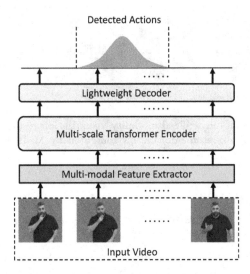

Fig. 3. The detailed structure of our framework in MSSL. We utilize the strong Transformer backbone to extract multi-modal features, followed by a lightweight CNN decoder to output detected actions.

tation $\mathbf{F} = \{\mathbf{F}^1, \mathbf{F}^2, \cdots, \mathbf{F}^L\}$. Then, our model decodes the feature pyramid $\mathbf{F}$ into the the sequence label $\hat{\mathbf{Y}} = \{\hat{\mathbf{y}}_1, \hat{\mathbf{y}}_2, \cdots, \hat{\mathbf{y}}_T\}$, where

$$\hat{\mathbf{y}}_i = (p_i, start_i, end_i),$$

the detailed workflow of our MSSL algorithm is illustrated in Fig. 3.

Multi-scale Transformer Encoder. We leverage the strong modeling capability of Transformer to perform localization similar to [35]. Specifically, it combines a multi-scale feature representation with local self-attention. We use three fc modules ($W_Q \in R^{D \times D_Q}$, $W_K \in R^{D \times D_K}$, $W_V \in R^{D \times D_V}$) to project features $\mathbf{F}$ to suitable space, getting the query, key, and value with $D_K = D_Q$. In more detail, we set $D_Q = D_K = D_V = 512$ in our experiment. The formulas are as follows,

$$\mathbf{F}^0 = concatenate[\mathbf{F}^r, \mathbf{F}^f, \mathbf{F}^p],$$
$$Q = \mathbf{F}^l \cdot W_Q, K = \mathbf{F}^l \cdot W_K, V = \mathbf{F}^l \cdot W_V. \tag{2}$$

Then, we use softmax to calculate the attention map and retrieve useful information from value features.

$$Attention(Q, K, V) = softmax(\frac{QK^T}{\sqrt{D_q}})V, \tag{3}$$

To jointly attend to information from different representation sub-spaces at different positions, researchers propose multi-head self-attention (MSA),

$$MultiHead(Q, K, V) = Concat(head_1, \cdots, head_n)W', \tag{4}$$
$$\text{where } head_i = Attention(Q, K, V).$$

Once self-attention is defined, to gather spatial data on multiple scales, we use a 2x downsampling module between every Transformer block, as follows:

$$\mathbf{F}^l = MSA(LN(\mathbf{F}^{l-1})) + \mathbf{F}^{l-1},$$
$$\mathbf{F}^l = MLP(LN(\mathbf{F}^l)) + \mathbf{F}^l, \tag{5}$$
$$\mathbf{F}^l = downsampling(\mathbf{F}^l),$$

where MLP is several fc modules with GELU introducing nonlinearity, LN is the abbreviation of layer normalization, and downsamlping is a depthwise 1D convolutional layer with kernel_size=1, stride=2 for its speed and fewer parameters. Finally, following by "actionformer" [35], we use another lightweight 1D convolutional network applying to each layer of the $\mathbf{F}^l$ and the parameters are shared across each layer.

Lightweight Decoder. decoder part is a lightweight 1D convolution network, including a classification head and regression head, classifying every moment in time and the corresponding action boundaries respectively. We use NMS as post-processing to get final proposals.

Objective Function. During training, L_{cls} is the focal loss [20] for sign action classification and L_{reg} is the generalized IoU loss [26] for distance regression,

$$L = \sum_{time} (\frac{1}{T}L_{cls} + \frac{\lambda_{reg}}{T_{pos}} \mathbb{1}L_{reg}), \tag{6}$$

where T is the length of the reference video frame, T_{pos} is the total number of positive samples, $\mathbb{1}$ is an indicator function denoting whether time step t is within action, and λ_{reg} is a parameter balancing two losses.

3.3   OSLWL Framework

In this section, we introduce our framework for the OSLWL task. Due to the diversity of appearances among different signers, we utilize the appearance-invariant features $\mathbf{F}^p$ from pose modality. Then we build the similarity graph to perform frame-wise matching between the query isolated sign video and the long reference video, as shown in Fig. 4.

Sign Instance Matching. We first calculate the cosine similarity between the pose feature of each frame in the query video and the reference video, and get a two-dimensional matrix $sims \in \mathbb{R}^{T_1 \times T_2}$, which is computed as follows,

$$sims[t_1, t_2] = \frac{F_{t_1}^T \cdot F_{t_2}}{\|F_{t_1}\|\|F_{t_2}\|}, t_1 \in \{1, 2, \cdots, T_1\}, t_2 \in \{1, 2, \cdots, T_2\}, \tag{7}$$

where F_{t_1} and F_{t_2} denote the frame feature from query and reference video, respectively. T_1 and T_2 denote the length of frames in the query and reference video, respectively. Then we create a graph. We use tuple (x, y) to represent each point in the graph, where x denotes query video frame and y denotes reference

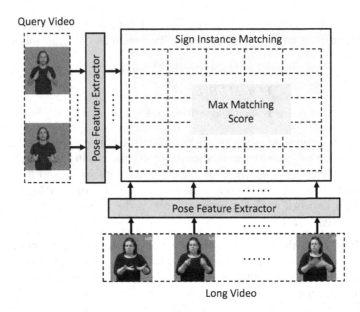

Fig. 4. The workflow diagram of OSLWL track. We use pose features to calculate cosine similarity and then use the longest path algorithm in graph theory to maximize the score.

video frame. All directed edge weights connected to node (x, y) are set to sims[x, y]. Our objective in this directed similarity graph is to find the longest weighted path to maximize the cosine similarity. This path will help us identify the portion of the query video and the reference video that are most similar. Additionally, for the sequence obtained by matching to be consistent with the query video sequence and reference video sequence, the path obtained by matching must not roll back and the time must be monotonically increasing. So, we connect a directed edge between two points (x_1, y_1) and (x_2, y_2) if they simultaneously satisfy the following four formulas.

$$
\begin{aligned}
x_1 &\leq x_2, \\
x_2 - x_1 &\leq max_{step}, \\
y_1 &\leq y_2, \\
y_2 - y_1 &\leq max_{step},
\end{aligned}
\tag{8}
$$

where we set $max_{step} = 4$. We artificially introduce the start node (0, 0) and the end node (T_1, T_2), which participate in the graph. Since the length of the reference video may be very long and there exist too many nodes in the graph, we utilize the pruning operation to decrease the temporal complexity. For each frame of the query, select the *top-k* = 16 frames with the highest cosine similarity in the reference and save them into the graph, otherwise remove the other edges. Finally, the longest weighted path in the graph is selected, and its start and end positions of the reference video (y min to y max) are extracted as a result of query and reference matching. In the experiments part, we will discuss the meaning of $max_{step}, top\text{-}k$ in detail.

Select Proper Features. It is well known that some sign languages are performed by sign hand pose to convey information. To this end, we adaptively utilize the proper parts of the pose feature for sign instance matching. Given a query video, first, we extract its pose information. Then since the pose feature includes three parts, *i.e.*, right hand, left hand and upper body, if OpenPose can't detect the left hand information, we delete the corresponding parts of the pose feature of the query and reference videos.

Extract Key Frames. The onset and offset of the query video usually contain a series of invalid frames, which are harmful to sign instance matching. To remove these invalid frames, we use the coordination of key points to detect the beginning and end of the query video. Specifically, given the query video, we select the continuous frames which satisfy the hands above the elbow joints as the effective duration T_1'. Further, we intercept a specific proportion of the effective duration as the final query video.

Use Polynomial Kernel. To enhance the discrimination of matching results, we utilize the polynomial kernel in the cosine similarity operation, which is computed as follows,

$$sims\_kernel[t_1, t_2] = sims^{\alpha}[t_1, t_2], \tag{9}$$

where α denotes the power coefficient of the cosine similarity. Instead of $sims$, we finally utilize $sims\_kernel$ in our experiment.

4 Experiments

4.1 Experiment Settings

Basic Settings. In MSSL, all networks were implemented by PyTorch and trained on NVIDIA RTX 3090. In OSLWL, we simply use CPU resources for feature matching. In MSSL, we utilize the Adam optimizer with a peak learning rate of 2e−4. The training lasts 100 epochs, with 10 epochs as the warm-up. During inference, we take the output for every time step, which is organized as the triplet sign action confidence score, onset, and offset of the action. These candidates are further processed via NMS to remove highly overlapping instances, which leads to the final localization outputs. Then, using the least squares method, we predict the length of the training set video and the number of actions in a linear fashion to determine their relationship.

Metrics. we use IoU of groundtruth interval and predicted video range to evaluate score per sign instance. Specifically, IoU threshold 't' will change from 0.2 to 0.8 in 0.05 steps. Given a specific reference video file 'n', query video file 'k', and query of sign 'i' to spot, we can calculate the True Positives, False Negatives,

Table 1. Effectiveness of the input modality on CSSL_MSSL validation dataset. The 'RGB', 'Pose' and 'Flow' denote the different input modalities. The final line marked with an asterisk in the table denotes that the sum of the three modalities is added directly rather than being concatenated.

RGB	Pose	Flow	F1_Score
✓			0.447
	✓		0.492
		✓	0.354
✓	✓		0.564
✓		✓	0.547
	✓	✓	0.525
✓	✓	✓	**0.594**
✓*	✓*	✓*	0.503

and so on as following formula,

$$IoU_i = \frac{Ret_i \cap Rel_i}{Ret_i \cup Rel_i},$$

$$TP = \sum_t \sum_n \sum_i \sum_k (IoU_i(k,n) >= t),$$

$$FP = (\sum_t \sum_n \sum_i \sum_k 1) - TP,$$

$$FN = (\sum_t \sum_n GT_n - TP),$$

$$Precision = TP/(TP + FP),$$

$$Recall = TP/(TP + FN),$$

$$F1 = 2 * (Precision * Recall)/(Precision + Recall),$$

$$(10)$$

where Ret_i is the prediction intervals of sign 'i' and Rel_i is the ground truth intervals of sign 'i'. Finally, we use F1-score (harmonic mean of precision and recall) to evaluate how good an algorithm is.

Data Augmentation. In MSSL, due to the small scale of the training data, we utilize data augmentation to strengthen the robustness of our framework. The data augmentation includes three aspects, *i.e.*, spatial, temporal, and feature levels. At the spatial level, we utilize random center crop operation for the RGB images and slightly disturb the pose coordination of the hand and body poses. At the temporal level, we utilize the temporal shift module proposed in [16] to enhance the temporal consistency. At the feature level, we add a random dropout to the input feature sequence with a probability of 0.5. Our framework achieves better performance with the above-mentioned augmentation.

Datasets. We conduct all experiments on the public sign spotting datasets.[1], *i.e.*, CSSL_MSSL and CSSL_OSLWL.

---

[1] https://chalearnlap.cvc.uab.cat/dataset/42/description/.

Table 2. Effectiveness of the input modality on CSSL_MSSL dataset. The 'Spatial', 'Temporal' and 'Feature' denote the different augmentation during the training procedure.

Spatial	Temporal	Feature	F1_Score
			0.532
✓			0.575
	✓		0.587
		✓	0.563
✓	✓	✓	**0.594**

Table 3. Ablation of the "power" parameter on CSSL_OSLWL dataset.

Power	F1_Score
1	0.629
2	0.644
3	0.644
4	**0.645**
5	0.642

Table 4. Ablation of the "max_step" parameter on CSSL_OSLWL dataset.

max_step	F1_Score
1	0.531
2	0.609
3	0.636
4	**0.645**
5	0.644

Table 5. Ablation of the "top-k" parameter on CSSL_OSLWL dataset.

Top-k	F1_score
12	0.636
13	0.637
14	0.640
15	0.643
16	**0.645**
17	0.640
18	0.636

Table 6. Ablation of the "percent" parameter on CSSL_OSLWL dataset.

Percent	F1_score
60%	0.629
65%	0.639
70%	0.645
75%	**0.645**
80%	0.643
85%	0.637
90%	0.625
95%	0.606

CSSL_MSSL. CSSL_MSSL is a dataset of Spanish sign language in the health domain, which contains samples of trimmed continuous sign language videos. The dataset includes 270 sign videos and 60 signs performed by 9 persons. All samples are split into three parts, *i.e.*, 112 for training, 80 for validation, and 78 for testing. Since the annotation of the test set is unavailable, we verify the performance of our proposed method on the validation set.

CSSL_OSLWL. CSSL_OSLWL is divided into two parts: query and reference. The query dataset contains 40 short videos of 40 isolated signs. The reference dataset is comprised of 1500 videos, performed by 5 people. For gender equality, both datasets have male and female signers.

4.2   Ablation on MSSL

In this section, we will perform several ablation studies for the data augmentation and input modalities of our proposed method in MSSL. As shown in Table 1, we experiment with different modalities. It is observed that the f1_score only reaches 0.49 using the single modality as input. We boost the performance by merging the features of three modalities, which demonstrates our assumption that multiple modalities could provide more sufficient information. Concatenate produced better results than sum when we tried to perform aggregation operations for the three modalities. We also validate the effectiveness of data augmentation during the training procedure in Table 2. Each of these data augmentations can enhance the robustness of our proposed method. The performance achieves the best result by applying three proposed data augmentations.

4.3   Ablation on OSLWL

In this section, we will perform several ablation studies for the hyper-parameters of our proposed method in OSLWL. There are about four hyper-parameters in our ablation, top-k in Table 5, max_step in Table 4, power in Table 3 and percent in Table 6. We also conduct ablation on Table 7, illustrating that removing outlier features can boost the performance. We will then explain what these four parameters mean.

For top-k, we want to reduce the number of nodes in the graph to reduce the time complexity of the algorithm. We keep the top-k frame with the highest cosine similarity in the long video for each frame of the short video to create a python tuple, and the remaining point pairs are removed from the graph.

For max_step, we will not link two nodes $(x_1, y_1), (x_2, y_2)$ in the graph, if

$$|x_2 - x_1| > max_{step} \text{ or } |y_2 - y_1| > max_{step}, \tag{11}$$

The optimal path is more continuous with a lower max_step value and the target path is slightly discrete when using a higher max_step. As observed in Table 4, performance is highest when the max_step is 4.

For power, by increasing the 'power' of cosine similarity, our algorithm's performance will be further enhanced. As observed in Table 3, the best performance reaches 0.645 when the 'power' is 4, and the performance decrease when the 'power' further increases.

For percent, once we get the onset and offset of the query videos using pose information, we extract 'percent' from the center frames as key frames, to let the algorithm focus on important frames. The performance ascends greatly once we reduce the number of key frames, achieving a 0.645 F1-score when 'percent' is 70%.

Grid search is used first to determine the general location of the optimal parameters, and line search is then used to obtain more precise results. In our best setting, the top-k, max_step, power, and percent are set to 16, 4, 4, and 70%, respectively. The ablation results for line search are as follows.

Table 7. Ablation of whether adaptively utilize the proper parts of pose feature on CSSL_OSLWL dataset, we conduct this experiments on best settings.

Full feature F1_Score	Remove left hand F1_Score
0.641	0.645

The tables above show that performance is significantly impacted by max_step. The performance drops to as low as 0.531 when it equals 1. With the increase of max_step, the performance ascends first and then descends. The maximum is reached when max_step equals 4. What's more, fine turning top-k, power, and percent have a positive impact on OSLWL's performance, although the performance improvement is not as good as max_step.

What's more, we also experiment using optical flow features with the best parameters described above. The resulting f1-score is 0.186, which is much lower than the result using the pose feature. We speculate that it is due to the optical flow feature being too noisy.

Additionally, we attempt to replace cosine similarity with other similarity functions like MSE. First, the smaller the MSE, the better. So, we will find a path with the least amount of edge weight. However, because all of MSE's edge weights are positive, when determining the shortest path, it generally includes just one step, and the outcome is unquestionably incorrect. For cosine similarity, the path weights are between $[-1, 1]$, both positive and negative, so the same problem as MSE will not be encountered.

5   Conclusion

In this work, we explore the suitable architectures in different settings of sign spotting. Specifically, the overall framework contains two stages, *i.e.,* feature extraction from multiple modalities and sign spotting for two different settings. Different from the previous methods focusing on uni-modal features, we fully exploit the complementary of multiple modalities, *i.e.,* RGB, optical flow, and pose. To this end, we propose a framework to merge the multiple modalities for multiple-shot supervised learning. Furthermore, we explore the sign spotting task with one-shot learning with a training-free method. Specifically, we introduce the sign instance matching algorithm to better localize the sign language action in the reference video. To validate the effectiveness of our approach, we participated in the Sign Spotting Challenge, organized by ECCV 2022. The competition includes two tracks, i.e., multiple-shot supervised learning (MSSL for track 1) and one-shot learning with weak labels (OSLWL for track 2). Finally, we achieve around 0.566 F1-score (rank 2) in track 1 and get first place with a 0.6 F1-score in track 2, which well validates the superiority of our proposed method.

5.1   Limitation and Future Work

MSSL Task. In MSSL task, we list three limitations. First, features used in our proposed method are not fine-tuned on the CSSL_MSSL dataset. We attempt to fine-tune the RGB feature using the CSSL_MSSL dataset, but sadly the scale of the CSSL_MSSL dataset is too small, making it simple to overfit. Second, the number of actions in the video is predicted using linear prediction, and there is some difference between the predicted and actual number of actions. Future methods may remedy this problem. Third, a straightforward concatenation method is used for multi-feature fusion. In the future, a well-designed network might produce better outcomes.

OSLWL Task. In OSLWL task, we also discover that the method in OSLWL task is sensitive to hyper-parameter, especially for parameter "max_step". Different parameters can affect performance by up to 10 points. Future research might concentrate on decreasing the algorithm's sensitivity to parameters.

Acknowledgement. This work was supported by the National Natural Science Foundation of China under Contract U20A20183. It was also supported by the GPU cluster built by MCC Lab of Information Science and Technology Institution, USTC.

References

1. Albanie, S., et al.: BSL-1K: scaling up co-articulated sign language recognition using mouthing cues. In: Vedaldi, A., Bischof, H., Brox, T., Frahm, J.-M. (eds.) ECCV 2020. LNCS, vol. 12356, pp. 35–53. Springer, Cham (2020). https://doi.org/10.1007/978-3-030-58621-8_3
2. Bai, Y., Wang, Y., Tong, Y., Yang, Y., Liu, Q., Liu, J.: Boundary content graph neural network for temporal action proposal generation. In: Vedaldi, A., Bischof, H., Brox, T., Frahm, J.-M. (eds.) ECCV 2020. LNCS, vol. 12373, pp. 121–137. Springer, Cham (2020). https://doi.org/10.1007/978-3-030-58604-1_8
3. Bull, H., Afouras, T., Varol, G., Albanie, S., Momeni, L., Zisserman, A.: Aligning subtitles in sign language videos. In: ICCV, pp. 11552–11561 (2021)
4. Cai, Y., et al.: Exploiting spatial-temporal relationships for 3D pose estimation via graph convolutional networks. In: ICCV, pp. 2272–2281 (2019)
5. Camgoz, N.C., Hadfield, S., Koller, O., Ney, H., Bowden, R.: Neural sign language translation. In: CVPR, pp. 7784–7793 (2018)
6. Carreira, J., Zisserman, A.: Quo Vadis, action recognition? A new model and the kinetics dataset. In: CVPR, pp. 6299–6308 (2017)
7. Chang, S., Wang, P., Wang, F., Li, H., Feng, J.: Augmented transformer with adaptive graph for temporal action proposal generation. arXiv (2021)
8. Chen, K., et al.: MMDetection: Open MMLab detection toolbox and benchmark. arXiv (2019)
9. MMP Contributors: OpenMMLab pose estimation toolbox and benchmark (2020). https://github.com/open-mmlab/mmpose
10. Douze, M., Jégou, H., Schmid, C., Pérez, P.: Compact video description for copy detection with precise temporal alignment. In: Daniilidis, K., Maragos, P., Paragios, N. (eds.) ECCV 2010. LNCS, vol. 6311, pp. 522–535. Springer, Heidelberg (2010). https://doi.org/10.1007/978-3-642-15549-9_38

11. Enriquez, M.V., Alba-Castro, J.L., Docio-Fernandez, L., Junior, J.C.S.J., Escalera, S.: ECCV 2022 sign spotting challenge: dataset, design and results. In: ECCVW (2022)
12. Gong, G., Zheng, L., Mu, Y.: Scale matters: temporal scale aggregation network for precise action localization in untrimmed videos. In: ICME, pp. 1–6 (2020)
13. Hu, H., Zhao, W., Zhou, W., Wang, Y., Li, H.: SignBERT: pre-training of hand-model-aware representation for sign language recognition. In: ICCV, pp. 11087–11096 (2021)
14. Jiang, Y.G., Wang, J.: Partial copy detection in videos: a benchmark and an evaluation of popular methods. TBD **2**(1), 32–42 (2016)
15. Li, D., et al.: TSPNet: hierarchical feature learning via temporal semantic pyramid for sign language translation. In: NeurIPS, vol. 33, pp. 12034–12045 (2020)
16. Lin, J., Gan, C., Han, S.: TSM: temporal shift module for efficient video understanding. In: ICCV, pp. 7083–7093 (2019)
17. Lin, T., Liu, X., Li, X., Ding, E., Wen, S.: BMN: boundary-matching network for temporal action proposal generation. In: ICCV, pp. 3889–3898 (2019)
18. Lin, T., Zhao, X., Shou, Z.: Single shot temporal action detection. In: ACM MM, pp. 988–996 (2017)
19. Lin, T., Zhao, X., Su, H., Wang, C., Yang, M.: BSN: boundary sensitive network for temporal action proposal generation. In: ECCV, pp. 3–19 (2018)
20. Lin, T.Y., Goyal, P., Girshick, R., He, K., Dollár, P.: Focal loss for dense object detection. In: ICCV, pp. 2980–2988 (2017)
21. Liu, W., et al.: SSD: single shot multibox detector. In: Leibe, B., Matas, J., Sebe, N., Welling, M. (eds.) ECCV 2016. LNCS, vol. 9905, pp. 21–37. Springer, Cham (2016). https://doi.org/10.1007/978-3-319-46448-0_2
22. Liu, Y., Ma, L., Zhang, Y., Liu, W., Chang, S.F.: Multi-granularity generator for temporal action proposal. In: CVPR, pp. 3604–3613 (2019)
23. Min, Y., Hao, A., Chai, X., Chen, X.: Visual alignment constraint for continuous sign language recognition. In: ICCV, pp. 11542–11551 (2021)
24. Momeni, L., Afouras, T., Stafylakis, T., Albanie, S., Zisserman, A.: Seeing wake words: audio-visual keyword spotting. arXiv (2020)
25. Ong, E.J., Koller, O., Pugeault, N., Bowden, R.: Sign spotting using hierarchical sequential patterns with temporal intervals. In: CVPR, pp. 1923–1930 (2014)
26. Rezatofighi, H., Tsoi, N., Gwak, J., Sadeghian, A., Reid, I., Savarese, S.: Generalized intersection over union: a metric and a loss for bounding box regression. In: CVPR, pp. 658–666 (2019)
27. Sincan, O.M., Keles, H.Y.: AUTSL: a large scale multi-modal Turkish sign language dataset and baseline methods. IEEE Access **8**, 181340–181355 (2020)
28. Stafylakis, T., Tzimiropoulos, G.: Zero-shot keyword spotting for visual speech recognition in-the-wild. In: ECCV, pp. 513–529 (2018)
29. Tan, H.K., Ngo, C.W., Hong, R., Chua, T.S.: Scalable detection of partial near-duplicate videos by visual-temporal consistency. In: ACM MM, pp. 145–154 (2009)
30. Tan, J., Tang, J., Wang, L., Wu, G.: Relaxed transformer decoders for direct action proposal generation. In: ICCV, pp. 13526–13535 (2021)
31. Varol, G., Momeni, L., Albanie, S., Afouras, T., Zisserman, A.: Read and attend: temporal localisation in sign language videos. In: CVPR, pp. 16857–16866 (2021)
32. Viitaniemi, V., Jantunen, T., Savolainen, L., Karppa, M., Laaksonen, J.: S-pot-a benchmark in spotting signs within continuous signing. In: LREC (2014)
33. Wang, L., Yang, H., Wu, W., Yao, H., Huang, H.: Temporal action proposal generation with transformers. arXiv (2021)

34. Xu, M., Zhao, C., Rojas, D.S., Thabet, A., Ghanem, B.: G-tad: sub-graph localization for temporal action detection. In: CVPR, pp. 10156–10165 (2020)
35. Zhang, C., Wu, J., Li, Y.: ActionFormer: localizing moments of actions with transformers. arXiv preprint arXiv:2202.07925 (2022)
36. Zhao, P., Xie, L., Ju, C., Zhang, Y., Wang, Y., Tian, Q.: Bottom-up temporal action localization with mutual regularization. In: Vedaldi, A., Bischof, H., Brox, T., Frahm, J.-M. (eds.) ECCV 2020. LNCS, vol. 12353, pp. 539–555. Springer, Cham (2020). https://doi.org/10.1007/978-3-030-58598-3_32

Sign Spotting via Multi-modal Fusion and Testing Time Transferring

Hongyu Fu[1,3], Chen Liu[1,2], Xingqun Qi[1,2], Beibei Lin[1], Lincheng Li[1], Li Zhang[3], and Xin Yu[2(✉)]

[1] NetEase Fuxi AI Lab, Hangzhou, China
[2] University of Technology Sydney, Sydney, Australia
xin.yu@uts.edu.au
[3] Tsinghua University, Beijing, China

Abstract. This work aims to locate a query isolated sign in a continuous sign video. In this task, the domain gap between the isolated and continuous sign videos often handicaps the localization performance. To address this issue, we propose a parallel multi-modal sign spotting framework. In a nutshell, our framework firstly takes advantage of multi-modal information (including RGB frames, 2D key-points and 3D key-points) to achieve representative sign features. The multi-modal features are employed to complement each other and thus compensate for the deficiency of a single modality, thus leading to informative representations for sign spotting. Moreover, we introduce a testing time top-k transferring technique into our framework to reduce the aforementioned domain gap. Concretely, we first compare the query sign with extracted sign video clips, and then update the feature of the query sign with the features of the top-k best matching video clips. In this manner, the updated query feature will exhibit a smaller domain gap with respect to continuous signs, facilitating feature matching in the following iterations. Experiments on the challenging OSLWL-Test-Set benchmark demonstrate that our method achieves superior performance (0.559 F1-score) compared to the baseline (0.395 F1-score). Our code is available at https://github.com/bb12346/OpenSLR.

Keywords: Sign language spotting · Multi-modal fusion framework · Pose-based feature extraction · RGB-based feature extraction · Top-K transferring technique

1 Introduction

Sign language spotting (SLS) aims to find accurate locations of the given isolated signs in continuous co-articulated sign language videos. As sign language is a common form of language communication used in the hearing-impaired community, sign spotting task plays a pivotal role in various real scene applications [7,8,19,20,24]. Inspired by the great achievements of deep neural networks, RGB image based SLS methods have achieved impressive progresses [6,9,10,18,22,37].

However, the domain gap between the isolated and continuous sign videos often limits the performance improvement for many RGB-based SLS approaches,

L. Karlinsky et al. (Eds.): ECCV 2022 Workshops, LNCS 13808, pp. 271–287, 2023.
https://doi.org/10.1007/978-3-031-25085-9_16

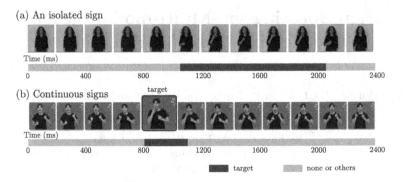

Fig. 1. The domain gap between isolated signs and continuous signs. (a) is an isolated query sign, while the target is the same sign cropped from continuous sign (b). It is obvious that the sign in (b) is performed much faster without hand up or hand down actions. In addition, one sign may be performed in many different ways, such as the signer in (b) performs while holding her left arm on.

as illustrated in Fig. 1. Pose-based SLS methods [1,3,4,15,34] have been proposed to leverage the changes of human joints for sign spotting. Pose-based SLS methods are roughly divided into 2D-pose based approaches and 3D-pose based approaches. Both of them need to extract the key joints of the signer from the given RGB image. However, estimating accurate poses in continuous sign videos is difficult as motion blur often exists in continuous sign videos. The blurry frames would lead to erroneous pose estimation.

To solve this problem, we propose a Multi-modal Sign Spotting Framework (MSSF) to generate more discriminative feature representations. Our MSSF is a dual-branch structure, including an RGB-based branch and a Pose-based branch. The pose-based branch pays more attention to the motion of different key points. To further capture the changes of key points, we introduce two types of poses, 2D key-points and 3D angle-axis. The 2D key-points describes the exact coordinates of each point in an image, while the 3D angle-axis represents the angle change of different points. The RGB-based branch focuses on extracting image details, such as holistic motions and hand shapes, and it also helps feature representation when pose-based methods suffer severe motion blur and fail to capture accurate joint poses. As a result, a total of three models, including 2D-Pose based SL-GCN [17], 3D-Pose based SL-GCN, and I3D-MLP [3], are employed to extract sign language features.

As shown in Fig. 1, the domain gap can be roughly summarized in two aspects: (i) a sign from the query set often contains the actions of hands up and hands down, while its counterpart in a continuous video does not contain such these actions. (ii) the difference of the same sign language instance among different persons. To address these problems, we first trim the isolated sign videos by clipping video frames before hands up and after hands down. Compared with the video before trimming, the trimmed sign videos will look closer to a sign instance in a continuous video (Fig. 1).

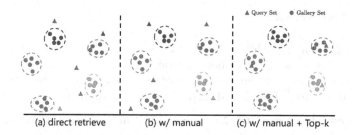

Fig. 2. Illustration of the manual trimming and testing-time Top-K transferring strategies. Triangles △ represent query features, circles ○ represent gallery features, and different colors represent different categories. (a) The feature distributions of isolated signs and continuous signs. (b) The feature distributions with manually trimming isolated sign videos, where the domain gap has been coarsely reduced. (c) The feature distribution after applying testing-time Top-K transferring.

Moreover, we propose a testing-time top-k transferring technique to alleviate the domain gap between the assigned isolated signs and continuous sign videos. To be specific, we first generate all feature representations from the query set. For each video in the test set, we utilize a sliding window to crop multiple clips. Then, all cropped clips are fed into our MSSF to generate feature representations. Next, we compare the feature representation of the query sign (specified by this video name) with all feature representations of the cropped video clips, and find the most similar clip to the query one. After all results have been retrieved, each sign from the query set can obtain many similar clips. We then sort the distances of all retrieved results in a ascending order and find the most similar K clips. Finally, each feature representation from the query set is updated by the average of the feature representations of the top-K retrieved clips. The updated feature representations will be used to retrieve the test data in the following iteration. We will conduct this top-K retrieval a few times to further reduce the domain gap in the retrieval.

Overall, the contributions of this paper are summarized as follows:

- We propose a multi-modal sign spotting framework which comprehensively extracts sign language features from RGB images, 2D-pose based and 3D angle-axis pose based joints, thus leading to representative sign features for sign spotting.
- We propose a testing-time Top-K transferring technique to address the domain gap between the gallery set and the query set to further improve the sign retrieval performance.
- Extensive experiments conducted on the OSLWL-Test-Set dataset demonstrate the priority of our method compared with the baseline model. Remarkably, our method surpasses the baseline provided by the challenge organizer by a large margin of 41.5% on the F1-score and achieves the runner-up in the sign spotting challenge in conjunction with ECCV 2022.

2   Related Work

Sign language spotting achieves significant progress in recent years due to powerful feature representations of deep neural networks. According to the different input data types, the sign-spotting methods are grouped into three classes, involving RGB-based, Pose-based, and Multi-modal approaches.

RGB-Based Sign Spotting Methods: RGB images are a widely used data type in sign language spotting tasks. However, informative feature representations plays a key role in sign spotting. To obtain effective feature representations, some studies first segment hand regions or remove face regions before extracting features [6,16,22,28]. Using colored gloves is a convenient way to attain better segment performance and solve hand occlusion problems [11,12,25]. Although the above methods improve sign spotting accuracy, wearing colored gloves is impractical in daily-life communications. Other studies use auxiliary information, such as hand motion speed, trajectory information, and skin colors, to attain high hand region segmentation quality [9,10,37]. The above methods work well in acquiring superb feature representation, but they may neglect the negative effects of the domain gap between isolated signs and continuous signs. The work [21] proposes a method to learn domain-invariant descriptors and coarsely align two domains, proving that the domain gap is another key factor affecting the quality of feature representations. In this work, we firstly crop the frames of the gallery videos to coarsely bridge the domain gap and then further propose a novel testing-time Top-K transferring technique to reduce the domain gap.

Pose-Based Sign Spotting Methods: A major challenge in sign language spotting tasks is exploring a distinguishable pattern and localizing it in a continuous sign video. Many studies prove that pose data is suitable for sign spotting tasks [4,30,34,36]. Nowadays, skeleton data is popular for sign language spotting tasks. It includes the body key points at bone junctions, such as shoulders, wrists, and elbows. Moreover, multi-modal data, such as bone motion and joint motion, can be obtained based on the skeleton joint data. Several methods [1,3,5,15] prove that collaboration of modalities promotes the spotting accuracy. Pose-based methods can reduce the impacts of interference of complex background and sign-irrelevant information [5]. However, videos captured in the wild is difficult to produce poses well due to occlusions, and fast movements. Besides, details of fingers are crucial for distinguishing similar signs, but poses may be erroneous when frames undergo severe blur. To compensate for the negative impact of inaccurate pose estimation on spotting performance, we introduce two feature extraction branches (RGB-based and pose-based) into our framework.

Multi-modal Sign Language Spotting Methods: Multi-modal methods leverage different data modalities in order to produce a complementary effect in sign feature extraction. The extracted comprehensive features would improve sign-spotting performance. To achieve better combination results, the work [14] designs a shared weights model for multi-modal scenarios. The method [32]

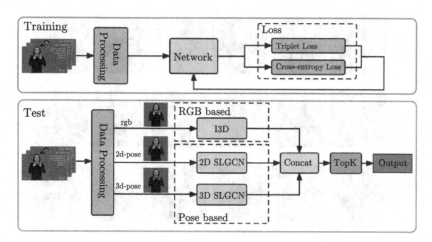

Fig. 3. The overview of our proposed method. Our network contains three branches, including 2D-Pose based SL-GCN, 3D-Pose based SL-GCN, and I3D-MLP, to extract sign language features. The network is trained on isolated signs with a combined loss, including a triplet loss and a cross-entropy loss. In particular, in the testing phase, we introduce a testing-time Top-K transferring to mitgate the domain gap between the isolated signs and continuous counterparts.

designs view-specific and view-independent modules to extract sign features and then combines disparate prediction results efficiently. Other studies design view-invariant representation learning frameworks to attain high-quality feature representation [33,38]. The approach [2] fuses multi-view feature representations together via a specifically designed super-vector. To acquire correlations among prediction from different modalities, [35] develop latent view distribution connections and proposes a post-fusion strategy. Inspired by the above studies, we fuse prediction results together via an ensemble strategy.

3 Proposed Method

The overview of the proposed method is shown in Fig. 3. First, we use different data processing operations to generate multi-modality data, including RGB images, 2D key points, 3D angle-axis pose. Then, we employ two branches, including RGB-based feature extraction and pose-based feature extraction, to generate multi-modal feature representations. The RGB-based feature extraction is built based on I3D-MLP [3], while the pose-based feature extraction is established by SL-GCN [17]. As a result, we totally develop three networks, including 2D-Pose based SL-GCN, 3D-Pose based SL-GCN, and RGB-based I3D-MLP. During the training phase, each network is trained separately. After feature extraction, a combined loss, consisting of a triplet loss and a cross-entropy loss [23], is proposed to train the network. In the inference phase, we employ our three models to extract visual features and then fuse them to generate multi-modal feature representations for

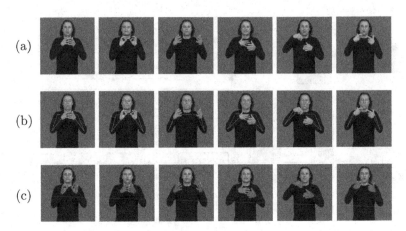

Fig. 4. Visualization of our employed modalities: (a) RGB image sequences; (b) 2D key-joints; (c) 3D key-joints. Here, we only visualize our extracted 3D keypoints via a 3D hand model (MANO) for illustration.

the sign spotting task. Moreover, we propose a novel top-k technique to alleviate the domain gap between the test set and the query set (Fig. 2).

3.1 Data Processing

Our framework takes multi-modal data as the input to generate discriminative feature representations. The multi-modal data includes three types: RGB images, 2D key points, 3D angle-axis poses (Fig. 4).

RGB Images. The original data provided by organizers is in MP4 format. Thus, we first use the opencv to extract RGB frames of each video. However, the person in different videos is misalignment. Hence, we further use HRNet [31] to get 133 key points of human pose, and then select the nose as the central landmark to crop the video frames. In this way, signers are in the center of images of 512 × 512 pixels.

2D Key Point. We use the HRNet [31] to estimate 133-point whole-body key points from the normalized RGB videos [17]. Then, we select 27 key points to construct a skeleton graph. Notably, for each key point, the tensor dimension is composed of 2-dimension spatial coordinates and 1-dimension confidence.

3D Angle-Axis Pose. We leverage the Frankmocap [29] to acquire the 54 angle-axis based 3D joints on whole-body from the normalized RGB videos. Afterwards, 39 upper body joints are selected to construct the 3D skeleton graph.

3.2 Multi-modal Feature Extraction

In this paper, we propose dual-branch structure, including RGB-based feature extraction and pose-based feature extraction, to extract sign language features from different multi-modal data types. To be specific, the RGB-based feature extraction is built based on I3D-MLP [3], while SL-GCN [17] is used to develop the pose-based feature extraction.

I3D-MLP. Inflated 3D CNN (I3D) [3] is one of the most common 3D CNN architecture in recent years. It is widely used to extract features from RGB input data. In this work, we utilize the I3D-MLP model from the Watch-Read-Lookup framework since the I3D has shown better performance than other 3D CNN architecture in sign spotting task [26]. The model mainly contains a two-stage architecture, including an I3D trunk and a three-layer MLP. The I3D network extracts 1024-dimensional feature vectors from RGB input frames and outputs a list of classification predictions. Then the 1024-dimensional feature vectors are fed into the MLP network for dimensionality reduction.

SL-GCN. Performing the pose skeleton extraction auxiliary task with sign language spotting is a straightforward way to enhance the spatial-structure information preservation. Such structure information facilitates adaptation of the extracted sign features between the source RGB images domain and the target domain. Taking this intrinsic observation into consideration, we further adopt the SL-GCN [17] model for structure-aware feature extraction. More specifically, we leverage the spatial temporary GCN proposed in [36] to model the signers upper body with the 2D-based skeleton and angle-axis 3D-based skeleton. Each joint is represented as a graph node in SL-GCN. Formally, the SL-GCN is expressed as:

$$F_{out} = D^{-\frac{1}{2}}(A + I)D^{-\frac{1}{2}}F_{in}W, \tag{1}$$

where the $D^{ii} = \sum_{j}(A^{ij}+I^{ij})$. A denotes the intra-body connections among each joint (i.e. 27 joints in 2D-based pose and 39 joints in 3D-based pose). The identity matrix I indicates the self-connections and W represents the trainable weight matrix. In practice, we implement the SL-GCN by conventional 2D convolution. For each input joint, we obtain the 256 dimension vector as the representative modality-aware feature.

3.3 Loss Function

Since the spotting task is similar to retrieval tasks, we present a combined loss, consisting of a triplet loss and a cross-entropy loss, to train the proposed network [23]. Given three sign language features Y_{p_i}, Y_{p_j} and Y_{n_k} corresponding to the sign samples p_i, p_j and n_k, respectively, where p_i and p_j are from the same class P while n_k comes from the class N. The triplet loss L_{cse} is defined as:

$$L_{tri} = Max((D(Y_{p_i}, Y_{p_j}) - D(Y_{p_i}, Y_{n_k}) + margin), 0), \tag{2}$$

where $D(d_1, d_2)$ is the Euclidean distance of features d_1 and d_2. $margin$ represents the margin threshold used in the triplet loss.

The cross-entropy loss can be formulated as:

$$L_{cse} = -\frac{1}{N}\sum_{i=1}^{N}log\frac{e^{W_{y_i}^T x_i + b_{y_i}}}{\sum_{j=1}^{n}e^{W_j^T x_i + b_j}}, \tag{3}$$

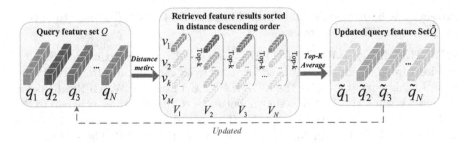

Fig. 5. Illustration of our proposed testing-time Top-K transferring technique. We firstly use the extracted query features from isolated sign videos to retrieve similar ones from the gallery continuous sign video clips. The retrieved sign features will be used to update the query sign features for the next round retrieval. This technique significantly reduces the feature domain gap between isolated and continuous sign videos.

where x_i is the feature of the i-th sample, and its label is y_i. Finally, the combined loss can be defined as:

$$L_{combined} = L_{tri} + L_{cse}. \tag{4}$$

3.4   Sign Spotting from Isolated Signs

For the spotting task, the dataset usually contains two subsets: the test set (unlabeled data) and the query set (labeled data). To spot the exact location of the sign instance in a provided video, we implement our method in the following steps:

First, we observe there is an obvious domain gap between isolated signs and continuous signs. Therefore, we trim the isolated sign videos by removing the video frames before hands up and after hands down. Then, we feed the query sign videos into our model to generate feature representations.

Second, for each video from the test set, we utilize a sliding window to crop video clips from the beginning to the end of a video. Then, we input all clips into the proposed model to extract feature representations. Here, the stride of the sliding window is 1.

Third, all feature representations from this video are used to calculate the Cosine Distance with respect to the corresponding feature representation of the sign provided in the query set. From the video name, we also know which sign to be spotted in a certain video. The clip that has the maximal similarity with the gallery sign representation are regarded as our retrieval result.

3.5   Top-K Transferring Technique

To alleviate the domain gap between the test set and the query set, we propose a Top-K transferring technique (Fig. 5). First, we denote $\mathbb{Q} = \{q_i | i = 1, 2, ..., N\}$ as all the query features from the query set, where q_i is the $i - th$ isolated sign. Note that each feature in the query set is combined from three model outputs.

After the third step in Sect. 3.4, we will obtain many retrieved results for each isolated sign. We sort the distances of all retrieved results and find the most similar K clips. For the isolated sign q_i, we denote $\mathbb{V}_i = \{v_i^j | j = 1, 2, ..., K\}$ as its most similar K clips' features in the test set, where v_i^j is the $j-th$ similar clips' features. Then, each feature representation from the query set will be updated by the average of the feature representations of the top-K retrieved clips. Here, the updated query set can be defined as $\widetilde{\mathbb{Q}} = \{\widetilde{q}_i | i = 1, 2, ..., N\}$, where $\widetilde{q}_i$ is a transferring sign feature of q_i. $\widetilde{q}_i$ can be formulated as:

$$\widetilde{q}_i = \frac{1}{K} \sum_{j=1}^{K} v_i^j. \tag{5}$$

Iteratively, the updated feature representation for a certain sign will be used to retrieve signs from continuous sign videos again.

4   Experiments

4.1   Dataset

We evaluate our framework on a continuous Spanish sign language video dataset in the health domain. The total length of the video in the dataset is 10 h. 10 people (7 deaf and 3 interpreters) conduct the continuous sign language, and they were asked to explain health contents in sign language after reading printed cards. Due to the abundant expressivity and naturalness, the reliance on presenters is very large. In this task, videos are partially annotated with the exact location of 100 signs, at 25 fps and each frame resolution is 1280×720.

The duration of each sign is also not uniform, as the fastest sign lasts about 120 ms (3 frames) up while the slowest one lasts 2 s (50 frames). The duration is not a Gaussian variable, with a mean duration of 520 ms (13 frames).

In detail, the dataset released for the ChaLearn ECCV2022 Challenge on Sign Spotting is divided into 5 splits and a query set. As shown in Table 1, Track1 is named MSSL (multiple shot supervised learning), and Track2 is named OSLWL (one-shot learning and weak labels). The total number of hand annotations exceeds 5000 in MSSL and 3000 in OSLWL. Our experiments mainly utilize MSSL-Train-set for training, while OSLWL-Query-Set and OSLWL-Test-Set for evaluation.

4.2   Implementation Details

In this paper, we propose a dual-branch structure to extract sign language features that includes RGB-based feature extraction and Pose-based feature extraction. Meanwhile, we introduce three data types to portray the human signs. Note that the data processing has been described in Sect. 3.1. As a result, a total of three networks are built to extract sign features. For the RGB-based feature extraction, we choose the I3D network [3] as our backbone and load pre-trained

Table 1. The ChaLearn ECCV 2022 sign spotting dataset has 5 parts containing continuous sign videos and a query set which includes isolated sign videos. There are 7 signers in this dataset. Moreover, different performing customs also contribute to the domain gap

Name	Duration	Signers	Number of signs
MSSL-Train-Set	Around 2.5 h	5 (p1, p2, p4, p5, p6)	60
MSSL-Val-Set	Around 1.5 h	4 (p1, p5, p7, p9)	60
MSSL-Test-Set	Around 1.5 h	4 (p1, p3, p5, p8)	60
OSLWL-Val-Set	1501 4-s videos	5 (p3, p4, p6, p9, p10)	40
OSLWL-Test-Set	607 4-s videos	6 (p1, p3, p4, p5, p8, p9)	40
OSLWL-Query-Set	40 short videos	2 (p0, p3)	40

model from [27]. For the Pose-based feature extraction, SL-GCN [17] is employed to extract pose-based features. The pre-trained model is also from [17]. Since the Pose-based feature extraction contains two data types, we separately train two networks, 2D-Pose based SL-GCN and 3D-Pose based SL-GCN. The main difference of both networks is the graph construction. The graph of the 2D-Pose is based on 27 key joints to build graphs, while the graph of the 3D-Pose is constructed by 39 angle-axis joints. As described in Sect. 3.3, we introduce a combined loss, including a triplet loss and cross-entropy loss, to train the proposed network. Hence, a Batch-ALL strategy is used for sampling in the training stage [13]. Specifically, the batch size includes $P \times K$ samples, where P is the number of sign instances and K is the number of samples for each sign instance. The margin in Eq. 2 is set to 0.2. For I3D, P and K are set to 6 and 2, respectively. For 2D-Pose based SL-GCN and 3D-Pose based SL-GCN, the parameters $P \times K$ are set to 12×4. The length of sequences in each batch is 16 frames. The optimizer of all experiments is Adam. The iteration of I3D, 2D-Pose based SL-GCN and 3D-Pose based SL-GCN is 100, 240 and 190, respectively.

4.3   Evaluation Metrics

We employ the F_1 score to evaluate the prediction performance produced by different methods. F_1 score takes harmonic values of *recall* and *precision* into account to validate the model spotting performance better.

In this task, we regard the IoU of the groundtruth interval and the predicted interval as the matching score of each sign instance. The IoU threshold is set from 0.2 to 0.8 in 0.5 steps to attain relatively slack locations. Positive instances and false instances are obtained by averaging from the 13 thresholds. Given a query of $sign_i$ and a gallery $videofile_n$, the specific calculation method of the IoU_i score is as follows:

$$IoU_i = \frac{Ret_i \cap Rel_i}{Ret_i \cup Rel_i}, \tag{6}$$

where Ret_i and Rel_i represent the prediction and groundtruth interval of $sign_i$ respectively. Noticeably, we regard a prediction instance as a positive sample only IoU_i satisfy greater than the threshold t, and IoU does not calculate if the retrieved and groundtruth classes do not match. For all Ret_i reported from each $videofile_n$ containing L_n groundtruth instances, the specific true positive, false positive, and false negative interval counts can be calculated as follows:

$$TP(t) = \sum_n \sum_i \sum_k IoU_i(k, n) \geq t, \tag{7}$$

$$FP(t) = (\sum_n \sum_i \sum_k 1) - TP(t), \tag{8}$$

$$FN(t) = (\sum_n^N L_n) - TP(t), \tag{9}$$

$$TP = \sum_t TP(t), \quad FP = \sum_t FP(t), \quad FN = \sum_t FN(t). \tag{10}$$

With the above data, $Precision$, $Recall$, and F_1 score are calculated by:

$$Precision = \frac{TP}{(TP + FP)}, \tag{11}$$

$$Recall = \frac{TP}{(TP + FN)}, \tag{12}$$

$$F1 = 2 \cdot \frac{Precision \cdot Recall}{Precision + Recall}. \tag{13}$$

4.4   Main Results

In this section, we firstly compare our method with some existing methods that our network is built on as well as the official baseline, and then conduct ablation study on the components of our method.

Comparison with the State-of-the-Art
In this work, our method is built upon BSL-I3D [27] and SL-GCN [17]. Note that the original SL-GCN only employs 2D key-joints rather than 3D poses for sign representation. As shown in Table 2, we can see that benefiting from the model ensemble and our newly proposed testing-time Top-K transferring, the spotting performance of our method is much higher than the state-of-the-art approaches. Moreover, our method outperforms the official baseline significantly. The visual results are also provided in Fig. 6.

Ablation Study

Model Ensemble. We propose a multi-modal sign spotting framework to generate comprehensive feature representations. To gain superior spotting performance, we explore various modal combination ways. The experimental results are shown in Table 2.

From the perspective of single model performance, 2D-Pose based SL-GCN achieves appealing performance, followed by the RGB-based I3D-MLP. Based

Table 2. The contributions of different models on the final performance. We evaluate the F1-score of different models on the OSLWL-Test-Set dataset. ↑ indicates the higher, the better. ✓ indicates the employment of a certain model or modality.

Diff. models			F_1 ↑
2D-based	3D-based	RGB-based	
✓			0.4692
	✓		0.4076
		✓	0.4249
✓	✓		0.4954
	✓	✓	0.5057
✓		✓	0.5503
✓	✓	✓	**0.5593**
Baseline			0.3951
SL-GCN [17]			0.4543
BSL [27]			0.4512
Ours			**0.5593**

Table 3. Impact of the proposed testing-time Top-K transferring on the final performance. We evaluate the performance using F1-score on the OSLWL-Test-Set dataset. "Trim" means manually removing video frames before hands up and after hands down; ✓ and × indicate the employment or removal of the manually trimming pre-processing, respectively. "Times" defines the iteration number of employing the Top-k transferring technique. "K" means the number of retrieved clips that are used to update feature representations. ↑ means the higher, the better.

Trim		×	✓	✓	✓	✓	✓	✓	✓	×
Top-K	Times	0	0	1	1	1	2	2	2	1
	K	–	–	[1]	[3]	[5]	[5, 1]	[5, 3]	[5, 5]	[5]
F_1 ↑		0.4602	0.5168	0.5272	0.5411	**0.5593**	0.5498	0.5538	0.5442	0.5205

on the assumption that multi-modal combinations can generate complementary results, we conduct plenty of experiments to explore the excellent combination way. As indicated in Table 2, all combination methods have a positive impact on spotting results, and the combination of three models (2D-Pose based SL-GCN, 3D-Pose based SL-GCN, and the RGB-based I3D-MLP) attains the highest F_1 score, with the 0.5593 F_1 score. Besides, we observe that any combination of two models achieves superior performance than any single model. In all pairwise combinations, the combination of the 2D-Pose based SL-GCN and the 3D-Pose based SL-GCN is the best one, of which F_1 score reaches 0.5503. The combination of the 3D Pose-based SL-GCN and the 2D-Pose based SL-GCN achieves 0.4954 F_1 score, which outperforms the 3D-Pose based SL-GCN by 5.6%. In contrast, the F_1 score of 3D-Pose based SL-GCN and the RGB-based I3D-MLP is 0.5057 that has an increase of 0.0981 F_1 score compared to the 3D-Pose based SL-GCN. The above results imply the positive complementary of combining various models. As a result, the combination of three models is the final setting in our sign language spotting framework.

Table 4. The impact of different sliding window frames which are used to crop video clips from gallery continuous sign videos on the final sign spotting performance. The performance is measured by F1-score on the OSLWL-Test-Set dataset. ↑ indicates the higher, the better. ✓ indicates the adopted frame number for the sliding windows.

The length of the sliding window					F_1 ↑
13 frames	14 frames	15 frames	16 frames	17 frames	
✓					0.5012
	✓				0.5369
		✓			0.5470
			✓		**0.5593**
				✓	0.5304
✓	✓			✓	0.5258
✓	✓	✓	✓	✓	0.5269

Top-K Transferring. We introduce a Top-K transferring strategy into our framework to diminish the gap between query signs and gallery signs. To attain superior spotting performance, we implement plenty of parameter adjustment experiments. All experiments are based on the three-model ensemble setting, and the length of sliding window is set as 16. By comparing the experimental results of the first and fifth rows in Table 3, using the Top-K strategy significantly raises the F_1 score by 21.5%.

In the experiments, we also adjust the number of feature representations (K) and the strategy iteration number $Times$. When the $Times$ set is 1, the F_1 score is in an increasing trend as the number of K increases, and $K = 5$ achieves a 0.5593 F_1 score. Based on the above results, we increase the number of $Times$. The experimental results prove that raising the number of $Times$ does not boost the sign spotting performance. This is mainly because the updated features selected by the Top-k strategy do not always match isolated sign instances in the query set. As the number of $Times$ increases, the deviation between the updated features and original sign features becomes larger. It degrades sign language spotting performance. As a result, $Times = 1$ is the optimal choice for our task. Besides, we also conduct experiments on using or not using a $Trim$ method. According to the results of the fifth and the last rows, clipping the video frames before hands up and after hands down facilitates the final spotting results. The $Trim$ method (Row 5) improves the F_1 score by about 7.5% compared to the untrimmed result (Row 9). The above results fully demonstrate the effectiveness of the Top-k strategy in reducing the domain gap between the query set and the gallery set.

Sliding Window. For each video of the gallery set, we set a sliding window strategy to crop video clips from the beginning to the end of a video. In all experiments, we use a three-model combination way to extract features and also introduce the Top-K strategy with the setting of $Times = 1$ and $K = 5$.

To attain a suitable window length setting, we conduct plenty of experiments on the OSLWL-Test-Set dataset. Considering the sign instance length distribution, we choose the length 13, 14, 15, 16, 17 as the basic length setting. As shown

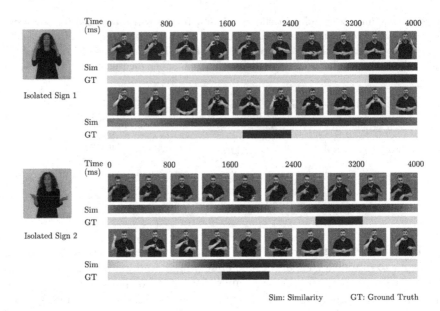

Fig. 6. Visualization of our retrieved results. As indicated by the similarity bar, the peak values predicted by our method are consistent with the ground-truth positions, demonstrating the effectiveness of our method.

in Table 4, the F_1 score is positively correlated with the increase of sliding window length within a certain range, and the F_1 score reaches the peak of 0.5593 when the window length is 16. A common consensus is that integrating the results obtained under various sliding window lengths facilitates sign spotting accuracy. To this end, we first integrate the results achieved by 13, 15, and 17 window lengths, and select the results with a minimum distance gap from the corresponding query sign instance. As the results show, there is a 6% decrease compared to using a sliding window length of 16. Although combining all window lengths slightly improve the F_1 score, there still is a gap with the results of setting 16 window length, which is 0.0324 lower. From the above results, we analyze the main reason is that the model tends to select features with comparatively low window length. All in all, we select a sliding window length of 16 as our final setting.

5   Conclusions

In this paper, we propose a dual-branch based multi-modal sign spotting framework. In our framework, we fully take advantage of the aggregated information from the various modalities by RGB-based branch and Pose-based branch. The aggregated information provides discriminative feature representations of the retrieved sign languages. Moreover, we present a testing-time top-k transferring technique to address the domain gap between isolated and continuous sign videos. In particular, this technique exploits an iterative retrieve strategy to update the extracted feature representations, leading to significant sign spotting performance improvement. As a result, our method achieves outperforms the baseline model by a large margin of 41.5% on the OSLWL-Test-Set dataset.

Acknowledgement. The research was supported by the Australian Research Council DP220100800 and Google Research Scholar Program.

References

1. Baradel, F., Wolf, C., Mille, J.: Human action recognition: pose-based attention draws focus to hands. In: Proceedings of the IEEE International Conference on Computer Vision Workshops, pp. 604–613 (2017)
2. Cai, Z., Wang, L., Peng, X., Qiao, Y.: Multi-view super vector for action recognition. In: Proceedings of the IEEE Conference on Computer Vision and Pattern Recognition, pp. 596–603 (2014)
3. Carreira, J., Zisserman, A.: Quo Vadis, action recognition? A new model and the kinetics dataset. In: Proceedings of the IEEE Conference on Computer Vision and Pattern Recognition, pp. 6299–6308 (2017)
4. Cheng, K., Zhang, Y., Cao, C., Shi, L., Cheng, J., Lu, H.: Decoupling GCN with DropGraph module for skeleton-based action recognition. In: Vedaldi, A., Bischof, H., Brox, T., Frahm, J.-M. (eds.) ECCV 2020. LNCS, vol. 12369, pp. 536–553. Springer, Cham (2020). https://doi.org/10.1007/978-3-030-58586-0_32
5. Choutas, V., Weinzaepfel, P., Revaud, J., Schmid, C.: Potion: pose motion representation for action recognition. In: Proceedings of the IEEE Conference on Computer Vision and Pattern Recognition, pp. 7024–7033 (2018)
6. Cihan Camgoz, N., Hadfield, S., Koller, O., Bowden, R.: SubUNets: end-to-end hand shape and continuous sign language recognition. In: Proceedings of the IEEE International Conference on Computer Vision, pp. 3056–3065 (2017)
7. Conneau, A., et al.: Unsupervised cross-lingual representation learning at scale. arXiv preprint arXiv:1911.02116 (2019)
8. Conneau, A., Lample, G.: Cross-lingual language model pretraining. In: Advances in Neural Information Processing Systems, vol. 32 (2019)
9. Cooper, H., Ong, E.J., Pugeault, N., Bowden, R.: Sign language recognition using sub-units. J. Mach. Learn. Res. **13**, 2205–2231 (2012)
10. Dardas, N.H., Georganas, N.D.: Real-time hand gesture detection and recognition using bag-of-features and support vector machine techniques. IEEE Trans. Instrum. Meas. **60**(11), 3592–3607 (2011)
11. Fels, S.S., Hinton, G.E.: Glove-talk: a neural network interface between a data-glove and a speech synthesizer. IEEE Trans. Neural Netw. **4**(1), 2–8 (1993)
12. Grobel, K., Assan, M.: Isolated sign language recognition using hidden Markov models. In: 1997 IEEE International Conference on Systems, Man, and Cybernetics. Computational Cybernetics and Simulation, vol. 1, pp. 162–167. IEEE (1997)
13. Hermans, A., Beyer, L., Leibe, B.: In defense of the triplet loss for person re-identification. arXiv preprint arXiv:1703.07737 (2017)
14. Hoffman, J., Gupta, S., Darrell, T.: Learning with side information through modality hallucination. In: Proceedings of the IEEE Conference on Computer Vision and Pattern Recognition, pp. 826–834 (2016)

15. Hu, J.F., Zheng, W.S., Pan, J., Lai, J., Zhang, J.: Deep bilinear learning for RGB-D action recognition. In: Proceedings of the European Conference on Computer Vision (ECCV), pp. 335–351 (2018)
16. Huang, J., Zhou, W., Zhang, Q., Li, H., Li, W.: Video-based sign language recognition without temporal segmentation. In: Proceedings of the AAAI Conference on Artificial Intelligence, vol. 32 (2018)
17. Jiang, S., Sun, B., Wang, L., Bai, Y., Li, K., Fu, Y.: Skeleton aware multi-modal sign language recognition. In: Proceedings of the IEEE/CVF Conference on Computer Vision and Pattern Recognition Workshops, pp. 3413–3423 (2021)
18. Li, D., Rodriguez, C., Yu, X., Li, H.: Word-level deep sign language recognition from video: a new large-scale dataset and methods comparison. In: Proceedings of the IEEE/CVF Winter Conference on Applications of Computer Vision, pp. 1459–1469 (2020)
19. Li, D., et al.: TSPNet: hierarchical feature learning via temporal semantic pyramid for sign language translation. In: Advances in Neural Information Processing Systems, vol. 33, pp. 12034–12045 (2020)
20. Li, D., Yu, X., Xu, C., Petersson, L., Li, H.: Transferring cross-domain knowledge for video sign language recognition. In: Proceedings of the IEEE/CVF Conference on Computer Vision and Pattern Recognition, pp. 6205–6214 (2020)
21. Li, D., Yu, X., Xu, C., Petersson, L., Li, H.: Transferring cross-domain knowledge for video sign language recognition. In: Proceedings of the IEEE/CVF Conference on Computer Vision and Pattern Recognition (CVPR) (2020)
22. Lim, K.M., Tan, A.W.C., Lee, C.P., Tan, S.C.: Isolated sign language recognition using convolutional neural network hand modelling and hand energy image. Multimed. Tools Appl. **78**(14), 19917–19944 (2019). https://doi.org/10.1007/s11042-019-7263-7
23. Lin, B., Zhang, S., Yu, X.: Gait recognition via effective global-local feature representation and local temporal aggregation. In: Proceedings of the IEEE/CVF International Conference on Computer Vision, pp. 14648–14656 (2021)
24. Liu, Y., et al.: Multilingual denoising pre-training for neural machine translation. Trans. Assoc. Comput. Linguist. **8**, 726–742 (2020)
25. Mehdi, S.A., Khan, Y.N.: Sign language recognition using sensor gloves. In: Proceedings of the 9th International Conference on Neural Information Processing, ICONIP 2002, vol. 5, pp. 2204–2206. IEEE (2002)
26. Momeni, L., Varol, G., Albanie, S., Afouras, T., Zisserman, A.: Watch, read and lookup: learning to spot signs from multiple supervisors. CoRR abs/2010.04002 (2020). https://arxiv.org/abs/2010.04002
27. Momeni, L., Varol, G., Albanie, S., Afouras, T., Zisserman, A.: Watch, read and lookup: learning to spot signs from multiple supervisors. In: Proceedings of the Asian Conference on Computer Vision (2020)
28. Parelli, M., Papadimitriou, K., Potamianos, G., Pavlakos, G., Maragos, P.: Exploiting 3D hand pose estimation in deep learning-based sign language recognition from RGB videos. In: Bartoli, A., Fusiello, A. (eds.) ECCV 2020. LNCS, vol. 12536, pp. 249–263. Springer, Cham (2020). https://doi.org/10.1007/978-3-030-66096-3_18
29. Rong, Y., Shiratori, T., Joo, H.: FrankMocap: a monocular 3D whole-body pose estimation system via regression and integration. In: IEEE International Conference on Computer Vision Workshops (2021)
30. Shi, L., Zhang, Y., Cheng, J., Lu, H.: Skeleton-based action recognition with multi-stream adaptive graph convolutional networks. IEEE Trans. Image Process. **29**, 9532–9545 (2020)

31. Sun, K., Xiao, B., Liu, D., Wang, J.: Deep high-resolution representation learning for human pose estimation. In: Proceedings of the IEEE/CVF Conference on Computer Vision and Pattern Recognition, pp. 5693–5703 (2019)

32. Wang, D., Ouyang, W., Li, W., Xu, D.: Dividing and aggregating network for multi-view action recognition. In: Proceedings of the European Conference on Computer Vision (ECCV), pp. 451–467 (2018)

33. Wang, H., Kläser, A., Schmid, C., Liu, C.L.: Dense trajectories and motion boundary descriptors for action recognition. Int. J. Comput. Vis. **103**(1), 60–79 (2013)

34. Wang, H., Wang, L.: Modeling temporal dynamics and spatial configurations of actions using two-stream recurrent neural networks. In: Proceedings of the IEEE Conference on Computer Vision and Pattern Recognition, pp. 499–508 (2017)

35. Wang, L., Ding, Z., Tao, Z., Liu, Y., Fu, Y.: Generative multi-view human action recognition. In: Proceedings of the IEEE/CVF International Conference on Computer Vision, pp. 6212–6221 (2019)

36. Yan, S., Xiong, Y., Lin, D.: Spatial temporal graph convolutional networks for skeleton-based action recognition. In: Thirty-Second AAAI Conference on Artificial Intelligence (2018)

37. Yang, Q.: Chinese sign language recognition based on video sequence appearance modeling. In: 2010 5th IEEE Conference on Industrial Electronics and Applications, pp. 1537–1542. IEEE (2010)

38. Zheng, J., Jiang, Z., Chellappa, R.: Cross-view action recognition via transferable dictionary learning. IEEE Trans. Image Process. **25**(6), 2542–2556 (2016)

W36 - A Challenge
for Out-of-Distribution Generalization
in Computer Vision

W36 - A Challenge for Out-of-Distribution Generalization in Computer Vision

Deep learning sparked a tremendous increase in the performance of computer vision systems over the past decade, under the implicit assumption that the training and test data are drawn independently and identically distributed (IID) from the same distribution. However, Deep Neural Networks (DNNs) are still far from reaching human-level performance at visual recognition tasks in real-world environments. The most important limitation of DNNs is that they fail to give reliable predictions in unseen or adverse viewing conditions, which would not fool a human observer, such as when objects have an unusual pose, texture, shape, or when objects occur in an unusual context or in challenging weather conditions. The lack of robustness of DNNs in such out-of-distribution (OOD) scenarios is generally acknowledged but largely remains unsolved, however, it needs to be overcome to make computer vision a reliable component of AI.

October 2022

Adam Kortylewski
Bingchen Zhao
Jiahao Wang
Shaozuo Yu
Siwei Yang
Dan Hendrycks
Oliver Zendel
Dawn Song
Alan Yuille

Domain-Conditioned Normalization for Test-Time Domain Generalization

Yuxuan Jiang[1]([✉]), Yanfeng Wang[1,2], Ruipeng Zhang[1], Qinwei Xu[1],
Ya Zhang[1,2], Xin Chen[3], and Qi Tian[4]

[1] Cooperative Medianet Innovation Center, Shanghai Jiao Tong University,
Shanghai, China
{g.e.m_jyx,wangyanfeng,zhangrp,qinweixu,ya_zhang}@sjtu.edu.cn
[2] Shanghai AI Laboratory, Shanghai, China
[3] Huawei Inc., Shenzhen, China
[4] Huawei Cloud & AI, Guangzhou, China
tian.qi1@huawei.com

Abstract. Domain generalization aims to train a robust model on multiple source domains that generalizes well to unseen target domains. While considerable attention has focused on training domain generalizable models, a few recent studies have shifted the attention to test time, *i.e.*, leveraging test samples for better target generalization. To this end, this paper proposes a novel test-time domain generalization method, Domain Conditioned Normalization (DCN), to infer the normalization statistics of the target domain from only a single test sample. In order to learn to predict the normalization statistics, DCN adopts a meta-learning framework and simulates the inference process of the normalization statistics at training. Extensive experimental results have shown that DCN brings consistent improvements to many state-of-the-art domain generalization methods on three widely adopted benchmarks.

Keywords: Domain generalization · Batch normalization · Test time adaptive · Distribution shift

1 Introduction

The performance of deep neural networks degrades drastically when the distribution of train (source) and test (target) data are different. In order to solve the problem of distribution shift, domain generalization (DG) aims to train a robust model on multiple source domains that generalizes well to arbitrary unseen target domains. Existing DG methods can be mainly divided into three categories: invariant representation learning which aims to learn a shared feature space across source domains, data augmentation which expands the source distributions, regularization techniques which are utilized to learn more semantic information from source domains.

Supplementary Information The online version contains supplementary material available at https://doi.org/10.1007/978-3-031-25085-9_17.

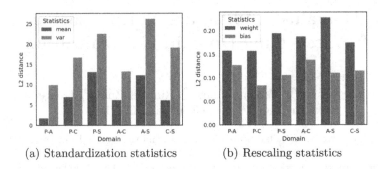

(a) Standardization statistics   (b) Rescaling statistics

Fig. 1. The difference in the first BN layer for (a) standardization statistics and (b) rescaling statistics. We train one model on all domains of PACS benchmark with different BN layers for each domain. The L2 distance is used to measure the statistics differences between every two domains.

Despite their success, most existing DG methods only focus on the training stage, *i.e.*, how to train a generalizable model based on source domains. In fact, DG is naturally an ill-posed problem due to information incompleteness. Knowledge of the target domain may greatly help the model generalization. Actually, the model always has access to at least one unlabeled test sample at inference which may provide important clues about the target domain. Some recent studies start to investigate how to make use of the test sample at inference to improve a model's generalizability to the target domain. TTT [33] and TENT [37] utilize target domain data to update model parameters by constructing an unsupervised loss. Dubey et al. [7] use some target samples to construct domain embeddings so that the classifier can make dynamic predictions based on it. T3A [13] computes pseudo-prototype representation for each class with past predicted target samples to modify the classifier. However, these methods have the following two limitations. (1) Updating parameters with target data increases the inference time significantly and may lead to catastrophic failure; (2) Batches of data are not necessarily available at inference and there is no guarantee that the domain labels of test samples are available.

In this paper, we propose a test-time DG method called Domain Conditioned Normalization (DCN), which estimates the normalization statistics for target domain at inference. DCN is based on the observation that the statistics (including standardization and rescaling statistics) in batch normalization (BN) [12] are unique for each domain and larger distribution shift is reflected by the increased difference in statistics, as shown in Fig. 1. Thus, to improve the domain generalizability of a model at test time, a straight-forward solution is to normalize the test data with the statistics of the domain it comes from. The challenge lies in we can assume access to only one test sample. Nevertheless, the good news is that, the source statistics are expected to contain essential semantic and style information shared by the source and target domains. We thus attempt to combine the instance statistics from a single test sample that reflect domain-specific information for the target domain, with the source statistics to infer the statistics of the target domain. The final inferred statistics is then used to normalize this test sample during inference.

In order to infer the target statistics, a meta-learning framework is adopted by DCN. In each meta-iteration, one source domain is chosen as the meta-target domain and the other source domains are considered as meta-source domains. DCN learns to infer the statistics of the meta-target domain by approximating the ground-truth meta-target statistics with the instance statistics of the meta-target sample and the statistics of meta-source domains. At testing, DCN infers the statistics of the target domain with a single test sample and the source statistics, which is used to replace the source statistics in the trained model for prediction. Our DCN only requires one forward pass of a single target sample, which avoids training at test time and requires no batched test data.

To validate the effectiveness of the proposed method, we experiment on several standard DG benchmarks, namely PACS [17], VLCS [35], OfficeHome [36]. Extensive experimental results have shown that DCN brings consistent improvements to the state-of-the-art DG approaches, indicating that the inferred target statistics obtained from a single test sample does help model generalize better across domains. We have also demonstrated that DCN becomes more effective as the distribution shift increases.

2 Related Work

2.1 Domain Generalization

The main goal of domain generalization (DG) is to learn a robust model from multiple available source domains that can generalize well to arbitrary unseen target domains [38,44]. Existing DG methods can be roughly categorized into three classes: invariant representation learning, data augmentation and regularization methods. Invariant representation learning aims to learn a shared feature space that retains the semantic information across source domains. Some work [19,22] learn this shared feature space in an adversarial training manner. Another work [29,31,32] align second-order statistics or gradients to learn the invariant representation. Data augmentation is also a common method for DG research. DDAIG [45] and FACT [41] generate new images to enlarge the training dataset. Different from augmentation at the image level, some methods [20,46,47] expand source distributions at the feature level. Another popular approach in DG is to use regularization techniques. Self-supervised objective is a popular choice as the regularization [2,4,16,39]. Another work [1,18] rely on meta-learning to simulate the domain shift during training. Furthermore, RSC [11] masks the maximum response gradient to learn more semantic information. These approaches only focus on source domains at the training staget and ignores the use of test data during inference.

2.2 Normalization in Neural Networks

Batch Normalization [12] is a common technique that optimizes the training of deep network by reducing internal covariate shift. However, normalizing the target domain data with the source statistics is sub-optimal because there is a distribution shift between source and target domains. To solve this problem, BIN [27] combines different standardization statistics with the learnable weights. ILM [14] learns the rescaling statistics for each sample by using neural networks. DSON [28] proposes to learn separate BIN for each source domain and ensemble the normalization for the target domain. MetaNorm [6] and ASR-Norm [8] learn the adaptive statistics based on a single sample. Different from all the methods above, our DCN uses both a single target sample and source statistics for adaptive normalization, so DCN can infer more accurate target statistics and achieve better generalization performance.

2.3 Adaptation and Generalization at Test Time

Recently, a lot of work has begun to study test-time adaptation. TTT [33] and TENT [37] update model parameters by training the target domain data with an unsupervised loss function. Apart from that, some work improves model generalization with test data while avoiding training at test time. BN-Test [26] computes statistics on batches of target samples to normalize them during inference. Dubey et al. [7] use very few test samples to construct the domain embedding for the target domain and use the domain embedding as a supplementary signal to make adaptive predictions. T3A [13] computes the representation templates with previous target samples and use these representation templates to modify the classifier. Compared with these methods, our DCN only requires a single test sample and avoids training during inference so that it avoids the increase of inference time and memory overhead.

3   Methodology

We here attempt to explore adaptive normalization at test time by inferring the target domain statistics with source domain statistics and a single target sample. In this section, we first describe the formal formulation of domain generalization and domain-specific batch normalization, followed by the details of our method, Domain Conditioned Normalization (DCN).

3.1   Preliminary

Domain Generalization. In domain generalization, the training dataset D_S consists of N_S source domains. Each source domain $D_d = \{(x_d^i, y_d^i)\}_{i=1}^{n_d}$ ($d \in \{1, \cdots, N_S\}$) contains n_d data and label pairs. There is also a target domain D_T, which is only available during testing. The goal of domain generalization is to train a model f_θ with D_S that generalizes well on the target domain D_T, where θ is the parameters of the model. A vanilla baseline is training with all source domain data with the empirical risk minimization (ERM) objective:

$$\min_\theta \frac{1}{N_S} \sum_{d=1}^{N_S} \frac{1}{n_d} \sum_{i=1}^{n_d} l(f_\theta(x_d^i), y_d^i), \tag{1}$$

where $l(\cdot)$ is the standard cross-entropy loss function for classification problems.

Domain Specific Batch Normalization. BN [12] is a widely used training technique in deep learning to reduce the internal covariate shift. The statistics in BN represent the characteristics of the training dataset. However, when present with multiple training domains, it is inappropriate to share the BN statistics among different domains considering each domain have its own characteristics. Therefore, in this paper, we employ DSBN [3] to separately store the domain-specific statistics for each source domain, which makes data from different domains go through different BN layers.

Given a batch of samples from domain d during training, the input feature maps of a normalization layer is denoted by $x_d \in \mathbb{R}^{N \times C \times H \times W}$, where N denotes the batch size, C denotes the number of channels, H and W denote height and width respectively. DSBN first uses the domain-specific batch mean μ_d^{bn} and variance σ_d^{bn} for standardization, then applies affine transformations on the standardized features x_d^{stan} with domain-specific rescaling parameters $\{\gamma_d, \beta_d\}$. The whole process is denoted as:

$$x_d^{stan} = \frac{x_d - \mu_d^{bn}}{\sqrt{\sigma_d^{bn} + \epsilon}}, \tag{2}$$

$$x_d^{dsbn} = \gamma_d \cdot x_d^{stan} + \beta_d, \tag{3}$$

where $\mu_d^{bn}, \sigma_d^{bn}, \gamma_d, \beta_d \in \mathbb{R}^C$ and ϵ is a small constant for numerical stability. The batch mean μ_d^{bn} and variance σ_d^{bn} are calculated as:

$$\mu_d^{bn} = \frac{\sum_{n,h,w} x_d}{N \cdot H \cdot W} \quad \text{and} \quad \sigma_d^{bn} = \frac{\sum_{n,h,w} (x_d - \mu_d^{bn})^2}{N \cdot H \cdot W}. \tag{4}$$

Following standard implementations [12], we acquire the domain-specific mean μ_d and variance σ_d of a source domain D_d through Exponential Moving Average (EMA). With the help of DSBN, we obtain the domain-specific normalization statistics (*i.e.*, standardization statistics $\{\mu_d, \sigma_d\}$ and rescaling statistics $\{\gamma_d, \beta_d\}$) of different source domains, which facilitates the training of DCN.

3.2 Domain Conditioned Normalization

Our approach aims to use a single target sample and source domain-specific statistics to infer the normalization statistics for an unseen target domain. To accomplish this, we adopt a meta-learning strategy to simulate such a target statistics inference task during each meta-iteration.

Specifically, within a meta-iteration, we randomly split the training domains D_S into two disjoint sets: one domain as meta-target and the others are meta-source domains. In the meta-train stage, we train the backbone network on meta-source domains and update the DSBN statistics of each meta-source domain by minimizing the cross-entropy loss. In the meta-test stage, we update the DSBN statistics of the meta-target domain in the same way to provide ground-truth meta-target statistics. Then we train the DCN modules which take the

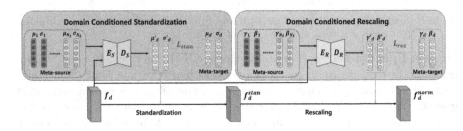

Fig. 2. Illustration of DCN for each layer when D_d is the meta-target domain. We use auto-encoders to infer statistics of D_d with input feature map f_d and the meta-source domains' statistics and use the inferred statistics to normalize f_d. L_{stan} and L_{res} are L2 loss between the inferred statistics and the ground truth statistics.

statistics of the meta-source domains and a single meta-target sample as inputs, to approximate the ground-truth meta-target statistics.

By looping over multiple meta-iterations, we hope the model would be able to infer the DSBN statistics on a novel target domain during testing. The learning process of DCN for a single layer is illustrated in Fig. 2, which can be further divided into domain conditioned standardization and domain conditioned rescaling. Next, we will describe these two parts in detail.

Domain Conditioned Standardization. The standardization statistics stand for the channel-wise mean μ and variance σ. To infer the meta-target standardization statistics, we adopt an auto-encoder structure, where both the encoder and decoder are composed of one fully-connected layer, followed by a ReLU activation. The predictions of the standardization auto-encoder are conditioned on the instance mean μ_d^i and variance σ_d^i of the meta-target sample x_d^i, as well as the standardization statistics $\{\mu_p, \sigma_p\}$ ($p \neq d$, $p \in \{1, \cdots, N_S\}$) of the meta-source domains. The instance mean μ_d^i and variance σ_d^i of x_d^i are calculated as:

$$\mu_d^i = \frac{\sum_{h,w} x_d^i}{H \cdot W} \quad \text{and} \quad \sigma_d^i = \frac{\sum_{h,w} (x_d^i - \mu_d^i)^2}{H \cdot W}. \tag{5}$$

Inspired from [15,42], we assume the standardization statistics of one domain to be a shifted version of those of another domain. Therefore, the meta-target standardization statistics can be inferred in the form of a linear combination of the meta-source standardization statistics and the instance statistics of x_d^i, which is specified as:

$$\mu_d' = \frac{1}{N_S - 1} \sum_{p \neq d} (w_{p,d}^i \cdot \mu_d^i + (1 - w_{p,d}^i) \cdot \mu_p), \tag{6}$$

$$\sigma_d' = \frac{1}{N_S - 1} \sum_{p \neq d} (w_{p,d}^i \cdot \sigma_d^i + (1 - w_{p,d}^i) \cdot \sigma_p). \tag{7}$$

where $w_{p,d}^i \in \mathbb{R}^C$ is the channel-wise linear combination weights learned from the standardization auto-encoders. Intuitively, different channels correspond to different degrees of transferability from the meta-source D_p to the meta-target D_d, so using a specific weight for each channel is a wiser choice.

To learn the combination weight $w_{p,d}^i$, we adopt a similar strategy as [21,40] by using the difference between the instance statistics $\{\mu_d^i, \sigma_d^i\}$ of x_d^i and the meta-source standardization statistics $\{\mu_p, \sigma_p\}$ as the inputs to the auto-encoder. The difference metric is calculated by:

$$M_{p,d}^i = \left| \frac{\mu_p}{\sqrt{\sigma_p + \epsilon}} - \frac{\mu_d^i}{\sqrt{\sigma_d^i + \epsilon}} \right|. \tag{8}$$

Intuitively, the channel-wise statistics difference $M_{p,d}^i$ can be viewed as an indicator of the domain transferability, where a small value of an entry in $M_{p,d}^i$ means the meta-source domain and the meta-target domain are similar in the corresponding channel, and vice versa. Such a transferability indicator allows the auto-encoder to learn a better trade-off between the meta-source standardization statistics and the instance statistics of x_d^i. Finally, the output of the decoder goes through a sigmoid activation to scale the learned weights into [0, 1].

After obtaining the inferred standardization statistics of D_d through Eq. (6) and Eq. (7), we train the standardization auto-encoder by minimizing the L2 distance with the ground-truth standardization statistics $\{\mu_d, \sigma_d\}$ obtained from DSBN on the meta-target domain:

$$L_{stan} = ||\mu_d' - \mu_d||_2^2 + ||\sigma_d' - \sigma_d||_2^2. \tag{9}$$

Domain Conditioned Rescaling. The rescaling statistics stand for the channel-wise affine parameters $\{\gamma, \beta\}$. Denote the rescaling statistics of the meta-target domain D_d and the meta-source domain D_p as $\{\gamma_d, \beta_d\}$ and $\{\gamma_p, \beta_p\}$, respectively. Similar as domain conditioned standardization, our goal is to learn a rescaling auto-encoder to infer the rescaling statistics of the meta-target domain D_d. The rescaling auto-encoder shares the same structure as the standardization auto-encoder, while its predictions are conditioned on the instance statistics $\{\mu_d^i, \sigma_d^i\}$ of the single meta-target sample x_d^i and the rescaling statistics $\{\gamma_p, \beta_p\}$ of the meta-source domain.

As stated in [10,23,34], the rescaling statistics $\{\gamma, \beta\}$ act like an attention mechanism that measures the contributions of each channel to the current domain. When transferring from the meta-source domain to the meta-target domain, the rescaling statistics of the domain-shared channels should be kept as the same since these channels make similar contributions on both domains. In contrast, the rescaling statistics of the non-shared channels should be calibrated to learn the meta-target domain-specific information. Therefore, we devise the learning strategy of the meta-target rescaling statistics in the manner of a scaling of the meta-source rescaling statistics. Specifically, we concatenate σ_d^i with γ_p and μ_d^i with β_p. Then we feed the two concatenated vectors into the rescaling auto-encoder and infer the meta-target rescaling statistics as:

$$\gamma_d' = \frac{1}{N_S - 1} \sum_{p \neq d} (g_r([\sigma_d^i, \gamma_p]) + 1) \cdot \gamma_p, \tag{10}$$

Algorithm 1. The Meta-learning Strategy for DCN

Input: Model f_θ, Total Iterations T, Source Domains D_S.
Output: Model f_θ, Statistics on D_S ($\{\mu, \sigma\}$, $\{\gamma, \beta\}$).
1: **for all** t in $1 \cdots T$ **do**
2:　　**Start a meta-iteration:**
3:　　　Random split D_S into meta-source domains D_p and meta-target domain D_d.
4:　　**Meta-train:**
5:　　　Forward and infer meta-source DSBN statistics $\{\mu_p, \sigma_p\}$, $\{\gamma_p, \beta_p\}$.
6:　　**Meta-test:**
7:　　　Random select one sample x_d^i from D_d.
8:　　　Obtain instance statistics $\{\mu_d^i, \sigma_d^i\}$ with Eq. (5).
9:　　　Infer standardization statistics with Eq. (6) and Eq. (7).
10:　　　Infer rescaling statistics with Eq. (10) and Eq. (11).
11:　　　Infer ground-truth meta-target DSBN statistics $\{\mu_d, \sigma_d\}$, $\{\gamma_d, \beta_d\}$.
12:　　　Compute the total loss with Eq. (13) and backward.
13: **end for**

$$\beta_d' = \frac{1}{N_S - 1} \sum_{p \neq d} (g_r([\mu_d^i, \beta_p]) + 1) \cdot \beta_p. \tag{11}$$

where a Tanh activation is further appended to scale the decoder outputs.

After acquiring the inferred rescaling statistics of the meta-target domain, we minimize its L2 distance with the ground truth $\{\gamma_d, \beta_d\}$ obtained from DSBN on the meta-target domain:

$$L_{res} = ||\gamma_d' - \gamma_d||_2^2 + ||\beta_d' - \beta_d||_2^2. \tag{12}$$

3.3 Training and Inference

Training. In our meta-learning process, the DCN modules attempt to approximate the meta-target DSBN statistics conditioned on only a single meta-target sample and the meta-source DSBN statistics. To obtain the meta-source DSBN statistics, we need to train the backbone network during the meta-train stage. Similarly, we should also train the backbone network during the meta-test stage to acquire reliable meta-target DSBN statistics as the ground truth for the approximation. Since the meta-source and meta-target domain are randomly selected in each meta-iteration and the whole training process goes through multiple meta-iterations, we could merge the training of DSBN on both the meta-source and meta-target domain into only on the meta-target domain. In this way, the meta-source DSBN statistics can be obtained with only a forward propagation without additional training, which results in a more simplified and efficient meta-learning process. The training in the meta-test stage is then composed of training the DSBN statistics with the classification loss and training the DCN modules to approximate DSBN with Eq. (9) and Eq. (12). To ensure the discriminality of DCN, we also add a classification loss on the features normalized by DCN. The total objective of the meta-test stage is then formulated as:

$$L = L_{cls} + \lambda(L'_{cls} + \bar{L}_{stan} + \bar{L}_{res}), \tag{13}$$

where L_{cls} and L'_{cls} are standard cross-entropy loss functions for training DSBN and DCN respectively, $\bar{L}_{stan}$ and $\bar{L}_{res}$ are the average of Eq. (9) and Eq. (12) for each DCN layer and λ are the balancing weight which is set as the gradient magnitudes of L_{cls}. The total training process is shown in Algorithm 1.

Inference. The inference process only adds a small amount of computation to the normalization operation and can achieve nearly the same speed as the ERM baseline method. When inferring a test example x_t in the target domain D_T, we first infer the normalization statistics of D_T with x_t and the domain-specific statistics of each source domain on the DCN normalization layers. We can obtain the inferred statistics through each source domain and we average them to normalize x_t. Next, we put the feature map of x_t normalized in this way into the classifier and get the final prediction result.

4   Experiments

In this section, we present both the quantitative and qualitative results of our method. Firstly, we describe the datasets and implementation details. Then, we compare our method with state-of-the-art methods to confirm the effectiveness of DCN. Furthermore, we conduct the ablation study to analyze the properties of our method.

4.1   Experiment Setup

Datasets. As our evaluation benchmarks. We use PACS [17], VLCS [35] and Office-Home [36]. Following [17], we apply the leave-one-domain-out protocol for all benchmarks, which means we leave one domain as the target domain and the other domains as source domains. For PACS and VLCS, we use the official split to conduct the experiment. For Office-Home, we split the dataset into 90% training set and 10% validation set randomly [17]. We select the best model on the validation set. All the reported results are conducted five times and averaged.

Implementation Details. For a fair comparison, we follow the implementations of [2,8,41]. We employ ResNet [9] as our backbone in all experiments and use the pretrained protocol in [17]. As for optimization, the backbone is trained with the SGD optimizer with the weight decay of 5e−4; the auto-encoders are trained with the Adam optimizer. We use a batch size of 32 and train the network for 30 epochs. The learning rate is 0.001 on PACS and Office-Home and 5e-5 on VLCS for the backbone, and the same for the auto-encoders except 0.0001 on PACS. Both of the learning rates are decayed by 0.1 at 80% of the total epochs. We also use some augmentations in RandAug [5] and implement the augmentation with probability 0.5, which makes the statistics of each source domain more robust.

Table 1. Comparison with different DG methods (%) using ResNet-18 and ResNet-50 on PACS [17] dataset. The best results are marked as bold. The asterisk means that ASR-Norm must be combined with RSC.

Backbone	Method	Photo	Art	Cartoon	Sketch	Avg.
ResNet-18	ERM	95.12	78.37	75.16	75.41	81.02
	BIN [27]	95.00	82.10	74.10	80.00	82.80
	EISNet [39]	95.93	81.89	76.44	74.33	82.15
	FACT [41]	95.15	85.37	78.38	79.15	84.51
	DSON [28]	95.87	84.67	77.65	82.23	85.11
	RSC [11]	95.99	83.43	80.31	80.85	85.15
	MetaNorm [6]	95.99	85.01	78.63	83.17	85.70
	ASR-Norm * [8]	96.10	84.80	**81.80**	82.60	86.30
	DCN (Ours)	**96.51**	**86.60**	81.47	**83.60**	**87.05**
ResNet-50	ERM	97.64	84.94	76.98	76.75	84.08
	EISNet [39]	97.11	86.64	81.53	78.07	85.84
	DSON [28]	95.99	87.04	80.62	82.90	86.64
	RSC [11]	**97.92**	87.89	82.16	83.85	87.83
	FACT [41]	96.75	**89.63**	81.77	84.46	88.15
	DCN (Ours)	96.82	89.16	**86.09**	**86.15**	**89.56**

For the dimension of auto-encoders in DCN, the encoders and decoders are $(C, C/2)$, $(C/2, C)$ in standardization stage and $(2C, C)$, (C, C) in rescaling stage (C denotes the channel number in that layer). Moreover, the decoder for γ and β are not shared. We take ϵ as 1e-5. For all experiments, only the BN in the first three residual blocks is replaced by DCN.

4.2 Comparison with State-of-the-Art Methods

In this section, we evaluate our method on PACS, VLCS and Office-Home. We compare our DCN with recent state-of-the-art DG methods to demonstrate the effectiveness of DCN.

PACS. We use ResNet-18 and ResNet-50 as our backbone to perform the evaluation on PACS. We compare with many normalization-based DG methods, such as BIN [27], DSON [28], Meta-Norm [6] and ASR-Norm [8]. We also compare with other SOTA methods, such as EISNet [39], RSC [11] and FACT [41]. The results are shown in Table 1. The ERM baseline can achieve good performance on the photo domain, because the photo domain is similar to the pretrained dataset ImageNet. However, ERM fails on art, cartoon and sketch domains due to the large distribution shift. When inferring target samples, our DCN takes full advantage of the single target sample's hint about the distribution of the target domain. Therefore, the improvement of DCN on ERM is more obvious when the distribution shift is larger.

Compared with the SOTA, our DCN significantly outperforms other DG methods. In ResNet-18, DCN achieves the best performance on three domains

Table 2. Comparison with different DG methods (%) using ResNet-18 on VLCS [35] dataset. The best performance is marked as **bold**.

Method	Caltech	LabelMe	Pascal	Sun	Avg.
ERM	91.86	61.81	67.48	68.77	72.48
JiGen [2]	96.17	62.06	70.93	71.40	75.14
MMLD [24]	97.01	62.20	73.01	72.49	76.18
RSC [11]	96.21	62.51	73.81	72.10	76.16
StableNet [43]	96.67	**65.36**	73.59	74.97	77.65
DCN (Ours)	**98.23**	62.11	**74.88**	**75.78**	**77.75**

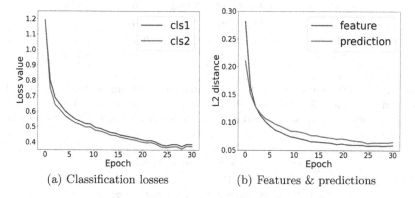

(a) Classification losses   (b) Features & predictions

Fig. 3. The analysis of the divergence between the two outputs. (a) compares the classification loss of the two outputs, and (b) compares the L2 distance on both the two feature maps before the classifier and the two prediction vectors. The ablation study is conducted on VLCS dataset with LabelMe as the target domain. (Color figure online)

and the second best in the other domain, resulting in the best average accuracy. In ResNet-50, DCN has an improvement of 3.93% and 1.69% over the second best on cartoon and sketch domains respectively. Moreover, DCN improves 1.41% over the SOTA method FACT on average accuracy. This results show the effectiveness of DCN when it is incorporated into different backbones.

VLCS. The results on VLCS are presented in Table 2. We compare with four DG methods: JiGen [2], MMLD [24], RSC [11] and StableNet [43]. Our method achieves the best performance on three domains and surpasses the latest StableNet [43] in terms of average performance. This shows that DCN can also perform well when the domain shift is small between source and target domains.

Office-Home. The results on Office-Home are presented in Table 3. Due to the variations being mainly background and viewpoint, the distribution discrepancy is small on Office-Home, and the best DG method lifts very little on Product and Real-World (0.5% in Product and 0.35% in Real-World). Therefore, ERM acts as a strong baseline and outperforms many DG methods. Our DCN focuses

Table 3. Comparision with different DG methods (%) using ResNet-18 on Office-Home [36] dataset. The best performance is marked as **bold**.

Method	Art	Clipart	Product	Real-World	Avg.
ERM	58.90	49.40	74.30	76.20	64.70
CCSA [25]	59.90	49.90	74.10	75.70	64.90
MMD-AAE [19]	56.50	47.30	72.10	74.80	62.70
CrossGrad [30]	58.40	49.40	73.90	75.80	64.40
JiGen [2]	53.04	47.51	71.47	72.79	61.20
DDAIG [45]	59.20	52.30	74.60	76.00	65.50
L2A-OT [46]	**60.60**	50.10	**74.80**	77.00	65.60
RSC [11]	58.42	47.90	71.63	74.54	63.12
FACT [41]	60.34	54.85	74.48	**76.55**	66.56
DCN (Ours)	59.83	**57.16**	73.78	75.63	**66.60**

Table 4. The ablation study of different components of our method on the PACS dataset with ResNet-18. DCR and DCS represent the model are trained with only the domain conditioned rescaling and domain conditioned standardization components respectively.

Method	Photo	Art	Cartoon	Sketch	Avg.
ERM	95.12	78.37	75.16	75.41	81.02
DCR	95.33	80.35	76.02	76.53	82.06
DCS	96.18	85.65	80.76	82.52	86.28
DCN	**96.51**	**86.60**	**81.47**	**83.60**	**87.05**

on inferring the normalization statistics for the target domain while the spatial features such as viewpoint are not reflected on the normalization statistics, so our method is slightly worse than ERM on Product and Real-World. However, our method improves 7.76% over ERM on Clipart, which is the hardest generalization task. It demonstrates once again that DCN is more effective when the distribution shift between source and target domains is larger. Moreover, DCN achieves comparable performance to the current SOTA method FACT. This again validates the effectiveness of our method.

4.3   Ablation Study

Impact of Different Components. Table 4 illustrates the effects of domain conditioned standardization (DCS) and domain conditioned rescaling (DCR) respectively. As shown in Tabel 4, the model only applying DCR has a small improvement on the baseline, while only applying DCS can improve the baseline by 5.26%. This is because the difference in standardization statistics between different domains is much larger than that in rescaling statistics. Furthermore, the model performs best when applying both DCR and DCS (i.e. DCN), which validates the necessity of both modules.

Table 5. The performance comparison of DCN in different locations of the network. The block1-4 means the first to last residual blocks in ResNet-18, respectively. The ablation study is conducted on PACS dataset.

Method	block1	block2	block3	block4	Photo	Art	Cartoon	Sketch	Avg.
Baseline	–	–	–	–	95.12	78.37	75.16	75.41	81.02
Model A	✓	–	–	–	95.65	83.57	77.86	81.09	84.54
Model B	✓	✓	–	–	96.35	85.79	79.86	82.43	86.11
Model C	–	✓	✓	–	96.47	85.55	79.61	81.70	85.83
Model D	✓	✓	✓	–	96.51	**86.6**	**81.47**	**83.6**	**87.05**
Model E	–	✓	✓	✓	96.59	84.03	78.84	81.37	85.21
Model F	✓	✓	✓	✓	**96.77**	84.72	80.33	81.98	85.95

The Divergence Between the Two Outputs. We analyse the divergence between the two outputs in different aspects. Figure 3(a) compares the classification loss of the two outputs. The blue line represents L_{cls} and the red line represents L'_{cls}. As shown in Fig. 3(a), the curves of the two classification losses are very close. Figure 3(b) compares the distance between the two feature maps before the classifier (denoted by the blue line) and the distance between the two prediction vectors (denoted by the red line). We can observe that both distances get smaller and smaller. These two figures demonstrate that the inferred statistics of the meta-target domain are reasonable and get closer to the ground truth statistics of the meta-target domain during training.

4.4 Further Analysis

Effect of where DCN is Used. We conduct an extensive ablation study to investigate the effect of where DCN is used. As shown is Table 5, only replacing BN with DCN in the first residual block can greatly improve the performance and get 3.52% raise than baseline. As the number of residual blocks using DCN increases, the generalization performance of the model is getting better and better. We can also notice that model F is worse than model D and model E is worse than model C. This is mainly because the normalization statistics in the last residual block are closely related to semantics, so the use of DCN in the last residual block may cause certain disturbances. Meanwhile, model B is better than model C and model D is better than model E. It means the statistics difference between source and target domains in the first residual block is larger, so DCN is more needed in block1 to infer the target statistics instead of using the source statistics during inference.

Weight Differences Between DCN Layers. Figure 4 shows the average weight (which is learned by the standardization auto-encoder) for each DCN layer when cartoon is the target domain on PACS dataset. It is obvious that the shallow DCN layers have a larger average weight than the deep DCN layers for all source domains. It confirms the intuition that shallow layers extract richer

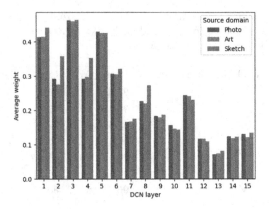

Fig. 4. Analysis of average weights (averaged on each channel) for different DCN layers when cartoon is the target domain on PACS dataset. The backbone is ResNet-18 and we only substitute the first 15 BN with DCN.

domain-specific features, so the shallow DCN layers rely more on a single target sample's instance information to explore the domain-specific information of the target domain. On the other hand, deep DCN layers rely more on source statistics because the deep layers extract more semantic features, which are similar between source and target domains.

5   Conclusions

In this paper, we propose a novel domain generalization method that can infer the normalization statistics of the target domain and use the inferred statistics to normalize test data during inference. We name our method Domain Conditioned Normalization (DCN). DCN simulates the inference process of the normalization statistics with a meta-learning framework during training, and uses the source statistics and one single target sample to infer the normalization statistics of the target domain with an optimization-free procedure during inference. Extensive experiments on three benchmarks demonstrate that our DCN can infer the reasonable target statistics so that it can achieve state-of-the-art performance for domain generalization. Furthermore, we conduct the ablation study to analyze the effects and characteristics of DCN.

Acknowledgement. This work is supported by the National Key R&D Program of China (No. 2019YFB1804304), 111 plan (No. BP0719010), and STCSM (No. 18DZ2270700, No. 21DZ1100100), and State Key Laboratory of UHD Video and Audio Production and Presentation.

References

1. Balaji, Y., Sankaranarayanan, S., Chellappa, R.: MetaReg: towards domain generalization using meta-regularization. In: Advances in Neural Information Processing Systems, vol. 31, pp. 998–1008 (2018)
2. Carlucci, F.M., D'Innocente, A., Bucci, S., Caputo, B., Tommasi, T.: Domain generalization by solving Jigsaw puzzles. In: Proceedings of the IEEE/CVF Conference on Computer Vision and Pattern Recognition, pp. 2229–2238 (2019)
3. Chang, W.G., You, T., Seo, S., Kwak, S., Han, B.: Domain-specific batch normalization for unsupervised domain adaptation. In: Proceedings of the IEEE/CVF Conference on Computer Vision and Pattern Recognition, pp. 7354–7362 (2019)
4. Chen, Y., Wang, Y., Pan, Y., Yao, T., Tian, X., Mei, T.: A style and semantic memory mechanism for domain generalization. In: Proceedings of the IEEE/CVF International Conference on Computer Vision, pp. 9164–9173 (2021)
5. Cubuk, E.D., Zoph, B., Shlens, J., Le, Q.V.: RandAugment: practical automated data augmentation with a reduced search space. In: Proceedings of the IEEE/CVF Conference on Computer Vision and Pattern Recognition Workshops, pp. 702–703 (2020)
6. Du, Y., Zhen, X., Shao, L., Snoek, C.G.: MetaNorm: learning to normalize few-shot batches across domains. In: International Conference on Learning Representations (2020)
7. Dubey, A., Ramanathan, V., Pentland, A., Mahajan, D.: Adaptive methods for real-world domain generalization. In: Proceedings of the IEEE/CVF Conference on Computer Vision and Pattern Recognition, pp. 14340–14349 (2021)
8. Fan, X., Wang, Q., Ke, J., Yang, F., Gong, B., Zhou, M.: Adversarially adaptive normalization for single domain generalization. In: Proceedings of the IEEE/CVF Conference on Computer Vision and Pattern Recognition, pp. 8208–8217 (2021)
9. He, K., Zhang, X., Ren, S., Sun, J.: Deep residual learning for image recognition. In: Proceedings of the IEEE Conference on Computer Vision and Pattern Recognition, pp. 770–778 (2016)
10. Hu, J., Shen, L., Sun, G.: Squeeze-and-excitation networks. In: Proceedings of the IEEE Conference on Computer Vision and Pattern Recognition, pp. 7132–7141 (2018)
11. Huang, Z., Wang, H., Xing, E.P., Huang, D.: Self-challenging improves cross-domain generalization. In: Vedaldi, A., Bischof, H., Brox, T., Frahm, J.-M. (eds.) ECCV 2020. LNCS, vol. 12347, pp. 124–140. Springer, Cham (2020). https://doi.org/10.1007/978-3-030-58536-5_8
12. Ioffe, S., Szegedy, C.: Batch normalization: accelerating deep network training by reducing internal covariate shift. In: International Conference on Machine Learning, pp. 448–456. PMLR (2015)
13. Iwasawa, Y., Matsuo, Y.: Test-time classifier adjustment module for model-agnostic domain generalization. In: Advances in Neural Information Processing Systems, vol. 34 (2021)
14. Jia, S., Chen, D.J., Chen, H.T.: Instance-level meta normalization. In: Proceedings of the IEEE/CVF Conference on Computer Vision and Pattern Recognition, pp. 4865–4873 (2019)
15. Khurana, A., Paul, S., Rai, P., Biswas, S., Aggarwal, G.: SITA: single image test-time adaptation. arXiv preprint arXiv:2112.02355 (2021)
16. Kim, D., Yoo, Y., Park, S., Kim, J., Lee, J.: SelfReg: self-supervised contrastive regularization for domain generalization. In: Proceedings of the IEEE/CVF International Conference on Computer Vision, pp. 9619–9628 (2021)

17. Li, D., Yang, Y., Song, Y.Z., Hospedales, T.M.: Deeper, broader and artier domain generalization. In: Proceedings of the IEEE International Conference on Computer Vision, pp. 5542–5550 (2017)

18. Li, D., Yang, Y., Song, Y.Z., Hospedales, T.M.: Learning to generalize: meta-learning for domain generalization. In: Thirty-Second AAAI Conference on Artificial Intelligence (2018)

19. Li, H., Pan, S.J., Wang, S., Kot, A.C.: Domain generalization with adversarial feature learning (2018)

20. Li, P., Li, D., Li, W., Gong, S., Fu, Y., Hospedales, T.M.: A simple feature augmentation for domain generalization. In: Proceedings of the IEEE/CVF International Conference on Computer Vision, pp. 8886–8895 (2021)

21. Li, S., Xie, B., Lin, Q., Liu, C.H., Huang, G., Wang, G.: Generalized domain conditioned adaptation network. IEEE Trans. Pattern Anal. Mach. Intell. (2021)

22. Li, Y., et al.: Deep domain generalization via conditional invariant adversarial networks. In: Proceedings of the European Conference on Computer Vision (ECCV), pp. 624–639 (2018)

23. Liang, S., Huang, Z., Liang, M., Yang, H.: Instance enhancement batch normalization: an adaptive regulator of batch noise. In: Proceedings of the AAAI Conference on Artificial Intelligence, pp. 4819–4827 (2020)

24. Matsuura, T., Harada, T.: Domain generalization using a mixture of multiple latent domains. In: Proceedings of the AAAI Conference on Artificial Intelligence, pp. 11749–11756 (2020)

25. Motiian, S., Piccirilli, M., Adjeroh, D.A., Doretto, G.: Unified deep supervised domain adaptation and generalization. In: Proceedings of the IEEE International Conference on Computer Vision, pp. 5715–5725 (2017)

26. Nado, Z., Padhy, S., Sculley, D., D'Amour, A., Lakshminarayanan, B., Snoek, J.: Evaluating prediction-time batch normalization for robustness under covariate shift. arXiv preprint arXiv:2006.10963 (2020)

27. Nam, H., Kim, H.E.: Batch-instance normalization for adaptively style-invariant neural networks. arXiv preprint arXiv:1805.07925 (2018)

28. Seo, S., Suh, Y., Kim, D., Kim, G., Han, J., Han, B.: Learning to optimize domain specific normalization for domain generalization. In: Vedaldi, A., Bischof, H., Brox, T., Frahm, J.-M. (eds.) ECCV 2020. LNCS, vol. 12367, pp. 68–83. Springer, Cham (2020). https://doi.org/10.1007/978-3-030-58542-6_5

29. Shahtalebi, S., Gagnon-Audet, J.C., Laleh, T., Faramarzi, M., Ahuja, K., Rish, I.: Sand-mask: an enhanced gradient masking strategy for the discovery of invariances in domain generalization. arXiv preprint arXiv:2106.02266 (2021)

30. Shankar, S., Piratla, V., Chakrabarti, S., Chaudhuri, S., Jyothi, P., Sarawagi, S.: Generalizing across domains via cross-gradient training. arXiv preprint arXiv:1804.10745 (2018)

31. Shi, Y., et al.: Gradient matching for domain generalization. arXiv preprint arXiv:2104.09937 (2021)

32. Sun, B., Saenko, K.: Deep CORAL: correlation alignment for deep domain adaptation. In: Hua, G., Jégou, H. (eds.) ECCV 2016. LNCS, vol. 9915, pp. 443–450. Springer, Cham (2016). https://doi.org/10.1007/978-3-319-49409-8_35

33. Sun, Y., Wang, X., Liu, Z., Miller, J., Efros, A., Hardt, M.: Test-time training with self-supervision for generalization under distribution shifts. In: International Conference on Machine Learning, pp. 9229–9248. PMLR (2020)

34. Tang, Z., Gao, Y., Zhu, Y., Zhang, Z., Li, M., Metaxas, D.N.: CrossNorm and SelfNorm for generalization under distribution shifts. In: Proceedings of the IEEE/CVF International Conference on Computer Vision, pp. 52–61 (2021)

35. Torralba, A., Efros, A.A.: Unbiased look at dataset bias. In: CVPR 2011, pp. 1521–1528. IEEE (2011)

36. Venkateswara, H., Eusebio, J., Chakraborty, S., Panchanathan, S.: Deep hashing network for unsupervised domain adaptation. In: Proceedings of the IEEE Conference on Computer Vision and Pattern Recognition, pp. 5018–5027 (2017)

37. Wang, D., Shelhamer, E., Liu, S., Olshausen, B., Darrell, T.: Tent: fully test-time adaptation by entropy minimization. arXiv preprint arXiv:2006.10726 (2020)

38. Wang, J., Lan, C., Liu, C., Ouyang, Y., Zeng, W., Qin, T.: Generalizing to unseen domains: a survey on domain generalization. arXiv preprint arXiv:2103.03097 (2021)

39. Wang, S., Yu, L., Li, C., Fu, C.-W., Heng, P.-A.: Learning from extrinsic and intrinsic supervisions for domain generalization. In: Vedaldi, A., Bischof, H., Brox, T., Frahm, J.-M. (eds.) ECCV 2020. LNCS, vol. 12354, pp. 159–176. Springer, Cham (2020). https://doi.org/10.1007/978-3-030-58545-7_10

40. Wang, X., Jin, Y., Long, M., Wang, J., Jordan, M.: Transferable normalization: towards improving transferability of deep neural networks. In: Advances in Neural Information Processing Systems (2019)

41. Xu, Q., Zhang, R., Zhang, Y., Wang, Y., Tian, Q.: A Fourier-based framework for domain generalization. In: Proceedings of the IEEE/CVF Conference on Computer Vision and Pattern Recognition, pp. 14383–14392 (2021)

42. You, F., Li, J., Zhao, Z.: Test-time batch statistics calibration for covariate shift. arXiv preprint arXiv:2110.04065 (2021)

43. Zhang, X., Cui, P., Xu, R., Zhou, L., He, Y., Shen, Z.: Deep stable learning for out-of-distribution generalization. In: Proceedings of the IEEE/CVF Conference on Computer Vision and Pattern Recognition, pp. 5372–5382 (2021)

44. Zhou, K., Liu, Z., Qiao, Y., Xiang, T., Loy, C.C.: Domain generalization: a survey. arXiv preprint arXiv:2103.02503 (2021)

45. Zhou, K., Yang, Y., Hospedales, T., Xiang, T.: Deep domain-adversarial image generation for domain generalisation. In: Proceedings of the AAAI Conference on Artificial Intelligence, pp. 13025–13032 (2020)

46. Zhou, K., Yang, Y., Hospedales, T., Xiang, T.: Learning to generate novel domains for domain generalization. In: Vedaldi, A., Bischof, H., Brox, T., Frahm, J.-M. (eds.) ECCV 2020. LNCS, vol. 12361, pp. 561–578. Springer, Cham (2020). https://doi.org/10.1007/978-3-030-58517-4_33

47. Zhou, K., Yang, Y., Qiao, Y., Xiang, T.: Domain generalization with mixstyle. arXiv preprint arXiv:2104.02008 (2021)

Unleashing the Potential of Adaptation Models via Go-getting Domain Labels

Xin Jin[1]([✉]), Tianyu He[2], Xu Shen[2], Songhua Wu[3], Tongliang Liu[3],
Jingwen Ye[4], Xinchao Wang[4], Jianqiang Huang[2], Zhibo Chen[5],
and Xian-Sheng Hua[2]

[1] Eastern Institute for Advanced Study, Ningbo, China
jinxin@eias.ac.cn
[2] Alibaba Group, Hangzhou, China
[3] The University of Sydney, Sydney, Australia
[4] National University of Singapore, Singapore, Singapore
[5] University of Science and Technology of China, Hefei, China

Abstract. In this paper, we propose an embarrassingly simple yet highly effective adversarial domain adaptation (ADA) method. We view ADA problem primarily from an optimization perspective and point out a fundamental dilemma, in that the real-world data often exhibits an imbalanced distribution where the large data clusters typically dominate and bias the adaptation process. Unlike prior works that either attempt loss re-weighting or data re-sampling for alleviating this defect, we introduce a new concept of go-getting domain labels (Go-labels) to replace the original immutable domain labels on the fly. The reason why call it as "Go-labels" is because "go-getting" means able to deal with new or difficult situations easily, like here Go-labels adaptively transfer the model attention from over-studied aligned data to those overlooked samples, which allows each sample to be well studied (*i.e.*, alleviating data imbalance influence) and fully unleashes the potential of adaption model. Albeit simple, this dynamic adversarial domain adaptation framework with Go-labels effectively addresses data imbalance issue and promotes adaptation. We demonstrate through theoretical insights, empirical results on real data as well as toy games that our method leads to efficient training without bells and whistles, while being robust to different backbones.

1 Introduction

Most deep models rely on huge amounts of labeled data and their learned features have proven brittle to data distribution shifts [58,68]. To mitigate the data discrepancy issue and reduce dataset bias, unsupervised domain adaptation (UDA) is extensively explored, which has access to labeled samples from a source domain and unlabeled data from a target domain. Its objective is to train a model that generalizes well to the target domain [8,10,14,15,19,24,25,28].

Supplementary Information The online version contains supplementary material available at https://doi.org/10.1007/978-3-031-25085-9_18.

As a mainstream branch of UDA, adversarial domain adaptation (ADA) approaches leverage a domain discriminator paired with a feature generator to adversarially learn a domain-invariant feature [9,11,15,35,50]. For the domain discriminator training, all source data are equally taken as one domain (*e.g.*, positive '1') while target data as another one (*e.g.*, negative '0') [11,15,35]. However, this fixed positive-negative separation neglects a fact that most real-world data exhibit imbalanced distributions [12,13]: the clusters with abundant examples (*i.e.*, large clusters) may **swamp** the clusters with few examples (*i.e.*, small clusters). Such imbalanceness contains two aspects, intra-class long-tailed distribution [34,44] and inter-class long-tailed distribution [55,64], and is widely existed in many UDA benchmarks. For example, in DomainNet [42], the "dog" class in the "clipart" domain has 70 image samples while has 782 image samples in the "real" domain. The majority "bike" samples (90%) in "Amazon" domain in Office31 [46] have no background scene (empty) while minority "bike" samples have real-world background instead.

On the other hand, deep neural networks (DNNs) typically learn simple patterns first before memorizing. In other words, DNN optimization is content-aware, taking advantage of patterns shared by multiple training examples [2]. Therefore, in the process of domain adaptation, the large domain clusters would dominate the optimization of domain discriminator, so that bias its decision boundary and hinder the effective adaptation. As shown in Fig. 1(a), only the large clusters of two domains (*i.e.*, two large circles) have been pulled close as the adaptation goes on, but those minority clusters (four small circles) are still under-aligned. This bias the optimization of domain discriminator so that misleads the feature extractor to learn unexpected domain-specific knowledge from large clusters. As a result, the adapted model still can not correctly classify these under-explored samples (marked by "misclassify").

In this paper, we attempt to design an optimization strategy to progressively take full advantage of both large and small data clusters across different domains,

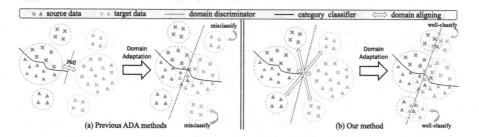

Fig. 1. Motivation illustration. ($\times$, $\triangle$) denote two different classes, and (blue, orange) color mean different domains. (a) Previous DA methods tend to be dominated by those large clusters and neglects small clusters, which will bias the domain discriminator optimization, leading to a sub-optimal adaptation accuracy. (b) Our method attempts to fully leverage both large and small data clusters for alignment, to enhance the domain-invariant representation learning, and thus achieving a better adaptation performance on the target set. (Color figure online)

like shown in Fig. 1(b). In this way, the domain-invariant representation learning could be gradually promoted, and the potential of adaptation model will be unleashed, leading a satisfied classification performance. Our study is different from existing methods that purely designed for long-tailed classification [22,69,71] in application scenarios and exhibits advantages in domain-agnostic representation learning. This problem is challenging, but valuable and meaningful for DA task.

There also exists few works have noticed the distribution imbalance issues in the domain adaptation task, and try to tackle it by re-weighting (IWAN [61,70]), data re-sampling (RADA [24]), or data augmentation (Domain Mixup [65]). Differently, our paper focuses on a more general imbalance setting, which contains two aspects of long-tailed intra-class and inter-class distribution. Besides, we try to achieve a high-powered optimization strategy to empower DA model study each sample well to promote distribution alignment without any cost increase.

To this end, we propose to replace the original immutable domain labels with an adjustable and importance-aware alternative, dubbed Go-getting Domain Labels (Go-labels). Its core idea is to adaptively reduce the importance of these dominated training data that have been aligned, and timely encourage the domain discriminator to pay more attention to those easy-to-miss minority clusters, which ensures each sample can be well studied. In the implementation, we assign a go-getting domain label (Go-label) to each sample according to its own optimization situation: If one sample has ambiguous domain predictions (*e.g.*, ∼0.5) when passing through domain discriminator, it means such sample has been well studied, or said, the learned feature w.r.t this sample has been domain-invariant. Then, we enforce a relaxation constraint on it through changing its groundtruth (*i.e.*, directly taking 0.5 as its new domain label), so as to reduce its optimization importance. Our contributions are summarized as follows,

- We revisit domain adaptation problem from an optimization perspective, and pinpoint the training defect caused by imbalanced data distributions issue.
- To alleviate this issue, we propose a novel concept of go-getting domain labels (Go-labels) to achieve a dynamic adaptation, which allows each sample to be well studied and reduces long-tailed influence, so as to promote domain alignment for DA without any increase in computational cost.
- As a byproduct, our work also provides a new perspective to understand the task of adaptation, and gives theoretical insights about the effectiveness of dynamic training strategy with Go-labels.

We thoroughly study the proposed Go-labels with several toy cases, and conduct experiments on multiple domain adaptation benchmarks, including Digit-Five, Office-31, Office-Home, VisDA-2017, and large-scale DomainNet, upon various baselines, to show it is effective and reasonable.

2   Related Work

Unsupervised Domain Adaptation. Recent UDA works focus on two mainstream branches, (1) moment matching and (2) adversarial training. The former works typically align features across domains by minimizing some distribution similarity metrics, such as Maximum Mean Discrepancy (MMD) [7,36,62] and second-/higher-order statistics [28,42,54]. Adversarial domain adaptation (ADA) methods have achieved superior performance and this paper also focuses on it. The pioneering works of DANN [15] and ADDA [59] both employ a domain discriminator to compete with a feature extractor in a two-player mini-max game. CDAN [35] improves this idea by conditioning domain discriminator on the information conveyed by the category classifier. MADA [41] uses multiple domain discriminators to capture multi-modal structures for fine-grained domain alignment. Recent GVB [11] gradually reduces the domain-specific characteristics in domain-invariant representations via a bridge layer between the generator and discriminator. MCD [49], STAR [37] and Symnet [72] all build an adversarial adaptation framework by leveraging the collision of multiple object classifiers. Unfortunately, all these methods ignore the imbalanced distribution issue in DA.

Imbalanced Domain Adaptation. Several prior works have noticed the distribution imbalance issues in domain-adversarial field, and provided rigorous analysis and explanations [23,26,55,64,74]. In particular, IWAN [70] leverages the idea of re-weighting for adaptation, and RADA [24] enhances the ability of domain discriminator in DA via sample re-sampling and augmentation. Besides, the works of [30,55,64] focus on the subpopulation shift issue (partial DA), where the source and target domains have imbalanced **label** distribution. Differently, our paper focuses on the more general covariate shift setting in DA, which contains two aspects of long-tailed intra-class and inter-class distribution. Such imbalanced problems are widely existed in the existing UDA benchmarks.

Adversarial Training. Our work is also related to the researches which aim to leverage or modify the discriminator output to further augment the standard GAN training [1,3,17,18,39,52,63]. Their core idea is to distill useful information from the discriminator to further regularize generator to obtain a better generation performance. Although our work shares a similar idea of enhancing adversarial training, the main contributions and target task are different.

3   Adversarial Domain Adaptation with Go-labels

3.1   Prior Knowledge Recap and Problem Definition

To be self-contained, we first simply review the problem formulation of adversarial domain adaptation (ADA). Taking classification task as example, we denote the source domain as $\mathcal{D}_S = \{(x_i^s, y_i^s, d_i^s)\}_{i=1}^{N_s}$ with N_s labeled samples covering C classes, $y_i^s \in [0, C-1]$. d_i^s is the domain label of each source sample and it always equals to '1' during the training [15,35]. The target domain is similarly denoted as $\mathcal{D}_T = \{x_j^t, d_i^t\}_{j=1}^{N_t}$ with N_t unlabeled samples that belong to the same C classes, d_i^t denotes the domain label of each target sample and it always equals to '0' so as to construct a '0–1' pair with source samples for adversarial optimization. Most ADA algorithms tend to learn domain-invariant representations, by adversarially training the feature extractor and domain discriminator in a minmax two-player game [11,15,21,35]. They typically use classification loss $\mathcal{L}_{cls}$ (*i.e.*, cross-entropy loss $\mathcal{L}_{ce}$) and domain adversarial loss $\mathcal{L}_{adv}$ (*i.e.*, binary cross-entropy loss $\mathcal{L}_{bce}$) for training,

$$
\mathcal{L}_{cls} = \frac{1}{N_s} \sum_{i=1}^{N_s} \mathcal{L}_{ce}(C(F(x_i^s)), y_i^s),
$$

$$
\mathcal{L}_{adv} = \frac{1}{N_s} \sum_{i=1}^{N_s} \mathcal{L}_{bce}(D(F(x_i^s)), d_i^s = 1) + \frac{1}{N_t} \sum_{i=1}^{N_t} \mathcal{L}_{bce}(D(F(x_i^t)), d_i^t = 0),
$$

(1)

where F, C, D represents the feature extractor, the category classifier, and the domain discriminator, respectively. They are shared across domains. The total optimization objective is described as $(\min_D \mathcal{L}_{adv} + \min_{F,C} \mathcal{L}_{cls} - \mathcal{L}_{adv})$. Note that, a gradient reversal layer (GRL) [15] is often used to connect feature extractor F and domain discriminator D to achieve the adversarial function by multiplying the gradient from D by a certain negative constant during the back-propagation to the feature extractor F.

Problem Definition of Imbalanced Data Distributions in DA. This paper focuses on the general covariate shift setting following [51,53] in the DA field, and assumes each domain presents an "imbalanced" data distributions. Suppose a source/target domain $\{(x_i, y_i)\}_{i=1}^n$ drawn i.i.d. from an imbalanced distribution $P(x, y)$. Such imbalanceness comprises two aspects: 1). the marginal distribution $P(y)$ of classes are likely long-tailed, i.e., inter-class long-tailed. 2). the data distribution within each class is also long-tailed, i.e., intra-class long-tailed distribution. We expect to learn a well adapted model $F(\cdot; \theta)$ with adversarial DA technique equipped with a domain discriminator $D(\cdot; \omega)$, to learn domain-invariant representations.

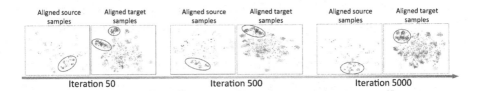

Fig. 2. Red and green points denote source and target domain data, respectively. The darker the color, the better the alignment, the more possible to be mis-classified by domain discriminator. (Color figure online)

Motivation Re-clarification. Here we look into whether the imbalanced data distribution issue actually hinders the effective ADA training, through a t-SNE [48] visualization results. This experiment is conducted on Office31 [46] (W→A setting) with the baseline of DANN [15]. We count the number of times each sample was **misclassified** by the domain discriminator during the DA training, and use this number as the color parameter. The darker the color, the better the alignment, the more possible to be mis-classified by domain discriminator. From Fig. 2, we see that, there obviously exists an imbalance situation with training going on, where some samples (surrounded by a blue circle) have been well aligned/studied by the domain discriminator (the darker the color, the better the alignment), but some samples are still under-studied or not aligned well. Therefore, treating those aligned and not aligned training data in different ways to promise each sample being well explored to alleviate imbalance influence is urgently required.

3.2 Proposed Go-getting Domain Labels

To alleviate the optimization difficulty caused by imbalanced data distributions and thus enhance the domain-invariant representation learning, we introduce a dynamic adversarial domain adaptation framework with the proposed go-labels: when calculating the domain adversarial loss on a mini-batch that contains both source and target domain samples, we replace the original immutable domain labels of samples (source as '1', target as '0') with an adjustable domain labels (*i.e.*, Go-labels) on the fly. In formula, we modify the domain adversarial loss $\mathcal{L}_{adv}$ of Eq. 1 to

$$\mathcal{L}_{adv} = \frac{1}{N_s} \sum_{i=1}^{N_s} \mathcal{L}_{bce}(D(F(x_i^s)), g_i^s) + \frac{1}{N_t} \sum_{i=1}^{N_t} \mathcal{L}_{bce}(D(F(x_i^t)), g_i^t), \quad (2)$$

where g_i^s and g_i^t are the updated go-getting domain labels for i-th source sample and i-th target sample in the mini-batch, they are **no longer** a fixed '1' or '0', but become adjustable and adaptive. Intuitively, a reliable metric to distinguish the well-aligned large cluster data and not aligned small cluster data is needed for the new updated domain labels assignment/decision.

Measurement of Alignment. The critic, domain discriminator D, can be seen as an online scoring function for data: one sample will receive a higher score ($\sim$1) if its extracted feature is close to the source distribution, and a lower score ($\sim$0) if its extracted feature is close to the target distribution. Thus, we directly take the predicted domain results of domain discriminator, denoted as $\widetilde{d^s}/\widetilde{d^t}$, as the alignment measurement metric for each source/target sample. For example, if the domain discriminator prefers to classify a source sample ($d_i^s = 1$) as target data, *i.e.*, $\widetilde{d_i^s} \to 0$, we believe the learned feature w.r.t this sample has been well aligned and is fake enough to fool domain discriminator. In this way, we could online distinguish the well-aligned and not aligned data during training.

Go-getting Domain Labels Update. In the implementation, we merge the alignment measurement (*i.e.*, well-aligned samples selection) and domain label update into a single step. Formally, we leverage a non-parametric mathematical rounding $Round(\cdot)$ to modify the original domain labels $d_i^s = 1$, $d_i^t = 0$ of i-th source, target sample according to their predicted domain results $\widetilde{d_i^s}$, $\widetilde{d_i^t}$:

$$g_i^s = \frac{d_i^s + Round(\widetilde{d_i^s})}{2}, \qquad g_i^t = \frac{d_i^t + Round(\widetilde{d_i^t})}{2} \qquad (3)$$

where go-getting domain labels of g_i^s, g_i^t are dynamic and adjustable, depending on the different domain prediction results $\widetilde{d_i^s}$, $\widetilde{d_i^t}$. The original domain label $d_i^s = 1$, $d_i^t = 0$ can be taken as groundtruth, and their intermediate decision boundary $(d_i^s + d_i^t)/2 = 0.5$ can be regarded as a threshold to automatically update go-getting domain labels through $Round(\cdot)$. That means that if the domain prediction result of a source sample is lower than the threshold of 0.5, i.e., $\widetilde{d_i^s} < 0.5$, we believe the learned feature w.r.t this sample has been well aligned and is fake enough to fool domain discriminator.

It can be see that the $Round(\cdot)$ function could keep the raw domain labels unchanged for those correctly classified samples by D. They have not been well aligned (i.e., $\widetilde{d_i^s} > 0.5$ and $\widetilde{d_i^t} < 0.5$). We only update the domain labels for these mis-classified well-aligned samples (i.e., $\widetilde{d_i^s} \leq 0.5$ and $\widetilde{d_i^t} \geq 0.5$), which reduces the optimization importance of these aligned training data and encourage the domain discriminator to pay more attention to those not aligned data.

Implementation in PyTorch. A simple PyTorch-like [40] pseudo-code snippet is shown below. The dynamic adversarial DA with go-getting domain labels (Go-labels) modification amounts simply to the addition of lines 9, 10 of the example code, which indicates its ease of implementation and generality.

```
1  # Extract features from source (s) and target (t) domain
2  feat_s, feat_t = Extractor(sample_s, sample_t)
3
4  # Get true domain labels and domain predictions
5  d_s, d_t = 1, 0
6  p_s, p_t = Domain_Discriminator(feat_s, feat_t)
7
8  # Get updated go-getting domain labels
9  g_s = (d_s + torch.Round(p_s.detach()) / 2.0
10 g_t = (d_t + torch.Round(p_t.detach()) / 2.0
11
12 # Compute adversarial loss with new go-getting domain labels
13 loss_adv = torch.BCELoss(p_s, g_s) + torch.BCELoss(p_t, g_t)
```

Discussion: Why use Rounding? Rounding-based dynamic domain labels *only* reduce the importance for these well-aligned (*i.e.*, mis-classified by discriminator) majority samples progressively, while keep unchanged for those not aligned minority data. This design makes the "dynamically change" of go-getting domain labels more "targeted". If no rounding, the real-valued soft Go-labels will be *always* affected by the probability scores of domain discriminator, even the discriminator has not yet been well-trained at early stage. In short, the physical meanings behind Go-labels is to **softly reduce** the importance for these dominated majority samples on the fly while **progressively** transferring optimization focus to those minority data.

3.3   Theoretical Insights of Go-labels

Many classic domain adaptation approaches typically bound/model the adapted target error by the sum of the *(1) source error* and *(2) a notion of distance between the source and the target distributions.* The classic generalization bound theory of the $\mathcal{H}$-divergence that based on the earlier work of [29] and used by [4,5,15] is obtained following theorem-1 in [6]:

$$\mathcal{R}_t(h) \leq \hat{\mathcal{R}}_s(h) + \frac{1}{2}d_{\mathcal{H}}(\hat{\mathcal{D}}_S^N, \hat{\mathcal{D}}_T^N) + C, \tag{4}$$

where C is a constant when such bound is achieved by hypothesis in $\mathcal{H}$. And $\hat{\mathcal{D}}_S^N$, $\hat{\mathcal{D}}_T^N$ denote the empirical distribution induced by sample of size N drawn from $\mathcal{D}_S$, $\mathcal{D}_T$ respectively. $\mathcal{R}_t$ denote the true risk on target domain, and $\hat{\mathcal{R}}_s$ denote the empirical risk on source domain.

Let $\{\mathbf{x}_i^s\}_{i=1}^N$, $\{\mathbf{x}_i^t\}_{i=1}^N$ be the samples in the empirical distributions $\hat{\mathcal{D}}_S$ and $\hat{\mathcal{D}}_T$ respectively. The empirical source risk can be written as $\hat{\mathcal{R}}_s(h) = \frac{1}{N}\sum_i^N \hat{\mathcal{R}}_{\mathbf{x}_i^s}(h)$.

Now, considering a *dynamic updated* source-target domain distributions $\hat{\mathcal{D}}_{dS}$ and $\hat{\mathcal{D}}_{dT}$ achieved by the proposed **adjustable** go-getting domain labels, which corresponds to relabeling the well-aligned samples (assuming that the number of selected well-aligned target samples is M), the new generalization bound for this updated data distribution can be modified as

$$\mathcal{R}_t(h) \leq \left(\frac{1}{N}\sum_i^N \hat{\mathcal{R}}_{\mathbf{x}_i^s}(h) + \frac{1}{M}\sum_j^M \hat{\mathcal{R}}_{\mathbf{x}_j^t}(h)\right) + \frac{1}{2}d_{\mathcal{H}\Delta\mathcal{H}}(\hat{\mathcal{D}}_{dS}^{N+M}, \hat{\mathcal{D}}_{dT}^{N-M}), \tag{5}$$

the first term on right becomes an **updated** source risk that could **re-energize** the object classifier optimization, and the second term becomes an **updated** domain discrepancy/divergency that could **re-energize** the domain discriminator optimization. They together unleash the potential of adaptation model. Besides, the risk of the target domain can be re-bounded by the risk of the updated source domain and the updated domain discrepancy, providing theoretical guarantees for the proposed approach. When $M = 0$, we get the original bound of Eq. (4). Hence, the original bound is in the feasible set of our optimization with Go-labels.

4 Experiments

4.1 Validation on Toy Problems

2D Random Point Classification. First, we observe the behavior of our method on toy problem of *2D random point classification*. We compared the class decision boundary of our method with *Baseline* obtained from the domain discriminator trained with immutable domain labels. To better evaluate adaptation performance of the trained model, we visualize source and target data separately. Experimental details are provided in **Supplementary**. We observe that the *Baseline* scheme is prone to miss the small tail cluster, especially when it is very closed to a large cluster belonged to the different class. In contrast, our method could better leverage both large/head and small/tail data clusters in the different domains to reduce discrepancy.

Inter-twinning Moons. Furthermore, we observe the behavior of Go-labels on toy problem of *inter-twinning moons* [15,49]. We compare our method with the model trained with source data only and DANN [15] in the Fig. 3. We observe that both baselines of *Source only* and *DANN* neglect the outlier samples. In contrast, our method not only gets a satisfactory classification boundary between two classes in the source domain, but also covers these minority tail data well and classifies them to the correct class. More details are presented in **Supplementary**.

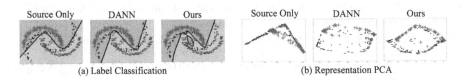

(a) Label Classification   (b) Representation PCA

Fig. 3. The second toy game of *inter-twinning moons*. Red "+", green "·", and black "." markers indicate the source positive samples (label 1), source negative samples (label 0), and target samples, respectively.

4.2 Experiments on the General UDA Benchmarks

Table 1. Classification accuracy (mean ± std %) of different schemes. We evaluate the effectiveness of Go-labels with different baselines, including DANN [15], CDAN [35], GVB [11], on the Digit-Five/Office31 datasets with Cov_3FC_2 [42]/ResNet-50 [20] as backbone. We re-implement all the baselines, thus the results are sightly different from the reported ones in the original papers.

(a) Comparison results on Digit-Five.

Method	mn → sv	mn → sy	sv → mn	sv → sy	sy → mn	sy → sv	Avg.
DANN [15]	23.2±0.5	40.0±0.3	71.0±0.3	84.6±0.1	93.6±0.4	84.7±0.3	66.2
+Go-labels	26.3±0.4	40.7±0.3	79.0±0.2	87.7±0.7	95.3±0.2	85.1±0.2	**69.0**
CDAN [35]	29.8±0.3	39.3±0.5	69.3±0.1	90.5±0.0	92.5±0.5	86.3±0.1	67.9
+Go-labels	28.1±0.5	41.3±0.3	78.6±0.0	90.6±0.0	95.5±0.5	86.4±0.1	**70.1**
GVB [11]	30.0±0.1	40.4±0.2	72.5±0.2	90.8±0.5	91.9±0.3	86.6±0.3	68.7
+Go-labels	30.3±0.1	42.1±0.2	79.6±0.1	90.9±0.5	95.9±0.3	87.2±0.0	**71.0**

(b) Comparison results on Office31.

Method	A → D	A → W	D → W	W → D	D → A	W → A	Avg.
DANN [15]	82.9±0.5	88.7±0.3	98.5±0.3	100±0.0	64.9±0.4	62.8±0.3	82.9
+Go-labels	89.9±0.4	92.4±0.3	98.9±0.2	100±0.0	71.6±0.0	68.3±0.2	**86.9**
CDAN [35]	92.2±0.3	93.1±0.5	98.7±0.1	100±0.0	72.8±0.5	70.1±0.0	87.8
+Go-labels	93.2±0.5	93.3±0.3	98.6±0.0	100±0.0	73.8±0.5	74.2±0.3	**88.9**
GVB [11]	94.8±0.1	92.2±0.3	94.5±0.3	100±0.0	75.3±0.2	73.2±0.3	88.3
+Go-labels	95.0±0.3	93.7±0.2	98.5±0.2	100±0.0	74.9±0.4	74.3±0.5	**89.4**

Datasets. Except for toy tasks, we also conduct experiments on the commonly-used domain adaptation (DA) datasets, including Digit-Five [14], Office31 [46], Office-Home [60], VisDA-2017 [43], and DomainNet [42]. These datasets cover various kinds of domain gaps, such as handwritten digit style discrepancy, office supplies imaging discrepancy, and synthetic↔real-world environment discrepancy. The data distribution imbalanced issue is also widely existed, and especially serious for the large-scale set, like DomainNet. The detailed introductions for each dataset can be found in **Supplementary**.

Implementation Details. As a plug-and-play optimization strategy, we apply our Go-labels on top of four representative ADA baselines, DANN [15], CDAN [35], GVB [11], and ASAN [45] for validation. DANN has been described in Sect. 3, and CDAN additionally conditions the domain discriminator on the information conveyed by the category classifier predictions (class likelihood). Recently-proposed GVB equips the adversarial adaptation framework with a gradually vanishing bridge, which reduces the transfer difficulty by reducing the domain-specific characteristics in representations. ASAN [45] integrates relevance spectral alignment and spectral normalization into CDAN. All reported results are obtained from the average of multiple runs (**Supplementary**).

Effectiveness of Go-getting Domain Labels. Our proposed Go-labels is generic and can be applied into most existing ADA frameworks, to alleviate the optimization difficulty caused by imbalanced domain data distributions, and thus enhance the domain-invariant representation learning. To prove that, we adopt three baselines, DANN [15], CDAN [35], GVB [11], and evaluate adaptation performance on Digit-Five and Office31, respectively. Table 1(a)(b) shows the comparison results, we observe that, regardless of the difference in framework design, our Go-labels (all *+Go-labels* schemes) consistently improves the accuracy of all three baselines on two datasets, *i.e.*, 2.8%/4.0%, 2.2%/1.1%, 2.3%/1.1% gains on average for DANN, CDAN, GVB, respectively on Digit-Five/Office31. With the help of Go-labels, each sample can be well explored in a dynamic way, resulting in better adaptation performance.

What Happens to Domain Discriminator When Training with Go-labels? For this experiment, we made statistics on the mis-classified cases of the domain discriminator during the training, and then visualize the *changing trend* in Fig. 4. There are two symmetrical mis-classified cases that need to be counted: mis-classify the raw source sample into the target domain or mis-classify the raw target sample into the source domain. Experiments are conducted on the Office31 and VisDA-2017 datasets, the compared baseline scheme is DANN [15]. As shown in Fig. 4, we observe that, the number of mis-classified cases by domain discriminator in our method is more than that in the baseline. We know that, 'mis-classified by domain discriminator' can be approximately equivalent to 'well-aligned'. Therefore, more 'mis-classified' samples by domain discriminator indicates that our method with Go-labels has a capability to align more samples, or said, could better cover those easy-to-miss minority clusters for alignment.

Loss Curve Comparison. Here we also show and compare the loss curves of domain discriminator for baseline DANN and our method. From Fig. 5, we can observe that the loss curve of baseline first drops quickly and gradually rises to near a constant as training progresses. In comparison, the domain discriminator loss curve of our method drops slowly, because more samples (including large and small cluster data) need to be studied/aligned during the training, which could in turn further drive better domain-invariant representations learning.

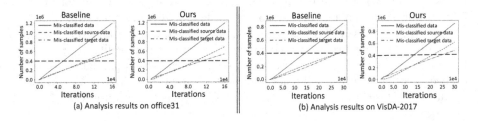

Fig. 4. Trend analysis of the mis-classified cases statistics for the domain discriminator in the training. Here, baseline is DANN [15] with ResNet-50 as backbone.

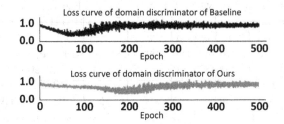

Fig. 5. Domain discriminator loss curves of baseline (DANN) and our method (DANN + Go-labels). Experiments are on the setting of W→A of Office31.

Why Not Directly Ignore Well-Aligned Data? The core idea of our dynamic adversarial domain adaptation with Go-labels is to transfer the model attention from over-studied aligned data to those overlooked samples progressively, so as to allow each sample to be well studied. Therefore, an intuitive alternative solution is to directly discard these over-aligned data, *e.g.*, simply zero out their gradients. We conduct this experiment on the Office31 based on DANN [15]. In Table 2, we see the scheme of *DANN + Zero Out* that directly discards these well-aligned samples is even inferior to *Baseline (DANN)* by 2.1% on average. This indicates that such **'hard and rude'** data filtering trick is sub-optimal because it may lose some important knowledge by mistake. Differently, our Go-labels training strategy could **softly** and **progressively** transfer the focus of optimization from the over-aligned samples to the under-explored data.

Comparison with Re-weighting Based Methods. As pointed in previous researches [27,33], the re-weighting schemes have the risks of over-fitting the tail data (by over-sampling) and also have the risk of under-fitting the global data distribution (by under-sampling), when data imbalance is extreme [76]. Besides, most sample re-weighting techniques [67] start re-weighting operation from the beginning of the entire training process. However, the non-converged feature extractor may affect the re-weighting decision, and cause unstable training. To prove that, we further compare our Go-labels with some sample (re)weighting based methods, including entropy-based re-weighting (+ E) [35], IWAN [70]. Entropy-based re-weighting (+ E) aims to prioritize the easy-to-transfer samples according to predictions of the category classifier to ease the entire adaptation optimization. IWAN [70] re-weights the source samples to exclude the outlier classes in the source domain. Table 2 shows the comparison results. We can observe that even all the sample re-weighting strategies bring performance gains, 3.0% for + E and 2.4% for + IWAN, but our Go-labels strategy still outperforms all competitors. In addition, our Go-labels is also complementary to these re-weighting techniques, the scheme of *DANN + E + Go-labels* still could achieve 1.6% gain in comparison with *DANN + E*.

Table 2. Comparison with gradient penalization and re-weighting related methods on Office31. The adopted baseline is DANN.

Method	A→D	A→W	D→W	W→D	D→A	W→A	Avg.
DANN [15]	82.9 ± 0.5	88.7 ± 0.3	98.5 ± 0.3	100 ± 0.0	64.9 ± 0.4	62.8 ± 0.3	82.9
+ Zero Out	84.3 ± 0.2	82.9 ± 0.1	98.2 ± 0.3	100 ± 0.0	58.0 ± 0.4	61.1 ± 0.5	80.8
+ E [35]	86.3 ± 0.1	91.0 ± 0.2	98.8 ± 0.3	100 ± 0.0	69.6 ± 0.3	69.8 ± 0.5	85.9
+ IWAN [70]	85.9 ± 0.1	91.9 ± 0.1	98.3 ± 0.2	100 ± 0.0	68.3 ± 0.4	67.5 ± 0.5	85.3
+ Go-labels	89.9 ± 0.4	92.4 ± 0.3	98.9 ± 0.2	100 ± 0.0	71.6 ± 0.0	68.3 ± 0.2	86.9
+ E + Go-labels	91.2 ± 0.2	91.4 ± 0.4	99.1 ± 0.1	100 ± 0.0	71.4 ± 0.1	71.9 ± 0.3	87.5

Go-labels is Well-Suited to DA Settings with Intra-class and Inter-class Imbalance. The results on DomainNet [42] can be taken as experimental evidence to prove this point. Because DomainNet has multiple domains, when testing the model adaptation ability on the certain target domain, the rest domains are mixuped as a large source domain. Such large source domain is seriously imbalanced, with both of intra-class and inter-class situations [55]. From the Table 3, we can observe that our Go-labels consistently achieves gains on the different sub-settings, which demonstrates it is always effective to DA settings with the different imbalances to some extents. We analyze that the go-getting labeling encourages the domain discriminator to learn well each sample to get a better source and target domain alignment. This in turn drives a better feature extractor to learn discriminative and domain-invariant features for all samples (they promote each other). Thus, a better feature extractor further improves the classifier and classification accuracy even the classes are still imbalanced.

Analysis About Rounding Operation in Go-labels. To validate the rounding design in Go-labels, we experimented with *real-valued soft* go-getting domain labels (based on the probability scores without rounding) for comparison. Actually, this is the initial version of our Go-labels. This scheme of using real-valued soft go-getting domain labels (built upon DANN) is inferior to our rounding version by 9.4% in average accuracy on Office31 (77.5% vs. 86.9%, baseline of DANN is 82.9%). We analyze such large drop is because that the real-valued soft go-getting domain labels of training samples are **always** affected by the probability scores of domain discriminator, even the discriminator has not yet been well-trained at early stage. On the contrary, our rounding-based Go-labels makes no influence for the entire optimization at the stage where the domain discriminator could clearly/correctly classify source-target sample. And, it **only** reduce the importance for these well-aligned (mis-classified by discriminator) majority samples progressively while keep unchanged for those not aligned minority data. In short, the rounding design makes Go-labels more robust.

Table 3. Classification accuracy on DomainNet. ResNet-101 as backbone.

Methods	Clipart	Infograph	Painting	Quickdraw	Real	Sketch	Average
MDAN [75]	60.3 ± 0.41	25.0 ± 0.43	50.3 ± 0.36	8.2 ± 1.92	61.5 ± 0.46	51.3 ± 0.58	42.8
M3SDA [42]	58.6 ± 0.53	26.0 ± 0.89	52.3 ± 0.55	6.3 ± 0.58	62.7 ± 0.51	49.5 ± 0.76	42.7
CMSS [67]	64.2 ± 0.18	$\mathbf{28.0 \pm 0.20}$	53.6 ± 0.39	16.0 ± 0.12	63.4 ± 0.21	$\mathbf{53.8 \pm 0.35}$	46.5
CDAN [35]	63.3 ± 0.21	23.2 ± 0.11	54.0 ± 0.34	16.8 ± 0.41	62.8 ± 0.14	50.9 ± 0.43	45.2
+ Go-labels	$\mathbf{65.7 \pm 0.22}$	25.7 ± 0.34	$\mathbf{55.6 \pm 0.21}$	$\mathbf{18.4 \pm 0.31}$	$\mathbf{63.6 \pm 0.28}$	53.6 ± 0.13	$\mathbf{47.1}$

Table 4. Performance (%) comparisons with the state-of-the-art UDA approaches on Office31. All experiments are based on ResNet-50 pre-trained on ImageNet.

Method	Venue	A→D	A→W	D→W	W→D	D→A	W→A	Avg.
DANCE [47]	NeurIPS'20	89.4 ± 0.1	88.6 ± 0.2	97.5 ± 0.4	$100.0 \pm .0$	69.5 ± 0.5	68.2 ± 0.2	85.5
Re-weight [61]	IJCAI'20	$91.7 \pm —$	$95.2 \pm —$	$98.6 \pm —$	$100.0 \pm -$	$74.5 \pm —$	$73.7 \pm —$	89.0
DADA [57]	AAAI'20	92.3 ± 0.1	93.9 ± 0.2	99.2 ± 0.1	$100.0 \pm .0$	74.4 ± 0.1	74.2 ± 0.1	89.0
MetaAlign [62]	CVPR'21	93.0 ± 0.5	94.5 ± 0.3	98.6 ± 0.0	$100.0 \pm .0$	75.0 ± 0.3	73.6 ± 0.0	89.2
FGDA [16]	ICCV'21	$93.3 \pm —$	$93.2 \pm —$	$99.1 \pm —$	$100.0 \pm .0$	$73.2 \pm —$	$72.7 \pm —$	88.6
SCDA [32]	ICCV'21	$94.2 \pm —$	$95.2 \pm —$	$98.7 \pm —$	$99.8 \pm —$	$75.7 \pm —$	$76.2 \pm —$	90.0
RADA [24]	ICCV'21	96.1 ± 0.4	96.2 ± 0.4	99.3 ± 0.1	$100.0 \pm .0$	77.5 ± 0.1	77.4 ± 0.3	91.1
CDAN + E (Baseline) [35]	NIPS'18	90.8 ± 0.3	94.0 ± 0.5	98.1 ± 0.3	$100.0 \pm .0$	72.4 ± 0.4	72.1 ± 0.3	87.9
CDAN + E + Go-labels	This work	94.2 ± 0.3	93.3 ± 0.1	99.0 ± 0.1	$100.0 \pm .0$	75.8 ± 0.1	75.2 ± 0.3	89.6
GVB (Baseline) [11]	CVPR'20	94.8 ± 0.1	92.2 ± 0.2	94.5 ± 0.2	$100.0 \pm .0$	75.3 ± 0.3	73.2 ± 0.4	88.3
GVB + Go-labels	This work	95.0 ± 0.3	93.7 ± 0.1	98.5 ± 0.1	$100.0 \pm .0$	74.9 ± 0.1	74.3 ± 0.3	89.4
ASAN (Baseline) [45]	ACCV'20	95.6 ± 0.4	$\mathbf{98.8 \pm 0.2}$	94.4 ± 0.9	$100.0 \pm .0$	74.7 ± 0.3	74.0 ± 0.9	90.0
ASAN + Go-labels	This work	$\mathbf{96.9 \pm 0.2}$	$\mathbf{98.8 \pm 0.1}$	99.1 ± 0.2	$100.0 \pm .0$	77.0 ± 0.3	75.9 ± 0.6	$\mathbf{91.3}$

4.3   Comparison with State-of-the-Arts

As a general technique, we insert our Go-labels into multiple DA algorithms to validate: CDAN with entropy regularization [35] (CDAN+E), GVB [11], ASAN [45], and RADA [24]. Table 4, Table 5 and Table 6 show the comparisons with the state-of-the-art approaches on Office31, Office-Home and VisDA-2017, respectively. For fair comparison, we report the results from their original papers if available, and we also report the results of the baseline schemes *GVB* and *CDAN+E* reproduced by our implementation. We find *GVB+Go-labels*, *CDAN+E+Go-labels*, and *ASAN+Go-labels* all outperform their corresponding baselines and also achieves the state-of-the-art performance on three datasets, and Go-labels is also more simple and efficient per without extra computation.

Table 5. Performance (%) comparisons with the state-of-the-art UDA approaches on Office-Home. All experiments are based on ResNet-50 pre-trained on ImageNet.

Method	Venue	Ar→Cl	Ar→Pr	Ar→Rw	Cl→Ar	Cl→Pr	Cl→Rw	Pr→Ar	Pr→Cl	Pr→Rw	Rw→Ar	Rw→Cl	Rw→Pr	Avg
BNM [10]	CVPR'20	52.3	73.9	80.0	63.3	72.9	**74.9**	61.7	49.5	79.7	70.5	53.6	82.2	67.9
DANCE [47]	NIPS'20	54.3	75.9	78.4	64.8	72.1	73.4	63.2	53.0	79.4	73.0	58.2	82.9	69.1
Reweight [61]	IJCAI'20	55.5	73.5	78.7	60.7	74.1	73.1	59.5	55.0	80.4	72.4	60.3	84.3	68.9
SRDC [56]	CVPR'20	52.3	76.3	81.0	**69.5**	76.2	78.0	**68.7**	53.8	81.7	**76.3**	57.1	85.0	71.3
CKB [38]	CVPR'21	54.2	74.1	77.5	64.6	72.2	71.0	64.5	53.4	78.7	72.6	58.4	82.8	68.7
TSA [31]	CVPR'21	57.6	75.8	80.7	64.3	76.3	75.1	66.7	55.7	81.2	75.7	61.9	83.8	71.2
MetaAg. [62]	CVPR'21	**59.3**	76.0	80.2	65.7	74.7	75.1	65.7	56.5	81.6	74.1	61.1	**85.2**	71.3
FGDA [16]	ICCV'21	52.3	**77.0**	78.2	64.6	75.5	73.7	64.0	49.5	80.7	70.1	52.3	81.6	68.3
SCDA [32]	ICCV'21	57.1	75.9	79.9	66.2	**76.7**	75.2	65.3	55.6	81.9	74.7	**62.6**	84.5	71.3
CDAN+E [35]	NIPS'18	55.6	72.5	77.9	62.1	71.2	73.4	61.2	52.6	80.6	73.1	55.5	81.4	68.1
+Go-labels	This work	56.0	74.4	78.2	63.9	72.7	72.0	63.7	54.1	81.7	73.3	59.6	83.0	69.4
ASAN [45]	ACCV'20	53.6	73.0	77.0	62.1	73.9	72.6	61.6	52.8	79.8	73.3	60.2	83.6	68.6
+Go-labels	This work	55.5	75.1	79.3	65.0	74.1	74.3	64.8	54.4	81.8	74.7	61.8	**85.2**	70.5
GVB [11]	CVPR'20	57.0	74.7	79.8	64.6	74.1	74.6	65.2	55.1	81.0	74.6	59.7	84.3	70.4
+Go-labels	This work	57.9	76.2	**81.1**	65.9	75.0	73.7	67.0	**56.6**	**82.9**	75.2	61.0	84.6	**71.4**

Table 6. Performance (%) comparisons with the state-of-the-art UDA approaches on VisDA-2017. All experiments are based on ResNet-50 pre-trained on ImageNet.

Method	Venue	Avg.
MDD [73]	ICML'19	74.61
SAFN [66]	ICCV'19	76.10
DANCE [47]	NeurIPS'20	70.20
CDAN + E (Baseline) [35]	NIPS'18	70.83
CDAN + E + Go-labels	This work	75.12
ASAN (Baseline) [45]	ACCV'20	72.34
ASAN + Go-labels	This work	75.21
GVB (Baseline) [11]	CVPR'20	75.34
GVB + Go-labels	This work	76.42
RADA (Baseline) [24]	ICCV'21	76.30
RADA + Go-labels	This work	**77.52**

5   Conclusion

We propose a simple plug-and-play technique dubbed go-getting domain labels (Go-labels) to achieve a dynamic adversarial domain adaptation framework, which effectively alleviates the imbalanced data distribution issue and significantly enhances the domain-invariant representation learning. Go-labels requires changing only two lines of code that yields non-trivial improvements across a wide variety of adversarial based UDA architectures. In fact, improvements of Go-labels come without bells and whistles on all domain adaptation benchmarks we evaluated, despite embarrassingly simple.

Acknowledgments. This work was supported in part by NSFC under Grant U1908209, 62021001 and the National Key Research and Development Program of China 2018AAA0101400. This work was also supported in part by the Advanced Research and Technology Innovation Centre (ARTIC), the National University of Singapore under Grant (project number: A-0005947-21-00).

References

1. Arjovsky, M., Chintala, S., Bottou, L.: Wasserstein generative adversarial networks. In: ICML, pp. 214–223. PMLR (2017)
2. Arpit, D., et al.: A closer look at memorization in deep networks. In: ICML, pp. 233–242. PMLR (2017)
3. Azadi, S., Olsson, C., Darrell, T., Goodfellow, I., Odena, A.: Discriminator rejection sampling. arXiv preprint arXiv:1810.06758 (2018)
4. Ben-David, S., Blitzer, J., Crammer, K., Kulesza, A., Pereira, F., Vaughan, J.W.: A theory of learning from different domains. Mach. Learn. **79**(1–2), 151–175 (2010)
5. Ben-David, S., Blitzer, J., Crammer, K., Pereira, F.: Analysis of representations for domain adaptation. In: NeurIPS, pp. 137–144 (2007)
6. Blitzer, J., Crammer, K., Kulesza, A., Pereira, F., Wortman, J.: Learning bounds for domain adaptation. In: NeurIPS, pp. 129–136 (2008)
7. Borgwardt, K.M., Gretton, A., Rasch, M.J., Kriegel, H.P., Schölkopf, B., Smola, A.J.: Integrating structured biological data by kernel maximum mean discrepancy. Bioinformatics **22**(14), e49–e57 (2006)
8. Chen, L., et al.: Reusing the task-specific classifier as a discriminator: discriminator-free adversarial domain adaptation. In: CVPR, pp. 7181–7190 (2022)
9. Chen, Q., Liu, Y., Wang, Z., Wassell, I., Chetty, K.: Re-weighted adversarial adaptation network for unsupervised domain adaptation. In: CVPR, pp. 7976–7985 (2018)
10. Cui, S., Wang, S., Zhuo, J., Li, L., Huang, Q., Tian, Q.: Towards discriminability and diversity: Batch nuclear-norm maximization under label insufficient situations. In: CVPR, pp. 3941–3950 (2020)
11. Cui, S., Wang, S., Zhuo, J., Su, C., Huang, Q., Tian, Q.: Gradually vanishing bridge for adversarial domain adaptation. In: CVPR, pp. 12455–12464 (2020)
12. Feldman, V.: Does learning require memorization? a short tale about a long tail. In: Proceedings of the 52nd Annual ACM SIGACT Symposium on Theory of Computing, pp. 954–959 (2020)
13. Feldman, V., Zhang, C.: What neural networks memorize and why: discovering the long tail via influence estimation. arXiv preprint arXiv:2008.03703 (2020)

14. Ganin, Y., Lempitsky, V.: Unsupervised domain adaptation by backpropagation. In: ICML, pp. 1180–1189. PMLR (2015)
15. Ganin, Y., et al.: Domain-adversarial training of neural networks. J. Mach. Learn. Res. **17**(1), 2030–2096 (2016)
16. Gao, Z., Zhang, S., Huang, K., Wang, Q., Zhong, C.: Gradient distribution alignment certificates better adversarial domain adaptation. In: ICCV, pp. 8937–8946 (2021)
17. Gulrajani, I., Ahmed, F., Arjovsky, M., Dumoulin, V., Courville, A.: Improved training of wasserstein gans. arXiv preprint arXiv:1704.00028 (2017)
18. Guo, T., et al.: On positive-unlabeled classification in gan. In: CVPR, pp. 8385–8393 (2020)
19. Haeusser, P., Frerix, T., Mordvintsev, A., Cremers, D.: Associative domain adaptation. In: ICCV, pp. 2765–2773 (2017)
20. He, K., Zhang, X., Ren, S., Sun, J.: Deep residual learning for image recognition. In: CVPR, pp. 770–778 (2016)
21. Hoffman, J., et al.: Cycada: cycle-consistent adversarial domain adaptation. In: ICML, pp. 1989–1998. PMLR (2018)
22. Huang, C., Li, Y., Loy, C.C., Tang, X.: Learning deep representation for imbalanced classification. In: CVPR, pp. 5375–5384 (2016)
23. Jiang, X., Lao, Q., Matwin, S., Havaei, M.: Implicit class-conditioned domain alignment for unsupervised domain adaptation. In: ICML, pp. 4816–4827. PMLR (2020)
24. Jin, X., Lan, C., Zeng, W., Chen, Z.: Re-energizing domain discriminator with sample relabeling for adversarial domain adaptation. In: ICCV (2021)
25. Jin, X., Lan, C., Zeng, W., Chen, Z.: Style normalization and restitution for domain generalization and adaptation. IEEE Trans. Multimedia **24**, 3636–3651 (2021)
26. Johansson, F.D., Sontag, D., Ranganath, R.: Support and invertibility in domain-invariant representations. In: AISTATS, pp. 527–536. PMLR (2019)
27. Kang, B., et al.: Decoupling representation and classifier for long-tailed recognition. In: ICLR (2019)
28. Kang, G., Jiang, L., Yang, Y., Hauptmann, A.G.: Contrastive adaptation network for unsupervised domain adaptation. In: CVPR, pp. 4893–4902 (2019)
29. Kifer, D., Ben-David, S., Gehrke, J.: Detecting change in data streams. In: VLDB, vol. 4, pp. 180–191. Toronto, Canada (2004)
30. Li, B., et al.: Rethinking distributional matching based domain adaptation. arXiv preprint arXiv:2006.13352 (2020)
31. Li, S., Xie, M., Gong, K., Liu, C.H., Wang, Y., Li, W.: Transferable semantic augmentation for domain adaptation. In: CVPR, pp. 11516–11525 (2021)
32. Li, S., et al.: Semantic concentration for domain adaptation. In: ICCV, pp. 9102–9111 (2021)
33. Li, Y., et al.: Overcoming classifier imbalance for long-tail object detection with balanced group softmax. In: CVPR, pp. 10991–11000 (2020)
34. Liu, J., Sun, Y., Han, C., Dou, Z., Li, W.: Deep representation learning on long-tailed data: a learnable embedding augmentation perspective. In: CVPR, pp. 2970–2979 (2020)
35. Long, M., Cao, Z., Wang, J., Jordan, M.I.: Conditional adversarial domain adaptation. In: NeurIPS, pp. 1640–1650 (2018)
36. Long, M., Zhu, H., Wang, J., Jordan, M.I.: Deep transfer learning with joint adaptation networks. In: ICML, pp. 2208–2217 (2017)
37. Lu, Z., Yang, Y., Zhu, X., Liu, C., Song, Y.Z., Xiang, T.: Stochastic classifiers for unsupervised domain adaptation. In: CVPR, pp. 9111–9120 (2020)

38. Luo, Y.W., Ren, C.X.: Conditional bures metric for domain adaptation. In: CVPR, pp. 13989–13998 (2021)
39. Nowozin, S., Cseke, B., Tomioka, R.: f-gan: training generative neural samplers using variational divergence minimization. In: NeurIPS (2016)
40. Paszke, A., et al.: Pytorch: an imperative style, high-performance deep learning library. arXiv preprint arXiv:1912.01703 (2019)
41. Pei, Z., Cao, Z., Long, M., Wang, J.: Multi-adversarial domain adaptation. In: AAAI, vol. 32 (2018)
42. Peng, X., Bai, Q., Xia, X., Huang, Z., Saenko, K., Wang, B.: Moment matching for multi-source domain adaptation. In: ICCV, pp. 1406–1415 (2019)
43. Peng, X., Usman, B., Kaushik, N., Hoffman, J., Wang, D., Saenko, K.: Visda: the visual domain adaptation challenge. arXiv preprint arXiv:1710.06924 (2017)
44. Peng, Z., Huang, W., Guo, Z., Zhang, X., Jiao, J., Ye, Q.: Long-tailed distribution adaptation. In: ACMMM, pp. 3275–3282 (2021)
45. Raab, C., Vath, P., Meier, P., Schleif, F.M.: Bridging adversarial and statistical domain transfer via spectral adaptation networks. In: ACCV (2020)
46. Saenko, K., Kulis, B., Fritz, M., Darrell, T.: Adapting visual category models to new domains. In: Daniilidis, K., Maragos, P., Paragios, N. (eds.) ECCV 2010. LNCS, vol. 6314, pp. 213–226. Springer, Heidelberg (2010). https://doi.org/10.1007/978-3-642-15561-1_16
47. Saito, K., Kim, D., Sclaroff, S., Saenko, K.: Universal domain adaptation through self supervision. In: NeurIPS (2020)
48. Saito, K., Ushiku, Y., Harada, T., Saenko, K.: Strong-weak distribution alignment for adaptive object detection. In: CVPR, pp. 6956–6965 (2019)
49. Saito, K., Watanabe, K., Ushiku, Y., Harada, T.: Maximum classifier discrepancy for unsupervised domain adaptation. In: CVPR, pp. 3723–3732 (2018)
50. Sankaranarayanan, S., Balaji, Y., Castillo, C.D., Chellappa, R.: Generate to adapt: aligning domains using generative adversarial networks. In: CVPR, pp. 8503–8512 (2018)
51. Shimodaira, H.: Improving predictive inference under covariate shift by weighting the log-likelihood function. J. Stat. Plan. Inference $90(2)$, 227–244 (2000)
52. Sinha, S., Zhao, Z., Alias Parth Goyal, A.G., Raffel, C.A., Odena, A.: Top-k training of gans: improving gan performance by throwing away bad samples. In: NeurIPS, vol. 33, pp. 14638–14649 (2020)
53. Sugiyama, M., Krauledat, M., Müller, K.R.: Covariate shift adaptation by importance weighted cross validation. J. Mach. Learn. Res. $8(5)$, 1–21 (2007)
54. Sun, B., Feng, J., Saenko, K.: Return of frustratingly easy domain adaptation. In: AAAI, vol. 30 (2016)
55. Tan, S., Peng, X., Saenko, K.: Class-imbalanced domain adaptation: an empirical odyssey. In: Bartoli, A., Fusiello, A. (eds.) ECCV 2020. LNCS, vol. 12535, pp. 585–602. Springer, Cham (2020). https://doi.org/10.1007/978-3-030-66415-2_38
56. Tang, H., Chen, K., Jia, K.: Unsupervised domain adaptation via structurally regularized deep clustering. In: CVPR, pp. 8725–8735 (2020)
57. Tang, H., Jia, K.: Discriminative adversarial domain adaptation. In: AAAI, vol. 34, pp. 5940–5947 (2020)
58. Torralba, A., Efros, A.A.: Unbiased look at dataset bias. In: CVPR 2011. pp. 1521–1528
59. Tzeng, E., Hoffman, J., Saenko, K., Darrell, T.: Adversarial discriminative domain adaptation. In: CVPR, pp. 7167–7176 (2017)
60. Venkateswara, H., Eusebio, J., Chakraborty, S., Panchanathan, S.: Deep hashing network for unsupervised domain adaptation. In: CVPR (2017)

61. Wang, S., Zhang, L.: Self-adaptive re-weighted adversarial domain adaptation. IJCAI (2020)
62. Wei, G., Lan, C., Zeng, W., Chen, Z.: Metaalign: coordinating domain alignment and classification for unsupervised domain adaptation. In: CVPR (2021)
63. Wu, Y., Donahue, J., Balduzzi, D., Simonyan, K., Lillicrap, T.: Logan: latent optimisation for generative adversarial networks. arXiv preprint arXiv:1912.00953 (2019)
64. Wu, Y., Winston, E., Kaushik, D., Lipton, Z.: Domain adaptation with asymmetrically-relaxed distribution alignment. In: ICML, pp. 6872–6881. PMLR (2019)
65. Xu, M., et al.: Adversarial domain adaptation with domain mixup. In: AAAI (2020)
66. Xu, R., Li, G., Yang, J., Lin, L.: Larger norm more transferable: an adaptive feature norm approach for unsupervised domain adaptation. In: ICCV, pp. 1426–1435 (2019)
67. Yang, L., Balaji, Y., Lim, S.-N., Shrivastava, A.: Curriculum manager for source selection in multi-source domain adaptation. In: Vedaldi, A., Bischof, H., Brox, T., Frahm, J.-M. (eds.) ECCV 2020. LNCS, vol. 12359, pp. 608–624. Springer, Cham (2020). https://doi.org/10.1007/978-3-030-58568-6_36
68. Yosinski, J., Clune, J., Bengio, Y., Lipson, H.: How transferable are features in deep neural networks? In: NeurIPS (2014)
69. You, C., Li, C., Robinson, D.P., Vidal, R.: Scalable exemplar-based subspace clustering on class-imbalanced data. In: ECCV, pp. 67–83 (2018)
70. Zhang, J., Ding, Z., Li, W., Ogunbona, P.: Importance weighted adversarial nets for partial domain adaptation. In: CVPR, pp. 8156–8164 (2018)
71. Zhang, X., Fang, Z., Wen, Y., Li, Z., Qiao, Y.: Range loss for deep face recognition with long-tailed training data. In: ICCV, pp. 5409–5418 (2017)
72. Zhang, Y., Tang, H., Jia, K., Tan, M.: Domain-symmetric networks for adversarial domain adaptation. In: CVPR, pp. 5031–5040 (2019)
73. Zhang, Y., Liu, T., Long, M., Jordan, M.I.: Bridging theory and algorithm for domain adaptation. In: ICML (2019)
74. Zhao, H., Des Combes, R.T., Zhang, K., Gordon, G.: On learning invariant representations for domain adaptation. In: ICML, pp. 7523–7532. PMLR (2019)
75. Zhao, H., Zhang, S., Wu, G., Moura, J.M., Costeira, J.P., Gordon, G.J.: Adversarial multiple source domain adaptation. In: NeurIPS, pp. 8559–8570 (2018)
76. Zhou, B., Cui, Q., Wei, X.S., Chen, Z.M.: Bbn: bilateral-branch network with cumulative learning for long-tailed visual recognition. In: CVPR, pp. 9719–9728 (2020)

ModSelect: Automatic Modality Selection for Synthetic-to-Real Domain Generalization

Zdravko Marinov$^{(\boxtimes)}$, Alina Roitberg, David Schneider, and Rainer Stiefelhagen

Institute for Anthropomatics and Robotics, Karlsruhe Institute of Technology,
Karlsruhe, Germany
{zdravko.marinov,alina.roitberg,david.schneider,
rainer.stiefelhagen}@kit.edu

Abstract. Modality selection is an important step when designing multimodal systems, especially in the case of cross-domain activity recognition as certain modalities are more robust to domain shift than others. However, selecting only the modalities which have a positive contribution requires a systematic approach. We tackle this problem by proposing an unsupervised modality selection method (ModSelect), which does not require any ground-truth labels. We determine the correlation between the predictions of multiple unimodal classifiers and the domain discrepancy between their embeddings. Then, we systematically compute modality selection thresholds, which select only modalities with a high correlation and low domain discrepancy. We show in our experiments that our method ModSelect chooses only modalities with positive contributions and consistently improves the performance on a SYNTHETIC→REAL domain adaptation benchmark, narrowing the domain gap.

Keywords: Modality selection · Domain generalization · Action recognition · Robust vision · Synthetic-to-real · Cross-domain

1 Introduction

Human activity analysis is vital for intuitive human-machine interaction, with applications ranging from driver assistance [55] to smart homes and assistive robotics [71]. Domain shifts, such as appearance changes, constitute a significant bottleneck for deploying such models in real-life. For example, while simulations are an excellent way of economical data collection, a SYNTHETIC→REAL domain shift leads to > **60%** drop in accuracy when recognizing daily living activities [70]. *Multimodality* is a way of mitigating this effect, since different types of data, such as RGB videos, optical flow and body poses, exhibit individual strengths and weaknesses. For example, models operating on body poses are

Supplementary Information The online version contains supplementary material available at https://doi.org/10.1007/978-3-031-25085-9_19.

less affected by appearance changes, as the relations between different joints are more stable given a good skeleton detector [22,23]. RGB videos, in contrast, are more sensitive to domain shifts [57,72] but also convenient since they cover the complete scene and video is the most ubiquitous modality [14,26,73,75].

Given the complementary nature of different data types (see Fig. 2), we believe, that multimodality has a strong potential for improving domain generalization of activity recognition models, but *which modalities to select* and *how to fuse the information* become important questions. Despite its high relevance for applications, the question of modality selection has been often overlooked in this field. The main goal of our work is to develop a systematic framework for studying the contribution of individual modalities in cross-domain human activity recognition. We specifically focus on the SYNTHETIC→REAL distributional shift [70], which opens new doors for economical data acquisition but comes with an especially large domain gap. We study five different modalities and examine how the prediction outcomes of multiple unimodal classifiers correlate as well as the domain discrepancy between their embeddings. We hope that our study will provide guidance for a better modality selection process in the future.

Contributions and Summary. We aim to make a step towards effective use of multimodality in the context of cross-domain activity recognition, which has been studied mostly for RGB videos in the past [14,26,73,75]. This work develops the modality selection framework ModSelect for quantifying the importance of individual data streams and can be summarized in two major contributions. (1) We propose a metric for quantifying the contribution of each modality for the specific task by calculating how the performance changes when the modality is included in the late fusion. Our new metric can be used by future research to justify decisions in modality selection. However, to estimate these performance changes, we use the ground-truth labels from the test data. (2) To detach ourselves from supervised labels, we propose to study the domain discrepancy between the embeddings and the correlation between the predictions of the unimodal classifiers of each modality. We use the discrepancy and the correlation to compute modality selection thresholds and show that these thresholds can be used to select only modalities with positive contributions w.r.t. our proposed metric in (1). Our unsupervised modality selection ModSelect can be applied in settings where no labels are present, e.g., in a multi-sensor setup deployed in unseen environments, where ModSelect would identify which sensors to trust.

2   Related Work

2.1   Multimodal Action Recognition

The usage of multimodal data represents a common technique in the field of action recognition, and is applied for both: increasing performance in supervised learning as well as unsupervised representation learning. Multimodal methods for action recognition include approaches which make use of video and audio

[3,4,50,62,64], optical flow [39,64], text [52] or pose information [22,23,65]. Such methods can be divided into lower level *early/feature* fusion which is based on merging latent space information from multiple modality streams [1,2,34,43, 56,63,84,94] and *late/score fusion* which combines the predictions of individual classifiers or representation encoders either with learned fusion modules [1,47, 61,76,77,93] or with rule-based algorithms.

For this work, we focus on the latter, since the variety of early fusion techniques and learned late fusion impedes a systematic comparison, while rule-based late fusion builds on few basic but successful techniques such as averaging single-modal scores [5,7,12,13,24,27,28,42,89], the max rule [5,27,45,67], product rule [25,27,42,45,48,67,78,81,91] or median rule. Ranking based solutions [20,30,31,66] like Borda count are less commonly used for action recognition but recognised in other fields of computer vision.

2.2 Modality Contribution Quantification

While the performance contribution of modalities has been analyzed in multiple previous works, e.g., by measuring the signal-to-noise ratio between modalities [83], determining class-wise modality contribution by learning an optimal linear modality combination [6,46] or extracting modality relations with threshold-based rules [80], in-depth analysis of modality contributions in the field of action recognition remains sparse and mostly limited to small ablation studies. Metrics to measure data distribution distances like Maximuim Mean Discrepancy (MMD) or Mean Pairwise Distance (MPD) have been applied in fields like domain adaptation. MMD is commonly used to estimate and to reduce domain shift [35,54,60,79] and can be adapted to be robust against class bias, e.g. in the form of weighted MMD [87], Mean Pairwise Distance (MPD) was applied to analyze semantic similarities of word embeddings, e.g., in [29]. In this work, we introduce a systematic approach for analyzing modality contributions in the context of cross-domain activity recognition, which, to the best of out knowledge, has not been addressed in the past.

2.3 Domain Generalization and Adaptation

Both domain generalization and domain adaptation present strategies to learn knowledge from a source domain which is transferable to a given target domain. While domain adaptation allows access to data from the target domain to fulfill this task, either paired with labels [16,59,69] or in the form of unsupervised domain adaptation [10,15,16,18,19,59,69,74], domain generalization assumes an unknown target domain and builds upon methods which condition a neural network to make use of features which are found to be more generalizable [88], apply heavy augmentations to increase robustness [90] or explore different methods of leveraging temporal data [90].

3 Approach

Our approach consists of three main steps. (1) We extract multiple modalities and train a unimodal action recognition classifier on each modality. Afterwards, we evaluate all possible combinations of the modalities with different late fusion methods. We define the action recognition task in Sect. 3.1, the datasets we use in Sect. 3.2, and the modality extraction and training in Sect. 3.3. (2) In Sect. 3.4, we determine which modalities lead to a performance gain based on our evaluation results from (1). This establishes a baseline for the (3) third step (Sect. 3.5), where we show how to systematically select these beneficial modalities in an unsupervised way with our framework ModSelect - without the need of labels nor evaluation results. We offer an optional notation table in the Supplementary for a better understanding of all of our equations.

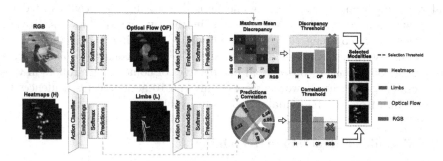

Fig. 1. ModSelect: our approach for unsupervised modality selection which uses predictions correlations and domain discrepancy.

We intentionally do not make use of learned late fusion techniques, such as [1,47,76,77,93], since such methods do not allow for comparing the contribution of individual modalities. Instead, a specific learned late fusion architecture could be better suited to some modalities in contrast to others, overshadowing a neutral evaluation. However, our work can be used to select modalities upon which such learned late fusion techniques can be designed.

3.1 Action Recognition Task

Our goal is to produce a systematic method for unsupervised modality selection in multimodal action recognition. More specifically, we focus on SYNTHETIC→REAL domain generalization to show the need for a modality selection approach when a large domain gap is present. In this scenario an action classifier is trained only on samples from a SYNTHETIC source domain $x_s \in X_s$ with action labels $y \in Y$. In domain generalization, the goal is to generalize to an unseen target domain X_t, without using any samples $x_t \in X_t$ from it during training. In our case, the target domain X_t consists of REAL data and the source and target data originate from distinct probability distributions $x_s \sim p_{synthetic}$ and $x_t \sim p_{real}$. The goal is to classify each instance x_t from the REAL target test domain X_t, which has a shared action label set Y with the training set. To achieve this, we use the synthetic Sims4Action dataset [70] for training and the real Toyota Smarthome (Toyota) [21] and ETRI-Activity3D-LivingLab (ETRI) [44] as two separate target test sets. We also evaluate our models on the Sims4Action official test split [70] in our additional SYNTHETIC→SYNTHETIC experiments.

3.2 Datasets

We focus on SYNTHETIC→REAL domain generalization between the synthetic Sims4Action [70] as a training dataset and the real Toyota Smarthome [21] and ETRI [44] as test datasets. Sims4Action consists of ten hours of video material recorded from the computer game Sims 4, covering 10 activities of daily living which have direct correspondences in the two REAL datasets. Toyota Smarthome [21] contains videos of 18 subjects performing 31 different everyday actions within a single apartment, and ETRI [44] consists of 50 subjects performing 55 actions recorded from perspectives of home service robots in various residential spaces. However, we use only the 10 action correspondences to Sims4Action from the REAL datasets for our evaluation.

3.3 Modality Extraction and Training

We leverage the multimodal nature of actions to extract additional modalities for our training data, such as body pose, movement dynamics, and object detections. To this end, we utilize the RGB videos from the synthetic Sims4Action [70] to produce four new modalities - heatmaps, limbs, optical flow, and object detections. An overview of all modalities can be seen in Fig. 2.

Heatmaps and Limbs. The heatmaps and limbs (H and L) are extracted via AlphaPose [32,51,86], which infers 17 joint locations of the human body. The heatmaps modality $h(x, y)$ at pixel (x, y) is obtained by stacking 2D Gaussian maps, which are centered at each joint location (x_i, y_i) and each map is weighted by its detection confidence c_i as shown in Eq. 1, where $\sigma = 6$.

$$h(x, y) := exp\left(\frac{-((x - x_i)^2 + (y - y_i)^2)}{2\sigma^2}\right) \cdot c_i \qquad (1)$$

The limbs modality is produced by connecting the joints with white lines and weighting each line by the smaller confidence of its endpoints. We weight both modalities by the detection confidences c_i so that uncertain and occluded body parts are dimmer and have a smaller contribution.

Optical Flow. The optical flow modality (OF) is estimated via the Gunnar-Farneback method [33]. The optical flow $of(x, y)$ at pixel (x, y) encodes the *magnitude* and *angle* of the pixel intensity changes between two frames in the *value* and *hue* components of the HSV color space. The *saturation* is used to adjust the visibility and we set it to its maximum value. The heatmaps, limbs, and optical flow are all *image-based* and are used as an input to models which usually utilize RGB images.

Object Detections. Our last modality (YOLO) consists of object detections obtained by YOLOv3 [68], which detects 80 different objects. Unlike the other modalities, we represent the detections as a vector, instead of an image. We show that such a simple representation achieves good domain generalization in our experiments. The YOLO modality for an image sample consists of a k-dimensional vector $\mathbf{v}$, where $\mathbf{v}[i]$ corresponds to the reciprocal Euclidean distance between the person's and the ith object's bounding box centers, and $k = 80$ is the number of detection classes. This way, objects closer to the person have a larger weight in $\mathbf{v}$ than ones which are further away. After computing the distances, $\mathbf{v}$ is normalized by its norm: $\mathbf{v} \leftarrow \mathbf{v}/\|\mathbf{v}\|$. We denote the set of all modalities as $\mathcal{M} := \{\text{H, L, OF, RGB, YOLO}\}$ and use the term $\mathcal{M}$ in our equations.

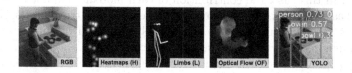

Fig. 2. Examples of all extracted modalities. Note: the YOLO modality is represented as a vector $\mathbf{v}$, which encodes distances to the person's detection (see Sect. 3.3).

Training. We train unimodal classifiers on each modality and evaluate all possible modality combinations with different late fusion methods. We utilize 3D-CNN models with the S3D backbone [85] for each one of the RGB, H, L, and OF modalities. The YOLO modality utilizes an MLP model as it is not image-based. We train all 5 action recognition models end-to-end on Sims4Action [70].

Evaluation. For our late fusion experiments, we combine the predictions of all unimodal classifiers at the class score level and obtain results for all $\sum_{i=1}^{5} \binom{i}{5} = 31$ modality combinations. We investigate 6 late fusion strategies - Sum, Squared Sum, Product, Maximum, Median, and Borda Count [8,40], which all operate on the class probability scores. Borda Count also uses the ranking of the class scores. For brevity, we refer to a late fusion of unimodal classifiers as a *multimodal classifier* and present our late fusion results in Sect. 4.1.

3.4   Quantification Study: Modality Contributions

In this section, we propose how to quantify the contributions of each modality based on the performance of the models on the target test sets. To this end, we propose a with-without metric, which computes the average difference $f(m)$ of the performance of a multimodal classifier *with* a modality m to the performance *without* it. Formally the contribution $f(m)$ of a modality m is defined as:

$$f(m) := \mathop{\mathbb{E}}_{C \in \mathcal{C}} [acc(C \cup \{m\}) - acc(C)] \tag{2}$$

where $\mathcal{C}$ is the set of all modality combinations, $acc(C)$ is the test accuracy of the multimodal classifier with the modality combination C, and $m \in \mathcal{M}$. We compute $f(m)$ for all modalities based on the late fusion results listed in Sect. 4.1. The contribution of each modality can be used to determine the modalities, which positively influence the performance on the test dataset $\mathcal{M}^+ \subseteq \mathcal{M}$.

3.5   ModSelect: Unsupervised Modality Selection

In this section, we introduce our method ModSelect for unsupervised modality selection. In this setting, we assume that we do not have any labels in the target test domain X_t. This is exactly the case for SYNTHETIC→REAL domain generalization, where a model trained on simulated data is deployed in real-world conditions. In this case, the contribution of each modality cannot be estimated with Eq. 2 as $acc(C)$ cannot be computed without ground-truth labels. Note that we do have labels in our test sets but we only use them in our quantification study and ignore them for our unsupervised experiments.

We propose ModSelect - a method for unsupervised modality selection based on the consensus of two metrics: (1) the correlation between the unimodal classifiers' predictions ρ and (2) the Maximum Mean Discrepancy (MMD) [36] between the classifiers' embeddings. We compute both metrics with our unimodal classifiers and propose how to systematically estimate modality selection thresholds. We show that the thresholds select the same modalities with positive contributions $\mathcal{M}^+$ as our quantification study in Sect. 3.4.

Correlation Metric. We define the predictions correlation vector $\boldsymbol{\rho}_{mn}$ between modalities m and n as:

$$\boldsymbol{\rho}_{mn} := \frac{\mathbb{E}[(\mathbf{z}_m - \boldsymbol{\mu}_m) \odot (\mathbf{z}_n - \boldsymbol{\mu}_n)]}{\boldsymbol{\sigma}_m \odot \boldsymbol{\sigma}_n} \tag{3}$$

where $\mathbf{z}_m, \mathbf{z}_n$ are the softmax class scores of the action classifiers trained on modalities m and n respectively, $(\boldsymbol{\mu}_m, \boldsymbol{\sigma}_m), (\boldsymbol{\mu}_n, \boldsymbol{\sigma}_n)$ are the mean and standard deviation vectors of $\mathbf{z}_m, \mathbf{z}_n$, and $\odot$ is the element-wise multiplication operator. We define the predictions correlation $\rho(m, n)$ between two modalities m, n as:

$$\rho(m, n) := \frac{1}{N} \sum_{i=1}^{N} \boldsymbol{\rho}_{mn}[i] \tag{4}$$

where $N = 10$ is the number of action classes.

MMD Metric. We show that the distance between the distributions of the embeddings of two unimodal classifiers can also be used to compute a modality selection threshold. The MMD metric [36] between two distributions P and Q over a set $\mathcal{X}$ is formally defined as:

$$MMD(P, Q) := ||\mathbb{E}_{X \sim P}[\varphi(X)] - \mathbb{E}_{Y \sim Q}[\varphi(Y)]||_{\mathcal{H}} \qquad (5)$$

where $\varphi : \mathcal{X} \mapsto \mathcal{H}$ is a feature map, and $\mathcal{H}$ is a reproducing kernel Hilbert space (RKHS) [9,36,37]. For our empirical calculation of MMD between the embeddings $\mathbf{h}_m, \mathbf{h}_n$ of two modalities m, n we set $\mathcal{X} = \mathcal{H} = \mathbb{R}^d$ and $\varphi(x) = x$:

$$MMD(m, n) := ||\mathbb{E}[\mathbf{h}_m] - \mathbb{E}[\mathbf{h}_n]|| \qquad (6)$$

where $\mathbf{h}_m, \mathbf{h}_n$ are the embeddings from the second-to-last linear layer of the action classifiers for modalities m and n respectively, and d is the embedding size. Note that using a linear feature map $\varphi(x) = x$ lets us determine only the discrepancy between the distributions' means. A linear mapping is sufficient to produce a good modality selection threshold, but one could also consider more complex alternatives, such as $\varphi(x) = (x, x^2)$ or a Gaussian kernel [38].

We make the following observations regarding both metrics for modality selection. Firstly, a high correlation between correct predictions is statistically more likely than a high correlation between wrong predictions, since there is only 1 correct class and $N-1$ possibilities for error. We believe that a stronger correlation between the predictions results in a higher performance. Secondly, unimodal classifiers should have a high agreement on easy samples and a disagreement on difficult cases [40]. A higher domain discrepancy between the classifiers' embeddings has been shown to indicate a lower agreement on their predictions [53,92], and hence, a decline in performance when fused. We therefore believe that good modalities are characterized by a low discrepancy and high correlation.

Modality Selection Thresholds. After computing $\rho(m, n)$ and $MMD(m, n)$ for all pairs $(m, n) \in \mathcal{M}^2$, we systematically calculate modality selection thresholds for each metric ρ and MMD. We consider two types of thresholds: (1) an aggregated threshold $\delta_{\mathbf{agg}}$, which selects a set of *individual modalities* $\mathcal{M}_{\mathbf{agg}} \subseteq \mathcal{M}$, and (2) a pairs-threshold $\delta_{\mathbf{pair}}$, which selects a set of *modality pairs* $\mathcal{C}_{\mathbf{pair}} \subseteq \mathcal{M}^2$.

Aggregated Threshold $\delta_{\mathbf{agg}}$. For the first threshold, we aggregate the ρ and MMD values for a modality m by averaging over all of its pairs:

$$\rho(m) := \frac{1}{|\mathcal{M}|} \sum_{n \in \mathcal{M}} \rho(m, n) \quad MMD(m) := \frac{1}{|\mathcal{M}|} \sum_{n \in \mathcal{M}} MMD(m, n) \qquad (7)$$

Thus, we produce the sets $A_\rho := \{\rho(m) | m \in \mathcal{M}\}$ and $A_{MMD} := \{MMD(m) | m \in \mathcal{M}\}$. A simple approach would be to use the mean or median as a threshold for A_ρ and A_{MMD}. However, such thresholds are sensitive to outliers (mean) or do not use all the information from the values (median). Additionally, one cannot tune the threshold with prior knowledge. To mitigate these issues, we propose to use the Winsorized Mean [41,82] $\mu_\lambda(A)$ for both sets, which is defined as:

$$\mu_\lambda(A) := \lambda a_\lambda + (1 - 2\lambda)\bar{a}_\lambda + \lambda a_{1-\lambda} \tag{8}$$

where a_λ is the λ-percentile of A, $\bar{a}_\lambda$ is the λ-trimmed mean of A, and $\lambda \in [0, 0.5]$ is a "trust" hyperparameter. A higher λ results in a lower contribution of edge values in A and a bigger trust in values near the center. Therefore, we set $\lambda = 0.2$ as we have 5 modalities and expect to trust at least 3. We compute two separate thresholds $\delta^\rho_{\mathbf{agg}} := \mu_{0.2}(A_\rho)$ and $\delta^{MMD}_{\mathbf{agg}} := \mu_{0.2}(A_{MMD})$ and select the modalities $\mathcal{M}_{\mathbf{agg}}$ as a consensus between the two metrics as:

$$\mathcal{M}_{\mathbf{agg}} := \{m \in \mathcal{M} | \rho(m) \geq \delta^\rho_{\mathbf{agg}} \vee MMD(m) \leq \delta^{MMD}_{\mathbf{agg}}\} \tag{9}$$

Pairs-Threshold $\delta_{\mathbf{pair}}$. The second type of selection threshold skips the aggregation step of $\delta_{\mathbf{agg}}$ and directly computes the Winsorized Means $\mu_{0.2}(\cdot)$ over the sets of all $\rho(m, n)$ and $MMD(m, n)$ values to obtain $\delta^\rho_{\mathbf{pair}}$ and $\delta^{MMD}_{\mathbf{pair}}$ respectively. This results in a selection of *modality pairs*, rather than individual modalities as in Eq. 9. In other words, the $\delta_{\mathbf{pair}}$ thresholds are suitable when one is searching for the *best pairs* of modalities, and $\delta_{\mathbf{agg}}$ for the *best individual* modalities. The selected modality pairs $\mathcal{C}_{\mathbf{pair}}$ with this method are:

$$\mathcal{C}_{\mathbf{pair}} := \{(m, n) \in \mathcal{M}^2 | \rho(m, n) \geq \delta^\rho_{\mathbf{pair}} \vee MMD(m, n) \leq \delta^{MMD}_{\mathbf{pair}}\} \tag{10}$$

Summary of our Approach. A summary of our unsupervised modality selection method ModSelect is illustrated in Fig. 1. We use the embeddings of unimodal action classifiers to compute the Maximum Mean Discrepancy (MMD) between all pairs of modalities. We also compute the correlation ρ between the predictions of all pairs of classifiers. We systematically estimate thresholds for MMD and ρ which discard certain modalities and select only modalities on which both metrics have a consensus. In the following Experiments 4 we show that the selected modalities $m \in \mathcal{M}$ with our method ModSelect are exactly the modalities with a positive contribution $f(m) > 0$ according to Eq. 2, although our unsupervised selection does not utilize any ground-truth labels.

4   Experiments

4.1   Late Fusion: Results

We evaluate our late fusion multimodal classifiers following the cross-subject protocol from [21] for Toyota, the inter-dataset protocol from [49] for ETRI, and the official test split for Sims4Action from [70]. We follow the original Sims4Action→Toyota evaluation protocol of [70] and utilize the mean-per-class accuracy (mPCA) as the number of samples per class are imbalanced in the Real test sets. The mPCA metric avoids bias towards overrepresented classes and is often used in unbalanced activity recognition datasets [11,21,55,70].

The results from our evaluation are displayed in Table 1. The domain gap of transferring to REAL data is apparent in the drastically lower performance, especially on the ETRI dataset. Combinations including the H, L, or RGB modalities exhibit the best performance on the Sims4Action dataset, whereas OF and YOLO are weaker. However, combinations including the RGB modality seem to have an overall lower performance on the REAL datasets, perhaps due to the large appearance change. Combinations with the YOLO modality show the best performance for both REAL test sets. Inspecting the results in Table 1 is tedious and prone to misinterpretation or confirmation bias [58]. It is also possible to overlook important tendencies. Hence, we show in Sect. 4.2 how our quantification study tackles these problems by systematically disentangling the modalities with a positive contribution $\mathcal{M}^+$ from the rest $\mathcal{M}^-$.

4.2   Quantification Study: Results

We use the results from Table 1 for the $acc(\cdot)$ term in Eq. 2 and compute the contribution of each modality $f(m)$. We do this for all late fusion strategies and all three test datasets and plot the results in Fig. 3. The SYNTHETIC→REAL domain gap is clearly seen in the substantial difference in the height of the bars in the test split of Sims4Action [70] compared to the REAL test datasets. The limbs and RGB modalities have the largest contribution on Sims4Action [70], followed by the heatmaps. The only modalities with negative contributions are the optical flow and YOLO, where YOLO reaches a drastic drop of over (-10%) for the Squared Sum and Maximum late fusion methods. We conclude that YOLO and optical flow have a negative contribution on Sims4Action [70].

The results on the REAL test datasets Toyota [21] and ETRI [44] show different tendencies. The contributions are smaller due to the domain shift, especially

Table 1. Results for the action classifiers trained on Sims4Action [70] in the mPCA metric. The late fusion results are averaged over the 6 fusion strategies discussed in Sect. 3.3. H: Heatmaps, L: Limbs, OF: Optical Flow.

	mPCA [%]						
	SYNTHETIC	REAL			SYNTHETIC	REAL	
Test Set	Sims4Action [70]	Toyota [21]	ETRI [44]	Test Set	Sims4Action [70]	Toyota [21]	ETRI [44]
Modalities				Modalities			
H	71.38	20.23	12.12	H L OF RGB	96.95	26.00	16.22
L	75.09	22.00	12.88	H L OF YOLO	90.05	28.98	20.27
OF	44.50	21.34	9.28	H L RGB YOLO	92.88	25.36	18.55
RGB	61.79	13.74	9.37	H OF RGB YOLO	91.14	26.10	17.20
YOLO	50.54	26.08	38.25	L OF RGB YOLO	92.56	26.12	18.56
H L	91.48	23.44	16.95	H L OF	94.03	26.70	17.26
H OF	89.86	25.97	15.10	H L RGB	95.72	23.25	15.78
H RGB	94.44	21.38	13.91	H OF RGB	95.41	23.73	13.79
L OF	90.84	25.60	15.90	L OF RGB	95.61	24.09	15.49
L RGB	94.42	20.58	15.46	H L YOLO	86.57	25.80	20.46
OF RGB	86.15	18.50	12.85	H OF YOLO	82.97	29.48	19.91
H YOLO	74.86	25.51	20.55	H RGB YOLO	88.79	23.27	17.77
L YOLO	80.25	24.77	19.91	L OF YOLO	86.31	28.51	20.68
OF YOLO	62.21	27.56	20.34	L RGB YOLO	90.59	23.29	19.09
RGB YOLO	76.23	17.29	19.41	OF RGB YOLO	82.37	21.15	16.92
H L OF RGB YOLO	94.27	27.52	18.53				

on the ETRI dataset. The RGB modality has explicitly negative contributions on both REAL datasets. We hypothesize that this is due to the appearance changes when transitioning from synthetic to real data. Apart from RGB, optical flow also has a consistently negative contribution on the ETRI dataset. An interesting observation is that the domain gap is much larger on ETRI than on Toyota. A reason for this might be that Roitberg et al. [70] design Sims4Action specifically as a SYNTHETIC→REAL domain adaptation benchmark to Toyota Smarthome [21], e.g., in Sims4Action the rooms are furnished the same way as in Toyota Smarthome. Our results indicate that optical flow has a negative contribution in ETRI, whereas RGB is negative in both REAL datasets. The average contribution of each modality over the 6 fusion methods is in Table 2(a).

4.3   Results from ModSelect: Unsupervised Modality Selection

Table 2(a) shows which modalities have a negative contribution on each target test dataset. However, to estimate these values we needed the performance, and hence, the labels for the target test sets. In this section, we show how to select only the modalities with a positive contribution without using any labels.

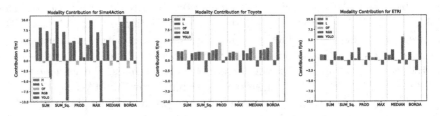

Fig. 3. Quantification study: quantification of the contribution of each modality for 6 late fusion methods and on three different test sets. The height of each bar corresponds to the contribution value $f(m)$ which is computed with Eq. 2.

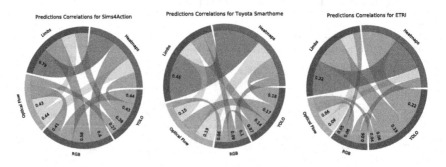

Fig. 4. Chord plots of the prediction correlations $\rho(m, n)$ for all modality pairs (m, n). The thickness of each arch corresponds to the correlation $\rho(m, n)$ between its two endpoint modalities m and n. Each value is computed according to Eq. 4.

Predictions Correlation. We utilize the predictions correlation metric $\rho(m, n)$ and compute it for all modality pairs using Eq. 4. The results for all datasets are illustrated in the chord diagrams in Fig. 4. The chord diagrams allow us to identify the same tendencies, which we observed in our quantification study. Each arch connects two modalities (m, n) and its thickness corresponds to the value $\rho(m, n)$. The YOLO modality has the weakest correlations on Sims4Action [70], depicted in the thinner green arches. Optical flow also exhibits weaker correlations compared to the heatmaps, limbs, and RGB. We see significantly thinner arches for the RGB modality in both REAL test datasets, and for optical flow in ETRI, which matches our results in Table 2(a).

However, simply inspecting the chord plots is not a systematic method for modality selection. Hence, we first compute the aggregated correlations $\rho(m)$ with Eq. 7, which constitute the set A_ρ. Then, we compute the aggregated threshold $\delta^\rho_{\text{agg}} := \mu_{0.2}(A_\rho)$ using the 0.2-Winsorized Mean [82] from Eq. 8. We compute these terms for each test dataset and obtain three thresholds. The aggregated correlations $\rho(m)$ and their thresholds δ^ρ_{agg} for each test dataset are illustrated in Table 2(b). The modalities underneath the thresholds are exactly the ones with negative contributions from Table 2(a).

Table 2. (a) Average contribution $f(m)$ over the 6 late fusion methods of each modality m. Negative contributions are colored in red. (b) Aggregated prediction correlation $\rho(m)$ values for each modality m on the three test datasets and the aggregated thresholds δ^ρ_{agg}. Values **below** the threshold are colored in red. (c) Aggregated $MMD(m)$ values for each modality m on the three test datasets and the aggregated thresholds $\delta^{MMD}_{\text{agg}}$. Values **above** the threshold are colored in red.

Test Dataset	(a) Contribution $f(m)$					(b) Aggregated $\rho(m)$						(c) Aggregated $MMD(m)$				
	H	L	OF	RGB	YOLO	H	L	OF	RGB	YOLO	δ^ρ_{agg}	H	L	OF	RGB	$\delta^{MMD}_{\text{agg}}$
Sims4Action [70]	4.37	8.86	-0.74	6.98	-2.57	0.57	0.55	0.38	0.50	0.37	0.40	9.49	8.12	13.07	9.92	10.15
Toyota [21]	2.14	2.46	2.90	-1.86	2.13	0.23	0.21	0.14	0.08	0.14	0.10	11.93	11.47	13.34	20.79	14.38
ETRI [44]	0.76	1.60	-0.13	-1.17	2.02	0.14	0.14	0.06	0.05	0.13	0.08	17.84	17.76	22.04	24.91	20.64

We also show that applying the pairs-threshold $\delta^\rho_{\text{pair}}$ leads to the same results. We skip the aggregation step of δ^ρ_{agg} and compose the A_ρ set out of the $\rho(m, n)$ values, i.e. we focus on *modality pairs* instead of *individual modalities*. We compute the threshold $\delta^\rho_{\text{pair}} := \mu_{0.2}(A_\rho)$ again with the 0.2-Winsorized Mean [82] from Eq. 8. The correlation values $\rho(m, n)$ as well as the accuracies of all bimodal action classifiers from Table 1 are shown in Fig. 5. The pairs-thresholds $\delta^\rho_{\text{pair}}$ for each test set are drawn as dashed lines and divide the modality pairs into two groups. The modality pairs below the thresholds are marked with an × so that it is possible to identify which pairs are selected.

For Sims4Action [70], optical flow (OF) and RGB show a negative $f(m)$ in Table 2(a) and are also discarded by δ^ρ_{agg}. Figure 5 shows that all modality pairs containing either OF or RGB are below the threshold, i.e. $\delta^\rho_{\text{pair}}$ discards the same modalities as δ^ρ_{agg} for Sims4Action. The same is true for the REAL test datasets, where all models containing RGB are discarded in Toyota and ETRI, as well as OF for ETRI. Another observation is that the majority of"peaks" in the yellow *accuracy* lines coincide with the peaks in the blue *correlation* lines. This result is in agreement with our theory that a high correlation between correct predictions is statistically more likely than a high correlation between wrong predictions, since there is one only correct class and multiple incorrect ones. Moreover, while we do achieve the same results with both thresholds, we recommend using $\delta^{\rho}_{\mathbf{agg}}$ when discarding an entire input modality, e.g. a faulty sensor in a multi-sensor setup, and $\delta^{\rho}_{\mathbf{pair}}$ when searching for the *best synergies* from all modality combinations.

Domain Discrepancy. The second metric we use to discern the contributing modalities from the rest is the Maximum Mean Discrepancy (MMD) [36] between the embeddings of the action classifiers. Note that the YOLO modality is not included in this experiment as its MLP model's embedding size is different that the other 4 *image-based* modalities. While MMD is widely used as a loss term for minimizing the domain gap between source and target domains [17,79,87], a large MMD is also associated with a decline in performance in fusion methods [53,92]. To utilize the MMD metrics to separate the modalities, we first compute $MMD(m, n)$ for all modality pairs and the aggregated discrepancies $MMD(m)$ with Eqs. 6 and 7. We then compute the pairs- $\delta^{MMD}_{\mathbf{pair}}$ and aggregated $\delta^{MMD}_{\mathbf{agg}}$ thresholds with the 0.2-Winsorized Mean [82] from Eq. 8 the same way as we did for the predictions' correlations ρ.

The $MMD(m, n)$ values for all modality pairs in the three test datasets are illustrated in Fig. 6. The higher discrepancy values are clearly apparent by their bright colors and contrast to the rest of the values. Optical flow has the largest values on Sims4Action, and RGB has the highest discrepancy on Toyota and ETRI. Optical flow also exhibits a high discrepancy on ETRI. Once again, we can see that the domain gap to the ETRI dataset is much larger, which is manifested in drastically higher $MMD(m, n)$ values. The pairs-thresholds $\delta^{MMD}_{\mathbf{pair}}$ for the three datasets are $\{13.68, 19.18, 27.50\}$ and separate exactly the same modality

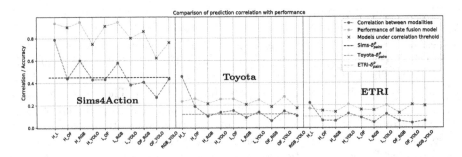

Fig. 5. Prediction correlations $\rho(m, n)$ between all modality pairs and the late fusion accuracy of the bi-modal action classifiers. The pairs-thresholds $\delta^{\rho}_{\mathbf{pair}}$ are depicted as dashed lines. Pairs under the thresholds are crossed out with an $\times$ in the yellow line. (Color figure online)

pairs as our quantification study and the $\delta^\rho_{\mathbf{pair}}$ thresholds, with the exception of the (H,OF) pair in ETRI. The aggregated discrepancies $MMD(m)$ for each modality and the aggregated thresholds are listed in Table 2(c), where the values above the thresholds are colored in red. The red values coincide exactly with the negative contributions $f(m)$ from our quantification study in Table 2(a).

ModSelect: Unsupervised Modality Selection. Finally, we select the modalities with either the aggregated $\delta_{\mathbf{agg}}$ or the pairs-thresholds $\delta_{\mathbf{pair}}$, by constructing the consensus between our two metrics ρ and MMD (see Eqs. 9 and 10). The selected modalities $\mathcal{M}_{\mathbf{agg}}$ with our aggregated thresholds $\delta_{\mathbf{agg}}$ and selected modality pairs $\mathcal{C}_{\mathbf{pair}}$ with our pairs-thresholds $\delta_{\mathbf{pair}}$ are listed in Table 3. The selected modalities $\mathcal{M}_{\mathbf{agg}}$ from our aggregated thresholds $\delta_{\mathbf{agg}}$ are exactly the ones with a positive contribution $\mathcal{M}^+$ in Table 2(a) from our quantification study in Sect. 4.2. The pairs-thresholds $\delta_{\mathbf{pair}}$ have selected only modality pairs $\mathcal{C}_{\mathbf{pair}}$ which are constituted out of modalities from $\mathcal{M}^+$, which means that $\mathcal{C}_{\mathbf{pair}}$ contains only pairs of modalities (m, n) with positive contributions, i.e. $f(m), f(n) > 0$. In other words, our proposed unsupervised modality selection is able to select only the modalities with positive contributions by utilizing the predictions correlation and MMD between the embeddings of the unimodal action classifiers, without the need of any ground-truth labels on the test datasets.

Impact on the Multimodal Accuracy. The impact of the unsupervised modality selection on the mean multimodal accuracy can be seen in Table 3. Selecting the modalities with our proposed thresholds leads to an average improvement of 5.2%, 3.6%, and 4.3% for Sims4Action [70], Toyota [21], and ETRI [44] respectively. This is a substantial improvement, given the low accuracies on the REAL test datasets due to the synthetic-to-real domain gap. These results confirm that our modality selection approach is able to discern between good and bad sources of information, even in the case of a large distributional shift.

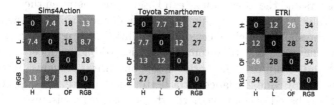

Fig. 6. Maximum Mean Discrepancy values $MMD(m, n)$ computed with Eq. 6 for all modality pairs (m, n). Warmer colors correspond to a higher discrepancy.

Table 3. Results from ModSelect: Selected modalities $\mathcal{M}_{agg}$ with δ_{agg} and selected modality pairs $\mathcal{C}_{pair}$ with δ_{pair} computed with Eqs. 9 and 10.

			Average multimodal accuracy	
Test dataset	$\mathcal{M}_{agg}$	$\mathcal{C}_{pair}$	All modalities $\mathcal{M}$	Ours: $\mathcal{M}_{agg}$
Sims4Action [70]	{H, L, RGB}	$\{(m,n) \in \mathcal{M}^2 \mid m \in \mathcal{M}^+ \wedge n \in \mathcal{M}^+\}$	85.7%	90.9% (+5.2%)
Toyota [21]	{H, L, OF, YOLO}		22.9%	26.5% (+3.6%)
ETRI [44]	{H, L, YOLO}		17.7%	22.0% (+4.3%)

5   Limitations and Conclusion

Limitations. A limitation of our work is that the contributions quantification metric relies on the evaluation results on the test datasets, i.e., the ground-truth labels are needed to calculate the metric. Moreover, the with-without metric $f(m)$ in Eq. 2 requires $\mathcal{O}(M * 2^{M-1})$ computations, where M is the number of modalities. However, one should note that in practice M is not too large, e.g., $M \leq 10$. Additionally, since our method is novel, it has only been tested on the task of cross-domain action recognition. To safely apply our method to other multimodal tasks, e.g., object recognition, future investigations are needed. Moreover, the overall accuracy is still relatively low for cross-domain activity recognition (<30%) and more research is needed for deployment-ready systems.

Conclusion. This is the first systematic study of modality selection in the context of cross-domain activity recognition, aimed at providing guidance for future work in multimodal domain generalization. Our experiments validate our assumption, that cross-domain activity recognition clearly benefits from multimodality, but not all modalities improve the recognition and a systematic modality selection is vital for achieving good results. We proposed a way to measure the contribution of each modality when it is included in a late fusion workflow. The contribution can be used to quantify the importance of each modality and to justify which sources of information are included in a multimodal framework. Our experiments indicate that the correlation between the predictions of unimodal classifiers and the Maximum Mean Discrepancy between their embeddings are both suitable metrics for unsupervised modality selection. The metrics allow to compute thresholds which select only modalities with positive contributions, which opens the possibility to automatically discard bad or uncertain sources of information and to improve the performance on unseen domains. We hope that our findings will provide guidance for a better modality selection process in the future, which is based on more structured and justified decisions.

Acknowledgements. This work was supported by the JuBot project sponsored by the Carl Zeiss Stiftung and Competence Center Karlsruhe for AI Systems Engineering (CC-KING) sponsored by the Ministry of Economic Affairs, Labour and Housing Baden-Württemberg.

References

1. Ahmad, Z., Khan, N.: Human action recognition using deep multilevel multimodal (M^2) fusion of depth and inertial sensors. IEEE Sens. J. **20**(3), 1445–1455 (2019)
2. Ahmad, Z., Khan, N.: CNN-based multistage gated average fusion (MGAF) for human action recognition using depth and inertial sensors. IEEE Sens. J. **21**(3), 3623–3634 (2020)
3. Alayrac, J.B., et al.: Self-supervised multimodal versatile networks. Adv. Neural Inf. Process. Syst. **33**, 25–37 (2020)
4. Alwassel, H., Mahajan, D., Korbar, B., Torresani, L., Ghanem, B., Tran, D.: Self-supervised learning by cross-modal audio-video clustering. Adv. Neural Inf. Process. Syst. **33**, 1–13 (2020)
5. Ardianto, S., Hang, H.M.: Multi-view and multi-modal action recognition with learned fusion. In: 2018 Asia-Pacific Signal and Information Processing Association Annual Summit and Conference (APSIPA ASC), pp. 1601–1604. IEEE (2018)
6. Atrey, P.K., Hossain, M.A., El Saddik, A., Kankanhalli, M.S.: Multimodal fusion for multimedia analysis: a survey. Multimedia Syst. **16**(6), 345–379 (2010)
7. Baradel, F., Wolf, C., Mille, J.: Human action recognition: pose-based attention draws focus to hands. In: IEEE International Conference on Computer Vision Workshops, pp. 604–613 (2017)
8. Black, D., et al.: The theory of committees and elections (1958)
9. Borgwardt, K.M., Gretton, A., Rasch, M.J., Kriegel, H.P., Schölkopf, B., Smola, A.J.: Integrating structured biological data by kernel maximum mean discrepancy. Bioinformatics **22**(14), e49–e57 (2006)
10. Busto, P.P., Iqbal, A., Gall, J.: Open set domain adaptation for image and action recognition. IEEE Trans. Pattern Anal. Mach. Intell. **42**(2), 413–429 (2018)
11. Caba Heilbron, F., Escorcia, V., Ghanem, B., Carlos Niebles, J.: Activitynet: a large-scale video benchmark for human activity understanding. In: Proceedings of the IEEE Conference on Computer Vision and Pattern Recognition, pp. 961–970 (2015)
12. Cai, J., Jiang, N., Han, X., Jia, K., Lu, J.: Jolo-gcn: mining joint-centered lightweight information for skeleton-based action recognition. In: Proceedings of the IEEE/CVF Winter Conference on Applications of Computer Vision, pp. 2735–2744 (2021)
13. Carreira, J., Zisserman, A.: Quo vadis, action recognition? a new model and the kinetics dataset. In: proceedings of the IEEE Conference on Computer Vision and Pattern Recognition. pp. 6299–6308 (2017)
14. Chaquet, J.M., Carmona, E.J., Fernández-Caballero, A.: A survey of video datasets for human action and activity recognition. Comput. Vision Image Underst. **117**(6), 633–659 (2013)
15. Chen, M.H., Kira, Z., AlRegib, G., Yoo, J., Chen, R., Zheng, J.: Temporal attentive alignment for large-scale video domain adaptation. In: Proceedings of the IEEE/CVF International Conference on Computer Vision, pp. 6321–6330 (2019)
16. Chen, M.H., Li, B., Bao, Y., AlRegib, G.: Action segmentation with mixed temporal domain adaptation. In: Proceedings of the IEEE/CVF Winter Conference on Applications of Computer Vision, pp. 605–614 (2020)
17. Chen, Y., Song, S., Li, S., Wu, C.: A graph embedding framework for maximum mean discrepancy-based domain adaptation algorithms. IEEE Trans. Image Process. **29**, 199–213 (2019)

18. Choi, J., Sharma, G., Chandraker, M., Huang, J.B.: Unsupervised and semi-supervised domain adaptation for action recognition from drones. In: Proceedings of the IEEE/CVF Winter Conference on Applications of Computer Vision, pp. 1717–1726 (2020)

19. Choi, J., Sharma, G., Schulter, S., Huang, J.-B.: Shuffle and attend: video domain adaptation. In: Vedaldi, A., Bischof, H., Brox, T., Frahm, J.-M. (eds.) ECCV 2020. LNCS, vol. 12357, pp. 678–695. Springer, Cham (2020). https://doi.org/10.1007/978-3-030-58610-2_40

20. Cormack, G.V., Clarke, C.L., Buettcher, S.: Reciprocal rank fusion outperforms condorcet and individual rank learning methods. In: International ACM SIGIR Conference on Research and Development in Information Retrieval, pp. 758–759 (2009)

21. Das, S., et al.: Toyota smarthome: real-world activities of daily living. In: Proceedings of the IEEE/CVF International Conference on Computer Vision, pp. 833–842 (2019)

22. Das, S., Dai, R., Yang, D., Bremond, F.: VPN++: rethinking video-pose embeddings for understanding activities of daily living. IEEE Trans. Pattern Anal. Mach. Intell. (2021)

23. Das, S., Sharma, S., Dai, R., Brémond, F., Thonnat, M.: VPN: learning video-pose embedding for activities of daily living. In: Vedaldi, A., Bischof, H., Brox, T., Frahm, J.-M. (eds.) ECCV 2020. LNCS, vol. 12354, pp. 72–90. Springer, Cham (2020). https://doi.org/10.1007/978-3-030-58545-7_5

24. Dawar, N., Kehtarnavaz, N.: A convolutional neural network-based sensor fusion system for monitoring transition movements in healthcare applications. In: 2018 IEEE 14th International Conference on Control and Automation (ICCA), pp. 482–485. IEEE (2018)

25. Dawar, N., Ostadabbas, S., Kehtarnavaz, N.: Data augmentation in deep learning-based fusion of depth and inertial sensing for action recognition. IEEE Sens. Lett. 3(1), 1–4 (2018)

26. Delaitre, V., Laptev, I., Sivic, J.: Recognizing human actions in still images: a study of bag-of-features and part-based representations. In: BMVC 2010–21st British Machine Vision Conference (2010)

27. Dhiman, C., Vishwakarma, D.K.: View-invariant deep architecture for human action recognition using two-stream motion and shape temporal dynamics. IEEE Trans. Image Process. 29, 3835–3844 (2020)

28. Duan, H., Zhao, Y., Chen, K., Shao, D., Lin, D., Dai, B.: Revisiting skeleton-based action recognition. arXiv preprint arXiv:2104.13586 (2021)

29. Elekes, Á., Schäler, M., Böhm, K.: On the various semantics of similarity in word embedding models. In: 2017 ACM/IEEE Joint Conference on Digital Libraries (JCDL), pp. 1–10. IEEE (2017)

30. Emerson, P.: The original borda count and partial voting. Social Choice Welfare 40(2), 353–358 (2013)

31. van Erp, M., Vuurpijl, L., Schomaker, L.: An overview and comparison of voting methods for pattern recognition. In: Proceedings Eighth International Workshop on Frontiers in Handwriting Recognition, pp. 195–200 (2002). https://doi.org/10.1109/IWFHR.2002.1030908

32. Fang, H.S., Xie, S., Tai, Y.W., Lu, C.: RMPE: regional multi-person pose estimation. In: ICCV (2017)

33. Farnebäck, G.: Two-frame motion estimation based on polynomial expansion. In: Bigun, J., Gustavsson, T. (eds.) SCIA 2003. LNCS, vol. 2749, pp. 363–370. Springer, Heidelberg (2003). https://doi.org/10.1007/3-540-45103-X_50

34. Gao, R., Oh, T.H., Grauman, K., Torresani, L.: Listen to look: action recognition by previewing audio. In: Proceedings of the IEEE/CVF Conference on Computer Vision and Pattern Recognition, pp. 10457–10467 (2020)
35. Ghifary, M., Kleijn, W.B., Zhang, M.: Domain adaptive neural networks for object recognition. In: Pham, D.-N., Park, S.-B. (eds.) PRICAI 2014. LNCS (LNAI), vol. 8862, pp. 898–904. Springer, Cham (2014). https://doi.org/10.1007/978-3-319-13560-1_76
36. Gretton, A., Borgwardt, K., Rasch, M., Schölkopf, B., Smola, A.: A kernel method for the two-sample-problem. Adv. Neural Inf. Process. Syst. **19** (2006)
37. Gretton, A., Borgwardt, K.M., Rasch, M.J., Schölkopf, B., Smola, A.: A kernel two-sample test. J. Mach. Learn. Res. **13**(1), 723–773 (2012)
38. Gretton, A., et al.: Optimal kernel choice for large-scale two-sample tests. Adv. Neural Inf. Process. Syst. **25** (2012)
39. Han, T., Xie, W., Zisserman, A.: Self-supervised co-training for video representation learning. In: NeurIPS (2020). http://arxiv.org/abs/2010.09709
40. Ho, T.K., Hull, J.J., Srihari, S.N.: Decision combination in multiple classifier systems. IEEE Trans. Pattern Anal. Mach. Intell. **16**(1), 66–75 (1994)
41. Huber, P.J.: Robust estimation of a location parameter. In: Breakthroughs in Statistics, pp. 492–518. Springer, Heidelberg (1992). https://doi.org/10.1007/978-1-4612-4380-9_35
42. Imran, J., Kumar, P.: Human action recognition using rgb-d sensor and deep convolutional neural networks. In: 2016 International Conference on Advances in Computing, Communications and Informatics (ICACCI), pp. 144–148. IEEE (2016)
43. Imran, J., Raman, B.: Evaluating fusion of rgb-d and inertial sensors for multimodal human action recognition. J. Ambient Intell. Hum. Comput. **11**(1), 189–208 (2020)
44. Jang, J., Kim, D., Park, C., Jang, M., Lee, J., Kim, J.: Etri-activity3d: a large-scale rgb-d dataset for robots to recognize daily activities of the elderly. In: 2020 IEEE/RSJ International Conference on Intelligent Robots and Systems (IROS), pp. 10990–10997. IEEE (2020)
45. Kamel, A., Sheng, B., Yang, P., Li, P., Shen, R., Feng, D.D.: Deep convolutional neural networks for human action recognition using depth maps and postures. IEEE Trans. Syst. Man Cybern. Syst. **49**(9), 1806–1819 (2018)
46. Kampman, O., Barezi, E.J., Bertero, D., Fung, P.: Investigating audio, visual, and text fusion methods for end-to-end automatic personality prediction. arXiv preprint arXiv:1805.00705 (2018)
47. Kazakos, E., Nagrani, A., Zisserman, A., Damen, D.: Epic-fusion: audio-visual temporal binding for egocentric action recognition. In: Proceedings of the IEEE/CVF International Conference on Computer Vision, pp. 5492–5501 (2019)
48. Khaire, P., Imran, J., Kumar, P.: Human activity recognition by fusion of RGB, depth, and skeletal data. In: Chaudhuri, B.B., Kankanhalli, M.S., Raman, B. (eds.) Proceedings of 2nd International Conference on Computer Vision & Image Processing. AISC, vol. 703, pp. 409–421. Springer, Singapore (2018). https://doi.org/10.1007/978-981-10-7895-8_32
49. Kim, D., Lee, I., Kim, D., Lee, S.: Action recognition using close-up of maximum activation and etri-activity3d livinglab dataset. Sensors **21**(20), 6774 (2021)
50. Korbar, B., Tran, D., Torresani, L.: Cooperative learning of audio and video models from self-supervised synchronization. In: Bengio, S., Wallach, H., Larochelle, H., Grauman, K., Cesa-Bianchi, N., Garnett, R. (eds.) Advances in Neural Information Processing Systems, vol. 31, pp. 7763–7774. Curran Associates, Inc. (2018)

51. Li, J., Wang, C., Zhu, H., Mao, Y., Fang, H.S., Lu, C.: Crowdpose: efficient crowded scenes pose estimation and a new benchmark. arXiv preprint arXiv:1812.00324 (2018)

52. Li, T., Wang, L.: Learning spatiotemporal features via video and text pair discrimination (2020)

53. Liang, T., Lin, G., Feng, L., Zhang, Y., Lv, F.: attention is not enough: mitigating the distribution discrepancy in asynchronous multimodal sequence fusion. In: Proceedings of the IEEE/CVF International Conference on Computer Vision, pp. 8148–8156 (2021)

54. Long, M., Wang, J., Ding, G., Sun, J., Yu, P.S.: Transfer feature learning with joint distribution adaptation. In: Proceedings of the IEEE International Conference on Computer Vision, pp. 2200–2207 (2013)

55. Martin, M., et al.: Drive&act: a multi-modal dataset for fine-grained driver behavior recognition in autonomous vehicles. In: Proceedings of the IEEE/CVF International Conference on Computer Vision, pp. 2801–2810 (2019)

56. Memmesheimer, R., Theisen, N., Paulus, D.: Gimme signals: discriminative signal encoding for multimodal activity recognition. In: 2020 IEEE/RSJ International Conference on Intelligent Robots and Systems (IROS), pp. 10394–10401. IEEE (2020)

57. Munro, J., Damen, D.: Multi-modal domain adaptation for fine-grained action recognition. In: Proceedings of the IEEE/CVF Conference on Computer Vision and Pattern Recognition, pp. 122–132 (2020)

58. Nickerson, R.S.: Confirmation bias: a ubiquitous phenomenon in many guises. Rev. Gen. Psychol. **2**(2), 175–220 (1998)

59. Pan, B., Cao, Z., Adeli, E., Niebles, J.C.: Adversarial cross-domain action recognition with co-attention. In: AAAI, vol. 34, pp. 11815–11822 (2020)

60. Pan, S.J., Tsang, I.W., Kwok, J.T., Yang, Q.: Domain adaptation via transfer component analysis. IEEE Trans. Neural Netw. **22**(2), 199–210 (2010)

61. Panda, R., et al.: AdaMML: adaptive multi-modal learning for efficient video recognition, pp. 7576–7585 (2021), https://openaccess.thecvf.com/content/ICCV2021/html/Panda_AdaMML_Adaptive_Multi-Modal_Learning_for_Efficient_Video_Recognition_ICCV_2021_paper.html

62. Patrick, M., Asano, Y., Fong, R., Henriques, J.F., Zweig, G., Vedaldi, A.: Multi-modal self-supervision from generalized data transformations. ArXiv abs/2003.04298 (2020)

63. Pham, C., Nguyen, L., Nguyen, A., Nguyen, N., Nguyen, V.-T.: Combining skeleton and accelerometer data for human fine-grained activity recognition and abnormal behaviour detection with deep temporal convolutional networks. Multimedia Tools and Applications **80**(19), 28919–28940 (2021). https://doi.org/10.1007/s11042-021-11058-w

64. Piergiovanni, A., Angelova, A., Ryoo, M.S.: Evolving losses for unsupervised video representation learning. In: Proceedings of the IEEE/CVF Conference on Computer Vision and Pattern Recognition, pp. 133–142 (2020)

65. Rai, N., Adeli, E., Lee, K.H., Gaidon, A., Niebles, J.C.: Cocon: cooperative-contrastive learning. In: Proceedings of the IEEE/CVF Conference on Computer Vision and Pattern Recognition. pp. 3384–3393 (2021)

66. Ramanathan, M., Kochanowicz, J., Thalmann, N.M.: Combining pose-invariant kinematic features and object context features for RGB-D action recognition. Int. J. Mach. Learn. Comput. **9**(1), 44–50 (2019)

67. Rani, S.S., Naidu, G.A., Shree, V.U.: Kinematic joint descriptor and depth motion descriptor with convolutional neural networks for human action recognition. Mater. Today: Proc. **37**, 3164–3173 (2021)
68. Redmon, J., Farhadi, A.: YOLOV3: an incremental improvement. arXiv preprint arXiv:1804.02767 (2018)
69. Reiß, S., Roitberg, A., Haurilet, M., Stiefelhagen, R.: Deep classification-driven domain adaptation for cross-modal driver behavior recognition. In: 2020 IEEE Intelligent Vehicles Symposium (IV), pp. 1042–1047. IEEE (2020)
70. Roitberg, A., Schneider, D., Djamal, A., Seibold, C., Reiß, S., Stiefelhagen, R.: Let's play for action: recognizing activities of daily living by learning from life simulation video games. In: 2021 IEEE/RSJ International Conference on Intelligent Robots and Systems (IROS), pp. 8563–8569. IEEE (2021)
71. Roitberg, A., Somani, N., Perzylo, A., Rickert, M., Knoll, A.: Multimodal human activity recognition for industrial manufacturing processes in robotic workcells. In: Proceedings of the 2015 ACM on International Conference on Multimodal Interaction, pp. 259–266 (2015)
72. Sankaranarayanan, S., Balaji, Y., Jain, A., Lim, S.N., Chellappa, R.: Learning from synthetic data: Addressing domain shift for semantic segmentation. In: Proceedings of the IEEE Conference on Computer Vision and Pattern Recognition, pp. 3752–3761 (2018)
73. Sharma, G., Jurie, F., Schmid, C.: Discriminative spatial saliency for image classification. In: 2012 IEEE Conference on Computer Vision and Pattern Recognition, pp. 3506–3513. IEEE (2012)
74. Song, X., et al.: Spatio-temporal contrastive domain adaptation for action recognition. In: Proceedings of the IEEE/CVF Conference on Computer Vision and Pattern Recognition, pp. 9787–9795 (2021)
75. Sun, Z., Ke, Q., Rahmani, H., Bennamoun, M., Wang, G., Liu, J.: Human action recognition from various data modalities: a review. arXiv preprint arXiv:2012.11866 (2020)
76. Wang, C., Yang, H., Meinel, C.: Exploring multimodal video representation for action recognition. In: International Joint Conference on Neural Networks, pp. 1924–1931. IEEE (2016)
77. Wang, L., Ding, Z., Tao, Z., Liu, Y., Fu, Y.: Generative multi-view human action recognition. In: Proceedings of the IEEE/CVF International Conference on Computer Vision, pp. 6212–6221 (2019)
78. Wang, P., Li, W., Gao, Z., Zhang, Y., Tang, C., Ogunbona, P.: Scene flow to action map: A new representation for rgb-d based action recognition with convolutional neural networks. In: IEEE Conference on Computer Vision and Pattern Recognition, pp. 595–604 (2017)
79. Wang, W., Li, H., Ding, Z., Wang, Z.: Rethink maximum mean discrepancy for domain adaptation. arXiv preprint arXiv:2007.00689 (2020)
80. Wang, X., He, J., Jin, Z., Yang, M., Wang, Y., Qu, H.: M2lens: visualizing and explaining multimodal models for sentiment analysis. IEEE Trans. Vis. Comput. Graph. **28**(1), 802–812 (2021)
81. Wei, H., Jafari, R., Kehtarnavaz, N.: Fusion of video and inertial sensing for deep learning-based human action recognition. Sensors **19**(17), 3680 (2019)
82. Wilcox, R.R., Keselman, H.: Modern robust data analysis methods: measures of central tendency. Psychol. Methods **8**(3), 254 (2003)
83. Wu, P., Liu, H., Li, X., Fan, T., Zhang, X.: A novel lip descriptor for audio-visual keyword spotting based on adaptive decision fusion. IEEE Trans. Multimedia **18**(3), 326–338 (2016)

84. Xiao, F., Lee, Y.J., Grauman, K., Malik, J., Feichtenhofer, C.: Audiovisual slowfast networks for video recognition. arXiv preprint arXiv:2001.08740 (2020)
85. Xie, S., Sun, C., Huang, J., Tu, Z., Murphy, K.: Rethinking spatiotemporal feature learning: speed-accuracy trade-offs in video classification. In: ECCV, pp. 305–321 (2018)
86. Xiu, Y., Li, J., Wang, H., Fang, Y., Lu, C.: Pose flow: efficient online pose tracking. In: BMVC (2018)
87. Yan, H., Ding, Y., Li, P., Wang, Q., Xu, Y., Zuo, W.: Mind the class weight bias: weighted maximum mean discrepancy for unsupervised domain adaptation. In: Proceedings of the IEEE Conference on Computer Vision and Pattern Recognition, pp. 2272–2281 (2017)
88. Yao, Z., Wang, Y., Wang, J., Yu, P., Long, M.: VIDEODG: generalizing temporal relations in videos to novel domains. IEEE Trans. Pattern Anal. Mach. Intell. **44**, 7989–8004 (2021)
89. Ye, J., Li, K., Qi, G.J., Hua, K.A.: Temporal order-preserving dynamic quantization for human action recognition from multimodal sensor streams. In: Proceedings of the 5th ACM on International Conference on Multimedia Retrieval, pp. 99–106 (2015)
90. Yi, C., Yang, S., Li, H., Tan, Y.P., Kot, A.: Benchmarking the robustness of spatial-temporal models against corruptions. arXiv preprint arXiv:2110.06513 (2021)
91. Zhao, C., Chen, M., Zhao, J., Wang, Q., Shen, Y.: 3D behavior recognition based on multi-modal deep space-time learning. Appl. Sci. **9**(4), 716 (2019)
92. Zheng, Y.: Methodologies for cross-domain data fusion: an overview. IEEE Trans. Big Data **1**(1), 16–34 (2015)
93. Zou, H., Yang, J., Prasanna Das, H., Liu, H., Zhou, Y., Spanos, C.J.: Wifi and vision multimodal learning for accurate and robust device-free human activity recognition. In: Proceedings of the IEEE/CVF Conference on Computer Vision and Pattern Recognition Workshops (2019)
94. Zou, Q., Wang, Y., Wang, Q., Zhao, Y., Li, Q.: Deep learning-based gait recognition using smartphones in the wild. IEEE Trans. Inf. Forensics Secur. **15**, 3197–3212 (2020)

Consistency Regularization for Domain Adaptation

Kian Boon Koh[1,2]($\boxtimes$) and Basura Fernando[1,2]

[1] Institute of High Performance Computing, A*STAR, Singapore, Singapore
kianboonkoh@gmail.com, fernando_basura@ihpc.a-star.edu.sg
[2] Centre for Frontier AI Research, A*STAR, Singapore, Singapore

Abstract. Collection of real world annotations for training semantic segmentation models is an expensive process. Unsupervised domain adaptation (UDA) tries to solve this problem by studying how more accessible data such as synthetic data can be used to train and adapt models to real world images without requiring their annotations. Recent UDA methods applies self-learning by training on pixel-wise classification loss using a student and teacher network. In this paper, we propose the addition of a consistency regularization term to semi-supervised UDA by modelling the inter-pixel relationship between elements in networks' output. We demonstrate the effectiveness of the proposed consistency regularization term by applying it to the state-of-the-art DAFormer framework and improving mIoU19 performance on the GTA5 to Cityscapes benchmark by 0.8 and mIou16 performance on the SYNTHIA to Cityscapes benchmark by 1.2.

1 Introduction

Semantic segmentation is a task which requires a lot of pixel level annotations and obtaining these annotations is expensive and time consuming. To overcome this issue, one solution is to obtain annotations from synthetic data such as Games (e.g. GTA5) and train models on these synthetic data for semantic segmentation. However, the problem is that even if modern synthetic data is near photo realistic, still there is a distribution mismatch between the synthetic data and real images. One solution is to develop models that can overcome this distribution mismatch between models that are trained on synthetic data and real data which is the topic of unsupervised domain adaptation (UDA).

UDA for semantic segmentation has made significant progress in recent years. One of the most recent method called DAFormer [10] obtained massive improvement over prior methods by using a Transformer architecture and self-training. However one of the challenges in self-training is that generated pseudo labels can be wrong and that may result in poor transfer of information from source domain to the target domain. Therefore, it is needed to further regularize the self-training learning process.

L. Karlinsky et al. (Eds.): ECCV 2022 Workshops, LNCS 13808, pp. 347–359, 2023.
https://doi.org/10.1007/978-3-031-25085-9_20

In this work, we present a new consistency regularization method based on correlation between pixel-wise class predictions. We enforce two models (teacher and student) to have similar inter-pixel similarity structure and by doing so we regularize the self-training process. This helps to improve the generalization of the student network as well as the teacher network allowing better transfer of information from the source domain to the target domain. We demonstrate its effectiveness by applying it to DAFormer and improving mIoU19 performance on the GTA5 to Cityscapes benchmark by 0.8 and mIou16 performance on the SYNTHIA to Cityscapes benchmark by 1.2. Implementation of our proposed method is available at our GitHub repository[1].

2   Related Work

2.1   Unsupervised Domain Adaptation

Domain adaptation is a field of techniques that aims to solve the domain shift problem, when data distributions experience change between datasets. UDA is a subset of the domain adaptation field that aims to utilize a labeled source domain to learn a model that performs well on an unlabeled target domain. Recent UDA methods can be grouped into either adversarial training or self-supervised learning (SSL) approaches. Adversarial training methods aim to reduce source and target distribution mismatch by aligning distributions at either the pixel [2,6,8] or intermediate feature level [9,22] using a generative adversarial network (GAN).

SSL methods allow models to be trained directly on the target domain by generating pseudo labels from the target domain. Recent advances focuses on improving the quality of pseudo labels using various approaches, such as using representative prototypes [29] or using more complex, Transformer-based network architecture [10]. It is also possible for methods to adopt a hybrid approach and use both adversarial training and SSL. Li et al. does so in their bidirectional learning framework [12]. Adversarial training is first used to obtain an image-to-image translation model and a segmentation model. Target domain pseudo labels are then generated from high confidence predictions, which are then subsequently used to fine tune the segmentation model. The improved segmentation model can then be used in the first adversarial stage to form a close loop.

2.2   Semantic Segmentation

Early methods on semantic segmentation problems were largely based on Fully Convolutional Network (FCN) [19], which typically follows an encoder-decoder architecture [1,17]. Further improvements were made by using dilated convolutions to overcome the loss of spatial resolution [26], and pyramid pooling [3,30] to enhance capturing of contextual information. Recent success of attention-based

---

[1] https://github.com/kw01sg/CRDA.

Transformers [23] in natural language processing has seen adaptations of Transformers for image segmentation [13,25] that were able to obtain state-of-the-art results.

2.3 Consistency Regularization

Consistency regularization is a regularization technique used to encourage networks to make consistent predictions that are invariant to perturbations. Tarvainen and Valpola improved model performance on the image classification problem by using a student and teacher network pair in their Mean Teacher model [21], where the weights of the teacher network are an exponential moving average (EMA) of the student network. Consistent predictions between the two networks are then promoted by optimizing a consistency loss between their predictions. Interpolation consistency training by Verma et al. [24] combines mixup [28] and the Mean Teacher model [21] to implement consistency regularization. During training, unlabelled samples are interpolated to create an augmented sample. Predictions by the student network on the augmented sample are then optimized to be consistent with interpolated predictions by the teacher network on the original non-interpolated samples. Kim et al. [11] uses cosine similarity in their consistency regularization method for semantic segmentation. They propose a structured consistency loss that optimizes predictions to be consistent in not only pixel-wise classification, but also inter-pixel relationship.

3 Our Method

Given source domain images $x_S \in X_S$ with their annotations (labels) $y_S \in Y_S$ and target domain images $x_T \in X_T$ without annotations (labels), we want to learn a network h that can correctly predict the annotations for target images X_T denoted by $\hat{Y}_T$. Typically, there is a mismatch in the joint probability distributions of source domain data $P(X_S, Y_S)$ and the target domain data $P(X_T, Y_T)$. Due to this mismatch or the gap between source and target domains, an image segmentation model h that is trained on the source data usually results in a low performance on target images. One common solution to address this issue is to use self-training as also done in the prior works such as DAFormer [10]. However, semi-supervised self-training methods could easily over-fit to source distribution and could generate inconsistent or wrong pseudo labels for the target domain images. To overcome this limitation, we propose the addition of consistency regularization to the DAFormer [10] framework during model training to further improve model performance. Next, we explain the overall training framework.

3.1 Overall Training

Overall training of the network is composed of three components: supervised training using source images, self-training using target images, and consistency regularization. Total loss $\mathcal{L}_{total}$ is given as

$$\mathcal{L}_{total} = \mathcal{L}_S + \mathcal{L}_T + \lambda_c \mathcal{L}_C \tag{1}$$

where $\mathcal{L}_S$ is supervised cross entropy loss using source images, $\mathcal{L}_T$ is self-trained cross entropy loss using pseudo labels, $\mathcal{L}_C$ is our consistency regularization term, and λ_c is a parameter we use to weigh $\mathcal{L}_C$. The following sections will present each of the losses in detail.

Supervised Training. Supervised training on the source domain is conducted using cross entropy loss for semantic segmentation. For a source image x_S and its annotation y_S, $\mathcal{L}_S$ can be defined as

$$\mathcal{L}_S(x_S, y_S) = -\frac{1}{HW} \sum_{j=1}^{H \times W} \sum_{c=1}^{C} y_S^{(j,c)} \log h(x_S)^{(j,c)} \qquad (2)$$

where C is the number of classes and H and W are the height and width of the segmentation output. The notation $y_S^{(j,c)}$ denotes the presence of class c at pixel location j (1 if present and 0 if not). Similarly, $h(x_S)^{(j,c)}$ denotes the predicted score for class c at pixel location j using model h for image x_S.

Self-training. Self-training uses a teacher network $f(; \phi)$ to produce pseudo labels on which the student network $h(; \theta)$ will be trained on. For a target image x_T, its pseudo label p_T is formally defined as

$$p_T^{(j,c)} = [\![c = \arg \max_{c'} f(x_T; \phi)^{(j,c')}]\!] \qquad (3)$$

where $[\![\cdot]\!]$ denotes the Iverson bracket.

We follow the Mean Teacher model [21] where the weights of the teacher network $f(; \phi)$ are the EMA of the weights of the student network $h(; \theta)$ after each training step t. The EMA weights used by the teacher model at training step t is formally defined as

$$\phi_{t+1} = \alpha \phi_t + (1 - \alpha) \theta_t \qquad (4)$$

where ϕ_{t+1} is the EMA of successive weights and α is a smoothing coefficient hyperparameter. It should also be noted that no gradients will be backpropagated into the teacher network from the student network.

A confidence estimate for the pseudo labels, defined as the ratio of pixels with maximum softmax probability exceeding a pre-defined threshold τ, is also used in the self-training loss. For a target image x_T, its confidence estimate q_T is formally defined as

$$q_T = \frac{\sum_{j=1}^{H \times W} [\max_{c'} f(x_T; \phi)^{j,c'} > \tau]}{HW} \qquad (5)$$

Self-training loss of the student network $\mathcal{L}_T$ for a target image x_T can thus be defined as

$$\mathcal{L}_T(x_T) = -\frac{1}{HW} \sum_{j=1}^{H \times W} \sum_{c=1}^{C} q_T \times p_T^{(j,c)} \times \log h(x_T; \theta)^{(j,c)} \qquad (6)$$

We follow DAFormer's [10] method of using non-augmented target images for the teacher network f to generate pseudo labels and augmented targeted images to train the student network h using Eq. 6. We also follow their usage of color jitter, Gaussian blur, and ClassMix [15] as data augmentations in our training process.

Consistency Regularization. As mentioned in Mean Teacher [21], cross entropy loss in Eq. 6 between predictions of the student model and pseudo labels (which are predictions from the teacher model) can be considered as a form of consistency regularization. However, different from classification problems, semantic segmentation problems have a property where pixel-wise class predictions are correlated with each other. Thus, we propose to further enhance consistency regularization by focusing on this inter-pixel relationship. Inspired by the method of Kim et al. [11], we use the inter-pixel cosine similarity of networks' predictions on target images to regularize the model. Formally, we define the similarity between pixel i and j class predictions on a target image x_T as

$$a_{i,j} = \frac{\mathbf{p}_i^T \mathbf{p}_j}{\|\mathbf{p}_i\| \cdot \|\mathbf{p}_j\|} \qquad (7)$$

where $a_{i,j}$ represents the cosine similarity between the prediction vector of the ith pixel and the prediction vector of the jth pixel. Note that the similarity between the probability vector $\mathbf{p}_i$ and $\mathbf{p}_j$ can also be computed using Kullback-Leibler (KL) divergence and cross entropy. We investigate these options in Sect. 4.3. The consistency regularization term, $\mathcal{L}_C$ can then be defined as the mean squared error (MSE) between the student network's similarity matrix and the teacher network's similarity matrix

$$\mathcal{L}_C = \frac{1}{(HW)^2} \sum_{i=1}^{H \times W} \sum_{j=1}^{H \times W} \|a_{i,j}^s - a_{i,j}^t\|^2 \qquad (8)$$

where $a_{i,j}^s$ is the similarity obtained from the student network and $a_{i,j}^t$ is the similarity obtained from the teacher network. We also follow the method of Kim et al. [11] to restrict the number of pixels used in the calculation of similarity matrices by performing a random sample of N_{pair} pixels for comparison. Thus, the consistency regularization in Eq. 8 is updated to the following equation

$$\mathcal{L}_C = \frac{1}{(N_{pair})^2} \sum_{i=1}^{N_{pair}} \sum_{j=1}^{N_{pair}} \|a_{i,j}^s - a_{i,j}^t\|^2 \qquad (9)$$

This term $\mathcal{L}_C$ is particularly useful for domain adaptation as it helps to minimize the divergence between the source representation and the target representation by enforcing a structural consistency in the image segmentation task.

4 Experiments

4.1 Implementation Details

Datasets. We use the Cityscapes street scenes dataset [5] as our target domain. Cityscapes contains 2975 training and 500 validation images with resolution of 2048×1024, and is labelled with 19 classes. For our source domain, we use the synthetic datasets GTA5 [16] and SYNTHIA [18]. GTA5 contains 24,966 images with resolution of 1914×1052, and is labelled with the same 19 classes as Cityscapes. For compatibility, we use a variant of SYNTHIA that is labelled with 16 of the 19 Cityscapes classes. It contains 9,400 images with resolution of 1280×760. Following DAFormer [10], we resize images from Cityscapes to 1024×512 pixels and images from GTA5 to 1280×720 pixels before training.

Network Architecture. Our implementation is based on DAFormer [10]. Previous UDA methods mostly used DeepLabV2 [4] or FCN8s [19] network architecture with ResNet [7] or VGG [20] backbone as their segmentation model. DAFormer proposes an updated UDA network architecture based on Transformers that was able to achieve state-of-the-art performance. They hypothesized that self-attention is more effective than convolutions in fostering the learning of domain-invariant features.

Training. We follow DAFormer [10] and train the network with AdamW [14], a learning rate of $\eta_{base} = 6 \times 10^{-5}$ for the encoder and 6×10^{-4} for the decoder, a weight decay of 0.01, linear learning rate warmup with $t_{warm} = 1500$, and linear decay. Images are randomly cropped to 512×512 and trained for 40,000 iterations on a batch size of 2 on a NVIDIA GeForce RTX 3090. We also adopt DAFormer's training strategy of rare class sampling and thing-class ImageNet feature distance to further improve results. For hyperparameters used in self-training, we follow DAFormer and set $\alpha = 0.99$ and $\tau = 0.968$. For hyperparameters used in consistency regularization, we set $N_{pair} = 512$, $\lambda_c = 1.0$ when calculating similarity using cosine similarity and $\lambda_c = 0.8 \times 10^{-3}$ when calculating similarity using KL divergence.

4.2 Results

Table 1. Comparison with other UDA methods on GTA5 to Cityscapes. Results for DAFormer and our method using cosine similarity are averaged over 6 random runs, while results for our method using KL Divergence are averaged over 3 random runs

Method	Road	Sidewalk	Build.	Wall	Fence	Pole	Tr.Light	Sign	Veget.	Terrain	Sky	Person	Rider	Car	Truck	Bus	Train	M.bike	Bike	mIoU19
BDL	91.0	44.7	84.2	34.6	27.6	30.2	36.0	36.0	85.0	43.6	83.0	58.6	31.6	83.3	35.3	49.7	3.3	28.8	35.6	48.5
ProDA	87.8	56.0	79.7	46.3	44.8	45.6	53.5	53.5	88.6	45.2	82.1	70.7	39.2	88.8	45.5	59.4	1.0	48.9	56.4	57.5
DAFormer	95.5	68.9	89.3	**53.2**	49.3	47.8	**55.5**	61.2	89.5	47.7	91.6	71.1	43.3	91.3	67.5	77.6	65.5	53.6	61.2	67.4
Ours (Cosine)	95.5	69.2	**89.5**	52.1	**49.6**	48.9	55.2	**62.1**	89.8	49.0	91.1	**71.7**	**45.1**	**91.7**	**70.0**	77.6	65.2	**56.6**	62.8	68.0
Ours (KL)	**96.1**	**71.6**	**89.5**	**53.2**	48.6	**49.5**	54.7	61.1	**90.0**	**49.4**	**91.7**	70.7	44.0	91.6	**70.0**	**78.1**	**68.9**	55.1	**62.9**	**68.2**

Table 2. Comparison with other UDA methods on SYNTHIA to Cityscapes. Results for DAFormer and our method using cosine similarity are averaged over 6 random runs, while results for our method using KL Divergence are averaged over 3 random runs

Method	Road	Sidewalk	Build.	Wall	Fence	Pole	Tr.Light	Sign	Veget.	Sky	Person	Rider	Car	Bus	M.bike	Bike	mIoU16	mIoU13
BDL	86.0	46.7	80.3	–	–	–	14.1	11.6	79.2	81.3	54.1	27.9	73.7	42.2	25.7	45.3	–	51.4
ProDA	87.8	45.7	84.6	37.1	0.6	44.0	54.6	37.0	88.1	84.4	74.2	24.3	88.2	51.1	40.5	40.5	55.5	62.0
DAFormer	80.5	37.6	87.9	**40.3**	9.1	**49.9**	**55.0**	51.8	85.9	88.4	73.7	**47.3**	87.1	58.1	53.0	61.0	60.4	66.7
Ours (Cosine)	86.3	44.2	**88.3**	39.2	7.5	49.2	54.7	**54.7**	**87.2**	**90.7**	**73.8**	**47.3**	**87.4**	55.9	53.7	60.7	61.3	68.1
Ours (KL)	**89.0**	**49.6**	88.1	**40.3**	7.3	49.2	53.5	52.1	87.0	88.0	**73.8**	46.4	87.1	**58.7**	**53.9**	**61.7**	**61.6**	**68.4**

| Image | DAFormer | Ours | Ground Truth |

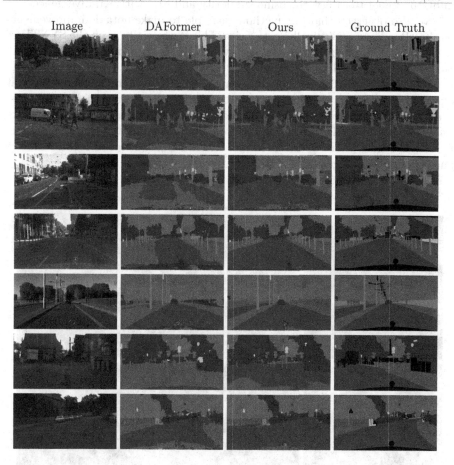

Fig. 1. Qualitative results comparing predictions on validation data of Cityscapes. From left: input image, predictions by DAFormer, predictions by our method, and the ground truth. The last row provides an example where DAFormer performed better compared to our method as it was able to correctly predict the sidewalks

We compare the results of our method against other state-of-the-art UDA segmentation methods such as BDL [12], ProDA [29] and DAFormer [10]. In Table 1, we present our experimental results on the GTA5 to Cityscapes problem. It can be observed that our method improved UDA performance from an mIoU19 of 67.4 to 68.0 when cosine similarity is used in Eq. 7 and 68.2 when KL divergence is used. Table 2 shows our experimental results on the SYNTHIA to Cityscapes problem. Similarly, our method improved performance from an mIoU16 of 60.4 to 61.3 with cosine similarity and 61.6 with KL divergence.

We also observed that our method was able to make notable improvements on the "Road" and "Sidewalk" categories. This is especially so on the SYNTHIA to Cityscapes problem, where we improved UDA performance on "Road" from 80.5 to 89.0 and "Sidewalk" from 37.6 to 49.6. We further verify this improvement in our qualitative analysis presented in Fig. 1, where we observed that our method had better recognition on the "Sidewalk" and "Road" categories. We attribute this improvement to our method's effectiveness in generating more accurate pseudo labels. We present pseudo labels generated during the training process in Fig. 2, where we observed more accurate pseudo labels for the "Road" and "Sidewalk" categories.

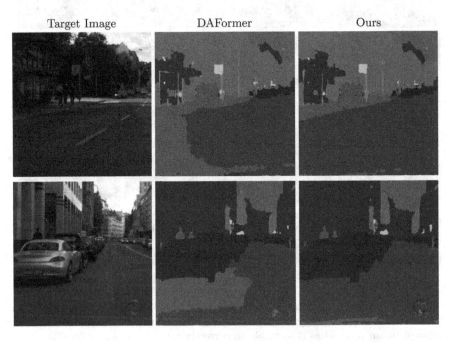

Fig. 2. Qualitative results on pseudo labels generated from the training data of Cityscapes. From left: target image, pseudo labels generated by DAFormer, and pseudo labels generated by our method using cosine similarity

It should be noted that experimental results obtained using the DAFormer method in Tables 1 and 2 were obtained by averaging 6 random runs using the official DAFormer implementation[2]. Even though we were unable to reproduce the exact numbers published in the DAFormer paper, we believe our experimental results for DAFormer are comparable.

4.3   Ablation Study

Table 3. Influence of N_{pair} on UDA performance. Results for all experiments were averaged over 3 random runs except for $N_{pair} = 512$, which was an average over 6 runs

N_{pair}	4	16	64	256	512	1024
mIoU16	**61.4**	60.8	**61.4**	61.1	61.3	60.6

Number of Pixels Sampled. We conducted additional experiments on SYN-THIA to Cityscapes to observe the effect of N_{pair} (from Eq. 9) on model performance. Theoretically, sampling more pixels for similarity calculation (i.e. a larger N_{pair}) allows us to have a more complete model of the inter-pixel relationship between predictions. However, empirical results in Table 3 suggests that N_{pair} does not have significant influence on UDA performance. We observe that very small samples, such as $N_{pair} = 4$, were able to obtain comparable results with larger sample sizes.

Additional experiments using $N_{pair} = 4$ were conducted to observe the locations of sampled pixels. Visualization of our ablation study is presented in Fig. 3. We found that after 40,000 training iterations, sampled pixels covered approximately 45.73% of the 512×512 images the network was trained on despite the small sampling size. This suggests that if a reasonable image coverage can be obtained during the training process, a small N_{pair} is sufficient to model the inter-pixel relationship between predictions, allowing us to minimize computational cost of our consistency regularization method. The influence of sampling coverage and sampling distribution on the effectiveness of consistency regularization is an interesting study that can be explored in the future.

Proximity of Sampled Pixels. Kim et al. adopted cutmix augmentation [27] in their consistency regularization method [11] to limit sampled pixel pairs to within a local region. They theorized that pixel pairs that are in close proximity to each other have high correlation, and hence have more effect on UDA performance. We tested this theory on SYNTHIA to Cityscapes by performing N_{box} crops and sampling N_{pair} pixels from each crop. This localizes sampled pixels and restricts them to have closer proximity. Sampled pixels are then used to compute inter-pixel similarity to obtain a $N_{box} \times N_{pair} \times N_{pair}$ similarity matrix

---

[2] https://github.com/lhoyer/DAFormer

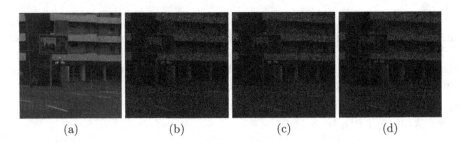

| (a) | (b) | (c) | (d) |

Fig. 3. Visualization of pixels sampled in experiments using $N_{pair} = 4$. (a) Target image cropped to 512×512; (b), (c) and (d) visualizes sampled pixels in three separate runs

Table 4. Influence of N_{box} and N_{pair} on UDA performance. Total number of sampled pixels i.e. $N_{box} \times N_{pair}$ is kept at 512 for a fair comparison

Crop size	N_{box}	N_{pair}	mIoU16
256	32	16	**61.2**
128	32	16	61.1
64	32	16	60.2

which is used for loss calculation in Eq. 8. We present the experimental results in Table 4 where three different crop sizes were varied to restrict the proximity of sampled pixels. We did not observe an improvement in UDA performance compared to results presented in Table 2, suggesting that proximity of sampled pixels perhaps may not be that influential for consistency regularization.

Measuring Inter-pixel Similarity. In Sect. 3.1, we adopted the method of Kim et al. to use cosine similarity in the measure of inter-pixel similarity [11]. In this section, we conducted additional experiments on SYNTHIA to Cityscapes to observe the influence different methods of measuring inter-pixel similarity have on UDA performance.

Table 5. Comparison of UDA performance using different methods to calculate inter-pixel similarity. We also provide the optimal λ_c obtained using hyperparameter tuning

Method	λ_c	mIoU16
Cosine similarity	1.0	61.3
Cross entropy	1.0×10^{-3}	61.2
KL divergence	0.8×10^{-3}	**61.6**

We tested the usage of cross entropy and KL divergence to measure inter-pixel similarity instead of cosine similarity in Eq. 7. Results from our empirical experiments are presented in Table 5. We observed that all three methods provided comparable results with each other, with KL divergence providing slightly better results.

5 Conclusion

In this work we presented a new consistency regularization method for UDA based on relationships between pixel-wise class predictions from semantic segmentation models. Using this technique we were able to improve the performance of the state-of-the-art DAFormer method. We also observed that even with smaller number of sampled pixel pairs N_{pair}, this regularization method was still able to be effective. Therefore, with minimal computational cost, we are able to improve the results of self-training methods for unsupervised domain adaptation.

Acknowledgment. This research is supported by the Centre for Frontier AI Research (CFAR) and Robotics-HTPO seed fund C211518008.

References

1. Badrinarayanan, V., Kendall, A., Cipolla, R.: SegNet: a deep convolutional encoder-decoder architecture for image segmentation. IEEE Trans. Pattern Anal. Mach. Intell. **39**, 2481–2495 (2017)
2. Bousmalis, K., Silberman, N., Dohan, D., Erhan, D., Krishnan, D.: Unsupervised pixel-level domain adaptation with generative adversarial networks. CoRR abs/1612.05424 (2016)
3. Chen, L., Papandreou, G., Kokkinos, I., Murphy, K., Yuille, A.L.: Deeplab: semantic image segmentation with deep convolutional nets, atrous convolution, and fully connected crfs. IEEE Trans. Pattern Anal. Mach. Intell. **40**(04), 834–848 (2018). https://doi.org/10.1109/TPAMI.2017.2699184
4. Chen, L.C., Papandreou, G., Kokkinos, I., Murphy, K., Yuille, A.: Deeplab: semantic image segmentation with deep convolutional nets, atrous convolution, and fully connected crfs. IEEE Trans. Pattern Anal. Mach. Intell. (2016). https://doi.org/10.1109/TPAMI.2017.2699184
5. Cordts, M., et al.: The cityscapes dataset for semantic urban scene understanding. In: Proceedingse IEEE Conference on Computer Vision and Pattern Recognition (CVPR) (2016)
6. Gong, R., Li, W., Chen, Y., Van Gool, L.: Dlow: domain flow for adaptation and generalization. In: 2019 IEEE/CVF Conference on Computer Vision and Pattern Recognition (CVPR), pp. 2472–2481 (2019). https://doi.org/10.1109/CVPR.2019.00258
7. He, K., Zhang, X., Ren, S., Sun, J.: Deep residual learning for image recognition. In: 2016 IEEE Conference on Computer Vision and Pattern Recognition (CVPR), pp. 770–778 (2016). https://doi.org/10.1109/CVPR.2016.90

8. Hoffman, J., et aal.: CyCADA: cycle-consistent adversarial domain adaptation. In: Dy, J., Krause, A. (eds.) Proceedings of the 35th International Conference on Machine Learning. Proceedings of Machine Learning Research, vol. 80, pp. 1989–1998. PMLR (2018). https://proceedings.mlr.press/v80/hoffman18a.html

9. Hoffman, J., Wang, D., Yu, F., Darrell, T.: FCNS in the wild: pixel-level adversarial and constraint-based adaptation (2016)

10. Hoyer, L., Dai, D., Gool, L.V.: Daformer: improving network architectures and training strategies for domain-adaptive semantic segmentation. CoRR **abs/2111.14887** (2021). https://arxiv.org/abs/2111.14887

11. Kim, J., Jang, J., Park, H.: Structured consistency loss for semi-supervised semantic segmentation. CoRR **abs/2001.04647** (2020). https://arxiv.org/abs/2001.04647

12. Li, Y., Yuan, L., Vasconcelos, N.: Bidirectional learning for domain adaptation of semantic segmentation. In: 2019 IEEE/CVF Conference on Computer Vision and Pattern Recognition (CVPR), pp. 6929–6938 (2019). https://doi.org/10.1109/CVPR.2019.00710

13. Liu, Z., et al.: Swin transformer: hierarchical vision transformer using shifted windows. In: Proceedings of the IEEE/CVF International Conference on Computer Vision (ICCV) (2021)

14. Loshchilov, I., Hutter, F.: Decoupled weight decay regularization. In: International Conference on Learning Representations (2019). https://openreview.net/forum?id=Bkg6RiCqY7

15. Olsson, V., Tranheden, W., Pinto, J., Svensson, L.: Classmix: segmentation-based data augmentation for semi-supervised learning. In: 2021 IEEE Winter Conference on Applications of Computer Vision (WACV), pp. 1368–1377 (2021). https://doi.org/10.1109/WACV48630.2021.00141

16. Richter, S.R., Vineet, V., Roth, S., Koltun, V.: Playing for data: ground truth from computer games. In: Leibe, B., Matas, J., Sebe, N., Welling, M. (eds.) ECCV 2016. LNCS, vol. 9906, pp. 102–118. Springer, Cham (2016). https://doi.org/10.1007/978-3-319-46475-6_7

17. Ronneberger, O., Fischer, P., Brox, T.: U-net: convolutional networks for biomedical image segmentation. In: Navab, N., Hornegger, J., Wells, W.M., Frangi, A.F. (eds.) MICCAI 2015. LNCS, vol. 9351, pp. 234–241. Springer, Cham (2015). https://doi.org/10.1007/978-3-319-24574-4_28

18. Ros, G., Sellart, L., Materzynska, J., Vazquez, D., Lopez, A.M.: The synthia dataset: a large collection of synthetic images for semantic segmentation of urban scenes. In: 2016 IEEE Conference on Computer Vision and Pattern Recognition (CVPR), pp. 3234–3243 (2016). https://doi.org/10.1109/CVPR.2016.352

19. Shelhamer, E., Long, J., Darrell, T.: Fully convolutional networks for semantic segmentation. IEEE Trans. Pattern Anal. Mach. Intell. **39**(4), 640–651 (2017). https://doi.org/10.1109/TPAMI.2016.2572683

20. Simonyan, K., Zisserman, A.: Very deep convolutional networks for large-scale image recognition. arXiv 1409.1556 (2014)

21. Tarvainen, A., Valpola, H.: Weight-averaged consistency targets improve semi-supervised deep learning results. CoRR abs/1703.01780 (2017)

22. Tzeng, E., Hoffman, J., Saenko, K., Darrell, T.: Adversarial discriminative domain adaptation. CoRR abs/1702.05464 (2017)

23. Vaswani, A., et al.: Attention is all you need. In: Guyon, I., et al. (eds.) Advances in Neural Information Processing Systems, vol. 30. Curran Associates, Inc. (2017). https://proceedings.neurips.cc/paper/2017/file/3f5ee243547dee91fbd053c1c4a845aa-Paper.pdf

24. Verma, V., Lamb, A., Kannala, J., Bengio, Y., Lopez-Paz, D.: Interpolation consistency training for semi-supervised learning. In: Proceedings of the Twenty-Eighth International Joint Conference on Artificial Intelligence, IJCAI-19, pp. 3635–3641. International Joint Conferences on Artificial Intelligence Organization (2019). https://doi.org/10.24963/ijcai.2019/504
25. Xie, E., Wang, W., Yu, Z., Anandkumar, A., Alvarez, J.M., Luo, P.: Segformer: simple and efficient design for semantic segmentation with transformers. arXiv preprint arXiv:2105.15203 (2021)
26. Yu, F., Koltun, V.: Multi-scale context aggregation by dilated convolutions. In: ICLR (2016)
27. Yun, S., Han, D., Oh, S.J., Chun, S., Choe, J., Yoo, Y.: Cutmix: regularization strategy to train strong classifiers with localizable features. In: Proceedings of the IEEE/CVF International Conference on Computer Vision (ICCV) (2019)
28. Zhang, H., Cisse, M., Dauphin, Y.N., Lopez-Paz, D.: mixup: beyond empirical risk minimization. In: International Conference on Learning Representations (2018). https://openreview.net/forum?id=r1Ddp1-Rb
29. Zhang, P., Zhang, B., Zhang, T., Chen, D., Wang, Y., Wen, F.: Prototypical pseudo label denoising and target structure learning for domain adaptive semantic segmentation. arXiv preprint arXiv:2101.10979 (2021)
30. Zhao, H., Shi, J., Qi, X., Wang, X., Jia, J.: Pyramid scene parsing network. In: 2017 IEEE Conference on Computer Vision and Pattern Recognition (CVPR), pp. 6230–6239 (2017). https://doi.org/10.1109/CVPR.2017.660

W37 - Vision With Biased or Scarce Data

W37 - Vision With Biased or Scarce Data

With the increasing appetite for data in data-driven methods, the issues of biased and scarce data have become a major bottleneck in developing generalizable and scalable computer vision solutions, as well as effective deployment of these solutions in real-world scenarios. To tackle these challenges, researchers from both academia and industry must collaborate and make progress in fundamental research and applied technologies. The organizing committee and keynote speakers of VBSD 2022 consist of experts from both academia and industry with rich experiences in designing and developing robust computer vision algorithms and tranferring them to real-world solutions. VBSD 2022 provides a focused venue to discuss and disseminate research related to bias and scarcity topics in computer vision.

October 2022   Kuan-Chuan Peng
                                                                     Ziyan Wu

CAT: Controllable Attribute Translation for Fair Facial Attribute Classification

Jiazhi Li[1,2,3](✉)(iD) and Wael Abd-Almageed[1,2,3](iD)

[1] USC Ming Hsieh Department of Electrical and Computer Engineering,
Los Angeles, USA
[2] USC Information Sciences Institute, Marina del Rey, USA
{jiazli,wamageed}@isi.edu
[3] Visual Intelligence and Multimedia Analytics Laboratory, Marina del Rey, USA

Abstract. As the social impact of visual recognition has been under scrutiny, several *protected-attribute* balanced datasets emerged to address *dataset bias* in imbalanced datasets. However, in facial attribute classification, *dataset bias* stems from both *protected attribute* level and facial attribute level, which makes it challenging to construct a multi-attribute-level balanced real dataset. To bridge the gap, we propose an effective pipeline to generate high-quality and sufficient facial images with desired facial attributes and supplement the original dataset to be a balanced dataset at both levels, which theoretically satisfies several fairness criteria. The effectiveness of our method is verified on sex classification and facial attribute classification by yielding comparable task performance as the original dataset and further improving fairness in a comprehensive fairness evaluation with a wide range of metrics. Furthermore, our method outperforms both resampling and balanced dataset construction to address *dataset bias*, and debiasing models to address *task bias*.

Keywords: Fairness · Synthetic face image generation

1 Introduction

Bias issues in machine learning methods involving *protected attributes* [7,46] (*e.g.*, sex and race) have lately garnered tremendous attention [5,19,56], such as in facial attribute classification [48,61] and sex classification [7] on face dataset. Many studies [7,61] show that one of the most obvious phenomena of bias issues in these downstream tasks is that the underlying learning algorithms yield uneven performance for different demographic groups and relatively worse performance for minorities. Such unfair performance across different demographic cohorts has hampered credibility of computer vision algorithms on face dataset from both individuals and the whole society, which is urgently to be resolved.

Ensuring fairness of visual recognition on face dataset, has aroused great interest in the machine learning [6,14,15,26,33,37,64,65,68]. Without loss of

---

Supplementary Information The online version contains supplementary material available at https://doi.org/10.1007/978-3-031-25085-9_21.

generality, there are two main challenges with respect to two distinct types of bias—(1) lack of generalized learned representations due to the spurious correlation between prediction target and confounding factors, which could include sensitive attribute (*e.g.*, sex) in training dataset, referred to as *sensitive attribute bias* and *task bias* [3,55,70], and (2) uneven performance across cohorts caused by insufficient samples for particular demographic groups in the existing dataset, referred to as *minority group bias* or *dataset bias* [1,49,53].

Mainstream debiasing methods [31, 55,70] focus on mitigating *task bias*, with less emphasis on tackling *dataset bias*. Furthermore, although these methods effectively address *task bias* [61], they are limited in addressing the issue of insufficient samples of minority demographics in the training dataset, as demonstrated in Table 3 with high bias scores. Instead, in the specific scenarios involving *dataset bias*, constructing a balanced dataset is a more appropriate way. For example, some work mitigate sex distribution differences or race distribution differences by either collecting fair dataset [29,67] or strategically resampling from existing datasets [6,37]. However, merely maintaining the balance across the *protected attributes* may not resolve the risk of *dataset bias* which may be implicitly derived

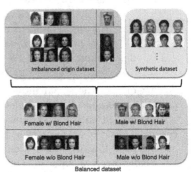

Fig. 1. As the original facial attribute dataset may be imbalanced at both *protected attribute* level (*e.g.*, sex) and facial attribute level (*e.g.*, blond hair), which induce *dataset bias*, we propose a pipeline to generate high-quality synthetic datasets with sufficient samples in a balanced distribution of both levels to rectify skew of the original dataset.

from imbalance on other facial attributes (*e.g.*, blond hair or chubby) [48]. Furthermore, due to the recourse limitations in academia to collect sufficient real images for minority groups, invasion of privacy while collecting more real data and the impossibility of covering all known and unknown confounding factors, it may be impractical to construct a balanced real dataset which is balanced across both *protected attributes* and all facial attributes so as to raise the research scenario to address *dataset bias*.

We therefore propose a method to iteratively generate high-quality synthetic images with desired facial attributes and combine the synthetic dataset with the original dataset to create a semi-synthetic balanced dataset by supplementing insufficient samples in both *protected attribute* level and facial attribute level for downstream tasks, as illustrated in Fig. 1. We use facial attribute classification and sex classification on face dataset as the proxy problem to discuss the effectiveness of our method instead of face recognition since the generation of synthetic face recognition ground truth relies on pre-trained models [23,66], which may be already biased involving *protected attributes* in debiasing face recognition literature. On the other hand, the generation of facial attribute ground truth of our synthetic dataset is bundled with synthetic images directly rather than rely on classifiers, as elaborated in Sect. 3. To construct semi-synthetic datasets which are balanced at both *protected attribute* level and facial attribute level, our method explicitly controls the facial attributes during image generation. Compared with debiasing models that focus on the process to learn fairly, our method to investigate training dataset to achieve fairness is super straightforward and effective, which is easier for generalization. Furthermore, compared with fair dataset collection restricted by long-tail distribution [71] and limited resources, the advantages of semi-synthetic dataset allow our method to generate sufficient face images with more diversity of facial attributes. Finally, different from data augmentation methods by adding slightly modified copies in existing datasets, we create a generative pipeline to first learn the facial attribute appearance of real data and then generate synthetic images with desired facial attributes. The key contributions of this paper can be summarized as follows:

- A pipeline to construct semi-synthetic face datasets to introduce fairness in facial attribute classification and sex classification.
- An investigation of the nature of *dataset bias* stemming from the facial attribute-level imbalance instead of the *protected attribute*-level imbalance.
- A comprehensive fairness evaluation for a wide range of debiasing techniques.

2 Related Work

Debiasing Face Datasets. Long tail distribution [71] leads to inequity towards minorities in face datasets [39,41], and several methods have been designed to address *dataset bias* by collecting fair datasets, such as PPB [8], RFW [59], UTK-Face [67] and FairFace [29]. Meanwhile, strategic resampling methods [6,37,58] attempt to balance the *appearance* of training data with respect to different demographic groups, referred to as *domain*. Since fair dataset and strategic resampling address bias issues before model training, in [9,28], they are also referred as pre-processing methods. On the other hand, to mitigate *task bias* from the perspective of model training, some debiasing models have been proposed. First, adversarial forgetting methods [3,16] manage to extract a new representation which only contains information for the recognition task and exclude the information of *protected attributes*. Second, domain adaptation methods [59,69] adapt learned recognition knowledge from the majority *domain* to the minority *domain*. Third, domain independent training methods [17,61] delicately design several separate classifiers in different demographic groups and garner an ensemble of classifiers by *representation sharing*.

Fairness Using GANs. Over the past couple years, Generative Adversarial Network (GAN) [18] is developed further in a tremendous number of work [10,24,27,43,45,60]. As pointed by [2], three main purposes of GAN are fidelity, diversity and privacy. Beside the mainstream developments, GAN has also been used to augment real datasets [21,50,72]. Some work [12,25] manipulate latent vectors to augment real datasets to learn disentangled representations. Moreover, Balakrishnan et al. [4] proposes an experimental method with face images generated by StyleGAN2 [30] to reveal correlation between *protected attributes* and fairness performance, which in nature is to study and measure bias instead of debiasing. While [32,62,63] use GANs to construct synthetic dataset to improve fairness, they conduct experiments on Census dataset, *e.g.*, Adult dataset (predict whether income exceeds $50K/yr based on census data), which is inapplicable for face dataset since label and data of census dataset are both census data and can be generated together, which is different from generating accurate labels together with generated images in face dataset.

Recently, a minority route of augmentation by GANs to introduce fairness [48,51,52] emerged. However, as pointed out by [48], some work [13,51,52] need to train a new GAN per bias since different facial attributes may yield different levels of bias. By contrast, V. Ramaswamy *et al.* [48] use a single GAN to construct a fair synthetic dataset by generating the complementary image with the same target label but the opposite *protected attribute* label corresponding to the randomly generated image. However, to assign the attribute labels, [48] relies on an attribute classifier trained on the original dataset, which may be harmful for fairness since the pre-trained classifier may be already biased by skew of the original dataset. By contrast, our method does not need the pre-trained classifier or significant resources of human annotations using Amazon Mechanical Turk as in [4] to assign labels for each generated image. As elaborated in Sect. 3, our method only needs a small set of images with labels as seeds and then construct the whole synthetic datasets automatically with labels, which can be directly used in the following performance and fairness evaluation pipeline. Furthermore, compared with [48] using additional images for the good recognition performance, our method achieves comparable performance and better fairness even with the same-size training dataset, as elaborated in Sect. 4.5.

3 Approach

As discussed in Sect. 2, some methods [13,51,52] focus on applying different data augmentation techniques to mitigate different types of bias. By contrast, we train a unified GAN on the original dataset to introduce fairness for both *protected attributes* and all facial attributes.

In the facial attribute classification task, given an attribute dataset $\mathcal{D}$ involving instances (x_i, y_i, z_i), the classification model H consumes an image $x_i \in \mathcal{X}$ annotated with a set of binary facial attributes $y_i \in \mathcal{Y}$ (*e.g.*, hair color) and *protected attributes* $z_i \in \mathcal{Z}$ (*e.g.*, sex), and produces the predicted facial attribute label $y_i' \in \mathcal{Y}$. As illustrated by conditions of three fairness criteria [20] defined with conditional probability, *dataset bias* may be induced from skew of training dataset. Specifically, $\mathcal{D}$ may naturally be imbalanced across the positive samples Z and the negative samples $\bar{Z}$ with respect to the *protected attribute* $\mathcal{Z}$ (if binary), *i.e.*, $|\mathcal{D}_Z| \neq |\mathcal{D}_{\bar{Z}}|$, on which the classification model H is trained may invalidate *demographic parity* [20] which requires that the prediction label Y' (if binary) and the *protected attribute* $\mathcal{Z}$ are independent, *i.e.*,

$$P(Y' = 1) = P(Y' = 1|Z) = P(Y' = 1|\bar{Z}). \tag{1}$$

Algorithm 1. *Controllable Attribute Translation* (CAT).

1: **initialize** A_Y^{intra} and B_Y^{inter} to be an empty set.
2: **for** $e_i \in \mathbb{R}^k$, $e_i \in S_Y$ **do**
3:      **for** $e_j \in \mathbb{R}^k$, $e_j \in S_Y$ **do**
4:          $A_{ij} = \{l \mid |e_i^l - e_j^l| < \texttt{intra\_threshold}, l \in [1,k]\}$
5:          $A_Y^{intra} \leftarrow A_Y^{intra} \cap A_{ij}$ ▷ Intra-class similarity.
6:      **end for**
7:      **for** $\bar{e}_j \in \mathbb{R}^k$, $\bar{e}_j \in S_{\bar{Y}}$ **do**
8:          $B_{ij} = \{l \mid |e_i^l - \bar{e}_j^l| > \texttt{inter\_threshold}, l \in [1,k]\}$
9:          $B_Y^{inter} \leftarrow B_Y^{inter} \cap B_{ij}$ ▷ Inter-class difference.
10:     **end for**
11: **end for**
12: **output** $C_Y \leftarrow A_Y^{intra} \cup B_Y^{inter}$

Furthermore, the original dataset $\mathcal{D}$ may be imbalanced at the facial attribute level across *protected attributes* for the facial attribute $\mathcal{Y}$, i.e., $|\mathcal{D}_Z^Y| \neq |\mathcal{D}_{\bar{Z}}^Y|$, where $|\mathcal{D}_Z^Y|$ is the number of positive samples Y among Z and $|\mathcal{D}_{\bar{Z}}^Y|$ is the number of Y among $\bar{Z}$ with respect to the facial attribute $\mathcal{Y}$, which invalidates *equal opportunity* [20], an another criterion that requires independence between the prediction label Y' and the *protected attribute* Z conditional on Y, i.e.,

$$P(Y' = 1 | Z, Y) = P(Y' = 1 | \bar{Z}, Y). \qquad (2)$$

Finally, *equalized odds* [20], which is a stronger criterion, requires that Y' and Z are independent conditional on both Y and $\bar{Y}$, i.e., with the additional constraint than *equal opportunity* as followed,

$$P(Y' = 1 | Z, \bar{Y}) = P(Y' = 1 | \bar{Z}, \bar{Y}), \qquad (3)$$

where $\bar{Y}$ represents the negative samples with respect to the facial attribute $\mathcal{Y}$. To mitigate the influence from skew of training dataset under the requirements of *equalized odds*, $\mathcal{D}$ should be balanced across Z for both Y and $\bar{Y}$, i.e., $|\mathcal{D}_Z^Y| = |\mathcal{D}_{\bar{Z}}^Y|$ and $|\mathcal{D}_Z^{\bar{Y}}| = |\mathcal{D}_{\bar{Z}}^{\bar{Y}}|$, where $|\mathcal{D}_Z^{\bar{Y}}|$ is the number of $\bar{Y}$ among Z and $|\mathcal{D}_{\bar{Z}}^{\bar{Y}}|$ is the number of $\bar{Y}$ among $\bar{Z}$ with respect to the facial attribute $\mathcal{Y}$.

However, even though prior methods that depend on fair dataset collection [29] or resampling [6,37] may satisfy *demographic parity* by balancing dataset across *protected attributes*, it is hard for these methods to satisfy *equal opportunity* and/or *equalized odds* constraints which require balance at the facial attribute level due to the lack of sufficient images which are balanced across one facial attribute, much less across multiple facial attributes in the multi-attribute classification task. Thus, to mitigate the effect of *dataset bias* on the trained model, we generate a semi-synthetic dataset $\mathcal{D}_{syn}$ which meets requirements of all three fairness criteria by controllably translating the representation of facial attributes in latent space to image space and generate sufficient images with desired facial attributes, referred to as *Controllable Attribute Translation* (CAT).

Given a k-dimensional latent space $\mathcal{E} \in \mathbb{R}^k$, a generative model $G : \mathcal{E} \to \mathcal{I}$ trained on real dataset $\mathcal{D}$ produces the synthetic image I_i from the latent vector

Fig. 2. Examples of generated female images and male images.

$e_i \in \mathcal{E}$. By human annotations denoted as F, we construct a **succinct** set of latent vectors with the facial attribute $\mathcal{Y}$, referred to as *attribute seeds* $S_Y \subset \mathcal{E}$, such that $\forall e_i \in S_Y, F(G(e_i)) = Y$. We refer to $\mathcal{Y}$ as Attribute of Interest (AOI)[1].

To stably generate images with AOI, we first study the common similarity of latent vectors among S_Y since I_i may yield same AOI and randomly yield other facial attributes, where $I_i = G(e_i), e_i \in S_Y$. Given two randomly picked latent vectors $e_i, e_j \in S_Y$, we traverse and aggregate the dimension indices where the absolute difference is smaller than a hyperparameter `intra_threshold`, *i.e.*, $A_{ij} = \{l \mid |e_i^l - e_j^l| < \texttt{intra\_threshold}, l \in [1, k]\}$, which are the dimensions representing the attribute similarity between e_i and e_j. Furthermore, since pairs of images generated by latent vectors from S_Y may be similar in other unwanted facial attributes, to purify the similarity of AOI and find predominate dimensions controlling AOI, we traverse all combination of pairs in S_Y to find A_{ij} of each pair and take the intersection across all A_{ij}, *i.e.*, $A_Y^{intra} = \bigcap_{e_i, e_j \in S_Y} A_{ij}$, which is the least common similarity, referred to as *intra-class similarity* of AOI, which effectively eliminates the similarity of other unwanted facial attributes.

To select more representative dimension indices for AOI, we compensate *intra-class similarity* with *inter-class difference*. We randomly pick $e_i \in S_Y$ and $\bar{e}_i \in S_{\bar{Y}}$, where $S_{\bar{Y}}$ is the set of latent vectors without AOI, such that $\forall \bar{e}_i \in S_{\bar{Y}}, F(G(\bar{e}_j)) = \bar{Y}$. Further, we aggregate the dimension indices where the absolute difference is greater than the other hyperparameter `inter_threshold`, *i.e.*, $B_{ij} = \{l \mid |e_i^l - \bar{e}_j^l| > \texttt{inter\_threshold}, l \in [1, k]\}$, which are the dimensions representing the attribute difference between e_i and $\bar{e}_j$. We then take the intersection across all attribute difference, $B_Y^{inter} = \bigcap_{e_i \in S_Y, \bar{e}_j \in S_{\bar{Y}}} B_{ij}$, which is the least common difference, referred to as *inter-class difference* of AOI.

Finally, we take $C_Y = A_Y^{intra} \cup B_Y^{inter}$ since the usage of B_Y^{inter} is to find the representative dimensions unseen by A_Y^{intra}, as elaborated in Sect. 4.5. A pseudo code of the method is shown in Algorithm 1, which can be extended for all other AOIs and all *protected attributes*. With *intra-class similarity* and *inter-class difference*, we can find the representative dimensions for both protected and facial attributes so that the generated images can properly capture the characteristics of AOI. In general, we translate the attribute appearance in images to the attribute representation in latent space.

To construct the synthetic dataset $\mathcal{D}_{syn}$, we controllably translate the found attribute representations in the latent space back to image space. We first

---

[1] AOI is set based on interests for different experiments.

Fig. 3. Examples of generated images with the *protected attribute* and a single assigned facial attribute, negative samples (**left**) and positive samples (**right**) in each 2×2 grid.

randomly simulate another set of latent vectors, referred to as *identity seeds* $S_{ID} \subset \mathcal{E}$ following standard normal distribution $N(0,1)$. Then, to assign a facial attribute Y, we perturb $e_{ID} \in S_{ID}$ to be $e_Y \in S_Y$ for the dimension indices in C_Y. In parallel, the label can be automatically assigned as Y. In practice, the latent space consists of R resolutions, and each resolution is a k-dimensional latent space. To further ensure the appearance of assigned facial attributes and mitigate the intervention from other facial attributes across resolution spectrum, although we do *not* assume chosen dimensions are independent between different attributes and the latent space is *not* disentangled by design, we subdivide the resolution spectrum to assign different kinds of facial attributes in different resolutions (*e.g.*, high-level facial attributes in lower resolutions and smaller scale facial attributes in higher resolutions), which is elaborated in Sect. 4.4.

By generating balanced synthetic dataset $\mathcal{D}_{syn}$ with desired attributes and supplementing the minority group of the original dataset with sufficient images at both *protected attribute* level and facial attribute level, we can ensure the requirements of all three fairness criteria in Eqs. (1) to (3). From the perspective of information theory, assuming the whole information contained in original dataset is fixed, *i.e.*, the best recognition performance trained on this dataset is upper-bounded, synthetic datasets generated by our method tries the best to transfer and express the recognition task-related information in a fairer way.

Advantages. By *intra-class similarity* and *inter-class difference*, we can accurately and effectively find a specific set of dimension indices representing AOI in the dimension spectrum so that we only need a small size of *attribute seeds* with annotations as starting seeds. Beside with the resolution separation, we have a two-dimensional separation in both latent space and resolution. Thus, we can assign multiple AOIs to one single identity seed and stably preserve randomness of other unassigned facial attributes, which endows synthetic dataset to be constructed under different facial attribute distribution with more freedom. Furthermore, we assign the attribute label to the generated image $I_i = G(e_i)$ naturally with the facial attribute $\mathcal{Y}$ represented by S_Y to which its latent vectors $e_i \in S_Y$ belongs, rather than rely on a shallow classifier.

4 Experimental Evaluation

We first investigate the nature of *dataset bias* by exploring each facial attribute and focus on the facial attributes which induce much bias. Furthermore, we empirically investigate the effectiveness of our method on two fashion tasks involving bias—(1) sex classification and (2) facial attribute classification. In sex classification, we concurrently manipulate multiple *femininity/masculinity attributes* to generate paired images with female and male appearance and construct sex-level balanced dataset. Meanwhile, in facial attribute classification, in the basis of sex-level balanced dataset, we further consider the other non-sex-related facial attributes to construct the semi-synthetic dataset which is balanced in terms of both sex and facial attribute. In both experiments, we first show that the classification performance of the model trained with synthetic datasets achieves comparable performance as the model trained on the original dataset. Then, we conduct a comprehensive fairness evaluation with a wide range of bias assessment metrics [11, 34, 42, 54, 57] to show that fairness has been improved after training with synthetic datasets compared to the model solely trained on the original dataset. Further, we demonstrate that our method outperforms both strategic resampling [6] and balanced dataset construction method with synthetic images [48] to address *dataset bias*, and several debiasing models [3, 61, 69] to address *task bias*. Finally, we present an ablation study to evaluate different factors influencing the performance of our method. Although we discuss sex as the *protected attribute* in this section, our method is general purpose and can be used for all *protected attributes* if the labels are available.

4.1 Attributes Study

Before presenting the improvement on fairness, we first conduct in-depth evaluation of *dataset bias* at the facial attribute level for CelebA dataset [39], which is a face dataset containing 202,599 images of celebrity faces and 40 binary attributes per image. Since our method produces a facial attribute-level balanced synthetic dataset, it is valuable to study *dataset bias* at facial attribute level instead of the general overall accuracy for all facial attributes. Following [48], we train a multitask facial attribute classifier with ResNet-50 [22] to recognize facial attributes on the sex-level balanced CelebA dataset. The main results are shown in Table 2 as baseline and full results are presented in the appendix.

As shown in Table 2 by results of baseline, solely balancing across sex does not guarantee fairness in facial attribute classification since even with balanced training across females and males, the imbalance of facial attribute across sex still exists. Although we know classification models tend to learn distinguishable representations from positive samples instead of negative samples [35], there are insufficient positive samples of some specific facial attributes in the minority group. Inspired by the categorization in [48] where they only categorize a subset of facial attributes, we summarize all 40 facial attributes in CelebA dataset into three groups—(1) *unbiased attributes* which do not yield much bias

Table 1. Performance and fairness comparison on sex classification.

		Number of training images			Classification accuracy ↑			Information leakage	Statistical dependence	
		Female	Male	Total	Female	Male	Overall	BA ↓	$dcor^2$ ↓	RLB ↓
Origin training dataset	Imbalanced	94509	68261	162770	**99.2**	97.7	**98.6**	−0.015	0.656	3.854
	Balanced	68261	68261	136522	**99.2**	97.6	98.5	−0.016	0.651	3.792
GAN-Debiasing [48]	Original	254509	228261	482770	99.1	97.4	98.4	**−0.018**	0.596	3.567
	Same size	68261	68261	136522	98.8	97.2	98.0	−0.017	0.583	3.578
	Supplement	94509	94509	189018	**99.2**	97.2	98.2	−0.016	0.574	3.574
Ours	Same size	68261	68261	136522	99.1	97.3	98.3	**−0.018**	0.592	3.742
	Supplement	94509	94509	189018	**99.2**	**97.8**	98.5	−0.016	**0.544**	**3.481**

(*i.e.*, the difference of AP between female and male is less than 5%), (2) *masculinity/femininity attributes*, which are considered as AOI in Sect. 4.3 to construct sex-level balanced dataset, and (3) non-sex-related facial attributes but inducing much bias even with sex-level balanced training, which are considered as AOI appending *masculinity/femininity attributes* in Sect. 4.4 to construct both sex-level and facial attribute-level balanced dataset.

4.2 Synthetic Attribute-Level Balanced Datasets

To construct the synthetic dataset $\mathcal{D}_{syn}$, we train the generative model solely on the training set of CelebA dataset containing 162,770 images to ensure the isolation from the testing set since the images generated by trained GAN are used for the following tasks. For the generative model choice to generate high-quality images (which is elaborated in appendix), we use StyleGAN2 [30] since *style mixing regularization* in StyleGAN2 facilitates our goal to assign different specific facial attributes and mitigate the interference from other unassigned facial attributes. We persist in training the generative model to generate more real images. As a standard metric, FID is used to evaluate sample qualities of generated images. FID of the presented results in the paper is 3.56, which is comparable to existing GANs [40, 44] trained on CelebA dataset and posted recently. Although the benchmark for FID of image generation on CelebA dataset is 2.71 [38], our objective is to utilize generated images to improve fairness of existing datasets rather than improve image quality of generated images.

To assign attributes, we use `intra_threshold` $= 2\sqrt{2}$ and `inter_threshold` $= \sqrt{2}$. In general, we empirically found that any settings for these two hyperparameters among the recommended range $[\sqrt{2}, 2\sqrt{2}]$ will not significantly affect the performance of downstream classification tasks, which confirms with our theoretical discussion in Sect. 4.5. Furthermore, for resolution spectrum separation, StyleGAN2 [30] have already provided reference to assign different attributes in different resolutions, *e.g.*, high-level attributes (hair style, face shape) in coarse spatial resolution ($4^2 - 8^2$), and racial appearance (colors of eyes, hair, skin) in fine spatial resolution ($16^2 - 1024^2$). The dimensionality of the latent space is 512 and the resolution of generated images is set to be 256^2 so that by [30] we generally have a 14-layer 512-dimensional latent space to assign attributes. Specifically, we leave 4^2 original for the basic identity construction. For resolution choices to assign attributes, we assign face shape attributes (Chubby, Big Nose,

Table 2. Performance and fairness comparison on facial attribute classification.

		Chubby	BigNose	WavyHair	PaleSkin	BlondHair	DoubleChin	PointyNose	BagsUnderEyes	HighCheekbones	Average
AP ↑	Baseline	**62.1**	**71.2**	**88.2**	**69.1**	92.0	**61.2**	**64.9**	**67.8**	95.4	**74.7**
	Resampling [6]	54.5	62.2	87.2	67.2	87.0	46.4	61.7	58.3	95.1	68.8
	GAN-Debiasing [48]	58.8	69.5	85.6	68.5	90.6	58.5	63.5	63.1	95.3	72.6
	Ours	60.5	66.5	86.8	69.0	**92.1**	58.6	61.4	62.9	**95.5**	72.6
DEO ↓	Baseline	28.7	35.6	33.0	11.9	10.8	26.7	32.8	20.1	10.8	23.4
	Resampling [6]	**15.0**	12.7	12.1	**6.2**	4.3	4.2	5.2	19.2	5.9	9.4
	GAN-Debiasing [48]	26.6	32.7	23.3	10.8	3.6	22.1	28.8	16.8	9.3	19.3
	Ours	21.2	**3.9**	**11.8**	9.3	**3.5**	**3.3**	**3.1**	**15.0**	**4.3**	**8.4**
BA ↓	Baseline	1.72	5.83	−3.27	1.19	0.12	0.46	4.78	2.27	−1.32	1.31
	Resampling [6]	1.38	**−8.69**	**−5.49**	−0.30	−3.37	**−1.73**	−3.21	**−9.44**	**−6.17**	**−4.11**
	GAN-Debiasing [48]	1.30	5.51	−3.60	0.20	0.50	0.46	3.71	1.14	−1.89	0.81
	Ours	**1.16**	−6.71	−5.10	**−0.44**	**−4.61**	−1.16	**−5.35**	−5.95	−4.27	−3.60
KL ↓	Baseline	0.39	0.60	0.42	0.08	0.46	0.30	0.32	0.27	0.12	0.33
	Resampling [6]	**0.18**	**0.13**	0.15	0.07	**0.05**	0.17	0.21	0.26	0.12	0.15
	GAN-Debiasing [48]	0.37	0.54	0.33	0.07	0.37	0.29	0.20	0.26	0.09	0.28
	Ours	0.33	0.13	**0.06**	**0.04**	0.34	**0.09**	**0.04**	**0.08**	**0.08**	**0.13**
dcor² ↓	Baseline	0.58	0.65	0.82	0.53	0.34	0.58	0.85	0.86	0.82	0.67
	Resampling [6]	0.58	0.59	0.77	0.47	0.69	0.53	0.83	0.65	0.56	0.63
	GAN-Debiasing [48]	0.35	0.54	0.50	0.08	**0.17**	0.31	0.39	0.43	0.33	0.34
	Ours	**0.18**	**0.31**	**0.38**	**0.06**	0.23	**0.17**	**0.24**	**0.25**	**0.30**	**0.23**
RLB ↓	Baseline	0.63	1.24	1.56	0.03	1.12	0.45	0.55	0.82	1.36	0.86
	Resampling [6]	0.58	1.13	1.45	0.03	1.03	0.44	0.50	0.63	0.96	0.75
	GAN-Debiasing [48]	0.34	1.14	1.42	**0.01**	0.86	0.31	0.49	0.61	0.69	0.65
	Ours	**0.28**	**1.11**	**1.30**	**0.01**	**0.73**	**0.30**	**0.09**	**0.43**	**0.39**	**0.52**

Pointy Nose, High Cheekbones and Double Chin) in 8^2, fine face shape attributes (Bags Under Eyes, Wavy Hair and Straight Hair) in 16^2, hair color attributes (Black Hair, Blond Hair, Brown Hair and Gray Hair) in $32^2 - 64^2$, and skin color attributes (Pale Skin) in $128^2 - 256^2$. Besides, we assign *masculinity/femininity attributes* in $8^2 - 16^2$. As shown in Fig. 2, the generated images are listed in pairs with highlighted *femininity/masculinity attributes*, and similar non-sex facial attributes (*e.g.*, skin color and hair color) and other photo settings (*e.g.*, lighting, pose and background). Besides, in Fig. 3, we controllably assign *femininity/masculinity attributes* and another facial attribute. Furthermore, in Fig. 4, we concurrently assign sex and multiple facial attributes and generally keep other unassigned facial attributes unchanged, which demonstrates the advantage that attribute assignment does not interfere with each other.

4.3 Sex Classification

As in [8,29], the imbalance of training dataset may hamper the accuracy of sex classification, we therefore validate the fairness improvement of our synthetic datasets on sex classification in this section. To help downstream classification models to well learn the difference between female and male (the useful information for sex classification), given an identity seed, we only modify *femininity/masculinity attributes* and leave other facial attributes unchanged, as shown in Fig. 2, which also ensures that there are sufficient samples with large diversity of facial attributes from identity seeds in both female group and male group.

Experiment Setup. ResNet-50 [22] is used as the backbone for sex classifier. The baseline model is trained with original CelebA dataset [39] under two settings—(1) `Original` to keep the training set of CelebA dataset fully original with the whole set of 162,770 images, and (2) `Balanced` to balance the training set across sex. For training with synthetic dataset, we also conduct two types of

Table 3. Comparison of debiasing models on facial attribute classification.

	mAP ↑			Information leakage ↓			Statistical dependence ↓	
	Female	Male	Overall	DEO	BA	KL	$dcor^2$ [54]	RLB [34]
Baseline	**75.3**	72.2	74.1	20.9	0.99	0.27	0.85	1.33
Adversarial forgetting [3]	73.2	70.4	72.1	18.8	0.61	0.25	0.51	1.25
Domain adaptation [69]	74.5	72.3	73.7	19.7	0.82	0.21	0.48	1.28
Domain independent [61]	74.6	**74.2**	**74.4**	20.8	0.55	0.27	0.37	1.12
Ours	73.0	73.2	73.1	**8.4**	**−3.26**	**0.16**	**0.24**	**0.65**

experiments—(1) **Same size** to combine one half of balanced original training set with a same size balanced synthetic dataset to be mixed dataset so that the number of mixed dataset is the same as the number of training images in the balanced original dataset, and (2) **Supplement** to supplement the lack of male images compared with female images in the original dataset so that the mixed dataset is balanced in the whole dataset scale. For comparison with the other attribute-level balance method [48], we conduct experiments under the original setting in their paper (the original CelebA dataset combined with the balanced synthetic dataset with 160,000 sex pairs of images) and two same settings of ours. The testing set is the original testing CelebA dataset for all experiments.

Evaluation Protocol. To demonstrate the effectiveness of our method on sex classification and facial attribute classification, we use accuracy and fairness as the main evaluation metrics. First, we compare the classification performance between the model trained on synthetic datasets and the original dataset to verify that the testing accuracy on the real testing dataset is preserved at the same level, which demonstrates that the generated images yield proper facial attributes in the metrics of the real dataset. Meanwhile, we use several fairness metrics for a comprehensive fairness comparison since different types of metrics assess bias in different directions. According to the taxonomy of bias assessment metrics in [34], we select several representative metrics in these categories. We select Bias Amplification (BA) [57,61] among *information leakage*-based metrics and two types of metrics based on *statistical dependence*, Distance Correlation $(dcor^2)$ [1,54] and Representation-Level Bias (RLB) [34] to increase the diversity of metrics and comprehensively evaluate fairness in different ways.

Results. As shown in Table 1, we can see the testing accuracy of synthetic dataset is kept at the same level under both two settings, which demonstrates that the generated images yield proper *femininity/masculinity attributes* in the original dataset. Furthermore, although the classification accuracy is already saturated, with additional images under the **Supplement** setting, we improve the accuracy in minority groups. On the other hand, for fairness evaluation compared with other datasets, our method further improves fairness. Although [48] uses more images than the original dataset, our method yields even better results in both accuracy and fairness metrics with fewer training images. We elaborate the discussion on the relation between the size of the synthetic dataset and performance of our method in the appendix.

Fig. 4. Examples of generated images with multiple assigned facial attributes.

4.4 Facial Attribute Classification

Having found that the remaining *dataset bias* of facial attribute classification on
CelebA dataset stems from imbalance at both *protected attribute* level and facial
attribute level as discussed in Sect. 4.1, we construct the synthetic dataset which
is balanced in both these two levels to study the effectiveness of our method
for all non-sex-related facial attributes, and compare our method with methods
based on sex-level balance [6] and facial attribute-level balance [48]. Furthermore,
we highlight the comparisons with debiasing models [3,61,69].

Experiment Setup. We train a multitask facial attribute classifier with
ResNet-50 [22]. The baseline is trained with sex-level balanced CelebA dataset
constructed from original CelebA dataset. Our method combines a portion of
original dataset without surplus samples of some facial attributes and the synthetic dataset which supplements insufficient samples of AOI so that the mixed
dataset is balanced at both sex level and facial attribute level, as illustrated
in Fig. 1.

Evaluation Protocol. We use Average Precision (AP) as the performance
metric. For bias assessment, beside BA, $dcor^2$ and RLB, Difference in Equal
Opportunity (DEO) [42,47] as a metric version of *equal opportunity* and KL-
divergence between score distribution (KL) [11,48] as a stronger notion stemming
from *equalized odds* are used to verify the achievements of Eqs. (2) and (3).

Results. In Table 2, we present one representative facial attribute for the mutually exclusive facial attributes, *e.g.*, Wavy Hair and Straight Hair, and full results
are presented in appendix. As AP is preserved at the same level for all AOIs,
our method outperforms other methods for Blond Hair and High Cheekbones. In
fairness comparison with baseline, all three methods introduce fairness under all
metrics in a same trend. As pointed by [34], $dcor^2$ and RLB, the metrics based
on *statistical dependence* are more consistent and stable. In this sense, although
strategically resampling [36] outperforms our method under BA, it yields second
worse performance on the metrics based on *statistical dependence*. Furthermore,
with better DEO and KL, we verify the achievement of Eqs. (2) and (3) in Sect. 3,
which is unreachable for sex-level balanced dataset and strategically resampling.
According to results of imbalanced training, due to the existence of the facial
attribute-level skew in the dataset, the classifier trained with these skew datasets
may be biased already. In this sense, compared with [48], our method is more
stable in all facial attributes since we assign label directly rather than rely on

such unreliable facial attribute classifier trained on original dataset as in [48], which may be harmful for fairness. Compared with Table 1, the range of two metrics based on *statistical dependence* in sex classification are clearly higher than the range in the facial attribute classification since they directly reveal the statistical dependence between learned representations to predict sex and sex labels themselves, which better assess the bias for sex classification.

We also conduct an overall comparison of average results over all non-sex-related AOIs (7 facial attributes in the paper and 7 mutually exclusive facial attributes in appendix) with several debiasing models based on adversarial forgetting [3], domain adaptation [69] and domain independent [61]. We train a ResNet-50 [22] to classify all non-sex-related AOIs on the original CelebA dataset as baseline model. For the multi-label classification, mean Average Precision (mAP) is used to evaluate the overall performance. In Table 3 of classification performance comparison, domain independent method outperforms other methods. In parallel, without the in-process debiasing techniques usage, our method yields comparable classification performance and outperforms other methods in all bias assessment metrics, which demonstrates a comprehensive fairness improvement. In general, there is a tradeoff between classification performance and fairness performance, *i.e.*, although domain independent method outperforms other methods in mAP, our method improves fairness mostly at an acceptable expense of classification performance. Furthermore, the proposed synthetic dataset is model-agnostic as compared under different backbones in appendix.

4.5   Ablation Study

Intra-class Similarity and *Inter-class Difference.* In this section, we will discuss *intra-class similarity* and *inter-class difference* involving two hyperparameter intra_threshold and inter_threshold. Considered the task to generate female and male images for which, given *masculinity attributes seeds* S_{male} and *femininity attributes seeds* S_{female}, *intra-class similarity* is deprecated when intra_threshold is small since there may be no dimension indice where $\forall e_i, e_j \in S_{male}, |e_i - e_j| <$ intra_threshold, *i.e.*, A_{male}^{intra} is empty. By contrast, *inter-class difference* will be deprecated when *inter_threshold* is large. For example, given an identity latent vector e_{ID} whose generated image is female, when *intra-class similarity* and *inter-class difference* are both deprecated, the generated image is same as the prime image generated from e_{ID}. As shown in Fig. 5a, without *intra-class similarity*, the generated image is apparently female. Further, when we apply *intra-class similarity* and deprecate *inter-class difference*, the generated image is transferred to be male with few *femininity attributes* (*e.g.*, hair) since the generated model has learned *masculinity attributes* but has not fully suppressed *femininity attributes*. On the other hand, if intra_threshold is large or inter_threshold is small, *i.e.*, A_{male}^{intra} or B_{male}^{inter} contains all dimension indices so that *intra-class similarity* or *inter-class difference* is overemphasized. As shown in Fig. 5b, compared with generated images under proper threshold settings, the generated images under strong threshold settings in the top row are cookie-cutter without the randomness from identity seeds. In the whole process, *intra-class similarity* contributes mostly and

(a) Deprecated.   (b) Overemphasized.

Fig. 5. The role of intra-class similarity and inter-class difference.

inter-class difference plays an auxiliary role. Since the random latent vector is simulated under standard normal distribution $N(0, 1)$ and the difference between two latent vectors follows normal distribution $N(0, \sqrt{2})$ where standard deviation is $\sqrt{2}$, the proper range for two thresholds is $[\sqrt{2}, 2\sqrt{2}]$ so that A_Y^{inter} may contain the dimension indices in the main lope representing attribute similarity among S_Y, and B_Y^{intra} may exclude the dimension indices in the main lope representing attribute similarity of S_Y and $S_{\bar{Y}}$, where $S_{\bar{Y}}$ is the set of latent vectors without *Attribute of Interest* $\mathcal{Y}$.

5   Conclusion

By the investigation on the imbalance at the facial attribute level, we can clearly find the root cause for the uneven performance across *protected attributes*. To address the imbalance, from the perspective of information theory, we try our best to mitigate the loss of information from original data (*i.e.*, comparable performance) and orderly express such information in a fairer way (*i.e.*, improved fairness) while the amount of whole information is fixed. In comparison with debiasing models, we have found fairness-performance tradeoff in model training, as fairness-efficiency tradeoff in the real world. Still, the valuable remark is that although the network trained with the proposed synthetic data does not outperform the debiasing models in recognition performance, it is impressive that synthetic images can achieve consistent performance with real data and further yields better fairness. Furthermore, in the image generation process, our method overcomes the difficulty to obtain labels by *intra-class similarity* and *inter-class difference* instead of relying on a shallow attribute classifier or significant extra human annotations. Therefore, our method can be scalably used in the model fully trained with synthetic datasets in the future. Besides, it is a good direction for future researchers to extend our method to other face datasets and introduce fairness for more facial attributes. Finally, our work presents a good example for the probability of the same-level performance for computer vision models partially or fully trained with synthetic dataset.

Acknowledgement. This research is based upon work supported in part by the Office of the Director of National Intelligence (ODNI), Intelligence Advanced Research Projects Activity (IARPA), via [2022-21102100007]. The views and conclusions contained herein are those of the authors and should not be interpreted as necessarily representing the official policies, either expressed or implied, of ODNI, IARPA, or the U.S. Government. The U.S. Government is authorized to reproduce and distribute reprints for governmental purposes notwithstanding any copyright annotation therein.

References

1. Adeli, E., et al.: Representation learning with statistical independence to mitigate bias. In: Proceedings of the IEEE/CVF Winter Conference on Applications of Computer Vision, pp. 2513–2523 (2021)
2. Alaa, A.M., van Breugel, B., Saveliev, E., van der Schaar, M.: How faithful is your synthetic data? Sample-level metrics for evaluating and auditing generative models. arXiv preprint arXiv:2102.08921 (2021)
3. Alvi, M., Zisserman, A., Nellåker, C.: Turning a blind eye: explicit removal of biases and variation from deep neural network embeddings. In: Proceedings of the European Conference on Computer Vision (ECCV) Workshops (2018)
4. Balakrishnan, G., Xiong, Y., Xia, W., Perona, P.: Towards causal benchmarking of biasin face analysis algorithms. In: Ratha, N.K., Patel, V.M., Chellappa, R. (eds.) Deep Learning-Based Face Analytics. ACVPR, pp. 327–359. Springer, Cham (2021). https://doi.org/10.1007/978-3-030-74697-1_15
5. Bellamy, R.K., et al.: AI fairness 360: an extensible toolkit for detecting, understanding, and mitigating unwanted algorithmic bias. arXiv preprint arXiv:1810.01943 (2018)
6. Bickel, S., Brückner, M., Scheffer, T.: Discriminative learning under covariate shift. J. Mach. Learn. Res. **10**(9) (2009)
7. Buolamwini, J., Gebru, T.: Gender shades: intersectional accuracy disparities in commercial gender classification. In: Friedler, S.A., Wilson, C. (eds.) Proceedings of the 1st Conference on Fairness, Accountability and Transparency. Proceedings of Machine Learning Research, vol. 81, pp. 77–91. PMLR (2018). http://proceedings.mlr.press/v81/buolamwini18a.html
8. Buolamwini, J., Raji, I.D.: Actionable auditing: investigating the impact of publicly naming biased performance results of commercial AI products. Conference on Artificial Intelligence, Ethics, and Society (2019)
9. Calmon, F.P., Wei, D., Vinzamuri, B., Ramamurthy, K.N., Varshney, K.R.: Optimized pre-processing for discrimination prevention. In: Proceedings of the 31st International Conference on Neural Information Processing Systems, pp. 3995–4004 (2017)
10. Chan, E.R., Monteiro, M., Kellnhofer, P., Wu, J., Wetzstein, G.: pi-GAN: periodic implicit generative adversarial networks for 3D-aware image synthesis. In: Proceedings of the IEEE/CVF Conference on Computer Vision and Pattern Recognition, pp. 5799–5809 (2021)
11. Chen, M., Wu, M.: Towards threshold invariant fair classification. In: Conference on Uncertainty in Artificial Intelligence, pp. 560–569. PMLR (2020)
12. Chen, X., Duan, Y., Houthooft, R., Schulman, J., Sutskever, I., Abbeel, P.: Infogan: Interpretable representation learning by information maximizing generative adversarial nets. In: Advances in Neural Information Processing Systems, vol. 29 (2016)
13. Choi, K., Grover, A., Singh, T., Shu, R., Ermon, S.: Fair generative modeling via weak supervision. In: International Conference on Machine Learning, pp. 1887–1898. PMLR (2020)
14. Das, A., Dantcheva, A., Bremond, F.: Mitigating bias in gender, age and ethnicity classification: a multi-task convolution neural network approach. In: Proceedings of the European Conference on Computer Vision (ECCV) Workshops (2018)
15. Dwork, C., Hardt, M., Pitassi, T., Reingold, O., Zemel, R.: Fairness through awareness. In: Proceedings of the 3rd Innovations in Theoretical Computer Science Conference, pp. 214–226 (2012)

16. Gong, S., Liu, X., Jain, A.K.: Jointly de-biasing face recognition and demographic attribute estimation. In: Vedaldi, A., Bischof, H., Brox, T., Frahm, J.-M. (eds.) ECCV 2020. LNCS, vol. 12374, pp. 330–347. Springer, Cham (2020). https://doi.org/10.1007/978-3-030-58526-6_20

17. Gong, S., Liu, X., Jain, A.K.: Mitigating face recognition bias via group adaptive classifier. In: Proceedings of the IEEE/CVF Conference on Computer Vision and Pattern Recognition, pp. 3414–3424 (2021)

18. Goodfellow, I., et al.: Generative adversarial nets. In: Advances in Neural Information Processing Systems, vol. 27 (2014)

19. Grother, P., Ngan, M., Hanaoka, K.: Face recognition vendor test (FVRT): part 3, demographic effects. National Institute of Standards and Technology (2019)

20. Hardt, M., Price, E., Srebro, N.: Equality of opportunity in supervised learning. In: Proceedings of the 30th International Conference on Neural Information Processing Systems, NIPS 2016, pp. 3323–3331. Curran Associates Inc., Red Hook (2016)

21. Hariharan, B., Girshick, R.: Low-shot visual recognition by shrinking and hallucinating features. In: Proceedings of the IEEE International Conference on Computer Vision, pp. 3018–3027 (2017)

22. He, K., Zhang, X., Ren, S., Sun, J.: Deep residual learning for image recognition. In: Proceedings of the IEEE Conference on Computer Vision and Pattern Recognition, pp. 770–778 (2016)

23. Hernandez-Ortega, J., Galbally, J., Fierrez, J., Beslay, L.: Biometric quality: review and application to face recognition with faceqnet. arXiv preprint arXiv:2006.03298 (2020)

24. Hudson, D.A., Zitnick, L.: Generative adversarial transformers. In: International Conference on Machine Learning, pp. 4487–4499. PMLR (2021)

25. Jaiswal, A., AbdAlmageed, W., Wu, Y., Natarajan, P.: Bidirectional conditional generative adversarial networks. In: Jawahar, C.V., Li, H., Mori, G., Schindler, K. (eds.) ACCV 2018. LNCS, vol. 11363, pp. 216–232. Springer, Cham (2019). https://doi.org/10.1007/978-3-030-20893-6_14

26. Jaiswal, A., Moyer, D., Ver Steeg, G., AbdAlmageed, W., Natarajan, P.: Invariant representations through adversarial forgetting. In: Proceedings of the AAAI Conference on Artificial Intelligence, vol. 34, pp. 4272–4279 (2020)

27. Jiang, Y., Chang, S., Wang, Z.: TransGAN: two pure transformers can make one strong GAN, and that can scale up. In: Advances in Neural Information Processing Systems, vol. 34 (2021)

28. Kamiran, F., Calders, T.: Data preprocessing techniques for classification without discrimination. Knowl. Inf. Syst. **33**(1), 1–33 (2012)

29. Karkkainen, K., Joo, J.: Fairface: face attribute dataset for balanced race, gender, and age for bias measurement and mitigation. In: Proceedings of the IEEE/CVF Winter Conference on Applications of Computer Vision, pp. 1548–1558 (2021)

30. Karras, T., Laine, S., Aittala, M., Hellsten, J., Lehtinen, J., Aila, T.: Analyzing and improving the image quality of stylegan. In: Proceedings of the IEEE/CVF Conference on Computer Vision and Pattern Recognition, pp. 8110–8119 (2020)

31. Kim, B., Kim, H., Kim, K., Kim, S., Kim, J.: Learning not to learn: Training deep neural networks with biased data. In: Proceedings of the IEEE/CVF Conference on Computer Vision and Pattern Recognition, pp. 9012–9020 (2019)

32. Kyono, T., van Breugel, B., Berrevoets, J., van der Schaar, M.: Decaf: generating fair synthetic data using causally-aware generative networks. In: NeurIPS (2021)

33. Leino, K., Fredrikson, M., Black, E., Sen, S., Datta, A.: Feature-wise bias amplification. In: International Conference on Learning Representations (2019). https://openreview.net/forum?id=S1ecm2C9K7

34. Li, J., Abd-Almageed, W.: Information-theoretic bias assessment of learned representations of pretrained face recognition. In: 2021 16th IEEE International Conference on Automatic Face and Gesture Recognition (FG 2021), pp. 1–8. IEEE (2021)
35. Li, X., Jia, X., Jing, X.Y.: Negative-aware training: be aware of negative samples. In: ECAI 2020, pp. 1269–1275. IOS Press (2020)
36. Li, Y., Vasconcelos, N.: Repair: removing representation bias by dataset resampling. In: Proceedings of the IEEE/CVF Conference on Computer Vision and Pattern Recognition, pp. 9572–9581 (2019)
37. Ling, C.X., Sheng, V.S.: Cost-sensitive learning and the class imbalance problem. In: Encyclopedia of Machine Learning 2011, pp. 231–235 (2008)
38. Liu, L., Ren, Y., Lin, Z., Zhao, Z.: Pseudo numerical methods for diffusion models on manifolds. In: International Conference on Learning Representations (2022). https://openreview.net/forum?id=PlKWVd2yBkY
39. Liu, Z., Luo, P., Wang, X., Tang, X.: Deep learning face attributes in the wild. In: Proceedings of the IEEE International Conference on Computer Vision, pp. 3730–3738 (2015)
40. Maggipinto, M., Terzi, M., Susto, G.A.: Introvac: introspective variational classifiers for learning interpretable latent subspaces. Eng. Appl. Artif. Intell. **109**, 104658 (2022)
41. Maze, B., et al.: IARPA Janus benchmark-C: face dataset and protocol. In: 2018 International Conference on Biometrics (ICB), pp. 158–165. IEEE (2018)
42. Morales, A., Fierrez, J., Vera-Rodriguez, R., Tolosana, R.: Sensitivenets: learning agnostic representations with application to face images. IEEE Trans. Pattern Anal. Mach. Intell. **43**(6), 2158–2164 (2020)
43. Niemeyer, M., Geiger, A.: Giraffe: representing scenes as compositional generative neural feature fields. In: Proceedings of the IEEE/CVF Conference on Computer Vision and Pattern Recognition, pp. 11453–11464 (2021)
44. Pandey, K., Mukherjee, A., Rai, P., Kumar, A.: Diffusevae: efficient, controllable and high-fidelity generation from low-dimensional latents. arXiv preprint arXiv:2201.00308 (2022)
45. Patashnik, O., Wu, Z., Shechtman, E., Cohen-Or, D., Lischinski, D.: Styleclip: text-driven manipulation of stylegan imagery. In: Proceedings of the IEEE/CVF International Conference on Computer Vision, pp. 2085–2094 (2021)
46. Pezdek, K., Blandón-Gitlin, I., Moore, C.: Children's face recognition memory: more evidence for the cross-race effect. J. Appl. Psychol. **88**(4), 760–3 (2003)
47. Quadrianto, N., Sharmanska, V., Thomas, O.: Discovering fair representations in the data domain. In: Proceedings of the IEEE/CVF Conference on Computer Vision and Pattern Recognition, pp. 8227–8236 (2019)
48. Ramaswamy, V.V., Kim, S.S., Russakovsky, O.: Fair attribute classification through latent space de-biasing. In: Proceedings of the IEEE/CVF Conference on Computer Vision and Pattern Recognition, pp. 9301–9310 (2021)
49. Robinson, J.P., Livitz, G., Henon, Y., Qin, C., Fu, Y., Timoner, S.: Face recognition: too bias, or not too bias? In: Proceedings of the IEEE/CVF Conference on Computer Vision and Pattern Recognition Workshops, p. 1 (2020)
50. Sandfort, V., Yan, K., Pickhardt, P.J., Summers, R.M.: Data augmentation using generative adversarial networks (CycleGAN) to improve generalizability in CT segmentation tasks. Sci. Rep. **9**(1), 1–9 (2019)
51. Sattigeri, P., Hoffman, S.C., Chenthamarakshan, V., Varshney, K.R.: Fairness GAN: generating datasets with fairness properties using a generative adversarial network. IBM J. Res. Dev. **63**(4/5), 3-1 (2019)

52. Sharmanska, V., Hendricks, L.A., Darrell, T., Quadrianto, N.: Contrastive examples for addressing the tyranny of the majority. arXiv preprint arXiv:2004.06524 (2020)
53. Stone, R.S., Ravikumar, N., Bulpitt, A.J., Hogg, D.C.: Epistemic uncertainty-weighted loss for visual bias mitigation. In: Proceedings of the IEEE/CVF Conference on Computer Vision and Pattern Recognition, pp. 2898–2905 (2022)
54. Székely, G.J., Rizzo, M.L., Bakirov, N.K., et al.: Measuring and testing dependence by correlation of distances. Ann. Stat. **35**(6), 2769–2794 (2007)
55. Tartaglione, E., Barbano, C.A., Grangetto, M.: End: entangling and disentangling deep representations for bias correction. In: Proceedings of the IEEE/CVF Conference on Computer Vision and Pattern Recognition, pp. 13508–13517 (2021)
56. Terhörst, P., et al.: A comprehensive study on face recognition biases beyond demographics. IEEE Trans. Technol. Soc. **3**(1), 16–30 (2021)
57. Wang, A., Russakovsky, O.: Directional bias amplification. arXiv preprint arXiv:2102.12594 (2021)
58. Wang, M., Deng, W.: Mitigate bias in face recognition using skewness-aware reinforcement learning. arXiv preprint arXiv:1911.10692 (2019)
59. Wang, M., Deng, W., Hu, J., Tao, X., Huang, Y.: Racial faces in the wild: Reducing racial bias by information maximization adaptation network. In: Proceedings of the IEEE/CVF International Conference on Computer Vision, pp. 692–702 (2019)
60. Wang, X., Xie, L., Dong, C., Shan, Y.: Real-ESRGAN: training real-world blind super-resolution with pure synthetic data. In: Proceedings of the IEEE/CVF International Conference on Computer Vision, pp. 1905–1914 (2021)
61. Wang, Z., et al.: Towards fairness in visual recognition: effective strategies for bias mitigation. In: Proceedings of the IEEE/CVF Conference on Computer Vision and Pattern Recognition, pp. 8919–8928 (2020)
62. Xu, D., Yuan, S., Zhang, L., Wu, X.: Fairgan: fairness-aware generative adversarial networks. In: 2018 IEEE International Conference on Big Data (Big Data), pp. 570–575. IEEE (2018)
63. Xu, D., Yuan, S., Zhang, L., Wu, X.: Fairgan+: achieving fair data generation and classification through generative adversarial nets. In: 2019 IEEE International Conference on Big Data (Big Data), pp. 1401–1406. IEEE (2019)
64. Yin, X., Yu, X., Sohn, K., Liu, X., Chandraker, M.: Feature transfer learning for face recognition with under-represented data. In: Proceedings of the IEEE/CVF Conference on Computer Vision and Pattern Recognition, pp. 5704–5713 (2019)
65. Zemel, R., Wu, Y., Swersky, K., Pitassi, T., Dwork, C.: Learning fair representations. In: Proceedings of the 30th International Conference on Machine Learning, ICML 2013, vol. 28, pp. III-325–III-333. JMLR.org (2013)
66. Zhang, H., Grimmer, M., Ramachandra, R., Raja, K., Busch, C.: On the applicability of synthetic data for face recognition. In: 2021 IEEE International Workshop on Biometrics and Forensics (IWBF), pp. 1–6. IEEE (2021)
67. Zhang, Z., Song, Y., Qi, H.: Age progression/regression by conditional adversarial autoencoder. In: Proceedings of the IEEE Conference on Computer Vision and Pattern Recognition, pp. 5810–5818 (2017)
68. Zhao, J., Yan, S., Feng, J.: Towards age-invariant face recognition. IEEE Trans. Pattern Anal. Mach. Intell. (2020)
69. Zhao, J., Wang, T., Yatskar, M., Ordonez, V., Chang, K.W.: Men also like shopping: reducing gender bias amplification using corpus-level constraints. In: Proceedings of the 2017 Conference on Empirical Methods in Natural Language Processing, pp. 2941–2951 (2017). https://www.aclweb.org/anthology/D17-1319

70. Zhu, W., Zheng, H., Liao, H., Li, W., Luo, J.: Learning bias-invariant representation by cross-sample mutual information minimization. In: Proceedings of the IEEE/CVF International Conference on Computer Vision, pp. 15002–15012 (2021)
71. Zhu, X., Anguelov, D., Ramanan, D.: Capturing long-tail distributions of object subcategories. In: 2014 IEEE Conference on Computer Vision and Pattern Recognition, pp. 915–922 (2014). https://doi.org/10.1109/CVPR.2014.122
72. Zhu, X., Liu, Y., Li, J., Wan, T., Qin, Z.: Emotion classification with data augmentation using generative adversarial networks. In: Phung, D., Tseng, V.S., Webb, G.I., Ho, B., Ganji, M., Rashidi, L. (eds.) PAKDD 2018. LNCS (LNAI), vol. 10939, pp. 349–360. Springer, Cham (2018). https://doi.org/10.1007/978-3-319-93040-4_28

Weakly Supervised Invariant Representation Learning via Disentangling Known and Unknown Nuisance Factors

Jiageng Zhu[1,2,3(✉)], Hanchen Xie[2,3], and Wael Abd-Almageed[1,2,3]

[1] USC Ming Hsieh Department of Electrical and Computer Engineering, Los Angeles, USA
[2] USC Information Sciences Institute, Marina del Rey, USA
{jiagengz,hanchenx,wamageed}@isi.edu
[3] Visual Intelligence and Multimedia Analytics Laboratory, Marina del Rey, USA

Abstract. Disentangled and invariant representations are two critical goals of representation learning and many approaches have been proposed to achieve either one of them. However, those two goals are actually complementary to each other so that we propose a framework to accomplish both of them simultaneously. We introduce a weakly supervised signal to learn disentangled representation which consists of three splits containing predictive, known nuisance and unknown nuisance information respectively. Furthermore, we incorporate contrastive method to enforce representation invariance. Experiments shows that the proposed method outperforms state-of-the-art (SOTA) methods on four standard benchmarks and shows that the proposed method can have better adversarial defense ability comparing to other methods without adversarial training.

1 Introduction

Robust representation learning which aims at preventing overfitting and increasing generality can benefit various down-stream tasks [12,17,29]. Typically, a DNN learns to encode a representation which contains all factors of variations of data, such as pose, expression, illumination, and age for face recognition, as well as other nuisance factors which are unknown or unlabelled. Disentangled representation learning and invariant representation learning are often used to address these challenges.

For disentangled representation learning, Bengio *et al.* [1] define disentangled representation which change in a given dimension corresponding to variation of one and only one generative factors of the input data. Although many unsupervised learning methods have been proposed [3,13,16], Locatello *et al.* [22] have shown both theoretically and empirically that the factor variants disentanglement is impossible without supervision or inductive bias. To this end, recent works have adopted the concept of semi-supervised learning [24] and weakly supervised learning [23]. On the other hand, Jaiswal *et al.* [14] take an invariant representation learning perspective in which they split representation z into two parts $z = [z_p, z_n]$, where z_p only contains predictive related information, and z_n merely contains nuisance factors.

Supplementary Information The online version contains supplementary material available at https://doi.org/10.1007/978-3-031-25085-9_22.

L. Karlinsky et al. (Eds.): ECCV 2022 Workshops, LNCS 13808, pp. 382–395, 2023.
https://doi.org/10.1007/978-3-031-25085-9_22

Invariant representation learning aims to learn to encode predictive latent factors which are invariant to nuisance factors in inputs [14, 25, 26, 28, 32]. By removing information of nuisance factors, invariant representation learning achieves good performance in challenges like adversarial attack [6] and out-of-distribution generalization [14]. Furthermore, invariant representation learning has also been studied in the reinforcement learning settings [5].

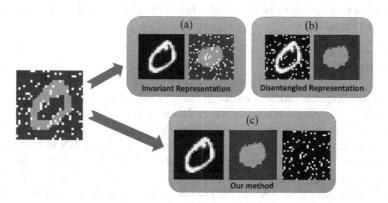

Fig. 1. Given an image with nuisance factors, (a): invariant representation learning splits predictive factors from all nuisance factors; (b): disentangled representation learning splits the known nuisance factors but leaving predictive and unknown nuisance factors; (c): Our method splits predictive, known nuisance and unknown nuisance factors simultaneously.

Despite the success of either disentangled or invariant representation learning methods, the relation between these two has not been thoroughly investigated. As shown in Fig. 1(a), invariant representation learning methods learn representations that maximize prediction accuracy by separating predictive factors from all other nuisance factors, while leaving the representations of both known and unknown nuisance factors entangled. Meanwhile, as illustrated in Fig. 1(b), although supervised disentangled representation methods can identify known nuisance factors, it fails to handle unknown nuisance factors, which may hurt downstream prediction tasks. Inspired by this observation, we propose a new training framework for seeking to achieve disentanglement and invariance of representation simultaneously. To split the known nuisance factors z_{nk} from predictive z_p and unknown nuisance factors z_{nu}, we introduce the weak supervision signals to achieve disentangled representation learning. To make predictive factors z_p independent of all nuisance factors z_n, we introduce a new invariant regularizer via reconstruction. The predictive factors from the same class are further aligned through contrastive loss to enforce invariance. Moreover, since our model achieve more robust representation comparing to other invariant models, our model is demonstrated to obtain better adversarial defense ability. In summary our main contributions are:

– Extending and combining both disentangled and invariant representation learning and proposing a novel approach to robust representation learning.

- Proposing a novel strategy for splitting the predictive, known nuisance factors and unknown nuisance factors, where mutual independence of those factors is achieved by the reconstruction step used during training.
- Outperforming state-of-the-art (SOTA) models on invariance tasks on standard benchmarks.
- Invariant latent representation trained using our method is also disentangled.
- Without using adversarial training, our model have better adversarial defense ability than other invariant models, which reflects that the generality of the model increases through our methods.

2   Related Work

Disentangled Representation Learning: Early works on disentangled representation learning aim at learning disentangled latent factors z by implementing an autoencoder framework [3,13,16]. Variational autoencoder (VAE) [17] is commonly used in disentanglement learning methods as basic framework. VAE uses DNN to map the high dimension input x to low dimension representation z. The latent representation z is then mapped to high dimension reconstruction $\hat{x}$. As shown in Eq. (1), the overall objective function to train VAE is the evidence lower bounds (ELBO) of likelihood $\log p_\theta(x_1, x_2, ...x_n)$, which contains two parts: quality of reconstruction and Kullback-Leibler divergence (D_{KL}) between distribution $q_\phi(z|x)$ and the assumed prior $p(z)$. Then, VAE uses the negative of ELBO, $L_{VAE} = -ELBO$, as loss function to update the parameters in the model.

$$L_{VAE} = -ELBO = -\sum_{i=1}^{N} \left[\mathbb{E}_{q_\phi(z|x^{(i)})}[\log p_\theta(x^{(i)}|z)] - D_{KL}(q_\phi(z|x^{(i)}||p(z)) \right] \quad (1)$$

Advanced methods based on VAE improve the disentanglement performance by implementing new disentanglement regularization. β-**VAE** [13] modifies the original VAE by adding a hyper-parameter β to balance the weights of reconstruction loss and D_{KL}. When $\beta > 1$, the model gains stronger disentanglement regularization. **AnnealedVAE** implements a dynamic algorithm to change the β from large to small value during training. **FactorVAE** [16] proposes to use a discriminator in order to distinguish between the joint distribution of latent factors $q(z)$ and multiplication of marginal distribution of every latent factor $\prod q(z_i)$. By using the discriminator, **FactorVAE** can automatically finds a better balance between reconstruction quality and disentangled representation. Compared to β-**VAE**, **DIP-VAE** [18] adds another regularization $D(q_\phi(z)||p(z))$ between the marginal distribution of latent factors $q_\phi(z) = \int q_\phi(z|x)p(x)dx$ and the prior $p(z)$ to further aid disentangled representation learning, where D can be any proper distance function such as mean square error. β-**TCVAE** proposed by [7] modifies the D_{KL} used in **VAE** into three part: *total correlation, index-coded mutual information* and *dimension-wise KL divergence*. To overcome the challenge proposed by [22], **AdaVAE** [23] purposely chooses pairs of inputs as supervision signal to learn representation disentanglement.

Invariant Representation Learning: The methods that aim at learning invariant representation can be classified into two groups: those methods that require annotations of nuisance factors [21, 25] and those that do not. A considerable number of approaches using nuisance factors annotations have been recently proposed. By implementing a regularizer which minimizes the Maximum Mean Discrepancy (MMD) on neural network (NN), The **NN+MMD** approach [21] removes affects of nuisance from predictive factors. On the basis of **NN+MMD**, The Variational Fair Autoencoder (**VFAE**) [25] uses special priors which encourage independence between nuisance factors and ideal invariant factors. The Controllable Adversarial Invariance (**CAI**) [32] approach applies the gradient reversal trick [9] which penalizes the model if latent representation has information of nuisance factors. **CVIB** [26] proposes a conditional form of Information Bottleneck (IB) and encourages the invariant representation learning by optimizing its variational bounds.

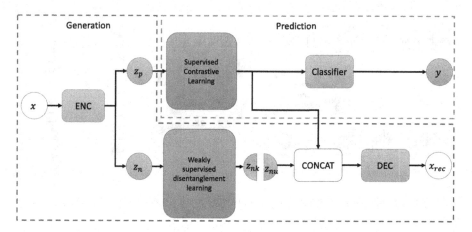

Fig. 2. Architecture of the model. Red box is the generation part and the blue box is the prediction part. Generation module aims at learning and splitting known nuisance factors z_{nk} and z_{nu}, and the prediction module aims at learning good predictive factors z_p. (Color figure online)

However, due to the constrains of demanding annotations, those methods take more effort to pre-process the data and encounter challenges when the annotations are inaccurate or insufficient. Comparing to annotation-eager approaches, annotation-free methods are easier to be implemented in practice. The Unsupervised Adversarial Invariance (**UAI**) [14] splits the latent factors into factors useful for prediction and nuisance factors. **UAI** encourages the independence of those two latent factors by incorporating competition between the prediction and the reconstruction objectives. **NN+DIM** [28] achieves invariant representation by using pairs of inputs and applying a neural network based mutual information estimator to minimize the mutual information between two shared representations. Furthermore, Sanchez et al. [28] employ a discriminator to distinguish the difference between shared representation and nuisance representation.

3 Learning Disentangled and Invariant Representation

3.1 Model Architecture

As illustrated in Fig. 2, the architecture of the proposed model contains two components: a generation module and a prediction module. Similar to VAE, the generation module performs the encoding-decoding task. However, it encodes the input x into latent factors z, $z = [z_p, z_n]$, where z_p represents the latent predictive factors that contains useful information for the prediction task, whereas z_n represents the latent nuisance factors and can be further divided into known latent factors z_{nk} and unknown nuisance factors z_{nu}.

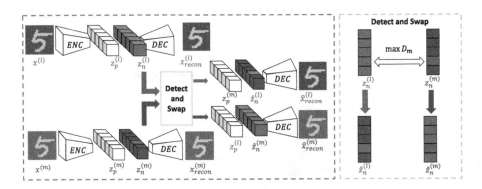

Fig. 3. Weakly supervised disentanglement representation learning for known nuisance factors z_{nk}

z_{nk} are discovered and separated from z_n via weakly supervised disentangled representation learning, where the joint distribution $p(z_{nk}) = \prod_i p(z_{nk_i})$. Since z_n is the split containing nuisance factor, after z_{nk} is identified, the remaining factors of z_n naturally result in unknown nuisance factors z_{nu}. Then, z_p and z_n are concatenated for generating reconstructions x_{rec} which are used to measure the quality of reconstruction. To enforce the independence between z_p and z_n, we add a regularizer using another reconstruction task, where the average mean and variance of predictive factors z_p are used to form new latent factors $\bar{z}_p$ and it will be discussed in Sect. 3.3. In the prediction module, we further incorporate contrastive loss to cluster the predictive latent factors belonging to the same class.

3.2 Learning Independent Known Nuisance Factors z_{nk}

As illustrated in Fig. 2, the known nuisance factors z_{nk} are discovered and separated from z_n, where $p(z_{nk}) = \prod_i p(z_{nk_i})$, since nuisance information is expected to be present only within z_n.

To fulfill the theoretical requirement of including supervision signal for disentangled representation learning as proven in [22], we use selected pairs of inputs $x^{(l)}$ and

$x^{(m)}$ as supervision signals, where only a few common generative factors are shared. As illustrated in Fig. 3, during training, the network encodes a pair of inputs $x^{(l)}$ and $x^{(m)}$ into two latent factors $z^{(l)} = [z_p^{(l)}, z_n^{(l)}]$ and $z^{(m)} = [z_p^{(m)}, z_n^{(m)}]$ respectively, which are then decoded to reconstruct $x_{rec}^{(l)}$ and $x_{rec}^{(m)}$. To encourage representation disentanglement, certain elements of $z_n^{(l)}$ and $z_n^{(m)}$ are *detected and swapped* to generate two new corresponding latent factors $\hat{z}^{(l)}$ and $\hat{z}^{(m)}$. The two new latent factors are then decoded to new reconstructions $\hat{x}_{rec}^{(l)}$ and $\hat{x}_{rec}^{(m)}$. By comparing $\hat{x}_{rec}$ with x_{rec}, the known nuisance factors z_{nk} are discovered and the elements of z_{nk} are enforced to be mutually independent with each other.

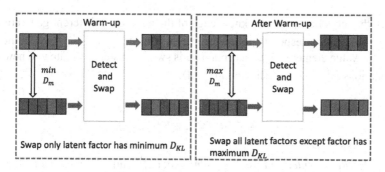

Fig. 4. In early training stages, small number of latent factors are swapped. the number of latent factors to be swapped increases gradually.

Selecting Image Pairs for Training and Latent Factors Assumptions: As mentioned by [13], the true world simulator using generative factors to generate x can be modeled as: $p(x|v, w) = Sim(v, w)$, where v is the generative factors and w is other nuisance factors. Inspired by this, we choose pairs of images by randomly selecting several generative factors to be the same and keeping the value of other generative factors to be random. Each image x has corresponding generative factors v, and the training pair is generated as follows: we first randomly select a sample $x^{(l)}$ whose generative factors are $v^{(l)} = [v_1, v_2, ...v_n]$. We then randomly change the value of k elements in $v^{(l)}$ to form a new generative factors $v^{(m)}$ and choose another sample $x^{(m)}$ according to $v^{(m)}$. During training, indices of different generative factors between $v^{(l)}$ and $v^{(m)}$, and the groundtruth value of all generative factors are not available to the model. The model is weakly supervised since it is trained with only the knowledge of the number of factors k that have changed. Ideally, if the model can learn a disentangled representation, the model will encode the image pair $x^{(l)}$ and $x^{(m)}$ to the corresponding representations $z^{(l)}$ and $z^{(m)}$ which have the characteristic shown in Eq. (2). We annotate the set of all different elements between $z^{(l)}$ and $z^{(m)}$ to be df_z and set of all latent factors to be d_z such that $df_z \subseteq d_z$.

$$p(z_{n_j}^{(l)}|\mathbf{x}^{(1)}) = p(z_{n_j}^{(m)}|\mathbf{x}^{(m)}); \ j \notin df_z$$
$$p(z_{n_i}^{(l)}|\mathbf{x}^{(1)}) \neq p(z_{n_i}^{(m)}|\mathbf{x}^{(m)}); \ i \in df_z$$
(2)

Detecting and Swapping the Distinct Latent Factors: VAE adopts the reparameterization to make posterior distribution $q_\theta(z|x)$ differentiable, where the posterior distribution of latent factors is commonly assumed to be a factorized multivariate Gaussian: $p(z|x) = q_\theta(z|x)$ [17]. By this assumption, we can directly measure the mutual information between the corresponding dimensions of the two latent representations $z^{(l)}$ and $z^{(m)}$ by measuring the divergence (D_m), which can be KL divergence (D_{KL}). We show the process of detecting distinct latent factors in Eq. (3), where a larger value of D_{KL} implies higher difference between the two corresponding latent factor distributions.

$$D_{KL}(q_\phi(z_i^{(l)}|x^{(l)})||q_\phi(z_i^{(m)}|x^{(m)})) = \frac{(\sigma_i^{(l)})^2 + (\mu_i^{(l)} - \mu_i^{(m)})^2}{2(\sigma_i^{(l)})^2} + log(\frac{\sigma_i^{(l)}}{\sigma_i^{(m)}}) - \frac{1}{2} \quad (3)$$

Since the model only has the knowledge of the number of different generative factors k, we swap all corresponding dimension elements of $z_n^{(l)}$ and $z_n^{(m)}$ except the top k highest D_m value elements. We incorporate this swapping step to create two new latent representations $\hat{z}_n^{(l)}$ and $\hat{z}_n^{(m)}$ shown in Eq. (4).

$$\begin{aligned} \hat{z}_{n_i}^{(l)} = z_{n_i}^{(m)}; \hat{z}_{n_i}^{(m)} = z_{n_i}^{(l)}; \ i \notin df_z \\ \hat{z}_{n_j}^{(l)} = z_{n_j}^{(l)}; \hat{z}_{n_j}^{(m)} = z_{n_j}^{(m)}; \ j \in df_z \end{aligned} \quad (4)$$

Disentangled Representation Loss Function: After $\hat{z}_n^{(l)}$ and $\hat{z}_n^{(m)}$ are obtained, they are concatenated with $z_p^{(m)}$ and $z_p^{(l)}$ respectively, to generate two new latent representations $\hat{z}^{(l)} = [z_p^{(m)}, \hat{z}_n^{(l)}]$ and $\hat{z}^{(m)} = [z_p^{(l)}, \hat{z}_n^{(m)}]$. $\hat{z}^{(l)}$ and $\hat{z}^{(m)}$ are decoded into new reconstructions $\hat{x}^{(l)}$ and $\hat{x}^{(m)}$. Since there are only k different generative factors between pair of images, ideally, after encoding the images, there should also be merely k pairs of different distributions on the latent representation space. By swapping other latent factors except them, the new representations $\hat{z}^{(l)}$ and $\hat{z}^{(m)}$ are the same with the original representations $z^{(l)}$ and $z^{(m)}$. Accordingly, the new reconstructions $\hat{x}_{rec}^{(l)}$ and $\hat{x}_{rec}^{(m)}$ should be identical to the original reconstruction $x_{rec}^{(l)}$ and $x_{rec}^{(m)}$. Therefore, we design the disentangled representation loss in Eq. (5), where D can be any suitable distance function *e.g.,* mean square error (MSE) or binary cross-entropy (BCE).

$$L = L_{VAE}(x_{rec}^{(l)}, z^{(l)}) + L_{VAE}(x_{rec}^{(m)}, z^{(m)}) + D(\hat{x}_{rec}^{(l)}, x_{rec}^{(l)}) + D(\hat{x}_{rec}^{(m)}, x_{rec}^{(m)}) \quad (5)$$

Training Strategies for Disentangled Representation Learning: To further improve the performance of disentangled representation learning, we design two strategies: *warmup by amount* and *warmup by difficulty*. Recalling that in the swapping step, the model needs to swap $|d_z| - k$ elements of latent representations. At beginning, exchanging too many latent factors will easily lead to mistakes. Therefore, in the first strategy, we gradually increase the number of latent factors being swapped from 1 to $|d_z| - k$. Further, to smoothly increase the training difficulty, we set the number of different generative factors to be 1 at the beginning and increase the number of different generative factors as training continues.

3.3 Learning Invariant Predictive Factors z_p

After we obtain the disentangled representation z_{nk}, the predictive factors z_p may still be entangled with z_{nu}. Therefore, we need to add other constraints to achieve fully invariant representation of z_p.

Making z_p Independent of z_n: As shown in [22], supervision signals need to be introduced for disentangled representation. Similarly, the independency of z_p and z_n also needs the help from a supervision signals as we discuss in Appendix. Luckily, for supervised training, a batch of samples naturally contains supervision signal. Similar to Eq. (2), the distribution of the representations z_p should be the same for the same class and can be shown in Eq. (6) where $C(x^{(l)})$ means the class of sample $x^{(l)}$.

$$p(z_p^{(l)}|\mathbf{x}^{(\mathbf{l})}) = p(z_p^{(m)}|\mathbf{x}^{(\mathbf{m})}); \; C(x^{(l)}) = C(x^{(m)})$$
$$p(z_p^{(l)}|\mathbf{x}^{(\mathbf{l})}) \neq p(z_p^{(m)}|\mathbf{x}^{(\mathbf{m})}); \; C(x^{(l)}) \neq C(x^{(m)})$$

(6)

Similar to the method we use for disentangled representation learning, we generate a new latent representation $\bar{z}_p$ and its corresponding reconstruction $\bar{x}_{rec-p}$. Then, we enforce the disentanglement between z_p and z_n by comparing the new reconstruction $\bar{x}_{rec-p}$ and x_{rec}. In contrast to the swapping method mentioned in Sect. 3.2, since the batch of samples used for training often contains more than two samples from the same class, the swapping method is hard to be implemented in this situation. Therefore, we generate the new latent representations $\bar{z}_p$ by calculating the average mean $\bar{\mu}_p$ and average variance $\bar{V}_p$ of the latent representations from the same class as shown in Eq. (7).

$$\bar{z}_p = \mathcal{N}(\bar{\mu}_p, \bar{V}_p); \; \bar{x}_{rec-p} = Decoder([\bar{z}_p, z_n])$$
$$\bar{\mu}_p = \frac{1}{|C|} \sum \mu_p^{(i)} \; ; \; \bar{V}_p = \frac{1}{|C|} \sum V_p^i; \; where \; \forall i \in C$$

(7)

We then generate the new reconstruction $\bar{x}_{rec-p}$ using the same decoder as in other reconstruction tasks and enforce the disentanglement of z_p and z_n by calculating the $D(x_{rec}, \bar{x}_{rec_p})$ and update the parameters of the model according to its gradient.

Constrastive Feature Alignment: To achieve invariant representation, we need to make sure the latent representation that is useful for prediction can also be clustered according to their corresponding classes. Even though the often used cross-entropy (CE) loss can accomplish similar goals, the direct goal of CE loss is to achieve logit-level alignment and change the representations distribution according to the logits, which does not guarantee the uniform distribution of features. Alternatively, we incorporate contrastive methods to ensure that representation/feature alignment can be accomplished effectively [31].

Similar to [15], we use supervised contrastive loss to achieve feature alignment and cluster the representations z_p according to their classes as shown in Eq. (8) where C is the set that contains samples from the same class and $y_p = y_i$.

$$\mathcal{L}_{sup} = \sum_{i \in I} \frac{-1}{|C|} \sum_{p \in C} \log \frac{\exp(z_i \cdot z_p / \tau)}{\sum\limits_{a \in A(i)} \exp(z_i \cdot z_a / \tau)}$$

(8)

The final loss function used to train the model, after adding the standard cross-entropy (CE) loss to train the classifier, is given by Eq. (9).

$$L = L_{CE}(x, y) + L_{VAE} + \alpha L_{disentangle} + \beta L_{Sup} + \gamma L_{Z_p}$$
$$L_{disentangle} = D(\hat{x}_{rec}^{(l)}, x_{rec}^{(l)}) + D(\hat{x}_{rec}^{(m)}, x_{rec}^{(m)}) \tag{9}$$
$$L_{Z_p} = D(\bar{x}_{rec-p}, x_{rec})$$

Table 1. Test average and worst accuracy results on Colored-MNIST, 3dShapes and MPI3D. **Bold, Black**: best result

Models	Colored-MNIST		3dShapes		MPI3D	
	Avg Acc	Worst Acc	Avg Acc	Worst Acc	Avg Acc	Worst Acc
Baseline	95.12 ± 2.42	66.17 ± 3.31	**98.87 ± 0.52**	96.89 ± 1.25	90.12 ± 3.13	87.89 ± 4.31
VFAE [25]	93.12 ± 3.07	65.54 ± 6.21	97.72 ± 0.81	93.34 ± 1.05	86.69 ± 3.12	82.43 ± 3.25
CAI [32]	93.56 ± 2.76	63.17 ± 5.61	97.62 ± 0.53	94.32 ± 0.89	86.63 ± 2.14	82.16 ± 5.83
CVIB [26]	93.31 ± 3.09	70.12 ± 4.77	97.11 ± 0.59	94.46 ± 0.90	87.04 ± 3.02	85.61 ± 2.08
UAI [14]	94.74 ± 2.19	74.25 ± 2.69	97.13 ± 1.02	95.21 ± 1.03	87.89 ± 4.23	83.01 ± 2.21
NN+DIM [28]	94.48 ± 2.35	80.25± 3.44	97.03 ± 1.07	96.02 ± 0.46	88.81 ± 1.37	82.01 ± 3.34
Our model	**97.96 ± 1.21**	**90.43 ± 2.79**	98.52 ± 0.51	**97.63 ± 0.72**	**91.32 ± 2.38**	**89.17 ± 2.69**

4   Experiments Evaluation

4.1   Benchmarks, Baselines and Metrics

The main objective of this work is to learn invariant representations and reduce overfitting to nuisance factors. Meanwhile, as a secondary objective, we also want to ensure that the learned representations are at least not less robust to adversarial attacks. Therefore, all models are evaluated on both invariant representation learning task and adversarial robustness task. We use four (4) dataset with different underlying factors of variations to evaluate the model:

– **Colored-MNIST** Colored-MNIST dataset is augmented version of MNIST [20] with two known nuisance factors: digit color and background color [28]. During training, the background color is chosen from three (3) colors and digit color is chosen from other six (6) colors. In test, we set the background color into three (3) new colors which is different from training set.
– **Rotation-Colored-MNIST** This dataset is further augmented version of Colored-MNIST. The background color and digit color setting is the same with the Colored-MNIST. This dataset further contains digits rotated to four (4) different angles $\Theta_{train} = \{0, \pm 22.5, \pm 45\}$. For test data, the rotation angles for digit is set to $\Theta_{test} = \{0, \pm 65, \pm 75\}$. The rotation angles are used as unknown nuisance factors.
– **3dShapes** [2] contains 480,000 RGB 64 × 64 × 3 images and the whole dataset has six (6) different generative factors. We choose object shape (four (4) classes) as the prediction task and only half number of object colors are used during training, and the remaining half of object color samples are used to evaluate performance of invariant representation.

– **MPI3D** [10] is a real-world dataset contains 1,036,800 RGB images and the whole dataset has seven (7) generative factors. Like 3dShapes, we choose object shape (six (6) classes) as the prediction target and half of object colors are used for training.

Prediction accuracy is used to evaluate the performance of invariant representation learning. Furthermore, we record both average test accuracy and worst-case test accuracy which was suggested by [27]. We find that using Eq. (9) directly does not guarantee good performance. This may be caused by inconsistent behavior of CE loss and supervised contrastive loss. Thus, we separately train the classifier using CE loss and use remaining part of total loss to train the rest of the model.

Table 2. Test average accuracy and worst accuracy results on Rotation-Colored-MNIST with different rotation angles. **Bold, Black**: best result

Models	Rotation-Colored-MNIST							
	Avg Acc	Worst Acc	Avg Acc	Worst Acc	Avg Acc	Worst Acc	Avg Acc	Worst Acc
	-75		-65		$+65$		$+75$	
Baseline	77.0 ± 1.3	62.3 ± 1.9	89.7 ± 1.2	77.5 ± 2.2	85.8 ± 1.2	65.8 ± 3.0	68.3 ± 2.2	49.9 ± 4.6
VFAE [25]	72.2 ± 2.4	58.9 ± 2.3	85.8 ± 1.7	74.4 ± 2.5	84.1 ± 2.1	64.6 ± 3.7	71.7 ± 1.3	48.0 ± 3.8
CAI [32]	74.9 ± 0.9	59.3 ± 3.9	86.5 ± 1.9	77.3 ± 2.0	84.2 ± 1.7	67.8 ± 1.9	64.7 ± 4.2	42.9 ± 3.7
CVIB [26]	76.1 ± 0.8	59.2 ± 3.0	88.6 ± 0.9	79.1 ± 1.2	85.6 ± 0.7	68.8 ± 2.9	72.2 ± 1.2	53.4 ± 2.6
UAI [14]	76.0 ± 1.7	61.1 ± 5.6	88.8 ± 0.7	80.0 ± 0.9	85.4 ± 1.6	68.2 ± 2.3	70.2 ± 0.9	51.1 ± 2.3
NN+DIM [28]	77.6 ± 2.6	69.2 ± 2.7	85.2 ± 3.4	76.3 ± 4.3	84.6 ± 3.1	66.7 ± 3.7	68.4 ± 3.1	53.2 ± 5.6
Our model	$\mathbf{81.0 \pm 2.1}$	$\mathbf{75.3 \pm 2.5}$	$\mathbf{90.8 \pm 1.6}$	$\mathbf{85.7 \pm 2.4}$	$\mathbf{87.3 \pm 2.5}$	$\mathbf{82.3 \pm 2.1}$	$\mathbf{73.2 \pm 2.3}$	$\mathbf{63.3 \pm 2.9}$

Meanwhile, the performance of representation disentanglement is also important for representation invariance since it can evaluate the invariance of latent factors representing known nuisance factors.

We adopt the following metrics to evaluate the performance of disentangled representation. All metrics range from 0 to 1, where 1 indicates that the latent factors are fully disentangled—(1) **Mutual Information Gap (MIG)** [7] evaluates the gap of top two highest mutual information between a latent factors and generative factors. (2) **Separated Attribute Predictability (SAP)** [18] measures the mean of the difference of perdition error between the top two most predictive latent factors. (3) **Interventional Robustness Score (IRS)** [30] evaluates reliance of a latent factor solely on generative factor regardless of other generative factors. (4) **FactorVAE (FVAE) score** [16] implements a majority vote classifier to predict the index of a fixed generative factor and take the prediction accuracy as the final score value. (5) **DCI-Disentanglement (DCI)** [8] calculates the entropy of the distribution obtained by normalizing among each dimension of the learned representation for predicting the value of a generative factor.

4.2 Comparison with Previous Work

We show invariance learning results which are the test average accuracy and worst accuracy in Tables 1 and 2. For Color-Rotation-MNIST dataset, since we rotate the test samples with $\theta \in \Theta_{test} = \{\pm65, \pm75\}$ and those angles are different with training rotation angles $\theta \in \Theta_{train} = \{0, \pm22.5, \pm45\}$, we record each average accuracy and worst accuracy under each rotation angles. The baseline model is the regular VGG16 model with no extra components for representation invariance. Our model largely outperforms prior work.

To compare the performance of disentanglement, we show the results of disentanglement representation learning in Table 3. Since the Color-MNIST and Rotation-Color-MNIST have only two generative factors, models for disentanglement learning tends to achieve nearly perfect disentanglement metric scores, which makes the results seems trivial. Therefore, we only record the disentanglement scores tested on **3dshapes** and **MPI3D** datasets.

Table 3. Disentanglement metrics on 3dShapes and MPI3D. **Bold, Black**: best result

Models	3dShapes					MPI3D				
	MIG	SAP	IRS	FVAE	DCI	MIG	SAP	IRS	FVAE	DCI
Unsupervised disentanglement learning										
β-VAE [13]	0.194	0.063	0.473	0.847	0.246	0.135	0.071	0.579	0.369	0.317
AnnealedVAE [3]	0.233	0.087	0.545	0.864	0.341	0.098	0.038	0.490	0.397	0.228
FactorVAE [16]	0.224	0.0440	0.630	0.792	0.304	0.092	0.031	0.529	0.379	0.164
DIP-VAE-I [18]	0.143	0.026	0.491	0.761	0.137	0.104	0.073	0.476	0.491	0.223
DIP-VAE-II [18]	0.137	0.020	0.424	0.742	0.083	0.131	0.075	0.509	0.544	0.244
β-TCVAE [7]	0.364	0.096	0.594	0.970	0.601	0.189	0.146	**0.636**	0.430	0.322
Weakly-supervised disentanglement learning										
Ada-ML-VAE [23]	0.509	0.127	0.620	0.996	0.940	0.240	0.074	0.576	0.476	0.285
Ada-GVAE [23]	0.569	0.150	0.708	0.996	**0.946**	0.269	0.215	0.604	0.589	0.401
Our model	**0.716**	**0.156**	**0.784**	0.996	0.919	**0.486**	**0.225**	0.615	**0.565**	**0.560**

To test the ability to defend adversarial attack without adversarial augmentation, we first train all models on **Colored-MNIST, CIFAR10** and **CIFAR100**. Then, we apply different adversarial attacks on those datasets. The adversarial attack types are: Fast Gradient Sign Method (FGSM) attack [11], Projected Gradient Descent (PGD) attack [19], and Carlini & Wagner (C&W) attack [4]. FSGM and PGD attacks results are shown in Fig. 5 and C&W attack results are included Appendix.

Table 4. Disentanglement metrics of with different training strategies applied to 3dShapes

Warmup by amount	Warmup by difficulty	MIG	SAP	IRS	FVAE	DCI
		0.492	0.096	0.661	0.902	0.697
✓		0.512	0.126	0.674	0.944	0.781
✓	✓	**0.716**	**0.156**	**0.784**	**0.996**	**0.919**

4.3 Ablation Study

Effectiveness of Training Strategies in Disentanglement Learning: To prove the effectiveness of the training strategies illustrated in Fig. 4, we compare results of three situations: (1) none of those strategies is used, (2) only *warmup by amount* strategy is used, and (3) both strategies are used. As shown in Table 4, using both training strategies clearly outperforms the others.

Table 5. Performance of different scheme on Colored-MNIST and Rotation-Colored-MNIST

Training scheme	Colored-MNIST		Rotation-Colored-MNIST (65)	
	Avg Acc	Worst Acc	Avg Acc	Worst Acc
L_{CE}	0.932	0.680	0.821	0.653
$L_{CE} + L_{contrastive}$	0.935	0.732	0.842	0.678
$L_{CE} \longleftrightarrow L_{contrastive}$	**0.980**	**0.904**	**0.873**	**0.823**

Separately Training the Classifier and the Rest of the Model: To prove the importance of the two-step training as mentioned in Sect. 4.1, we compare the results of training the entire model together versus separately training classifier and other parts. Further, we also record the results of our model which does not use contrastive loss for feature-level alignment. As shown in Table 5, either only using CE loss (L_{CE}) or training the whole

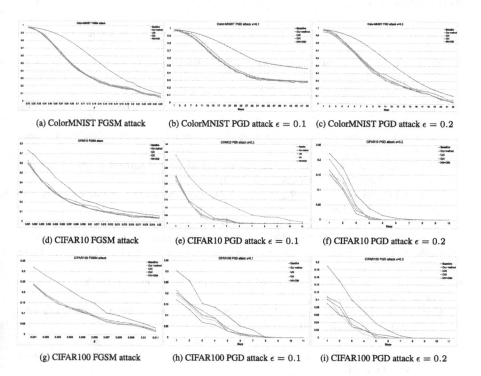

(a) ColorMNIST FGSM attack   (b) ColorMNIST PGD attack $\epsilon = 0.1$ (c) ColorMNIST PGD attack $\epsilon = 0.2$

(d) CIFAR10 FGSM attack   (e) CIFAR10 PGD attack $\epsilon = 0.1$ (f) CIFAR10 PGD attack $\epsilon = 0.2$

(g) CIFAR100 FGSM attack   (h) CIFAR100 PGD attack $\epsilon = 0.1$ (i) CIFAR100 PGD attack $\epsilon = 0.2$

Fig. 5. FGSM attacks and PGD attacks results

together ($L_{CE} + L_{contrastive}$) will harm the performance of the model. By separately training the classifier and other parts ($L_{CE} \longleftrightarrow L_{contrastive}$), the framework has the best results for both average and worst accuracy.

5 Conclusion

In this work, we extend the ideas of representation disentanglement and representation invariance by combining them to achieve both goals at the same time. By introducing contrastive loss and new invariant regularization loss, we make predictive factor z_p to be more invariant to nuisance and increase both average and worst accuracy on invariant learning tasks. Furthermore, we demonstrate that simultaneously achieving invariant and disentangled representation can increase the performance of adversarial defense comparing to merely using invariant representation learning.

Acknowledgement. This research is based upon work supported by the Defense Advanced Research Projects Agency (DARPA), under cooperative agreement number HR00112020009. The views and conclusions contained herein should not be interpreted as necessarily representing the official policies or endorsements, either expressed or implied, of DARPA or the U.S. Government. The U.S. Government is authorized to reproduce and distribute reprints for governmental purposes notwithstanding any copyright notation thereon.

References

1. Bengio, Y., Courville, A., Vincent, P.: Representation learning: a review and new perspectives (2014)
2. Burgess, C., Kim, H.: 3D shapes dataset (2018). https://github.com/deepmind/3dshapes-dataset/
3. Burgess, C.P., et al.: Understanding disentangling in β-VAE (2018)
4. Carlini, N., Wagner, D.: Towards evaluating the robustness of neural networks. In: 2017 IEEE Symposium on Security and Privacy (SP) (2017)
5. Castro, P.S.: Scalable methods for computing state similarity in deterministic Markov decision processes. In: AAAI (2020)
6. Chen, J., Konrad, J., Ishwar, P.: A cyclically-trained adversarial network for invariant representation learning. In: CVPR Workshops (2020)
7. Chen, R.T.Q., Li, X., Grosse, R., Duvenaud, D.: Isolating sources of disentanglement in variational autoencoders (2019)
8. Eastwood, C., Williams, C.K.I.: A framework for the quantitative evaluation of disentangled representations. In: ICLR (2018). https://openreview.net/forum?id=By-7dz-AZ
9. Ganin, Y., et al.: Domain-adversarial training of neural networks. J. Mach. Learn. Res. **17**(1), 2096–2130 (2016)
10. Gondal, M.W., et al.: On the transfer of inductive bias from simulation to the real world: a new disentanglement dataset (2019)
11. Goodfellow, I.J., Shlens, J., Szegedy, C.: Explaining and harnessing adversarial examples (2014). http://arxiv.org/abs/1412.6572, cite arxiv:1412.6572
12. He, K., Zhang, X., Ren, S., Sun, J.: Deep residual learning for image recognition (2015)
13. Higgins, I., et al.: β-VAE: learning basic visual concepts with a constrained variational framework. In: ICLR (2017)

14. Jaiswal, A., Wu, R.Y., Abd-Almageed, W., Natarajan, P.: Unsupervised adversarial invariance. In: Advances in Neural Information Processing Systems, vol. 31 (2018)

15. Khosla, P., et al.: Supervised contrastive learning. CoRR **abs/2004.11362** (2020). https:// arxiv.org/abs/2004.11362

16. Kim, H., Mnih, A.: Disentangling by factorising. In: ICML (2018). http://proceedings.mlr. press/v80/kim18b.html

17. Kingma, D.P., Welling, M.: Auto-encoding variational bayes. In: ICLR (2014)

18. Kumar, A., Sattigeri, P., Balakrishnan, A.: Variational inference of disentangled latent concepts from unlabeled observations. In: ICLR (2018)

19. Kurakin, A., Goodfellow, I.J., Bengio, S.: Adversarial machine learning at scale. CoRR **abs/1611.01236** (2016). http://arxiv.org/abs/1611.01236

20. LeCun, Y., Cortes, C.: MNIST handwritten digit database (2010). http://yann.lecun.com/ exdb/mnist/

21. Li, Y., Swersky, K., Zemel, R.: Learning unbiased features. arXiv preprint arXiv:1412.5244 (2014)

22. Locatello, F., et al.: Challenging common assumptions in the unsupervised learning of disentangled representations (2019)

23. Locatello, F., Poole, B., Raetsch, G., Schölkopf, B., Bachem, O., Tschannen, M.: Weakly-supervised disentanglement without compromises. In: ICML (2020). http://proceedings.mlr. press/v119/locatello20a.html

24. Locatello, F., et al.: Disentangling factors of variations using few labels. In: ICLR (2020). https://openreview.net/forum?id=SygagpEKwB

25. Louizos, C., Swersky, K., Li, Y., Welling, M., Zeme, R.: The variational fair autoencoder. In: ICLR (2016)

26. Moyer, D., Gao, S., Brekelmans, R., Galstyan, A., Ver Steeg, G.: Invariant representations without adversarial training. In: Advances in Neural Information Processing Systems, vol. 31 (2018)

27. Sagawa, S., Koh, P.W., Hashimoto, T.B., Liang, P.: Distributionally robust neural networks. In: ICLR (2020). https://openreview.net/forum?id=ryxGuJrFvS

28. Sanchez, E.H., Serrurier, M., Ortner, M.: Learning disentangled representations via mutual information estimation. In: Vedaldi, A., Bischof, H., Brox, T., Frahm, J.-M. (eds.) ECCV 2020. LNCS, vol. 12367, pp. 205–221. Springer, Cham (2020). https://doi.org/10.1007/978-3-030-58542-6_13

29. van Steenkiste, S., Locatello, F., Schmidhuber, J., Bachem, O.: Are disentangled representations helpful for abstract visual reasoning? CoRR **abs/1905.12506** (2019). http://arxiv.org/ abs/1905.12506

30. Suter, R., Ðorđe Miladinović, Schölkopf, B., Bauer, S.: Robustly disentangled causal mechanisms: validating deep representations for interventional robustness (2019)

31. Wang, F., Liu, H.: Understanding the behaviour of contrastive loss. In: CVPR (2021)

32. Xie, Q., Dai, Z., Du, Y., Hovy, E., Neubig, G.: Controllable invariance through adversarial feature learning. In: Advances in Neural Information Processing Systems, vol. 30 (2017)

Learning Visual Explanations for DCNN-Based Image Classifiers Using an Attention Mechanism

Ioanna Gkartzonika[ID], Nikolaos Gkalelis$^{(\boxtimes)}$[ID], and Vasileios Mezaris[ID]

CERTH-ITI, 6th Km Charilaou-Thermi Road, P.O. BOX 60361, Thessaloniki, Greece
{gkartzoni,gkalelis,bmezaris}@iti.gr

Abstract. In this paper two new learning-based eXplainable AI (XAI) methods for deep convolutional neural network (DCNN) image classifiers, called L-CAM-Fm and L-CAM-Img, are proposed. Both methods use an attention mechanism that is inserted in the original (frozen) DCNN and is trained to derive class activation maps (CAMs) from the last convolutional layer's feature maps. During training, CAMs are applied to the feature maps (L-CAM-Fm) or the input image (L-CAM-Img) forcing the attention mechanism to learn the image regions explaining the DCNN's outcome. Experimental evaluation on ImageNet shows that the proposed methods achieve competitive results while requiring a single forward pass at the inference stage. Moreover, based on the derived explanations a comprehensive qualitative analysis is performed providing valuable insight for understanding the reasons behind classification errors, including possible dataset biases affecting the trained classifier (Source code is made publicly available at: https://github.com/bmezaris/L-CAM).

Keywords: Explainable AI · XAI · Image classification · Class activation map · Deep convolutional neural networks · Attention · Bias

1 Introduction

During the last years, we are witnessing a breakthrough performance of DCNN image classifiers. However, the widespread commercial adoption of these methods is still hindered by the difficulty of users to attain some kind of explanations concerning the DCNN decisions. This lack of DCNN transparency affects especially the adoption of this technology in safety-critical applications as in the medical, security and self-driving vehicles industries, where a wrong DCNN decision may have serious implications. To this end, there is great demand for developing eXplainable AI (XAI) methods [6,13,14,16–18,21,23,24,34].

A category of XAI approaches for DCNN image classifiers that is currently getting increasing attention concerns methods that provide a visual explanation depicting the regions of the input image that contribute the most to the decision of the classifier. We should note that these approaches differ from methods used in weakly supervised learning tasks such as weakly supervised object localization and segmentation, where the goal is to locate the region of the target object and

L. Karlinsky et al. (Eds.): ECCV 2022 Workshops, LNCS 13808, pp. 396–411, 2023.
https://doi.org/10.1007/978-3-031-25085-9_23

not the image regions that contribute to the prediction of the classifier [15,19,30]. This for instance can be seen in the explanation examples of the various figures in our experimental evaluation section (Sect. 4.4), where often the focus region produced by the XAI approach does not coincide with the region of the object instance corresponding to the class label of the image.

Gradient-based class activation map (CAM) [8,11,27,28,36] and perturbation-based [22,26,31,35] approaches have shown promising explanation performance. Given an input image and its inferred class label, these methods generate a CAM, which is rescaled to the image size providing the so-called saliency map (SM); the SM indicates the image regions that the DCNN has focused on in order to infer this class. However, these methods are either based on backpropagating gradients [8,27,28], producing suboptimal SMs due to the well-known gradient problems [4], or require many forward passes at the inference stage [22,26,31,35], thus introducing significant computational overhead. Furthermore, the training dataset is not exploited in the exploration of the internal mechanisms concerning the decision process of the classifier. To this end, two new learning-based CAM methods are proposed, called L-CAM-Fm and L-CAM-Img, which utilize an appropriate loss function to train an attention mechanism [5] for generating visual explanations. Both methods can be used to generate explanations for arbitrary DCNN classifiers, are gradient-free and during inference require only one forward pass to derive a CAM and generate the respective SM of an input image. Experimental evaluation using VGG-16 and ResNet-50 backbones on ImageNet shows the efficacy of the proposed approaches in terms of both explainability performance and computational efficiency. Moreover, an extensive qualitative analysis using the generated SMs to explain misclassification errors leads to interesting conclusions including, among others, possible biases in the classifier's training dataset. In summary, the contributions of this paper are:

- We present the first, to the best of our knowledge, learning-based CAM framework for explaining image classifiers; this materializes into two XAI methods, L-CAM-Fm and L-CAM-Img.
- An appropriate loss function, consisting of the cross-entropy loss and an average and total variation CAM loss components, is employed during training, forcing the attention mechanism to extract CAMs of low energy that constitute good explanations.

The paper structure is: Related work and the proposed methods are presented in Sects. 2 and 3, while, experiments and conclusions in Sects. 4 and 5.

2   Related Work

We discuss here visual XAI approaches that are mostly related to ours. For a more comprehensive survey the reader is referred to [6,7,13,25].

Gradient-based CAM approaches (Fig. 1a) calculate a weight for each feature map of the last convolutional layer using the gradients backpropagated from the output; and, derive the CAM as the weighted sum of the feature maps [8,27,28].

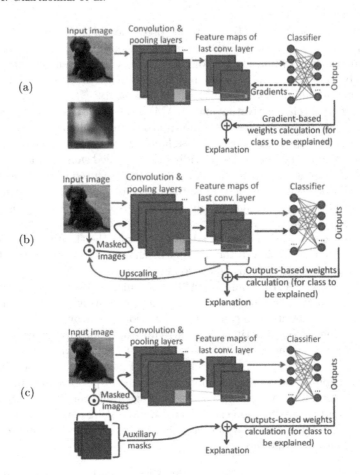

Fig. 1. Inference stage for literature XAI approaches: (a) Gradient-based CAM methods, (b) Perturbation-based methods with feature maps, and, (c) Perturbation-based methods with auxiliary masks. Blue components and arrows denote the (frozen) DCNN classifier and the flow of the original input image through its layers. Red arrows, indicating the flow of modified information such as masked images, and components in red, are introduced by the respective approach to derive the model decision's explanation (SMs). (Color figure online)

Grad-CAM [28] calculates the importance of each feature map by the gradients flowing from the output layer into the last convolutional layer. In [8], Grad-CAM++ utilizes a weighted combination of the positive partial derivatives of the last convolutional layer. Integrated Grad-CAM [27] introduces Integrated Gradient [32] to further improve the CAM's quality.

Perturbation-based approaches are gradient-free [22,26,31,35]. Both Score-CAM [35] and SIDU [21] derive the weight of each feature map by forward passing perturbed copies of the input image. Similarly, RISE [22] generates randomly masked versions of the input image to compute the aggregation weights. In [26], SISE selects feature maps at various depths, generates the so-called attribution masks and combines them using their classification scores. In [31], a fraction of the feature maps are adaptively selected by ADA-SISE, reducing the computational complexity of SISE. The general architecture of SIDU, Score-CAM, SISE and ADA-SISE is shown in Fig. 1b, while the respective architecture for RISE is depicted in Fig. 1c. The form of explanation (SM) produced by all methods is shown right below the input image in Fig. 1a.

3 Proposed Method

3.1 Problem Formulation

Let f be a DCNN model trained to categorize images to one of R different classes. Suppose an input image $\mathbf{X} \in \mathbb{R}^{W \times H \times C}$ that passes through f producing a model-truth label $y \in \{1, \ldots, R\}$, i.e. the top-1 class label inferred by f, and K feature maps extracted from f's last convolutional layer,

$$\mathbf{A} \in \mathbb{R}^{P \times Q \times K}, \tag{1}$$

where, W, H, C and P, Q, K, are the width, height and number of channels of $\mathbf{X}$ and $\mathbf{A}$, respectively, and $\mathbf{A}_{:,:,k}$ is the kth feature map. Given the above formulation, the goal of CAM-based methods is to derive an activation map from the K feature maps, the so-called CAM, and based on it generate the respective SM, visualizing the salient image regions that explain f's decision.

3.2 Training the Attention Mechanism

Consider a training set of R classes (the same classes that were used to train f), where each image $\mathbf{X}$ in the dataset is associated with a model-truth label y. This dataset is used to train an attention mechanism $g()$,

$$\mathbf{L}^{(y)} = g(y, \mathbf{A}), \tag{2}$$

where $\mathbf{L}^{(y)} \in \mathbb{R}^{P \times Q}$ is the CAM produced for a specified $\mathbf{X}$ and y. Specifically, the attention mechanism is implemented as follows

$$g(y, \mathbf{A}) = \sum_{k=1}^{K} w_k^{(y)} \mathbf{A}_{:,:,k} + b^{(y)} \mathbf{J}, \tag{3}$$

where, the weight matrix $\mathbf{W} = [\mathbf{w}^{(1)}, \ldots, \mathbf{w}^{(R)}]^T \in \mathbb{R}^{R \times K}$ and bias vector $\mathbf{b} = [b^{(1)}, \ldots, b^{(R)}]^T \in \mathbb{R}^R$ are the parameters of the attention mechanism, the transpose of vector $\mathbf{w}^{(r)} = [w_1^{(r)}, \ldots, w_K^{(r)}]^T \in \mathbb{R}^K$ is the rth row of $\mathbf{W}$, $w_k^{(r)} \in \mathbb{R}$ is the kth element of $\mathbf{w}^{(r)}$, and $\mathbf{J} \in \mathbb{R}^{P \times Q}$ is an all-ones matrix. That is, the model-truth label y at the input of $g()$ is used to select the class-specific weight vector and bias term from the yth row of $\mathbf{W}$ and $\mathbf{b}$, respectively.

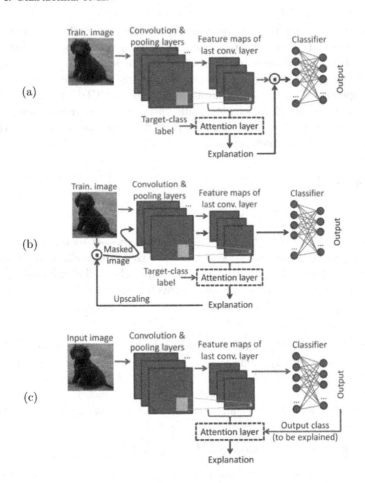

Fig. 2. Network architectures of the proposed approaches: (a) L-CAM-Fm training, (b) L-CAM-Img training, (c) L-CAM-Fm/-Img inference.

To learn the parameters of the attention mechanism, two different approaches, called L-CAM-Fm and L-CAM-Img, are proposed, with the corresponding network architectures shown in Figs. 2a and 2b. In both architectures, the attention mechanism is placed at the output of the last convolutional layer of the DCNN and the elements of the derived CAM are normalized to [0,1] using the element-wise sigmoid function $\sigma()$. In L-CAM-Fm, the CAM produced by the attention mechanism is used as a self-attention mask to re-weight the elements of the feature maps, i.e.,

$$\mathbf{A}_{:,:,k} \leftarrow \mathbf{A}_{:,:,k} \odot \sigma(\mathbf{L}^{(y)}), k = 1, \ldots, K, \tag{4}$$

where $\odot$ denotes element-wise multiplication. Contrarily, in L-CAM-Img the derived CAM is upscaled and applied to each channel of the input image,

$$\mathbf{X}_{:,:,c} \leftarrow \mathbf{X}_{:,:,c} \odot \theta(\sigma(\mathbf{L}^{(y)})), c = 1, \ldots, C, \qquad (5)$$

where, $\theta() : \mathbb{R}^{P \times Q} \rightarrow \mathbb{R}^{W \times H}$ is the upscaling operator (e.g. bilinear interpolation) and $\mathbf{X}_{:,:,c}$ is the cth channel of $\mathbf{X}$.

The overall architecture is trained end-to-end using an iterative gradient descent algorithm, where the attention mechanism's weights are updated at every iteration, while the weights of f remain fixed to their original values. The following loss function is used during training for both L-CAM-Fm and L-CAM-Img,

$$\lambda_1 TV(\sigma(\mathbf{L}^{(y)})) + \lambda_2 AV(\sigma(\mathbf{L}^{(y)})) + \lambda_3 CE(y, u), \qquad (6)$$

where, $CE(,)$ is the cross-entropy loss, u is the confidence score for class y derived using the L-CAM-Fm or -Img network, $\lambda_1, \lambda_2, \lambda_3$, are regularization parameters, and $AV()$, $TV()$ are the average and total variation operator, respectively. For the two latter operators we use the definitions presented in [9],

$$AV(\mathbf{S}) = \frac{1}{PQ} \sum_{p,q} (s_{p,q})^{\lambda_4}, \qquad (7)$$

$$TV(\mathbf{S}) = \sum_{p,q} [(s_{p,q} - s_{p,q+1})^2 + (s_{p,q} - s_{p+1,q})^2], \qquad (8)$$

where, $s_{p,q}$ is the element at the pth row and qth column of any tensor $\mathbf{S} \in \mathbb{R}^{P \times Q}$ and λ_4 is a fourth regularization parameter. The $AV()$ and $TV()$ are used to reduce the energy (i.e., the number of nonzero or high-valed elements) and remove spurious/noise areas in the SM, respectively. Intuitively, the above components, in synergy with the CE loss, guide the DCNN to learn more informative, fine-grained SMs, i.e., SMs that contain only a few high-valued elements corresponding to the image regions contributing mostly to classifier's decision. We should note that although f's weights are kept frozen, the gradients backpropagate through it and train effectively the attention mechanism, as explained for instance in [33]. Thus, the attention mechanism is forced to learn a transformation of the feature maps so that the CAM retains the regions of the input image that best explain f's decision.

3.3 Inference of Model Decision's Explanation

During the inference stage, the procedure to derive the CAM of a test image is the same for both L-CAM-Fm and L-CAM-Img (see Fig. 2c). That is, the test image is forward-passed through the DCNN to produce the corresponding feature maps and the model-truth label, which are then forwarded to the trained attention mechanism for computing the CAM (Eqs. (2), (3)). Similarly to [8,28], the explanation (SM) $\mathbf{V} \in \mathbb{R}^{W \times H}$ is then derived by

$$\mathbf{V}^{(y)} = \theta(\varsigma(\mathbf{L}^{(y)})), \qquad (9)$$

where, $\theta()$ is an upscaling operator to the input image size (e.g. bilinear interpolation), $\varsigma()$ is the min-max normalization operator, i.e., transform each element $l^{(y)}$ of $\mathbf{L}^{(y)}$ using, $l^{(y)} \leftarrow (l^{(y)} - \min(\mathbf{L}^{(y)}))/(\max(\mathbf{L}^{(y)}) - \min(\mathbf{L}^{(y)}))$, and, $\min(\mathbf{L}^{(y)})$, $\max(\mathbf{L}^{(y)})$ are the smallest and largest element of $\mathbf{L}^{(y)}$, respectively.

4   Experiments

4.1   Dataset

ImageNet [10], which is among the most popular datasets in the visual XAI domain, was selected for the experiments. It contains $R = 1000$ classes, 1.3 million images for training and 50K images for testing. Due to the prohibitively high computational cost of perturbation-based approaches that are used for experimental comparison, only 2000 randomly-selected testing images are used for evaluation, following an evaluation protocol similar to [35].

4.2   Experimental Setup

The proposed L-CAM-Fm and L-CAM-Img are compared against the top-performing approaches in the literature for which publicly-available code is provided, i.e., Grad-CAM [28], Grad-CAM++ [8], Score-CAM [35], and RISE [26] (using the implementations of [2] for the first three and of [1] for the fourth). Two sets of experiments are conducted with respect to the employed DCNN classifier, i.e., one using VGG-16 [20] and another using ResNet-50 [12]. In both cases, pretrained models from the PyTorch model zoo [3] are used.

The proposed approaches are trained using the loss of Eq. (6) with stochastic gradient descent, batch size 64 and learning rate 10^{-4}. The learning rate decay factor per epoch and total number of epochs are 0.75, 7 for the VGG-16 experiment and 0.95, 25 for the ResNet-50 one. The loss regularization parameters (Eqs. (6), (7)) are chosen empirically using the training set in order to minimize the total loss (Eq. (6)) and at the same time bring the different loss components at the same order of magnitude (thus ensuring that all of them contribute similarly to the loss function): $\lambda_1 = 0.01$, $\lambda_2 = 2$, $\lambda_3 = 1.5$, $\lambda_4 = 0.3$. We should note that in all experiments the proposed methods exhibit a quite stable performance with respect to the above optimization parameters. During training, each image is rescaled and normalized as done during training of the original DCNN classifier, i.e., its shorter side is scaled to 256 pixels, then random-cropped to $W \times H \times C$, where, $W = H = 224$ and $C = 3$ (the three RGB channels) and normalized to zero mean and unit variance. The same preprocessing is used during testing, except that center-cropping is applied. The size $P \times Q \times K$ of the feature maps tensor at the last convolutional layer of the DCNN is $P = Q = 14$, $K = 512$ and $P = Q = 7$, $K = 2048$ for VGG-16 and ResNet-50, respectively. For the compared CAM approaches the SM of an input image is derived as follows [8,28]: the derived CAM is normalized to [0,1] using the min-max operator and transformed to the size of the input image using bilinear interpolation (Eq. 9). In contrary, for Score-CAM and RISE, as proposed in their corresponding papers [26,35], the opposite procedure is followed to derive the SM, i.e., bilinear interpolation to the input image's size and then min-max normalization.

Table 1. Evaluation results for a VGG-16 (upper half) and ResNet-50 (lower half) backbone classifier using 2000 randomly-selected testing images of ImageNet. The best and 2nd-best performance for a given evaluation measure are shown in bold and underline, respectively.

	AD(100%)↓	IC(100%)↑	AD(50%)↓	IC(50%)↑	AD(15%)↓	IC(15%)↑	#FW↓
Grad-CAM [28]	32.12	22.1	58.65	9.5	84.15	2.2	1
Grad-CAM++ [8]	30.75	22.05	54.11	11.15	82.72	3.15	1
Score-CAM [35]	27.75	22.8	45.6	14.1	<u>75.7</u>	<u>4.3</u>	512
RISE [22]	**8.74**	**51.3**	<u>42.42</u>	<u>17.55</u>	78.7	**4.45**	4000
L-CAM-Fm*	20.63	31.05	51.34	13.45	82.4	3.05	1
L-CAM-Fm	16.47	35.4	47	14.45	79.39	3.65	1
L-CAM-Img*	18.01	37.2	50.88	12.05	82.1	3	1
L-CAM-Img	12.96	<u>41.25</u>	45.56	14.9	78.14	4.2	1
L-CAM-Img†	<u>12.15</u>	40.95	**37.37**	**20.25**	**74.23**	**4.45**	1
Grad-CAM [28]	13.61	38.1	29.28	23.05	**78.61**	3.4	1
Grad-CAM++ [8]	13.63	37.95	30.37	23.45	79.58	3.4	1
Score-CAM [35]	**11.01**	39.55	**26.8**	**24.75**	78.72	3.6	2048
RISE [22]	11.12	**46.15**	36.31	21.55	82.05	3.2	8000
L-CAM-Fm*	14.44	35.45	32.18	20.5	80.66	2.9	1
L-CAM-Fm	12.16	40.2	29.44	23.4	<u>78.64</u>	**4.1**	1
L-CAM-Img*	15.93	32.8	39.9	14.85	84.67	2.25	1
L-CAM-Img	<u>11.09</u>	<u>43.75</u>	<u>29.12</u>	<u>24.1</u>	79.41	<u>3.9</u>	1

4.3 Evaluation Measures

Two widely used evaluation measures, Average Drop (AD) and Increase in Confidence (IC) [8], are used in the experimental evaluation,

$$AD = \sum_{i=1}^{\Upsilon} \frac{max(0, f(\mathbf{X}_i) - f(\mathbf{X}_i \odot \phi_\nu(\mathbf{V}_i)))}{\Upsilon f(\mathbf{X}_i)} 100, \tag{10}$$

$$IC = \sum_{i=1}^{\Upsilon} \frac{\delta(f(\mathbf{X}_i \odot \phi_\nu(\mathbf{V}_i)) > f(\mathbf{X}_i))}{\Upsilon} 100, \tag{11}$$

where, $f()$ is the original DCNN classifier, $\phi_\nu()$ is a threshold function to select the ν percent higher-valued pixels of the SM [11,35], $\delta()$ returns 1 when the input condition is satisfied and zero otherwise, Υ is the number of test images, $\mathbf{X}_i$ is the ith test image and $\mathbf{V}_i$ is the respective SM produced by the XAI method under evaluation. Intuitively, both measures assess the pixel-wise contribution to the classification confidence score. Specifically, AD is the average model's confidence score drop when the masked test images are used, while, IC is the portion of

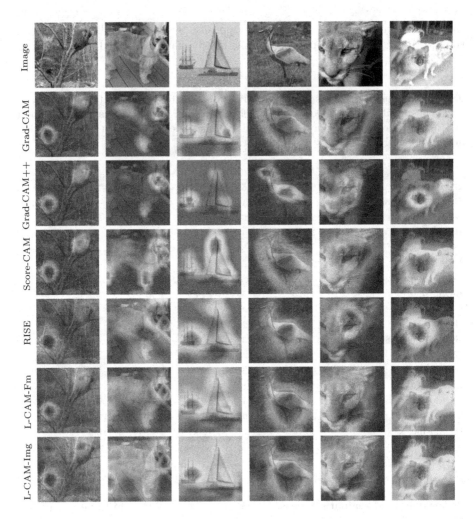

Fig. 3. Visualization of SMs with $\nu = 100\%$ from various XAI methods superimposed on the original input image to produce class-specific visual explanations for the VGG-16 (columns 1 to 3) and ResNet-50 (columns 4 to 6) backbones.

test images for which the model's confidence score increased when the masked images are used. Lower AD and higher IC indicate a better explanation.

4.4 Results

Comparisons and Ablation Study: The evaluation results in terms of $AD(\nu)$ and $IC(\nu)$ for different thresholds ν at $\phi_\nu()$, i.e., $\nu = 100\%, 50\%$ and 15%, are depicted in the upper and lower half of Table 1 for VGG-16 and ResNet-50, respectively. As an ablation study, we also report results for the proposed

| Image | Pug | Tiger cat | Image | soccer ball | Maltese |

Fig. 4. Two examples of using class-specific SMs (superimposed on the input image) produced by L-CAM-Img† on VGG-16 with $\nu = 100\%$ for classes "pug" and "tiger cat" (left) and classes "soccer ball" and "Maltese" (right).

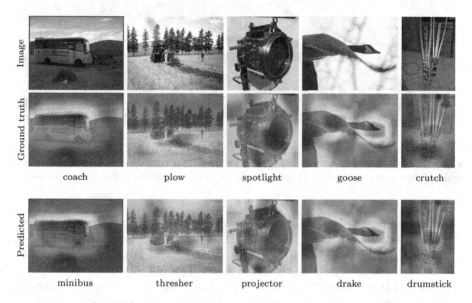

Fig. 5. Illustration of images and class-specific SMs (superimposed on the input image) whose ground truth and predicted labels are highly correlated.

methods when trained using only the CE loss, denoted as L-CAM-Fm* and L-CAM-Img*. The number of forward passes, #FW, needed to compute the SM for an input image at the inference stage, is also shown at the last column of this table. We should note that the auxiliary masks used by RISE (Fig. 1c) in the VGG-16 experiment are of size 7×7 [22] (which contrasts to the other approaches that use 14×14 feature maps for this experiment). For a fair comparison, we performed an additional experiment with the 7×7 feature maps after the last max pooling layer of VGG-16 using our L-CAM-Img, denoted as L-CAM-Img†. The results of this experiment are reported in the last row of the upper half of Table 1, under the L-CAM-Img's results (i.e. the ones obtained using the 14×14 feature maps). Moreover, qualitative results for the SMs produced by the different methods for six sample input images are shown in Fig. 3, while class-specific SM results for two images containing instances of two different classes are provided in Fig. 4. From the obtained results we observe the following:

Fig. 6. Illustration of images and class-specific SMs (superimposed on the input image), which (although single-labeled) contain instances of two different ImageNet classes.

i) The proposed L-CAM-Img generally outperforms the gradient-based approaches and is comparable in AD, IC scores to the perturbation-based approaches Score-CAM, RISE, yet contrarily to the latter requires only one FW instead of 512-8000 at the inference stage.

ii) L-CAM-Img† using 7×7 feature maps achieves the best performance in VGG-16; our approach is learning-based and, as the experiments showed, it is easier for it to learn the combination of the feature maps in the lower-dimensional space. This is consistent with the typical behavior of learning methods when working with high-dimensional data that may lay in a low-dimensional manifold (which is often the case with images), i.e. the curse of dimensionality.

iii) L-CAM-Img outperforms L-CAM-Fm, but the latter still generally outperforms the gradient-based approaches.

iv) The proposed approaches provide smooth SMs focusing on important regions of the image, as illustrated in Fig. 3 (and also shown from the very good results obtained for $\nu = 15\%$ in Table 1) and can produce class-specific explanations, as depicted in the examples of Fig. 4.

v) From the ablation study of employing only the CE loss (L-CAM-Fm*, L-CAM-Img*), we conclude that incorporating the two additional terms in the loss function (Eq. (6)) is very beneficial, especially for smaller values of ν.

Predicted SM = Ground truth SM ("rugby ball")

bobsled   football helmet

Fig. 7. Images and SMs (superimposed on the input image) from the category "rugby ball'. We observe that the classifier mostly learns the environment where the rugby activity takes place, e.g., football field, rugby players, playing rugby, rather than the rugby ball itself. In the absence of these clues the classifier fails to categorize correctly the image, as shown in the examples of the last two columns.

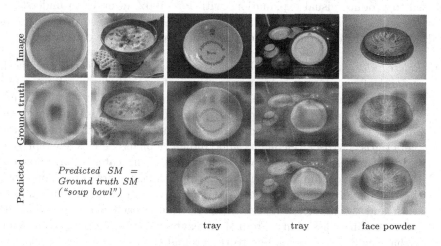

Predicted SM = Ground truth SM ("soup bowl")

tray   tray   face powder

Fig. 8. Images and SMs (superimposed on the input image) from the category "soup bowl". We see that the classifier has learned to classify to this category the soup bowls when they contain food; contrarily, empty soup bowls are miscategorized to other classes such as tray and face powder.

Qualitative Analysis and Discussion: In the following, a qualitative study is performed using the proposed L-CAM-Img[†]. Specifically, the proposed approach

Image

Ground truth

Predicted

Predicted SM =
Ground truth SM
("sunglass")

shoe shop   loudspeaker   seatbelt

Fig. 9. Images and SMs (superimposed on the input image) from the category "sunglass". We see that the classifier tends to focus on the sunglasses and the surrounding human face region; when the relevant human face region does not appear in the image or is occluded, the classifier infers the wrong category (e.g. shoe shop, loudspeaker).

is used to produce visual explanations with $\nu = 100\%$ in order to understand why the VGG-16 classifier may fail to categorize a test image correctly. To this end, we group the different classification error cases into three categories:

i) *Related classes*: Some ImageNet classes are very close to each other both semantically and/or in appearance. For instance, there are classes such as "maillot" and "bikini", "seacoast" and "promontory", "schooner" and "yawl", "cap" and "coffee mug", and others. The same is also true for many animals, e.g., "miniature poodle" and "toy poodle", "panther" and "panthera tigris", "African elephant" and "Indian elephant", etc. A few representative examples of images belonging to classes of this category are shown in Fig. 5. In each column of this figure, the input image is presented along with the SM (superimposed on the input image) corresponding to the ground truth and the predicted labels. Moreover, under each SM we provide the corresponding class name. From these examples, an interesting conclusion is that the SMs corresponding to the ground truth and predicted class are similar, i.e., in both cases the classifier focuses on the same image regions to infer the label of the image.

ii) *Multilabel images*: ImageNet is a single label dataset, i.e., each image is annotated with only one class label. However, some images may contain instances belonging to more than one ImageNet class. For instance, we have identified images containing together instances of the classes "screw" and "screwdriver", "warplane" and "aircraft carrier", "pier" and "boathouse", and other. For these cases, the classifier may correctly detect the instance of a class that is visible but does not correspond to the label of the image, which is considered a classification error. A few such examples are shown in Fig. 6. In contrast to the previous classification error category (related classes), we observe that now the SMs of the ground truth and predicted class differ significantly and usually identify a different region of the image.

iii) *Class bias*: Finally, we performed an analysis of the classification results in order to discover possible biases on specific ImageNet classes and understand how these biases affect the classifier decisions. To this end, representative examples from three classes are depicted in Figs. 7, 8, 9. From Fig. 7 we observe that the classifier has difficulty inferring the "soup bowl" label when an empty soup bowl is depicted. This finding, together with a visual inspection of the positive training samples for this class, which reveals that the training set is dominated with images depicting soup bowls filled with food, indicates that the classifier has in fact learned to detect mostly the "soup bowl filled with food" class instead of the more general "soup bowl". Similarly, from the results shown in Figs. 8 and 9 we get clear indications that in place of target class "rugby ball" the classifier to a large extent has learned to detect a broader "rugby game" class; and in place of target class "sunglass" the classifier has learned to detect the narrower "human face wearing sunglasses" class: when the sunglasses are shown but the human face is not visible, the classifier fails.

5   Conclusion

Two new visual XAI methods were presented, which, in contrast to current approaches, train an attention mechanism to produce explanations. We showed that it is possible to learn the feature maps' weights for deriving very good CAM-based explanations. We also performed a qualitative study using the explanations produced by our approach to shed light on the reasons why an image is misclassified, obtaining interesting conclusions, including the discovery of possible biases in the training set. As future work we plan to utilize the explanation masks for automatic bias detection, e.g., extending the work presented in [29].

Acknowledgments. This work was supported by the EU Horizon 2020 programme under grant agreements H2020-101021866 CRiTERIA and H2020-951911 AI4Media.

References

1. RISE implementation. https://github.com/eclique/RISE. Accessed 01 Feb 2022
2. Score-CAM with pytorch. https://github.com/yiskw713/ScoreCAM. Accessed 01 Feb 2022
3. TORCHVISION.MODELS.   https://pytorch.org/vision/stable/models.html. Accessed 01 Feb 2022

4. Adebayo, J., Gilmer, J., Muelly, M., Goodfellow, I., Hardt, M., Kim, B.: Sanity checks for saliency maps. In: Proceedings of NIPS, Montréal, Canada, pp. 9525–9536 (2018)

5. Bahdanau, D., Cho, K., Bengio, Y.: Neural machine translation by jointly learning to align and translate. In: Proceedings of ICLR, San Diego, CA, USA, pp. 2921–2929 (2015)

6. Bai, X., et al.: Explainable deep learning for efficient and robust pattern recognition: a survey of recent developments. Pattern Recognit. **120**, 108102 (2021)

7. Barredo Arrieta, A., et al.: Explainable artificial intelligence (XAI): concepts, taxonomies, opportunities and challenges toward responsible AI. Inf. Fusion **58**, 82–115 (2020)

8. Chattopadhay, A., Sarkar, A., Howlader, P., Balasubramanian, V.N.: Grad-CAM++: generalized gradient-based visual explanations for deep convolutional networks. In: Proceedings of IEEE WACV, Lake Tahoe, NV, USA, pp. 839–847 (2018)

9. Dabkowski, P., Gal, Y.: Real time image saliency for black box classifiers. In: Proceedings of NIPS, Long Beach, California, USA, pp. 6970–6979 (2017)

10. Deng, J., Dong, W., Socher, R., Li, L.J., Li, K., Fei-Fei, L.: ImageNet: a large-scale hierarchical image database. In: Proceedings of IEEE CVPR, Miami, FL, USA, pp. 248–255 (2009)

11. Desai, S., Ramaswamy, H.G.: Ablation-CAM: visual explanations for deep convolutional network via gradient-free localization. In: Proceedings of IEEE WACV, Snowmass Village, CO, USA, pp. 972–980 (2020)

12. He, K., Zhang, X., Ren, S., Sun, J.: Deep residual learning for image recognition. In: Proceedings of IEEE CVPR, Las Vegas, NV, USA, pp. 770–778 (2016)

13. Holzinger, A., Goebel, R., Fong, R., Moon, T., Müller, K.R., Samek, W.: XxAI - beyond explainable artificial intelligence. In: Proceedings of ICMLW, Vienna, Austria, pp. 3–10 (2020)

14. Hu, B., Vasu, B., Hoogs, A.: X-MIR: explainable medical image retrieval. In: Proceedings of WACV, Waikoloa, HI, USA, pp. 440–450 (2022)

15. Jiang, P.T., Zhang, C.B., Hou, Q., Cheng, M.M., Wei, Y.: LayerCAM: exploring hierarchical class activation maps for localization. IEEE Trans. Image Process. **30**, 5875–5888 (2021)

16. Jung, D., Lee, J., Yi, J., Yoon, S.: iCaps: an interpretable classifier via disentangled capsule networks. In: Vedaldi, A., Bischof, H., Brox, T., Frahm, J.-M. (eds.) ECCV 2020. LNCS, vol. 12364, pp. 314–330. Springer, Cham (2020). https://doi.org/10.1007/978-3-030-58529-7_19

17. Jung, S., Byun, J., Shim, K., Hwang, S., Kim, C.: Understanding VQA for negative answers through visual and linguistic inference. In: Proceedings of IEEE ICIP, Virtual Event/Anchorage, Alaska, USA, pp. 2873–2877 (2021)

18. Kim, J., Rohrbach, A., Darrell, T., Canny, J., Akata, Z.: Textual explanations for self-driving vehicles. In: Proceedings of ECCV, Munich, Germany, pp. 577–593 (2018)

19. Li, K., Wu, Z., Peng, K.C., Ernst, J., Fu, Y.: Tell me where to look: guided attention inference network. In: Proceedings of IEEE CVPR, Salt Lake City, UT, USA, pp. 9215–9223 (2018)

20. Liu, S., Deng, W.: Very deep convolutional neural network based image classification using small training sample size. In: Proceedings of ACPR, Kuala Lumpur, Malaysia, pp. 730–734 (2015)

21. Muddamsetty, S.M., Mohammad, N.S.J., Moeslund, T.B.: SIDU: similarity difference and uniqueness method for explainable AI. In: Proceedings of IEEE ICIP, Virtual Event, pp. 3269–3273 (2020)
22. Petsiuk, V., Das, A., Saenko, K.: RISE: randomized input sampling for explanation of black-box models. In: Proceedings of BMVC, Newcastle, UK (2018)
23. Plummer, B.A., Vasileva, M.I., Petsiuk, V., Saenko, K., Forsyth, D.: Why do these match? Explaining the behavior of image similarity models. In: Vedaldi, A., Bischof, H., Brox, T., Frahm, J.-M. (eds.) ECCV 2020. LNCS, vol. 12356, pp. 652–669. Springer, Cham (2020). https://doi.org/10.1007/978-3-030-58621-8_38
24. Prabhushankar, M., Kwon, G., Temel, D., AlRegib, G.: Contrastive explanations in neural networks. In: Proceedings of IEEE ICIP, Virtual Event, pp. 3289–3293 (2020)
25. Samek, W., Montavon, G., Vedaldi, A., Hansen, L.K., Müller, K. (eds.): Explainable AI: Interpreting, Explaining and Visualizing Deep Learning. Lecture Notes in Computer Science, vol. 11700. Springer, Cham (2019). https://doi.org/10.1007/978-3-030-28954-6
26. Sattarzadeh, S., et al.: Explaining convolutional neural networks through attribution-based input sampling and block-wise feature aggregation. In: Proceedings of AAAI, Virtual Event, pp. 11639–11647 (2021)
27. Sattarzadeh, S., Sudhakar, M., Plataniotis, K.N., Jang, J., Jeong, Y., Kim, H.: Integrated Grad-CAM: sensitivity-aware visual explanation of deep convolutional networks via integrated gradient-based scoring. In: Proceedings of IEEE ICASSP, Toronto, ON, Canada, pp. 1775–1779 (2021)
28. Selvaraju, R.R., Cogswell, M., Das, A., Vedantam, R., Parikh, D., Batra, D.: Grad-CAM: visual explanations from deep networks via gradient-based localization. In: Proceedings of IEEE ICCV, Venice, Italy, pp. 618–626 (2017)
29. Serna, I., Peña, A., Morales, A., Fiérrez, J.: InsideBias: measuring bias in deep networks and application to face gender biometrics. In: Proceedings of IEEE ICPR, Virtual Event/Milan, Italy, pp. 3720–3727 (2020)
30. Shi, X., Khademi, S., Li, Y., van Gemert, J.: Zoom-CAM: Generating fine-grained pixel annotations from image labels. In: Proceedings of IEEE ICPR, Virtual Event/Milan, Italy, pp. 10289–10296 (2020)
31. Sudhakar, M., Sattarzadeh, S., Plataniotis, K.N., Jang, J., Jeong, Y., Kim, H.: Ada-SISE: adaptive semantic input sampling for efficient explanation of convolutional neural networks. In: Proceedings of IEEE ICASSP, Toronto, ON, Canada, pp. 1715–1719 (2021)
32. Sundararajan, M., Taly, A., Yan, Q.: Axiomatic attribution for deep networks. In: Proceedings of ICML, Sydney, NSW, Australia, vol. 70, pp. 3319–3328 (2017)
33. Tsimpoukelli, M., Menick, J.L., Cabi, S., Eslami, S.M.A., Vinyals, O., Hill, F.: Multimodal few-shot learning with frozen language models. In: Proceedings of NIPS, Virtual Event, vol. 34, pp. 200–212 (2021)
34. Uehara, K., Murakawa, M., Nosato, H., Sakanashi, H.: Multi-scale explainable feature learning for pathological image analysis using convolutional neural networks. In: Proceedings of IEEE ICIP, Virtual Event, pp. 1931–1935 (2020)
35. Wang, H., et al.: Score-CAM: score-weighted visual explanations for convolutional neural networks. In: Proceedings of IEEE/CVF CVPRW, Virtual Event, pp. 111–119 (2020)
36. Zhou, B., Khosla, A., Lapedriza, À., Oliva, A., Torralba, A.: Learning deep features for discriminative localization. In: Proceedings of IEEE CVPR, Las Vegas, NV, USA, pp. 2921–2929 (2016)

Self-supervised Orientation-Guided Deep Network for Segmentation of Carbon Nanotubes in SEM Imagery

Nguyen P. Nguyen[1], Ramakrishna Surya[2], Matthew Maschmann[2], Prasad Calyam[1], Kannappan Palaniappan[1], and Filiz Bunyak[1(✉)]

[1] Department of Electrical Engineering and Computer Science, Columbia, USA
npntz3@mail.missouri.edu, {calyam,pal,bunyak}@missouri.edu
[2] Department of Mechanical and Aerospace Engineering,
University of Missouri-Columbia, Columbia, MO, USA
rst7b@mail.missouri.edu, maschmannm@missouri.edu

Abstract. Electron microscopy images of carbon nanotube (CNT) forests are difficult to segment due to the long and thin nature of the CNTs; density of the CNT forests resulting in CNTs touching, crossing, and occluding each other; and low signal-to-noise ratio electron microscopy imagery. In addition, due to image complexity, it is not feasible to prepare training segmentation masks. In this paper, we propose *CNTSegNet*, a dual loss, orientation-guided, self-supervised, deep learning network for CNT forest segmentation in scanning electron microscopy (SEM) images. Our training labels consist of weak segmentation labels produced by intensity thresholding of the raw SEM images and self labels produced by estimating orientation distribution of CNTs in these raw images. The proposed network extends a U-net-like encoder-decoder architecture with a novel two-component loss function. The first component is dice loss computed between the predicted segmentation maps and the weak segmentation labels. The second component is mean squared error (MSE) loss measuring the difference between the orientation histogram of the predicted segmentation map and the original raw image. Weighted sum of these two loss functions is used to train the proposed CNTSegNet network. The dice loss forces the network to perform background-foreground segmentation using local intensity features. The MSE loss guides the network with global orientation features and leads to refined segmentation results. The proposed system needs only a few-shot dataset for training. Thanks to it's self-supervised nature, it can easily be adapted to new datasets.

Keywords: Semantic segmentation · Self-supervised learning · Carbon nanotubes (cnt) · Electron microscopy

1 Introduction

Carbon nanotubes (CNTs), discovered in 1991, [20] have an intriguing combination of mechanical, thermal, electrical, and chemical properties [13,21].

© The Author(s), under exclusive license to Springer Nature Switzerland AG 2023
L. Karlinsky et al. (Eds.): ECCV 2022 Workshops, LNCS 13808, pp. 412–428, 2023.
https://doi.org/10.1007/978-3-031-25085-9_24

The unique physical properties of CNTs are a result of the hexagonal sp2-bonded graphene sheets that comprise their walls. Single-walled CNTs (SWNTs) may exhibit metallic or semiconducting properties depending on their chirality. SWNT transistor devices fabricated with sub-10 nm channel lengths have exhibited high current density of 2.4 mA/micron and low operating voltage of 0.5 V. Multi-walled CNTs (MWNTs) are metallic in nature, with diameters ranging from approximately 2–40 nm. MWNTs have been spun together to form yarns [43] that are electrically conductive and strong, yet capable of being tied into a knot.

Fig. 1. Carbon nanotube (CNT) pillar imaged using scanning electron microscope (SEM). (A-left) Full pillar view. (B-right) Zoomed side view of the CNT pillar.

Isolated, individual CNTs are difficult to synthesize and are often impractical for device-level integration. CNTs are more frequently grown as CNT forests – high density CNT populations synthesized on a support substrate. Crowding within a growing CNT population forces CNTs to orient vertically within a forest, normal to the growth substrate. Interactions between contacting CNTs generate persistent attractive van der Waals bonds which resist mechanical loads generated within the forest during synthesis. Individual CNTs within a forest in response to mechanical loading, leading to a CNT forest morphology that resembles an open-cell foam. The properties of the CNT forests are vastly diminished when compared to that of an individual CNT. For example the elastic modulus of an individual CNT is in the order of 1 TPa, while the compressive elastic modulus of a CNT forest may be as low as 10 MPa [32] – similar to that of natural rubber. The deformation mechanisms of compressed CNT forests are highly variable [7, 8, 18, 31–33, 39, 42] and are thought to result from variations in CNT forest morphology generated during cooperative synthesis [3, 28, 37]. CNT forests are candidates for the dry spinning of conductive, high-strength fibers[23, 43], piezoresistive sensing [29, 30, 36], electrochemical energy storage [9, 12], and thermal interface materials [10, 11]. Testing for physical properties of CNT forests often requires destruction of the forest which prevents further data collection. A method to determine physical properties of CNT forests indirectly using images of the said forest would help overcome the problem (Fig. 1). Thus a thorough analysis of CNT images is a critical step in determining the CNT forests' physical characteristics. The data obtained could then be used to determine how growth parameters can be modified to obtain a CNT forest with favorable properties. CNT image analytics aims to quantify CNT attributes such as orientation, linearity, density, diameter etc. The first step towards CNT feature characterization is segmentation. Earlier works on CNT image analytics relied on classical image processing approaches. In [14] thresholding was used to produce partial CNT masks to determining CNT diameters. In [41] class-entropy maximization was used to segment CNT images with modest magnification levels (800X-4000X). [40] thresholded image pixels into three classes: background, CNT, and uncertain areas. Feature vectors of class background and CNT extracted from small image patches were used to train a multi layer perceptron neural network. The network then classified pixels of uncertain area as either background or CNTs. However, this strategy was only effective with extremely sparse, non-overlapping CNTs in small patches. In [15,16] synthetic CNT forest images obtained from physics-based simulation have been analyzed using machine learning approaches to predict mechanical properties. While not developed for CNT image segmentation, recent works on detection and segmentation of other curvilinear structures such as fibers may be of interest for CNT image analytics. In [27] a 3D deep neural network pipeline was proposed for segmentation of short and thick glass fibers with acceptable density levels. The network was trained with a combination of synthetic data and real CT scan data with associated ground truths. In [4] an improved pipeline with deep center regression and geometric clustering was proposed for this type of glass fiber data.

As shown in Fig. 1B, our dataset contains long, thin, and dense CNT fibers. Clustering or thresholding-based, unsupervised segmentation methods lead to limited success. Because of data complexity and ambiguity, manual labeling is not feasible. Thus, there is a shortage of high-quality datasets with associated labels that can enable use of supervised learning based approaches. Self-supervising learning [22] has emerged as an approach to learn good representations from unlabeled data and to perform fine-tuning with labeled features at the down-stream tasks.

In this paper, we present a self-supervised deep learning network for segmentation of CNT forests in scanning electron microscopy (SEM) images. The imaged CNT forests were grown using an in-situ SEM synthesis process based on chemical vapor deposition (CVD) [26]. The proposed deep segmentation network relies on two complementary training labels. The first label, intensity thresholded raw input image, serves as a weak label that leads the network to perform binary CNT segmentation. The second label, CNT orientation histogram calculated directly from the raw input image, constraints the segmentation process by enforcing the network to preserve orientation characteristics of the original image. Experimental results demonstrate refined segmentation results without supervision and need for manual image annotation.

2   Methods

In order to enable characterization of CNT properties within dense CNT forests, we have developed a self-supervised segmentation method. According to [22], self-supervised learning trains a model by using pseudo labels that are generated automatically without the requirement for human annotations. The training procedure consists of two steps: a pretext task and a downstream task. Feature representation is learned in the pretext task first, whereas model adaptation is completed in the downstream task. The downstream task also evaluates the quality of features learned by the pretext task. Our proposed system consists of a novel deep neural network with two complementary loss functions which correspond to the pretext task and the downstream task. The following subsections describe the network architecture, loss functions, and generation of training labels.

2.1   Network Architecture

We have built a self-supervised deep neural network with two complementary loss functions named CNTSegNet. The architecture of this network is similar to the classical U-Net [38] architecture with the encoder and decoder branches. Input to the CNTSegNet consists of a single channel 2D grayscale image. Output of the CNTSegNet consists of a binary, single channel, 2D image. During training and testing, SEM images of CNT forests are fed to the network to generate binary segmentation masks of CNTs. The network's encoder uses the ResNet-34 model [17] as the backbone. It's a fully convolutional deep neural network with shortcut connections to learn residual features. The encoder includes three main layers with 16, 32, and 64 filters respectively. Figure 2 depicts the network architecture and the network training process. The network generates a pixel-wise class likelihood map which is binarized at the level of 0.7 to produce a binary segmentation mask. Orientation histogram for the prediction is computed from the generated class likelihood map. The proposed network is trained with weighted sum of two complementary loss functions described below.

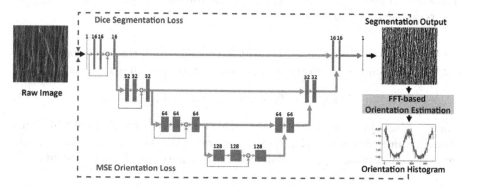

Fig. 2. Network architecture and training pipeline for the proposed dual loss and self-supervised network CNTSegNet. The network is based on an encoder-decoder architecture similar to U-Net segmentation network [17, 19, 38], but includes a second loss function and involves a self-supervised training scheme.

A) CNT Segmentation Loss: The first loss component aims to drive the network to perform segmentation prediction to match a given binary segmentation mask. Given a prediction mask $\text{Mask}_{\text{pred}}$ and a training mask $\text{Mask}_{\text{train}}$, dice loss is computed using the following equation

$$\text{Loss}_{\text{Dice}}(\text{Mask}_{\text{pred}}, \text{Mask}_{\text{train}}) = 1 - 2 \times \frac{|\text{Mask}_{\text{pred}} \cap \text{Mask}_{\text{train}}|}{|\text{Mask}_{\text{pred}}| + |\text{Mask}_{\text{train}}|} \quad (1)$$

Dice loss is a pixel-wise function that matches local image features in spatial domain. In this case, the dice loss by itself is not sufficient to generate reliable segmentation masks since the network is trained with automatically generated coarse weak labels rather than precise ground truth segmentation masks. This weakly supervised learning step plays the role of a pretext task.

B) CNT Orientation Loss: To compensate for weak labels, we introduced a second loss component. This second loss aims to drive the network to refine the predicted segmentation masks by enforcing the output to preserve the orientation patterns of the input. CNT forest orientation patterns is encoded using orientation histograms calculated without human annotation using the frequency domain methods proposed in [6,24] and briefly described in Sect. 2.2. Given the input and output orientation histograms h_{in} and h_{pred}, orientation loss is computed as the following equation

$$\text{Loss}_{\text{MSE}}(h_{\text{pred}}, h_{\text{in}}) = \sum_{b=1}^{n} (h_{pred}(b) - h_{in}(b))^2 \quad (2)$$

where n and b refer to number of bins in the histogram and index for an histogram bin. Optimizing this orientation loss will push the arrangement of foreground pixels in the segmentation mask towards the orientation of corresponding pixels in the raw images. This operation plays the role of a downstream task in self-supervised learning approach.

C) Total Loss: The proposed network is trained with a total loss computed as the weighted sum of the dice segmentation and MSE orientation losses

$$\text{Loss}_{\text{Total}} = k_1 \times \text{Loss}_{\text{Dice}} + k_2 \times \text{Loss}_{\text{MSE}} \quad (3)$$

where k_1 and k_2 refer to scalar weights. This total loss combines local and global image features extracted using spatial and frequency domain operations to preserve CNT morphology, particularly orientation patterns, during segmentation.

2.2   Generation of Training Labels

The two sets of training labels, the weak CNT segmentation masks and the CNT orientation histograms are generated as follows to enable self-supervised network training.

A) Weak CNT Segmentation Masks: To produce weak segmentation labels we explored two thresholding-based unsupervised methods. The first method, Otsu [35], is a global thresholding strategy that calculates an "ideal" threshold by maximizing the inter-class variance between background and foreground classes. The results of the Otsu approach are shown in the second columns of Fig. 3. Because of the increasing illumination, CNTs in the lower image regions are prominent while in the upper regions fade into the background. The second method, adaptive thresholding, generates improved segmentation results if the background intensity varies widely. Adaptive thresholding determines threshold values in local regions by computing the weighted mean of the local neighborhood minus an offset value [2]. The third column of Fig. 3 illustrates the outcomes of adaptive thresholding. These segmentation masks provide a greater level of detail. For training of the proposed deep neural network, we used weak segmentation masks generated using adaptive thresholding.

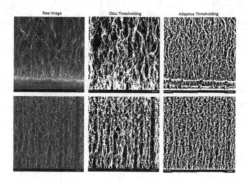

Fig. 3. Thresholding of CNT forest SEM images. Raw SEM image (first column), Otsu thresholding [35] (second column), adaptive thresholding [2] (last column).

B) CNT Forest Orientation Histograms: CNT forests' physical properties are strongly affected by orientation and alignment of the CNTs forming them. Alignment of the CNTs can enhance the mechanical, thermal, and electrical characteristics. Birefringence and linear dichroism, fluorescence polarization, and polarized Raman spectroscopy are some of the most commonly used methods to evaluate CNT alignment. [6,24].

In this study, we employed an image-based CNT forest orientation distribution estimation approach using radial sum method described in [6,24]. As the orientation feature is extracted by summing operation, it is possible to incorporate this feature to the computational pipeline and perform back-propagation to train our deep neural network. This approach can estimate the orientation of CNTs in either raw images or their binary masks. Figure 4 illustrates the processes required to estimate the orientation distribution of CNTs. Initially, the input image (raw image or binary mask) (Fig. 4A-B) is transformed into

Fourier space (Fig. 4D). The output is masked by a circle (Fig. 4E) divided into a number of bins (e.g. 360 bins, corresponding to 360°C). The total intensity in each bin indicates count of image pixels of that associated angle (Fig. 4C). This histogram represents the practical orientation distribution of CNTs in the input image. It's possible to fit this practical distribution as a mixture of several theoretical distributions as shown in Fig. 4F.

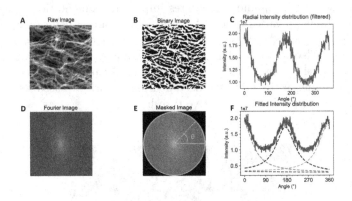

Fig. 4. CNT forest orientation distribution estimation using robust frequency-domain orientation estimation method described in [6,24].

3   Experimental Results

3.1   Datasets

The CNT forests used in this study were grown using chemical vapor deposition (CVD) [26] and were imaged using a FEI Quanta environmental SEM. The collected images had a resolution of 1536 × 1094 pixels and were acquired with a pixel dwell time of 10 μs. A CNT forest can grow up to a height of several millimeters and often has different morphology at different locations of the forest pillar [5], thus it is important to collect images from different locations to obtain data encapsulating all the morphologies in the CNT forest. SEM images in this study were collected at a magnification of 50,000X. Care was taken to prevent a large overlap between images. 110 image patches of size of 768 × 768 were used, with 34 of them going into the training set and the rest 76 patches going into the test set.

3.2   Training Process

We trained the proposed deep neural network with the dice and MSE loss functions separately and with the combined (dice + MSE) loss function to explore the effects of each loss component. During training, we maintained the same

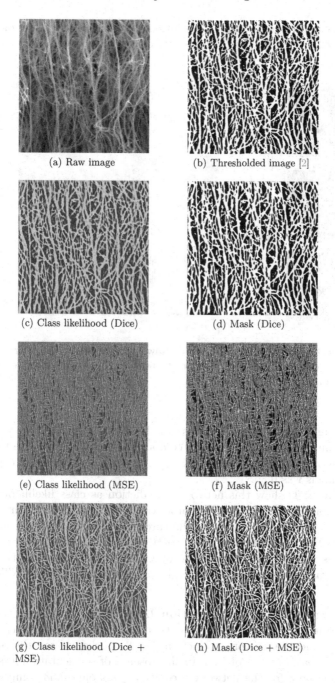

(a) Raw image

(b) Thresholded image [2]

(c) Class likelihood (Dice)

(d) Mask (Dice)

(e) Class likelihood (MSE)

(f) Mask (MSE)

(g) Class likelihood (Dice + MSE)

(h) Mask (Dice + MSE)

Fig. 5. Segmentation results for CNTSegNet with different loss functions. Raw SEM image of a sample CNT forest (a). Thresholded image [2] used as a weak segmentation label (b). Class likelihood maps and binary masks predicted using only the dice loss (c-d), only the MSE loss applied to orientation histogram (e-f), the combined dice and MSE loss functions (g-h).

learning rate and the same type of optimizer but varied the number of iterations for each loss function due to their distinct convergence processes.

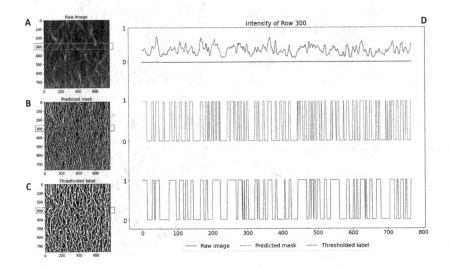

Fig. 6. Inspection of intensity profiles. (a) Raw SEM image; (b) mask predicted using CNTSegNet; (c) weak-label used to train the network; (d) associated intensity profiles at a sample image row.

A) Training with Dice Segmentation Loss. The network was first trained with just the dice segmentation loss component for 10 epochs. We used Adam optimizer with a learning rate of 5e-4. The final average dice score compared to the weak labels was 0.86.

Figure 5C & D show this network's prediction as class likelihood map and as binary mask. The binary mask produced by the network is smoother and has fewer tiny debris compared to the thresholded image due to the effects of convolution filters and upsampling layers of the network. However, since the network was trained with thresholded images rather than ground truth masks, prediction was coarse resulting in merging of many neighboring CNTs.

B) Training with MSE Orientation Loss. In an other task, the network was trained using just the MSE loss between the orientation vectors of the raw input image and the network prediction. For 8 epochs, we ran the Adam optimizer with a learning rate of 5e-4. In the absence of segmentation loss, the final average dice score for the network predictions was only 0.53 compared to the weak segmentation labels. This is due to the fact that the orientation vector includes just the summed global features. The loss function lacked the necessary local (pixel-wise) information needed to drive the segmentation mask.

Figure 5E & F show this network's prediction as class likelihood map and as binary mask. As expected, this approach dropped the dice scores significantly from 0.86 to 0.53.

C) Training with Two-Component Loss Function. In the final task, we trained the proposed network with a combined loss function consisting of dice segmentation and MSE orientation losses. We utilized an Adam optimizer with a learning rate of 5e-4. In the first 6 epochs, we initialized the segmentation output by training only with the dice loss. To regularize this output mask, for the following 3 epochs, we continued to train the network with a weighted sum of the dice segmentation and MSE orientation losses. As the maximum value of the dice loss is 1 but the magnitude of the orientation loss reaches much larger values at the levels of 1e+7 (as seen in Fig. 4C & F), we set the weight of the dice loss to 0.6, and the weight of the MSE loss to 1e-7 for a more balanced influence. This scheme resulted in an average dice score of 0.84 compared to the weak labels.

Figure 5G & H show the prediction class likelihood map and associated binary mask obtained using this dual loss network. In this scheme, dice loss guides the network to match the segmentation mask of the weak labels, while the MSE loss regularizes the segmentation process and guides the network to preserve the orientation properties of its input resulting in finer segmentation details.

Figure 6 aims to provide further visual insight. We plotted single-row intensity profiles (row=300) for a sample raw SEM image, the corresponding weak segmentation label used for network training, and the binary mask predicted using the proposed CNTSegNet network. The intensity peaks in the original signal correspond to individual CNTs. These plots show that the weak label tends to merge the neighboring peaks in the original signal resulting in wider/thicker foreground blocks. We can observe that the CNTSegNet prediction can better detect these intensity peaks resulting in narrower/thinner foreground blocks and more refined segmentation masks.

3.3 Segmentation Evaluation

We conducted inference on the test set after training the proposed CNTSegNet network with the combined loss function. We compared the performance of the CNTSegNet to three other segmentation methods, adaptive intensity thresholding [2], k-means clustering [1], and a recent unsupervised deep learning-based segmentation method [25]. K-means [1] is an unsupervised clustering method that partitions a dataset into K clusters where each data item belongs to the cluster with the closest centroid. We used k-means clustering to cluster the intensity levels into two clusters corresponding to CNTs and background. [25] is a novel unsupervised convolutional neural network using differentiable feature clustering to enable unsupervised image segmentation. The network includes a normalizing function to generate a response map from the network output, and an argument max function to select a cluster for each pixel. For a single

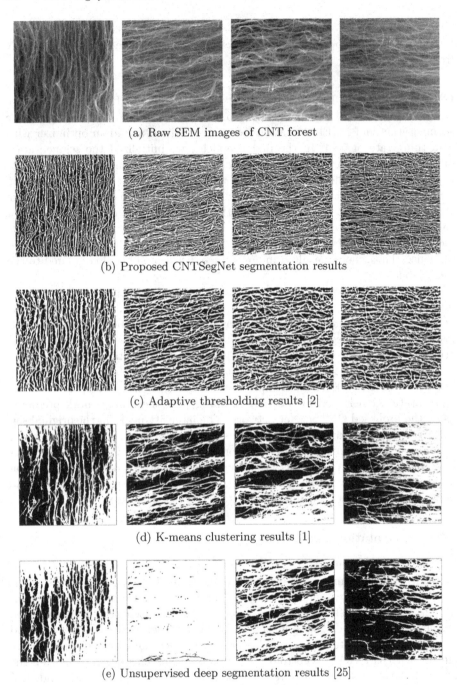

(a) Raw SEM images of CNT forest

(b) Proposed CNTSegNet segmentation results

(c) Adaptive thresholding results [2]

(d) K-means clustering results [1]

(e) Unsupervised deep segmentation results [25]

Fig. 7. Segmentation results for four sample images (columns 1–4). (a) Raw SEM images. (b) Segmentation masks obtained using the proposed CNTSegNet network. (c) Adaptive thresholding results [2]. (d) K-means clustering results [1]. (e) Unsupervised deep segmentation results using [25].

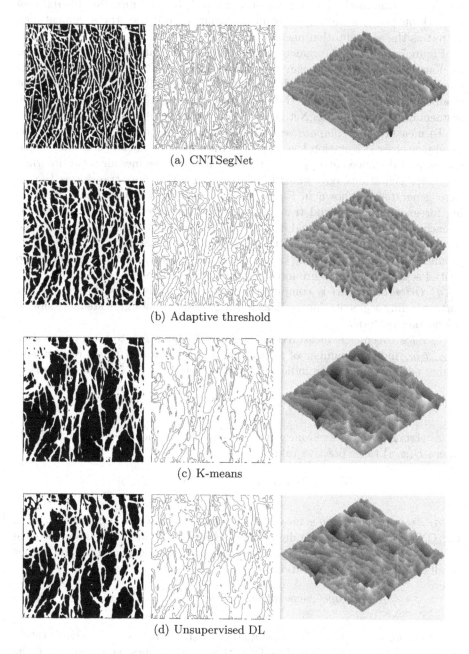

(a) CNTSegNet

(b) Adaptive threshold

(c) K-means

(d) Unsupervised DL

Fig. 8. Segmentation masks (left) and associated edge maps (middle), associated signed distance transform maps (right) for the proposed CNTSegNet network, adaptive thresholding [2], k-means clustering[1], and unsupervised deep segmentation [25] results.

test image, this method first trains the network to minimize the difference of network output and the argument-max output, then it uses the argument-max output as the segmentation mask.

Figure 7 depicts the segmentation results of four aforementioned methods for four sample images. It can be observed that CNTSegNet results in more refined segmentation masks with better recall of the individual CNTs compared to all three methods. Compared to k-means clustering [1] and unsupervised deep segmentation [25], CNTSegNet is also more robust to illumination variations.

To measure and compare segmentation quality we computed five unsupervised measures: orientation loss, edge coverage, average thickness, average separation, and distance entropy. In order to compute these measures two intermediate representations, edge maps $E(x, y)$ and signed distance transforms $D(x, y)$ were generated as shown in Fig. 8. An edge map is a binary image identifying foreground-background transitions. Signed distance transform [34] assigns to each pixel of the foreground its distance to the closest background point, and to each pixel of the background the opposite of its distance to the closest foreground point. Signed distance transform can be used to measure thickness and spatial separations of foreground structures in an image.

a) Orientation loss is computed between the original raw image and the corresponding segmentation mask as in Eq. 2. Lower values indicate more similar orientation patterns.

b) Edge coverage is measured as the ratio of edge pixels to total image area. n_{FG}, n_{BG} indicate number of foreground and background pixels respectively. Higher edge coverage is an indication of more details in the segmentation mask.

$$Edge\ coverage = \frac{100}{n_{FG} + n_{BG}} \sum_{x,y} E(x, y) \qquad (4)$$

c) Average thickness is measured as average distance on foreground pixels where $D(x, y)$ takes positive values.

$$Average\ thickness = \frac{1}{n_{FG}} \sum_{D(x,y)>0} D(x, y) \qquad (5)$$

d) Average separation is measured as average distance on background pixels. For CNT forest images containing dense clusters of thin CNTs, lower average thickness and lower average separation indicate finer segmentation and higher recall of CNTs in a segmentation mask.

$$Average\ \ separation = -\frac{1}{n_{BG}} \sum_{D(x,y)<0} D(x, y) \qquad (6)$$

e) Distance entropy is the last statistic we utilized to assess our segmentation outcomes. Entropy, which derives from thermal dynamics, is a measure of the disorder and uncertainty of a piece of information in information theory. It is computed by multiplying an event likelihood by its log probability

$$Distance\ entropy = -\sum_{x} p_D(x) log\ p_D(x) \qquad (7)$$

for our case p_D refers to probability distribution of signed distance transform. Lower distance entropy indicates lower variations in CNT thickness and CNT separation which is an indication of good segmentation for CNT forest images consisting of dense layouts of CNTs with similar diameters. Table 1 presents results of these unsupervised segmentation quality measures for the proposed CNTSegNet method and compared adaptive intensity thresholding [2], k-means clustering [1], and unsupervised deep learning-based segmentation [25] methods. The table indicates that the proposed CNTSegNet outperforms the compared methods in terms of all measures.

Table 1. Unsupervised evaluation of segmentation quality for the proposed CNTSeg-Net method and compared adaptive intensity thresholding [2], k-means clustering [1], and unsupervised deep learning-based segmentation [25] methods. The values indicate average for 76 test images. Underlined number in each row indicates the best result.

Measurement	CNTSegNet	Adaptive Threshold [2]	K-Means [1]	Unsup. DL [25]
Orientation loss ↓	9922	9977	10009	10038
Edge coverage ↑	15.64%	10.85%	9.37%	8.46%
Average thickness ↓	1.9965	2.8043	3.8478	7.9640
Average separation ↓	2.4233	3.2337	5.4858	11.2432
Distance entropy ↓	4.1677	4.9932	5.6466	5.7731

4   Conclusions

In this paper, we proposed a self-supervised deep neural network, CNTSegNet, with two complementary loss functions for segmentation of CNT forests in SEM imagery. Despite lack of supervision, the proposed network was able to generate more detailed segmentation masks indicated by various unsupervised segmentation quality measures. The network was also able to better preserve orientation characteristics as indicated by lower orientation losses. This was an important feature since CNT forest physical properties are strongly affected by orientation and alignment of CNTs forming them. Thanks to it's self-supervised nature, the proposed network is highly suitable for complex segmentation tasks where manual annotation is not practical or even feasible. The proposed network can easily be retrained using new datasets to improve performance or to adapt to new image characteristics. This study is our first step towards effective quantification of CNT forest characteristics from SEM imagery. Imaging and automated image analysis will be critical steps towards our ultimate goal of human out of the loop material discovery.

Acknowledgement. This work was partially supported by the National Science Foundation under award number CMMI-2026847. Any opinions, findings, and conclusions or recommendations expressed in this publication are those of the author(s) and do not necessarily reflect the views of the National Science Foundation.

References

1. OpenCV: K-Means Clustering in OpenCV (Aug 2022). https://docs.opencv.org/3.4/d1/d5c/tutorial_py_kmeans_opencv.html. (Accessed 12 Aug 2022)
2. Thresholding—skimage v0.13.1 docs (Jun 2022). http://devdoc.net/python/scikit-image-doc-0.13.1/auto_examples/xx_applications/plot_thresholding.html#id4. (Accessed 28 Jun 2022)
3. Abadi, P.P.S.S., et al.: Reversible tailoring of mechanical properties of carbon nanotube forests by immersing in solvents. Carbon **69**, 178–187 (2014)
4. Aguilar, C., Comer, M., Hanhan, I., Agyei, R., Sangid, M.: 3D fiber segmentation with deep center regression and geometric clustering. In: 2021 IEEE/CVF Conference on Computer Vision and Pattern Recognition Workshops (CVPRW), pp. 3741–3749. IEEE (June 2021). https://doi.org/10.1109/CVPRW53098.2021.00415
5. Bedewy, M., Meshot, E.R., Guo, H., Verploegen, E.A., Lu, W., Hart, A.J.: Collective mechanism for the evolution and self-termination of vertically aligned carbon nanotube growth. J. Phys. Chem. C **113**(48), 20576–20582 (2009)
6. Brandley, E., Greenhalgh, E.S., Shaffer, M.S.P., Li, Q.: Mapping carbon nanotube orientation by fast fourier transform of scanning electron micrographs. Carbon **137**, 78–87 (2018). https://doi.org/10.1016/j.carbon.2018.04.063
7. Brieland-Shoultz, A., Tawfick, S., Park, S.J., Bedewy, M., Maschmann, M.R., Baur, J.W., Hart, A.J.: Scaling the stiffness, strength, and toughness of ceramic-coated nanotube foams into the structural regime. Adv. Func. Mater. **24**(36), 5728–5735 (2014)
8. Cao, A., Dickrell, P.L., Sawyer, W.G., Ghasemi-Nejhad, M.N., Ajayan, P.M.: Super-compressible foamlike carbon nanotube films. Science **310**(5752), 1307–1310 (2005)
9. Carter, R., Davis, B., Oakes, L., Maschmann, M.R., Pint, C.L.: A high areal capacity lithium-sulfur battery cathode prepared by site-selective vapor infiltration of hierarchical carbon nanotube arrays. Nanoscale **9**(39), 15018–15026 (2017)
10. Cola, B.A., Xu, J., Cheng, C., Xu, X., Fisher, T.S., Hu, H.: Photoacoustic characterization of carbon nanotube array thermal interfaces. J. Appl. Phys. **101**(5), 054313 (2007)
11. Cola, B.A., Xu, X., Fisher, T.S.: Increased real contact in thermal interfaces: A carbon nanotube/foil material. Appl. Phys. Lett. **90**(9), 093513 (2007)
12. Davis, B.F., Yan, X., Muralidharan, N., Oakes, L., Pint, C.L., Maschmann, M.R.: Electrically conductive hierarchical carbon nanotube networks with tunable mechanical response. ACS Appli. Mater. Interfaces **8**(41), 28004–28011 (2016)
13. De Volder, M.F.L., Tawfick, S.H., Baughman, R.H., Hart, A.J.: Carbon nanotubes: present and future commercial applications. Science **339**(6119), 535–539 (2013). https://doi.org/10.1126/science.1222453
14. Gommes, C., et al.: Image analysis characterization of multi-walled carbon nanotubes. Carbon **41**(13), 2561–2572 (2003). https://doi.org/10.1016/S0008-6223(03)00375-0
15. Hajilounezhad, T., Bao, R., Palaniappan, K., Bunyak, F., Calyam, P., Maschmann, M.R.: Predicting carbon nanotube forest attributes and mechanical properties using simulated images and deep learning. NPJ Comput. Mater. **7**(1), 1–11 (2021)
16. Hajilounezhad, T., et al.: Exploration of carbon nanotube forest synthesis-structure relationships using physics-based simulation and machine learning. In: IEEE Applied Imagery Pattern Recognition Workshop (AIPR), pp. 1–8 (2019)

17. He, K., Zhang, X., Ren, S., Sun, J.: Deep residual learning for image recognition. In: 2016 IEEE Conference on Computer Vision and Pattern Recognition (CVPR), pp. 770–778 (2016). https://doi.org/10.1109/CVPR.2016.90

18. Hines, R., Hajilounezhad, T., Love-Baker, C., Koerner, G., Maschmann, M.R.: Growth and mechanics of heterogeneous, 3d carbon nanotube forest microstructures formed by sequential selective-area synthesis. ACS Appli. Mater. Interfaces **12**(15), 17893–17900 (2020)

19. Iakubovskii, P.: Segmentation models pytorch. www.github.com/qubvel/segmentation_models.pytorch (2019)

20. Iijima, S.: Helical microtubules of graphitic carbon. Nature **354**(6348), 56–58 (1991). https://doi.org/10.1038/354056a0

21. Iijima, S.: Carbon nanotubes: past, present, and future. Phys. B **323**(1), 1–5 (2002). https://doi.org/10.1016/S0921-4526(02)00869-4

22. Jing, L., Tian, Y.: Self-supervised visual feature learning with deep neural networks: a survey. IEEE Trans. Pattern Anal. Mach. Intell. **43**(11), 4037–4058 (2020). https://doi.org/10.1109/TPAMI.2020.2992393

23. Jung, Y., Cho, Y.S., Lee, J.W., Oh, J.Y., Park, C.R.: How can we make carbon nanotube yarn stronger? Compos. Sci. Technol. **166**, 95–108 (2018)

24. Kaniyoor, A., Gspann, T.S., Mizen, J.E., Elliott, J.A.: Quantifying alignment in carbon nanotube yarns and similar two-dimensional anisotropic systems. J. Appl. Polym. Sci. **138**(37), 50939 (2021). https://doi.org/10.1002/app.50939

25. Kim, W., Kanezaki, A., Tanaka, M.: Unsupervised learning of image segmentation based on differentiable feature clustering. IEEE Trans. Image Process. **29**, 8055–8068 (2020). https://doi.org/10.1109/TIP.2020.3011269

26. Koerner, G., Surya, R., Palaniappan, K., Calyam, P., Bunyak, F., Maschmann, M.R.: In-situ scanning electron microscope chemical vapor deposition as a platform for nanomanufacturing insights. In: ASME International Mechanical Engineering Congress and Exposition. vol. 85567, p. V02BT02A052 (2021)

27. Konopczyński, T., Kröger, T., Zheng, L., Hesser, J.: Instance Segmentation of Fibers from Low Resolution CT Scans via 3D Deep Embedding Learning. arXiv (Jan 2019). 10.48550/arXiv. 1901.01034

28. Maschmann, M.R.: Integrated simulation of active carbon nanotube forest growth and mechanical compression. Carbon **86**, 26–37 (2015)

29. Maschmann, M.R., Dickinson, B., Ehlert, G.J., Baur, J.W.: Force sensitive carbon nanotube arrays for biologically inspired airflow sensing. Smart Mater. Struct. **21**(9), 094024 (2012)

30. Maschmann, M.R., et al.: Bioinspired carbon nanotube fuzzy fiber hair sensor for air-flow detection. Adv. Mater. **26**(20), 3230–3234 (2014)

31. Maschmann, M.R.: Visualizing strain evolution and coordinated buckling within cnt arrays by in situ digital image correlation. Adv. Func. Mater. **22**(22), 4686–4695 (2012)

32. Maschmann, M.R., Zhang, Q., Du, F., Dai, L., Baur, J.: Length dependent foam-like mechanical response of axially indented vertically oriented carbon nanotube arrays. Carbon **49**(2), 386–397 (2011)

33. Maschmann, M.R., Zhang, Q., Wheeler, R., Du, F., Dai, L., Baur, J.: In situ sem observation of column-like and foam-like cnt array nanoindentation. ACS Appli. Mater. Interfaces **3**(3), 648–653 (2011)

34. Maurer, C.R., Qi, R., Raghavan, V.: A linear time algorithm for computing exact euclidean distance transforms of binary images in arbitrary dimensions. IEEE Trans. Pattern Anal. Mach. Intell. **25**(2), 265–270 (2003)

35. Otsu, N.: A threshold selection method from gray-level histograms. IEEE Trans. Syst. Man Cybern. **9**(1), 62–66 (1979). https://doi.org/10.1109/TSMC.1979. 4310076
36. Park, M., et al.: Effects of a carbon nanotube layer on electrical contact resistance between copper substrates. Nanotechnology **17**(9), 2294 (2006)
37. Pathak, S., et al.: Local relative density modulates failure and strength in vertically aligned carbon nanotubes. ACS Nano **7**(10), 8593–8604 (2013)
38. Ronneberger, O., Fischer, P., Brox, T.: U-Net: convolutional networks for biomedical image segmentation. In: Navab, N., Hornegger, J., Wells, W.M., Frangi, A.F. (eds.) MICCAI 2015. LNCS, vol. 9351, pp. 234–241. Springer, Cham (2015). https://doi.org/10.1007/978-3-319-24574-4_28
39. Tawfick, S., et al.: Mechanics of capillary forming of aligned carbon nanotube assemblies. Langmuir **29**(17), 5190–5198 (2013)
40. Trujillo, M.C.R., Alarcón, T.E., Dalmau, O.S., Zamudio Ojeda, A.: Segmentation of carbon nanotube images through an artificial neural network. Soft. Comput. **21**(3), 611–625 (2016). https://doi.org/10.1007/s00500-016-2426-1
41. Wortmann, T., Fatikow, S.: Carbon nanotube detection by scanning electron microscopy. In: Proceedings of the Eleventh IAPR Conference on Machine Vision Applications, MVA 2009 (2009)
42. Zbib, A.A., Mesarovic, S.D., Lilleodden, E.T., McClain, D., Jiao, J., Bahr, D.F.: The coordinated buckling of carbon nanotube turfs under uniform compression. Nanotechnology **19**(17), 175704 (2008). https://doi.org/10.1088/0957-4484/19/17/175704
43. Zhang, M., Atkinson, K.R., Baughman, R.H.: Multifunctional carbon nanotube yarns by downsizing an ancient technology. Science **306**(5700), 1358–1361 (2004)

W38 - Visual Object Tracking Challenge

W38 - Visual Object Tracking Challenge

The VOT challenges provide the tracking community with a precisely defined and repeatable way of comparing short-term trackers and long-term trackers as well as a common platform for discussing the evaluation and advancements made in the field of visual tracking. Following nine highly successful VOT challenges, the 10th Visual Object Tracking Challenge VOT2022 was held in spring of 2022 (challenge closed) hosting 7 subchallenges. This workshop includes results presentations, winning tracker talks, a keynote and contributed papers talks.

October 2022

Matej Kristan
Aleš Leonardis
Jiří Matas
Hyung Jin Chang
Joni-Kristian Kämäräinen
Roman Pflugfelder
Luka Čehovin Zajc
Alan Lukežič
Gustavo Fernández
Michael Felsberg
Martin Danelljan

The Tenth Visual Object Tracking VOT2022 Challenge Results

Matej Kristan[1], Aleš Leonardis[2(✉)], Jiří Matas[3], Michael Felsberg[4], Roman Pflugfelder[5,6,7,8], Joni-Kristian Kämäräinen[9], Hyung Jin Chang[2], Martin Danelljan[10], Luka Čehovin Zajc[1], Alan Lukežič[1], Ondrej Drbohlav[3], Johanna Björklund[11], Yushan Zhang[4], Zhongqun Zhang[2], Song Yan[9], Wenyan Yang[9], Dingding Cai[9], Christoph Mayer[10], Gustavo Fernández[5], Kang Ben[18], Goutam Bhat[10], Hong Chang[24], Guangqi Chen[16], Jiaye Chen[26], Shengyong Chen[43], Xilin Chen[24], Xin Chen[18], Xiuyi Chen[13], Yiwei Chen[35], Yu-Hsi Chen[12], Zhixing Chen[16], Yangming Cheng[55], Angelo Ciaramella[47], Yutao Cui[30], Benjamin Džubur[1], Mohana Murali Dasari[22], Qili Deng[16], Debajyoti Dhar[39], Shangzhe Di[14], Emanuel Di Nardo[46,47], Daniel K. Du[16], Matteo Dunnhofer[51], Heng Fan[48], Zhenhua Feng[50], Zhihong Fu[16], Shang Gao[41], Rama Krishna Gorthi[22], Eric Granger[27], Q. H. Gu[15], Himanshu Gupta[19], Jianfeng He[49], Keji He[13], Yan Huang[13], Deepak Jangid[19], Rongrong Ji[53], Cheng Jiang[30], Yingjie Jiang[26], Felix Järemo Lawin[4], Ze Kang[26], Madhu Kiran[27], Josef Kittler[50], Simiao Lai[18], Xiangyuan Lan[32], Dongwook Lee[34], Hyunjeong Lee[34], Seohyung Lee[34], Hui Li[26], Ming Li[17], Wangkai Li[49], Xi Li[55], Xianxian Li[20], Xiao Li[16], Zhe Li[41], Liting Lin[37], Haibin Ling[40], Bo Liu[25], Chang Liu[18], Si Liu[23], Huchuan Lu[18], Rafael M. O. Cruz[27], Bingpeng Ma[44], Chao Ma[36], Jie Ma[21], Yinchao Ma[49], Niki Martinel[51], Alireza Memarmoghadam[45], Christian Micheloni[51], Payman Moallem[45], Le Thanh Nguyen-Meidine[27], Siyang Pan[35], ChangBeom Park[34], Danda Paudel[10], Matthieu Paul[10], Houwen Peng[28], Andreas Robinson[4], Litu Rout[39], Shiguang Shan[24], Kristian Simonato[51], Tianhui Song[30], Xiaoning Song[26], Chao Sun[55], Jingna Sun[16], Zhangyong Tang[26], Radu Timofte[10,52], Chi-Yi Tsai[42], Luc Van Gool[10], Om Prakash Verma[19], Dong Wang[18], Fei Wang[49], Liang Wang[13], Liangliang Wang[16], Lijun Wang[18], Limin Wang[30], Qiang Wang[35], Gangshan Wu[30], Jinlin Wu[13], Xiaojun Wu[26], Fei Xie[38], Tianyang Xu[26], Wei Xu[16], Yong Xu[37], Yuanyou Xu[55], Wanli Xue[43], Zizheng Xun[14], Bin Yan[18], Dawei Yang[49], Jinyu Yang[41], Wankou Yang[38], Xiaoyun Yang[33], Yi Yang[55], Yichun Yang[30], Zongxin Yang[55], Botao Ye[24], Fisher Yu[10], Hongyuan Yu[13], Jiaqian Yu[35], Qianjin Yu[49], Weichen Yu[13], Kang Ze[26], Jiang Zhai[38], Chengwei Zhang[17], Chunhu Zhang[36], Kaihua Zhang[29], Tianzhu Zhang[49], Wenkang Zhang[38], Zhibin Zhang[43], Zhipeng Zhang[31], Jie Zhao[18], Shaochuan Zhao[26], Feng Zheng[41], Haixia Zheng[54], Min Zheng[16], Bineng Zhong[20], Jiawen Zhu[18], Xuefeng Zhu[26], and Yueting Zhuang[55]

[1] University of Ljubljana, Ljubljana, Slovenia
matej.kristan@fri.uni-lj.si

L. Karlinsky et al. (Eds.): ECCV 2022 Workshops, LNCS 13808, pp. 431–460, 2023.
https://doi.org/10.1007/978-3-031-25085-9_25

2 University of Birmingham, Birmingham, UK
a.leonardis@cs.bham.ac.uk
3 Czech Technical University, Prague, Czech Republic
4 Linköping University, Linkping, Sweden
5 Austrian Institute of Technology, Seibersdorf, Austria
6 TU Vienna, Vienna, Austria
7 TU Munich, Munich, Germany
8 Technion Israel Institute of Technology, Haifa, Israel
9 Tampere University, Tampere, Finland
10 ETH Zurich, Zurich, Switzerland
11 Umeå University, Umea, Sweden
12 Academia Sinica, New Taipei, Taiwan
13 AI School, Beijing, China
14 Beihang University, Beijing, China
15 Beijing Jiaotong University, Beijing, China
16 ByteDance, Beijing, China
17 Dalian Maritime University, Dalian, China
18 Dalian University of Technology, Dalian, China
19 Dr B R Ambedkar National Institute of Technology Jalandhar, Jalandhar, India
20 Guangxi Normal University, Guilin, China
21 Huaqiao University, Quanzhou, China
22 Indian Institute of Technology, Mumbai, India
23 Institute of Artificial Intelligence, Beijing, China
24 Institute of Computing Technology, Chinese Academy of Sciences, Beijing, China
25 JD Finance America Corporation, Mountain View, USA
26 Jiangnan University, Wuxi, China
27 LIVIA-École de technologie supérieure, Montreal, Canada
28 Microsoft Research Asia, Beijing, China
29 Nanjing University of Information Science and Technology, Nanjing, China
30 Nanjing University, Nanjing, China
31 NLP, Beijing, China
32 Peng Cheng Laboratory, Shenzhen, China
33 Remark AI, Las Vegas, China
34 Samsung Advanced Institute of Technology (SAIT), Yongin-si, Korea
35 Samsung R&D Institute China Beijing (SRCB), Beijing, China
36 Shanghai Jiao Tong University, Shanghai, China
37 South China University of Technology, Beijing, China
38 Southeast University, Nanjin, China
39 Space Applications Centre, Ahmedabad, India
40 Stony Brook University, New York, USA
41 Sustech, Shenzhen, China
42 Tamkang University, New Taipei, Taiwan
43 Tianjin University of Technology, Tianjin, China
44 University of Chinese Academy of Sciences, Beijing, China
45 University of Isfahan, Isfahan, Iran
46 University of Milan, Milano, Italy
47 University of Naples Parthenope, Naples, Italy
48 University of North Texas, Denton, USA
49 University of Science and Technology of China, Hefei, China
50 University of Surrey, Guildford, UK

[51] University of Udine, Udine, Italy
[52] University of Wurzburg, Wrzburg, Germany
[53] Xiamen University, Xiamen, China
[54] Xian Jiaotong University, Xian, China
[55] Zhejiang University, Hangzhou, China

Abstract. The Visual Object Tracking challenge VOT2022 is the tenth annual tracker benchmarking activity organized by the VOT initiative. Results of 93 entries are presented; many are state-of-the-art trackers published at major computer vision conferences or in journals in recent years. The VOT2022 challenge was composed of seven sub-challenges focusing on different tracking domains: (i) VOT-STs2022 challenge focused on short-term tracking in RGB by segmentation, (ii) VOT-STb2022 challenge focused on short-term tracking in RGB by bounding boxes, (iii) VOT-RTs2022 challenge focused on "real-time" short-term tracking in RGB by segmentation, (iv) VOT-RTb2022 challenge focused on "real-time" short-term tracking in RGB by bounding boxes, (v) VOT-LT2022 focused on long-term tracking, namely coping with target disappearance and reappearance, (vi) VOT-RGBD2022 challenge focused on short-term tracking in RGB and depth imagery, and (vii) VOT-D2022 challenge focused on short-term tracking in depth-only imagery. New datasets were introduced in VOT-LT2022 and VOT-RGBD2022, VOT-ST2022 dataset was refreshed, and a training dataset was introduced for VOT-LT2022. The source code for most of the trackers, the datasets, the evaluation kit and the results are publicly available at the challenge website (http://votchallenge.net).

Keywords: Visual Object Tracking challenge · VOT · Short-term tracking · Long-term tracking · Performance evaluation

1   Introduction

A decade ago, the Visual Object Tracking (VOT) initiative was founded in response to the lack of standardised performance evaluation in visual object tracking. To facilitate the development of this highly active computer vision field, the first VOT2013 challenge [13] was organized in conjunction with ICCV2013. Encouraged by the strong interest of the emerging community, eight VOT challenges have been organized since, with the results presented at the accompanying workshops at major computer vision conferences: ECCV2014 (VOT2014 [14]), ICCV2015 (VOT2015 [12]), ECCV2016 (VOT2016 [10]), ICCV2017 (VOT2017 [9]), ECCV2018 (VOT2018 [8]), ICCV2019 (VOT2019 [6]), ECCV2020 (VOT2020 [7]), ICCV2021 (VOT2021 [11]). The VOT challenge is now the main annual tracking performance evaluation event in computer vision.

The primary mission of the VOT initiative has been the promotion of the development of general trackers for single-camera, single-target, model-free, causal tracking. For nearly a decade the VOT has thus been a community-driven forum for gradual development and in-situ testing of performance evaluation protocols, dataset development and exploration of the tracking challenges landscape. The VOT2013 [13] started with a single short-term tracking challenge; VOT-ST. In VOT2014 [14] the VOT-TIR challenge was added to explore tracking in thermal imagery. In VOT2017 [9] the real-time tracking challenge VOT-RT was established to promote tracking speed and computational efficiency in parallel to robustness. Long-term tracking challenge VOT-LT was introduced in VOT2018 [8] and a year later in VOT2019 [6], multi-modal (RGB+thermal and RGB+depth) tracking challenges VOT-RGBT and VOT-RGBD were introduced.

Particular attention has been put on the development of informative performance evaluation measures. Two basic weakly correlated performance measures were introduced in VOT2013 [13] to evaluate the tracking accuracy and robustness of short-term trackers. A ranking-based methodology to identify the top performers was also proposed but was abandoned in VOT2015 [12] in favor of a more principled and interpretable combination of the primary scores in form of the expected average overlap score EAO. For the first seven VOT challenges, the measures were calculated under a reset-based protocol, in which a tracker is reset upon drifting off the target. This protocol was replaced in VOT2020 [7] by the anchor-based evaluation protocol that produces the most stable performance evaluation results compared to related protocols, yet inherits the benefits from the reset-based protocol. Similarly, a performance evaluation protocol and measures tailored for long-term tracking have been developed [16] and applied first in VOT2018 [8]. These measures have consistently shown good evaluation capabilities for long-term trackers.

Several datasets have been developed over the years. A dataset creation and maintenance protocol has been established for the main short-term tracking challenge to produce datasets which are sufficiently small for practical evaluation yet include a variety of challenging tracking situations for in-depth analysis. In VOT2017 [9], a sequestered dataset for identification of the short-term tracking challenge winner was introduced. This dataset has been refreshed along with the public versions over the years. Alongside, datasets specialized for long-term tracking, RGB+thermal and RGBD tracking were constructed and gradually updated.

In most of the VOT challenges, the trackers are required to report the target position as an axis-aligned bounding box. While this is a reasonable target state encoding, the VOT short-term tracking challenge gradually explored more detailed pose encodings to push the bar on tracking accuracy and expand the range of applications. Thus rotated bounding boxes were introduced in VOT2014 [14]. To reduce human annotation bias, VOT2016 [10] introduced fitting rotated bounding boxes to semi-automatically segmented objects in each frame. In VOT2020 [7] bounding boxes were abandoned and the short-term trackers are required to provide full target segmentation (the VOT-ST dataset was accordingly re-annotated to ensure high ground truth accuracy) – with this move, the VOT short-term tracking challenge has started narrowing the gap between visual object tracking and the related field of video object segmentation. The remaining challenges (VOT-LT, VOT-RGBD, VOT-RGBT) maintain axis-aligned target annotation.

This paper presents the tenth edition of the VOT challenges – the VOT2022 challenge. After two years of virtual editions due to the global Covid19 pandemic, the 10th anniversary of VOT was organized in a mixed form with in-person and online attendance, in conjunction with the ECCV2022 Visual Object Tracking VOT2022 Workshop. In the following, we overview the challenge and participation requirements.

1.1 The VOT2022 Challenge

The evaluation toolkit and the datasets are provided by the VOT2022 organizers. The challenges opened in the first week of April and closed on May 3rd. The winners of individual challenges were identified in late June, but not publicly disclosed. The results were presented at the ECCV2022 VOT2022 workshop on 24th October. The VOT2022 challenge contained seven challenges:

1. **VOT-STs2022 challenge** addressed short-term tracking by target segmentation in RGB images.
2. **VOT-STb2022 challenge** addressed short-term tracking by bounding boxes in RGB images.
3. **VOT-RTs2022 challenge** addressed the same class of trackers as VOT-STs2022, except that the trackers had to process the sequences in real-time.
4. **VOT-RTb2022 challenge** addressed the same class of trackers as VOT-STb2022, except that the trackers had to process the sequences in real-time.
5. **VOT-LT2022 challenge** addressed long-term tracking by bounding boxes in RGB images.
6. **VOT-RGBD2022 challenge** addressed short-term tracking by bounding boxes in RGB+depth (RGBD) imagery.
7. **VOT-D2022 challenge** addressed short-term tracking by bounding boxes in depth map images.

The authors participating in the challenge were required to integrate their tracker into the VOT2022 evaluation kit, which automatically performed a set of standardized experiments. The results were analyzed according to the VOT2022 evaluation methodology.

Participants were encouraged to submit their own new or previously published trackers as well as modified versions of third-party trackers. In the latter case, modifications had to be significant enough for acceptance. Participants were expected to submit a single set of results per tracker If a participant coauthored several submissions with a similar design, only the top performer from this *cluster* was considered to compete in the final top-performer ranking and winner identification.

Each submission was accompanied by a short abstract describing the tracker, which was used for the short tracker descriptions in Appendix [5] – the authors were asked to provide a clear description useful to the readers of the VOT2022 results report. In addition, participants filled out a questionnaire on the VOT submission page to categorize their tracker according to various design properties. Authors were encouraged to submit their tracker integrated into a Singularity container provided by VOT, which allows result reproduction and aids potential further evaluation. The participants with sufficiently well-performing submissions who contributed to the text for this paper and agreed to make their tracker code publicly available from the VOT page (or upon request) were offered co-authorship of this results paper. The committee reserved the right to disqualify any tracker that, by their judgement, attempted to cheat the evaluation protocols or failed in the post-hoc evaluation.

Methods considered for prizes in the VOT2022 challenge were not allowed to be trained on certain datasets (OTB, VOT, ALOV, UAV123, NUSPRO, Temple-Color and RGBT234), except for VOT-LT2022, where the VOT-LT2021 dataset was allowed. For GOT10k, a list of 1k prohibited sequences was created in VOT2019, while the remaining 9k+ sequences were allowed for learning. The reason was that part of the GOT10k was used in the VOT-ST2022 dataset.

The use of class labels specific to VOT was not allowed (i.e., identifying a target class in each sequence and applying pre-trained class-specific trackers was not allowed). The organizers of VOT2022 were allowed to participate in the challenge but were not eligible to win. Further details are available from the challenge homepage[1].

VOT2022 goes beyond previous challenges by updating the datasets in VOT-ST2022 and VOT-RT2022, introducing a training dataset as well as a sequestered dataset in the VOT-RGBD2022 challenge, introducing a depth-only tracking challenge VOT-D2022 and a new challenging VOT-LT2022 tracking dataset. The Python VOT evaluation toolkit was updated as well.

The remainder of this report is structured as follows. Section 2 describes the performance evaluation protocols, Sect. 3 describes the individual challenges, Sect. 4.5 overviews the results and conclusions are drawn in Sect. 5. Short descriptions of the tested trackers are available in Appendix [5].

2   Performance Evaluation Protocol

Since VOT2018, the VOT challenges adopt the following definitions from [16] to distinguish between short-term and long-term trackers:

- **Short-term tracker** (ST_0). The target position is reported at each frame. The tracker does not implement target re-detection and does not explicitly detect occlusion.
- **Short-term tracker with conservative updating** (ST_1). The target position is reported at each frame. Target re-detection is not implemented, but tracking robustness is increased by selectively updating the visual model depending on a tracking confidence estimation mechanism.

---

[1] https://www.votchallenge.net/vot2022/participation.html.

- **Pseudo long-term tracker** (LT_0). The target position is not reported in frames when the target is predicted not visible. The tracker does not implement explicit target re-detection but uses an internal mechanism to identify and report tracking failure.
- **Re-detecting long-term tracker** (LT_1). The target position is not reported in frames when the target is predicted not visible. The tracker detects tracking failure and implements explicit target re-detection.

Since the two classes of trackers make distinct assumptions on target presence, separate performance measures and evaluation protocols were designed in VOT to probe the tracking properties.

2.1 The Short-Term Evaluation Protocols

The short-term performance evaluation protocol entails initializing the tracker at several frames in the sequence, called the anchor points, which are spaced approximately 50 frames apart. The tracker is run from each anchor - in the first half of the sequences in the forward direction, for anchors in the second half backwards, till the first frame. Performance is evaluated by two basic measures *accuracy* (A) and *robustness* (R).

Accuracy is the average overlap on frames before tracking failure, averaged over all sub-sequences. Robustness is the percentage of successfully tracked sub-sequence frames, averaged over all sub-sequences. Tracking failure is defined as the frame at which the overlap between the ground truth and predicted target position dropped below 0.1 and did not increase above this during the next 10 frames. This definition allows short-term failure recovery in short-term trackers. The primary performance measure is the expected average overlap EAO, which is a principled combination of tracking accuracy and robustness. Please see [7] for further details on the VOT short-term tracking performance measures.

2.2 The Long-Term Evaluation Protocol

The long-term performance evaluation protocol follows the protocol proposed in [16] and entails initializing the tracker in the first frame of the sequence and running it until the end of the sequence. The tracker is required to report the target position in each frame along with a score that reflects the certainty that the target is present at that position. Performance is measured by two basic measures called the tracking precision (Pr) and the tracking recall (Re), while the overall performance is summarized by the tracking F-measure.

The performance measures depend on the target presence certainty threshold, thus the performance can be visualized by the tracking precision-recall and tracking F-measure plots obtained by computing these scores for all thresholds. The final values of Pr, Re and F-measure are obtained by selecting the certainty threshold that maximizes tracker-specific F-measure. This avoids all manually-set thresholds in the primary performance measures.

3   Description of Individual Challenges

3.1   VOT-ST2022 Challenge Outline

This challenge addressed RGB tracking in a short-term tracking setup. The initial VOT challenges required target prediction in form of bounding boxes, while a transition to segmentation output requirement has been made in VOT2020. Nevertheless, to support the very much active community that develops bounding box prediction trackers, the bounding box challenge is re-introduced in VOT2022. Thus the VOT-ST2022 ran two subchallenges: the main segmentation-based short-term tracking challenge VOT-STs2022, and the legacy bounding-box-based short-term tracking challenge VOT-STb2022.

The Dataset. Results of the VOT2021 showed that the dataset was not saturated [11], thus the public dataset has been only refreshed by the addition of two sequences which include new challenging scenarios not present in previous VOT datasets: (i) a transparent deforming object and (ii) a flat object with significant out of plane rotations (see Fig. 1). The sequestered dataset has been updated with two sequences matching the public dataset extension.

Fig. 1. Two sequences with new challenging scenarios were added to the VOT-ST2022 public dataset. In the sequence 'bubble' the bubble has to be tracked, while in the sequence 'tennis' the racquet is the target object.

The new sequences were frame-by-frame semi-automatically segmented to provide the segmentation ground truth for the main VOT-STs2022 subchallenge. For the legacy VOT-STb2022 subchallenge, the target position was annotated in all sequences by fitting axis-aligned bounding boxes to the target segmentation masks. Per-frame visual attributes were semi-automatically assigned to the new sequences following the VOT attribute annotation protocol. In particular, each frame was annotated by the following visual attributes: (i) occlusion, (ii) illumination change, (iii) motion change, (iv) size change, (v) camera motion.

Winner Identification Protocol. The VOT-STs2022 winner was identified as follows. Trackers were ranked according to the EAO measure on the public dataset. The top five ranked trackers were then re-run by the VOT2022 committee on the sequestered dataset. The top-ranked tracker on the sequestered dataset not submitted by the VOT2022 committee members is the winner. The same protocol was used to identify the winner of the legacy short-term challenge VOT-STb2022.

3.2 VOT-RT2022 Challenge Outline

This challenge addressed *real-time* RGB tracking in a short-term tracking setup. The dataset was the same as in the VOT-ST2022 challenge, but the evaluation protocol was modified to emphasize the real-time component in tracking performance. In particular, the VOT-RT2022 challenge requires predisetcting bounding boxes faster or equal to the video frame rate. The toolkit sends images to the tracker via the Trax protocol [21] at 20fps. If the tracker does not respond in time, the last reported bounding box is assumed as the reported tracker output at the available frame (zero-order hold dynamic model). The same performance evaluation protocol as in VOT-ST2022 is then applied. As in VOT-ST2022, two realtime subchallenges were considered: the main segmentation-based realtime subchallenge VOT-RTs2022 and the legacy bounding-box-based realtime subchallenge VOT-RTb2022.

Winner Identification Protocol. All trackers are ranked on the public RGB short-term tracking dataset with respect to the EAO measure. The winner of the main VOT-RTs2022 subchallenge was identified as the top-ranked tracker not submitted by the VOT2022 committee members. The same methodology was applied to identify the winner of the VOT-RTb2022 challenge.

3.3 VOT-LT2022 Challenge Outline

This challenge addressed RGB tracking in a long-term tracking setup and is a continuation of the VOT-LT2021 challenge. We adopt the definitions from [16], which are used to position the trackers on the short-term/long-term spectrum. A long-term performance evaluation protocol and measures from Sect. 2.2 were used to evaluate tracking performance on VOT-LT2022. Compared to VOT-LT2021, a significant change is a new dataset described in the following.

The Dataset. The new VOT-LT dataset contains 50 sequences, carefully selected to obtain a dataset with long sequences containing many target disappearances. The LTB50 [16], which was used in VOT-LT2021, is the training set this year. The new VOT-LT dataset contains 50 challenging sequences of diverse objects (persons, cars, motorcycles, bicycles, boats, animals, etc.) with a total length of 168,282 frames. The sequence resolution is 1280×720. Each sequence contains on average 10 long-term target disappearances, each lasting on average 52 frames. An overview of the dataset is shown in Fig. 2.

The targets are annotated by axis-aligned bounding boxes. Sequences are annotated by the following visual attributes: (i) full occlusion, (ii) out-of-view, (iii) partial occlusion, (iv) camera motion, (v) fast motion, (vi) scale change, (vii) aspect ratio change, (viii) viewpoint change, (ix) similar objects. Note this is per-sequence, not per-frame annotation and a sequence can be annotated by several attributes. Compared with LTB50, the new VOT-LT dataset is more challenging in small objects, similar objects, fast motion, and full/partial occlusions.

Winner Identification Protocol. The VOT-LT2022 winner was identified as follows. Trackers were ranked according to the tracking F-score on the new LT dataset (no sequestered dataset available). The top-ranked tracker on the

Fig. 2. The new VOT-LT dataset - a frame selected from each sequence. Name and length (top), visual attributes (bottom left): (O) full occlusion, (V) out-of-view, (P) partial occlusion, (C) camera motion, (F) fast motion, (S) scale change, (A) aspect ratio change, (W) viewpoint change, (I) similar objects. The dataset is highly diverse in attributes and target types and contains many target disappearances.

dataset not submitted by the VOT2022 committee members is the winner of the VOT-LT2022 challenge.

3.4   VOT-RGBD2022 Challenge Outline

The first RGBD (RGB and Depth) challenge was introduced to VOT 2019 and the two first challenges were based on the same public dataset, CDTB [15], which consists of 80 sequences where the target momentarily disappears or is fully occluded. In VOT 2021, the CDTB dataset was replaced with new sequences captured with an Intel RealSense 415 RGBD camera that provides spatially aligned RGB and depth frames. The 2021 dataset contained 80 public and 50 sequestered test sequences. The main motivation for the new dataset was to make it more challenging in the sense that sometimes depth cue is more informative and sometimes RGB. Moreover, separate training and test sequences were provided to allow method fine-tuning with dataset-specific data. More details about the dataset and its properties can be found from [25]. The two major changes as compared to the previous years' RGBD tracks are that 1) the challenge is now a short-term (ST) tracking challenge and 2) the challenge is divided into RGBD and depth-only (D) tracks in order to better understand how much depth contributes to RGBD tracking, i.e. complementarity of the two modalities.

The main motivation to switch from the long-term evaluation to short-term evaluation is that in the long-term setting the target disappearance played an important role and many of the proposed RGBD trackers used the depth channel to assist in occlusion detection, but otherwise the cue was omitted. Now, the two tracks, RGBD and D, provide information about the complementary properties of color texture and depth. It is noteworthy that the RGBD and D challenges use otherwise exactly the same data.

The Dataset. Inspired by the recent work on depth-only tracking [26], we converted the long-term sequences from the CDTB dataset used in the first two VOT-RGBD challenges and DepthTrack used in the latest challenge, to short-term sequences. We converted all 80 sequences from CDTB and 50 test sequences of DepthTrack. Since the DepthTrack training sequences were not used they can be used in training learning-based trackers. The short-term sequences were manually checked and sequences with poor depth information or other errors were removed. Finally, 127 sequences were selected and published on the VOT Web site. See Fig. 3 for example frames.

Fig. 3. Samples from the RGBD and D challenge sequences. The first two from the left are from the CDTB sequences and the next two from DepthTrack-test sequences.

VOT-D2022. The data for the VOT-D2022 challenge is exactly the same as for VOT-RGBD except that the RGB frames are removed.

Winner Identification Protocol. The VOT-RGBD2022 and VOT-D2022 winners were identified as follows. Trackers were ranked according to the EAO measure on the public dataset and the top-ranked tracker on the public dataset not submitted by the VOT2022 committee members is the winner. The same protocol was used to identify the winners of both the VOT-RGBD and VOT-D challenges.

4   The VOT2022 Challenge Results

This section summarizes the trackers submitted, results analysis and winner identification for each of the VOT2022 challenges. Due to page limit, we provide the appendix with more detailed descriptions of the submitted trackers in the supplementary material [5]. For browsing convenience, we also compiled a version of the paper with the appendix included – please see the VOT2022 resutls page[2] for this verison.

4.1   The VOT-STs2022 Challenge Results

The VOT-STs2022 challenge tested 31 trackers, including the baselines contributed by the VOT committee. Each submission included the binaries or source code that allowed verification of the results if required. In the following, we briefly overview the entries and provide the references to original papers in the Appendix [5] where available.

Of the participating trackers, 13 trackers (42%) were categorized as ST_0, 14 trackers (45%) as ST_1, and 4 (13%) as LT_0. 81% applied discriminative and 19% applied generative models. Most trackers (81%) used a holistic model, while 19% of the participating trackers used part-based models. Most trackers (75%) applied an equally probably displacement within a region centered at the current

---

[2] https://www.votchallenge.net/vot2022.

position[3] or a random walk dynamic model (25%). 42% of trackers localized the target in a single stage, while the rest applied several stages, typically involving approximate target localization and position refinement. Most of the trackers (84%) use deep features. The majority of the submissions (72%) localized the target by segmentation, while the rest reported a bounding box.

The trackers were based on various tracking principles. 11 trackers were based on classical or deep discriminative correlation filters (RTS, ATOM_AR, DiMP_AR, KYS_AR, PrDiMP_AR, CSRDCF, D3Sv2, SuperDiMP_AR, KCF, LWL, LWL-B2S), 2 trackers were based purely on Siamese correlation (SiamFC, SiamUSCMix), 14 trackers were based on transformers (DAMT, DAMT-Mask, DGformer, Linker, MixFormerM, MS_AOT, OSTrackSTS, SwinT, SRA-TransTS, TransLL, TransT, transt_ar, TransT_M, and TRASFUSTm), two were deformable parts trackers (ANT and LGT), a meanshift tracker (ASMS), and a video-object segmentation method adapted to tracking (STM).

In summary, we observe a significant increase in a new class of trackers identified in VOT2021 – the transformers. In fact, 47% of trackers are now from this class, 41% of trackers apply discriminative correlation filters, while 6% apply classical siamese correlation networks.

Results. The results are summarized in the AR-raw plots and EAO plots in Fig. 4 and in Table 9. The top ten trackers according to the primary EAO measure (Fig. 4) are MS_AOT, DAMTMask, MixFormerM, OSTrackSTS, Linker, SRATransTS, TransT_M, DGformer, TransLL and LWL-B2S. Nine of the top trackers apply transformers as the core tracking methodology and one applied deep DCFs. Seven apply a two-stage target localization, meaning that they first localize the target by a bounding box and the segment the target withing the bounding box with a separate network (two of these apply Alpha-Refine [24] – the winner of VOT-RT2020 challenge). Three of the top 10 trackers are single-stage, meaning that they directly segment the target. Four of the trackers are apply elements (or are extensions) of MixFormers [3], four extend TransT [2] and three apply ViT [4].

The top tracker on the public set according to EAO is MS_AOT, which is based on the recent transformer-based video object segmentation AOT [28]. For normal-sized objects, the tracker acts as a single-stage segmentation method. For tiny objects, the tracker works in a two-stage regime in which the object is first localized by bounding box using MixFormer [3] and then segmented by the AOT.

The second-best tracker is DAMTMask, which is build on top of Mix-Former [3] and SuperDiMP [1], and applied a two-stage target localization and segmentation approach. The target location is predicted by RepPoints [27] and a MixFormer-like head is implemented to predict the segmentation mask.

---

[3] The target was sought in a window centered at its estimated position in the previous frame. This is the simplest dynamic model that assumes all positions within a search region containing the target have an equal prior probability.

The third-best tracker is MixFormerM, a two-stage tracker which uses a new mixed attention module for simultaneous feature extraction and target information fusion.

The three top performers in EAO are among the top three performers in accuracy (A) and robustness (R) measures as well (Table 9). While these trackers are comparable in target localization accuracy, MS_AOT stands out by its remarkable robustness (Fig. 4).

Table 1. VOT-STs2022 tracking difficulty with respect to the following visual attributes: camera motion (CM), illumination change (IC), motion change (MC), occlusion (OC) and size change (SC).

	CM	IC	OC	SC	MC
Accuracy	0.62	0.62	0.52	0.64	0.63
Robustness	0.79	0.75	0.68	0.78	0.76

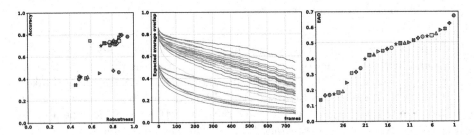

Fig. 4. The VOT-STs2022 AR-raw plots generated by sequence pooling (left) and EAO curves (center) and the VOT-STs2022 expected average overlap graph with trackers ranked from right to left. The right-most tracker is the top-performing according to the VOT-STs2022 expected average overlap values. The dashed horizontal line denotes the average performance of three state-of-the-art trackers published in 2021/2022 at major computer vision venues. These trackers are denoted by gray circle in the bottom part of the graph. See Table 9 for the tracker labels.

Three of the tested trackers have been published in major computer vision journals and conferences in the last two years (2021/2022). These trackers are indicated in Fig. 4, along with their average performance (EAO = 0.504), which constitutes the VOT2022 state-of-the-art bound. Approximately 32% of the submissions exceed this bound.

The per-attribute robustness analysis is shown in Fig. 5 for individual trackers. The overall top performers remain at the top of per-attribute ranks as well. MS_AOT achieves top robustness in all attributes. According to the median failure over each attribute (Table 1) the most challenging attribute remains occlusion. The drop on this attribute is consistent for all trackers (Fig. 5).

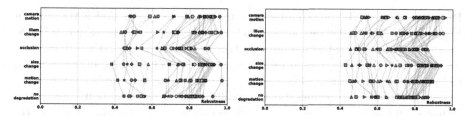

Fig. 5. Robustness with respect to the visual attributes in VOT-STs2022 challenge (left) and in the VOT-STb2022 challenge (right). See Table 9 and Table 10 for VOT-STs2022 and VOT-STb2022 tracker labels, respectively.

The VOT-STs2022 Challenge Winner. The top five trackers from the baseline experiment (Table 9) were re-run on the sequestered dataset. Their scores obtained on the sequestered dataset are shown in Table 2. The top tracker according to the EAO is MS_AOT and is thus the VOT-STs2022 challenge winner.

Table 2. The top five trackers from Table 9 re-ranked on the VOT-STs2022 sequestered dataset.

	Tracker	EAO	A	R
1	MS_AOT	0.565	0.823	0.906
2	DAMTMask	0.513	0.846	0.830
3	OSTrackSTS	0.500	0.822	0.845
4	MixFormerM	0.497	0.844	0.819
5	Linker	0.492	0.829	0.830

4.2 The VOT-STb2022 Challenge Results

The VOT-STb2022 challenge tested 41 trackers, including the baselines contributed by the VOT committee. Each submission included the binaries or source code that allowed verification of the results if required. In the following, we briefly overview the entries and provide the references to original papers in the Appendix [5] where available. The trackers were based on various tracking principles. 13 trackers were based on classical or deep discriminative correlation filters (SuperFus, TCLCFcpp, KCF, D3Sv2, DiMP, ATOM, CSRDCF, SuperDiMP, PrDiMP, FSC2F, oceancycle, DeepTCLCF, KYS), 4 trackers were based purely on Siamese correlation (NfS, SiamUSCMix, SiamVGGpp, SiamFC), 19 trackers were based on transformers (TransT_M, TransT, ADOTstb, GOANET, DAMT, tomp, TransLL, APMT_MR, APMT_RT, DGformer, Linker_B, Mix-Former, ViTCRT, MixFormerL, OSTrackSTB, SRATransT, vittrack, SwinTrack, SBT), one ensamble-based (TRASFUST), one was based on meta-learning (ReptileFPN), one was scale-adaptive mean-shift tracker (ASMS), and two were part-based generative trackers (ANT and LGT).

Results. The results are summarized in the AR-raw plots and EAO plots in Fig. 6, and in Table 10. The top ten trackers according to the primary EAO measure (Fig. 6) are DAMT, MixFormerL, OSTrackSTB, APMT_MR, MixFormer, APMT_RT, ADOTstb, SRATransT, Linker_B, TransT_M. Like in the segmentation tracking challenge VOT-STs2022, all top ten trackers apply transformers. In fact, seven of the top trackers are modifications of segmentation-based counterparts, ranked among the top ten trackers on the VOT-STs2022: MixFormerL, DAMT, OSTrackSTB, MixFormer, SRATransT, Linker, TransT.

All three top-ranked trackers on the public dataset according to EAO, are counterparts of the top-ranked trackers on the main segmentation challenge VOT-STs2022. The two top performers, with equal EAO are MixFormerL and DAMT. MixFormerL, is a counterpart of the tracker ranked third on VOT-STs2022, while DAMT is a counterpart of the second-ranked tracker on VOT-STs2022. The two trackers excel in different tracking properties. DAMT is more robust than MixFormerL, while MixformerL is delivers a more accurate target estimation than DAMT. The third-best ranked tracker is OSTrackSTB is a counterpart of the fourth-best ranked tracker on VOT-STs2022.

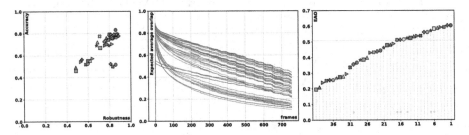

Fig. 6. The VOT-STb2022 AR-raw plots generated by sequence pooling (left) and EAO curves (center) and the VOT-STb2022 expected average overlap graph with trackers ranked from right to left. The right-most tracker is the top-performing according to the VOT-STs2022 expected average overlap values. The dashed horizontal line denotes the average performance of ten state-of-the-art trackers published in 2021/2022 at major computer vision venues. These trackers are denoted by gray circle in the bottom part of the graph. See Table 10 the tracker labels.

Seven of the tested trackers have been published in major computer vision journals and conferences in the last two years (2021/2022). These trackers are indicated in Fig. 6, along with their average performance (EAO=0.484), which constitutes the VOT2022 state-of-the-art bound. Approximately 43.9% of the submissions exceed this bound.

The per-attribute robustness analysis is shown in Fig. 5 for individual trackers. The overall top performers remain at the top of per-attribute ranks as well, however, none of the trackers consistently outperforms the rest in all attributes. According to the median failure over each attribute (Table 3) the most challenging attribute remains occlusion. The drop on this attribute is consistent for all trackers (Fig. 5).

Table 3. VOT-STb2022 tracking difficulty with respect to the following visual attributes: camera motion (CM), illumination change (IC), motion change (MC), occlusion (OC) and size change (SC).

	CM	IC	OC	SC	MC
Accuracy	0.68	0.63	0.55	0.66	0.65
Robustness	0.79	0.74	0.69	0.77	0.74

The VOT-STb2022 Challenge Winner. Top trackers from the baseline experiment (Table 10) were re-run on the sequestered dataset. Since some of the top trackers were variations of the same tracker, the VOT committee selected only the top-performing variant as a representative to be run on the sequestered dataset. Note that there are several ways to specify the ground truth against which the predicted bounding boxes from the trackers can be evaluated. The most straight-forward way is to fit bounding boxes to the ground truth masks (as done in the public evaluation). However, the most accurate ground truth target location specification is actually a segmentation mask and the predicted bounding box from the tracker can be considered as its parametric approximation. We thus inspected the tracker performance for winner identification along the bounding box ground truth specification and along the segmentation mask ground truth specification.

The scores using the bounding box ground truth are shown in Table 4, while the scores using the segmentation mask ground truth are shown in Table 5. We observe that the tracker ranks remain the same across the two ground truth specifications, except from the top two, who switch ranks. For this reason, both top-performers are determined as the winners of the VOT-STb2022 challenge, each in its category. The winner of the VOT-STb2022 challenge in the bounding box ground truth category is OSTrackSTB, while the winner in the segmentation mask ground truth category is APMT_MR.

Table 4. The top five trackers from Table 10 re-ranked on the VOT-STb2022 sequestered dataset using the bounding box ground truth.

	Tracker	EAO	A	R
1	OSTrackSTB	0.523	0.800	0.881
2	APMT_MR	0.508	0.800	0.862
3	MixFormerL	0.500	0.837	0.825
4	ADOTstb	0.499	0.812	0.840
5	DAMT	0.479	0.804	0.826

Table 5. The top five trackers from Table 10 re-ranked on the VOT-STb2022 sequestered dataset using the segmentation masks as ground truth.

	Tracker	EAO	A	R
1	APMT_MR	0.322	0.528	0.845
2	OSTrackSTB	0.309	0.517	0.839
3	MixFormerL	0.306	0.542	0.803
4	ADOTstb	0.301	0.532	0.806
5	DAMT	0.289	0.503	0.807

4.3   The VOT-RTs2022 Challenge Results

The trackers that entered the VOT-STs2022 challenge were also run on the VOT-RTs2022 challenge. Thus the statistics of submitted trackers were the same as in VOT-ST2022. For details please see Sect. 4.2.

Results. The EAO scores and AR-raw plots for the trackers participating in the VOT-RTs2022 challenge are shown in Fig. 7 and Table 9. The top ten segmentation-based real-time trackers are MS_AOT, OSTrackSTS, SRA-TransTS, TransT_M, DGformer, MixFormerM, TransLL, TransT and Linker and RTS.

Nine of the top ten trackers are based on transformers. Nine trackers are ranked among to top 10 on the VOT-STs2022 challenge: MS_AOT, OSTrackSTS, SRATransTS, TransT_M, DGformer, MixFormerM, TransLL, Linker and rts, while TransT is a variation of TransT_M. The top-ranked tracker on realtime challenge according to EAO is MS_AOT, which is also the top-performer on the VOT-STs2022 public datast, the second-best is OSTrackSTS, which ranks fourth on VOT-STs2022 and the third is SRATransTS, which ranks seventh on VOT-STs2022. This indicates significant advancement in the field of visual object tracking since the inception of the VOT realtime challenges, indicating that the speed limitation of modern robust trackers has been confidently breached by transformers.

Three of the tested trackers have been published in major computer vision journals and conferences in the last two years (2021/2022). These trackers are indicated in Fig. 7, along with their average performance (EAO = 0.422), which constitutes the VOT2022 state-of-the-art bound. Approximately 45.2% of the submissions exceed this bound.

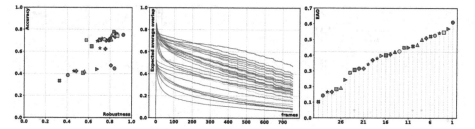

Fig. 7. The VOT-RTs2022 AR plot (left), the EAO curves (center) and the EAO plot (right). The dashed horizontal line denotes the average performance of seven state-of-the-art trackers published in 2021/2022 at major computer vision venues. These trackers are denoted by gray circle in the bottom part of the graph.

The VOT-RTs2022 Challenge Winner. According to the EAO results in Table 9, the top performer and the winner of the segmentation-based real-time tracking challenge VOT-RTs2022 is MS_AOT.

4.4 The VOT-RTb2022 Challenge Results

The trackers that entered the VOT-STb2022 challenge were also run on the VOT-RTb2022 challenge. Thus the statistics of submitted trackers were the same as in VOT-STb2022. For details please see Sect. 4.1 and [5].

Results. The EAO scores and AR-raw plots for the trackers participating in the VOT-RTb2022 challenge are shown in Fig. 8 and Table 10. The top ten bounding-box-based real-time trackers are OSTrackSTB, APMT_RT, Mix-Former, APMT_MR, SRATransT, DAMT, TransT_M, vittrack, SBT, TransT. All of these are based on transformers. Seven are among the top ten performers on the public dataset in VOT-STb2022: OSTrackSTB, APMT_RT, Mix-Former, APMT_MR, SRATransT, DAMT and TransT_M. Thus, similarly to VOT-RTs2022, results here show that performance is minimally compromised if at all on account of speed in transformer-based tracking.

The top-performer according to the EAO on the public dataset is OSTrack-STB, which is based on the recent OSTrack [29] and uses a ViT [4] backbone. This tracker is ranked third on VOT-STb2022. The second and the third-best trackers on VOT-RTb2022 are APMT_RT and MixFormer, which are ranked fourth and fifth on VOT-STb2022.

Note that 7 of the tested trackers have been published in major computer vision journals and conferences in the last two years (2021/2022). These trackers are indicated in Fig. 8, along with their average performance (EAO=0.421), which constitutes the VOT2022 state-of-the-art bound. Approximately 53.7% of the submissions exceed this bound.

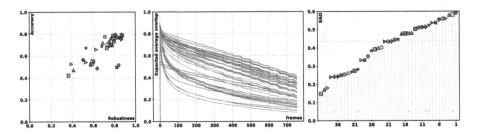

Fig. 8. The VOT-RTb2022 AR plot (left), the EAO curves (center) and the EAO plot (right). The dashed horizontal line denotes the average performance of ten state-of-the-art trackers published in 2021/2022 at major computer vision venues. These trackers are denoted by gray circle in the bottom part of the graph.

The VOT-RTb2022 Challenge Winner. According to the EAO results in Table 10, the top performer and the winner of the bounding-box-based real-time tracking challenge VOT-RTb2022 is OSTrackSTB.

4.5   The VOT-LT2022 Challenge Results

Trackers Submitted. The VOT-LT2022 challenge received 7 valid entries. The VOT2022 committee contributed additional trackers SuperDiMP and KeepTrack as baselines; thus 9 trackers were considered in the challenge. In the following, we briefly overview the entries and provide the references to original papers in [5] where available.

All participating trackers were categorized as ST_1 according to the ST-LT taxonomy from Sect. 2 in that they implemented explicit target re-detection. All trackers were based on convolutional neural networks. Four trackers applied Transformer architecture akin to STARK [23] for target localization (CoCoLoT, mixLT, mlpLT, and VITKT_M). Particularly, VITKT_M is based purely on a Transformer-backbone [20] for feature extraction. Four trackers applied SuperDiMP structure [1] as their basic tracker (ADiMPLT, mixLT, mlpLT, SuperDiMP). Three trackers selected KeepTrack [18] as their auxiliary tracker due to its robustness to distractors (CoCoLoT, VITKT_M, KeepTrack). One tracker is based on MixFormer [3] to design a long-term tracker that focuses on target recapture (HuntFormer). One tracker extends the D3Sv2 [17] short-term tracker with long-term capabilities (D3SLT). Four trackers combined different tracking methods and switched them based on their tracking scores (CoCoLoT, D3SLT, mixLT, mlpLT, VITKT_M). Among them, two trackers use an online real-time MDNet-based [19] verifier to determine the tracking score (CoCoLoT, D3SLT).

Table 6. List of trackers that participated in the VOT-LT2022 challenge along with their performance scores (Pr, Re, F-score) and ST/LT categorization.

Tracker	Pr	Re	F-Score	Year
●VITKT_M	0.629①	0.604②	0.617①	2022
✚mixLT	0.608②	0.592③	0.600②	2022
✖HuntFormer	0.586	0.610①	0.598③	2022
▶CoCoLoT	0.591③	0.577	0.584	2022
▲mlpLT	0.568	0.562	0.565	2022
☐KeepTrack	0.572	0.550	0.561	2022
★D3SLT	0.520	0.516	0.518	2022
●Super_DiMP	0.510	0.496	0.503	2022
✚ADiMPLT	0.489	0.514	0.501	2022

Results. The overall performance is summarized in Fig. 9 and Table 6. The top-three performers are VITKT_M, mixLT_LT and HuntFormer. VITKT_M obtains the highest F-score (0.617) in 2022, while last year winner (mlpLT) obtains 0.565. Since the new VOT-LT dataset is more challenging, it should be noted that the average F-Score of these trackers decreased by 11.4% than last year. All the results are based on the submitted numbers, but these were verified by running the codes multiple times. The VITKT_M is composed of a Transformer-based tracker VitTrack, an auxiliary tracker KeepTrack and a motion module. Specifically, the master tracker VitTrack is a Transformer-based tracker composed of a backbone network, a corner prediction head and a classification head. Besides, a simple motion module is used to predict the target current state according to the temporal trajector. When scores of VitTrack and KeepTrack are all lower than a threshold, and the target moves abnormally, this motion module is triggered to predict the current state.

The mixLT architecture is a progressive fusion of multiple trackers, mainly STARK [23] and SuperDiMP. Specifically, it first fuses the results of two trackers, STARK-ST50 and STARK-ST101. The states of two trackers are then corrected based on the fusion resuls. SuperDiMP controlled by meta-updater is introduced for further fusion between dissimilar trackers, in order to improve the robustness of long-term tracking. The final tracking result is determined according to the confidences of the trackers over several frames, and another tracker correction is performed.

Based on MixFormer, the tracker HuntFormer propose an effective motion prediction model that provides a reliable search region for the tracker to recapture the target. Meanwhile, we propose a novel soft-threshold-based dynamic memory update model, which keeps a set of reliable target templates in the memory that can be used to match the target position in the search region. The two modules cooperate with each other, which greatly improves the recapture ability of the tracker.

The VITKT_M achieves an overall best F-score and significantly surpasses mixLT (by 1.7%) and MixFormer (by 1.9%). All of these methods are based on Transformer. Two similar trackers, VITKT_M and VITKT were submitted by one team. The only difference is that the VITKT is a more concise version than VITKT_M without the motion module. When ablating the motion module (VITKT), the F-score decreases by 1.2%. Since VITKT is a minor variant of VITKT_M, we only keep VITKT_M in our ranking.

The VOT-LT2022 Challenge Winner. According to the F-score in Table 6, the top-performing tracker is VITKT_M, closely followed by mixLT and Hunt-Former. Thus the winner of the VOT-LT 2021 challenge is VITKT_M.

4.6   The VOT-RGBD2022 Challenge Results

Eight trackers were submitted to the 2022 RGBD challenge: DMTracker, keep_track, MixForRGBD, OSTrack, ProMix, SAMF, SBT_RGBD and SPT.

All trackers are based on the popular deep learning-based tracker architectures that have performed well in the previous years VOT RGB challenges. The new deep architecture for this year is MixFormer [3] that is in multiple submissions (MixForRGBD, ProMix and SAMF). The main difference between the submitted trackers is how they fuse the two modalities, depth and RGB, and in their training prodedures. Some teams submitted multiple trackers, but since their architectures are different they were all accepted.

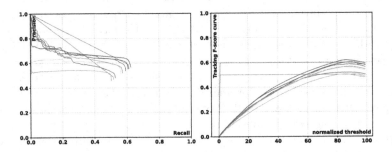

Fig. 9. VOT-LT2022 challenge average tracking precision-recall curves (left) and the corresponding F-score curves (right). Tracker labels are sorted according to maximum of the F-score (see Table 6).

Results. The Expected Average Overlap (EAO), Accuracy (A) and Robustness (R) metrics of the submitted and a number of additional trackers are shown in Table 7. The two best performing trackers, MixForRGBD and SAMF, are distinctively better than the next ones. The six best performing trackers are this year submissions while the DepthTrack database baseline, DeT_DiMP50_Max, is the seventh. The two RGB trackers perform the worst as was expected.

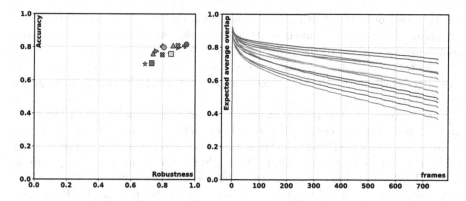

Fig. 10. The VOT-RGBD2022 AR plot (left) and the EAO curves (right).

The VOT-RGBD2022 Challenge Winner. The results in Fig. 10 show that MixForRGBD and SAMF perform very similarly and are clearly better than the rest. Still, MixForRGBD obtains the best EAO score and is thus the winner of the VOT-RGBD2022 challenge.

Table 7. Results for the eight submitted VOT-RGBD2022 trackers. For comparison, the table also includes the results for the three best performing RGBD trackers from VOT2020 (ATCAIS) and VOT2021 (STARK_RGBD and DRefine), two strong baseline RGB trackers from the previous years (DiMP and ATOM) and the baseline RGBD tracker from the DepthTrack dataset (DeT_DiMP50_Max [25]).

Tracker	EAO	A	R
1. ●MixForRGBD	0.779①	0.816	0.946
2. ♣SAMF	0.762②	0.807	0.936
3. ✖OSTrack	0.729③	0.808	0.894
4. ▶ProMix	0.722	0.798	0.900
5. ▲SBT_RGBD	0.708	0.809	0.864
6. □DMTracker	0.658	0.758	0.851
7. ★DeT_DiMP50_Max	0.657	0.760	0.845
8. ○SPT	0.651	0.798	0.851
9. ♣STARK_RGBD	0.647	0.803	0.798
10. ✖keep_track	0.606	0.753	0.797
11. ▶DRefine	0.592	0.775	0.760
12. ▲ATCAIS	0.559	0.761	0.739
13. ○DiMP	0.534	0.703	0.731
14. ★ATOM	0.505	0.698	0.688

4.7 The VOT-D2022 Challenge Results

The VOT-D2022 challenge uses the same 127 short-time tracking sequences as the above RGBD2022 challenge, but in the D (depth-only) challenge the trackers are provided only the depth map frames. This challenge was added as it was intriguing to study how much RGB adds to the depth cue and what is the complementary power of the two modalities.

The total of six trackers were submitted to the depth-only challenge. The submitted trackers are: CoDeT, MixFormerD, OSTrack_D, RSDiMP, SBT_Depth and UpDoT.

Not surprisingly the D-only challenge attracted submissions from the same groups that also participated the RGBD challenge. For example, CoDeT is a D-only version of DMTRacker, MixFormerD of MixForRGBD, OSTrack_D of OSTrack, and SBT_Depth of SBT_RGB. RSDiMP is from the same group as the SPT RGBD tracker, but the two architectures are different. The authors of CoDeT also submitted UpDoT which corresponds to standard DiMP trained with two different versions of depth data.

Results. The computed performance metrics for the D (depth-only) trackers are in Table 8 and the corresponding graphs in Fig. 11. From the results we can see that the depth-only variants of the best performing RGBD trackers also perform well in the D-only challenge (MixFormerRGBD → MixFormerD and OSTrack

Table 8. Results for the six submitted VOT-D2022 trackers. For comparison, the table also includes the results for the recent dept-only tracker DOT [26] and RGB DiMP that was trained with RGB but tested with colormap converted depth images.

Tracker	EAO	A	R
1. ●MixFormerD	0.600①	0.758	0.806
2. ✚RSDiMP	0.573②	0.734	0.759
3. ✖OSTrack_D	0.568③	0.735	0.774
4. ▶DOT	0.469	0.672	0.673
5. ▲SBT_Depth	0.462	0.756	0.571
6. ☐UpDoT	0.439	0.652	0.627
7. ★CoDeT	0.372	0.597	0.594
8. ●DiMP	0.336	0.623	0.496

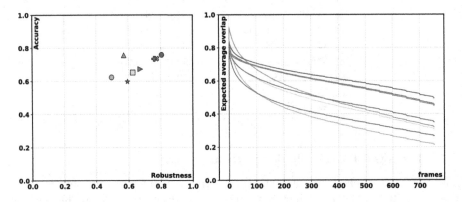

Fig. 11. The VOT-D2022 AR plot (left) and the EAO curves (right).

→ OSTrack_D). The only dedicated D-only tracker submitted to the D-only challenge and which does not have an RGBD counterpart, RSDiMP, obtains the second best EAO score. Overall the three best methods, MixFormerD, RSDiMP and OSTrack_D, perform almost on par and are distinctively better than the rest. Therefore, these three trackers are good starting points to understand how to effectively use the depth channel in tracking.

Table 9. Results for VOT-STs2022 and VOTs-RT2022 challenges. Expected average overlap (EAO), accuracy and robustness are shown. For reference, a no-reset average overlap AO [22] is shown under *Unsupervised*.

| | baseline | | | realtime | | | unsupervised |
| | EAO | AR | | EAO | AR | | Avg. acc. |
Tracker	EAO	A	R	EAO	A	R	AUC
MS_AOT	0.673①	0.781③	0.944①	0.610①	0.751③	0.921①	0.734②
DAMTMask	0.624②	0.796②	0.891②	0.369	0.623	0.756	0.765①
MixFormerM	0.589③	0.799①	0.878③	0.521	0.778①	0.834	0.722③
OSTrackSTS	0.581	0.775	0.867	0.569②	0.766②	0.860	0.665
Linker	0.559	0.772	0.861	0.467	0.709	0.811	0.697
SRATransTS	0.547	0.743	0.866	0.547③	0.743	0.866②	0.673
TransT_M	0.542	0.743	0.865	0.542	0.742	0.865③	0.667
DGformer	0.538	0.744	0.861	0.527	0.744	0.850	0.668
TransLL	0.530	0.735	0.861	0.509	0.733	0.846	0.649
LWL_B2S	0.516	0.736	0.831	0.458	0.715	0.800	0.616
rts	0.502	0.710	0.843	0.450	0.698	0.802	0.592
TransT	0.500	0.749	0.815	0.500	0.749	0.815	0.611
D3Sv2	0.497	0.713	0.827	0.307	0.648	0.627	0.553
TRASFUSTm	0.491	0.740	0.805	0.342	0.631	0.705	0.616
SuperDiMP_AR	0.468	0.734	0.773	0.426	0.711	0.746	0.580
LWL	0.461	0.721	0.798	0.406	0.699	0.763	0.582
SiamUSCMix	0.449	0.702	0.776	0.449	0.702	0.776	0.502
KYS_AR	0.445	0.722	0.749	0.397	0.702	0.708	0.574
DiMP_AR	0.426	0.723	0.719	0.422	0.718	0.724	0.547
PrDiMP_AR	0.422	0.723	0.724	0.400	0.707	0.706	0.581
ATOM_AR	0.398	0.699	0.691	0.380	0.695	0.681	0.505
DAMT	0.339	0.459	0.861	0.315	0.447	0.841	0.434
transt_ar	0.315	0.474	0.809	0.315	0.474	0.809	0.397
STM	0.308	0.745	0.589	0.288	0.703	0.577	0.455
SwinT	0.247	0.452	0.679	0.243	0.440	0.680	0.369
ASMS	0.193	0.419	0.565	0.192	0.419	0.561	0.256
CSRDCF	0.184	0.406	0.550	0.183	0.406	0.548	0.236
SiamFC	0.174	0.417	0.502	0.165	0.423	0.462	0.232
ANT	0.169	0.399	0.482	0.142	0.386	0.407	0.224
KCF	0.165	0.423	0.487	0.165	0.423	0.486	0.162
LGT	0.137	0.345	0.451	0.102	0.335	0.331	0.162

Table 10. Results for VOT-STb2022 and VOTb-RT2022 challenges. Expected average overlap (EAO), accuracy and robustness are shown. For reference, a no-reset average overlap AO [22] is shown under *Unsupervised*.

Tracker	baseline EAO	A	R	realtime EAO	A	R	unsupervised Avg. acc. AUC
●MixFormerL	0.602①	0.831①	0.859	0.512	0.792②	0.814	0.708
✚DAMT	0.602②	0.776	0.887①	0.554	0.752	0.866	0.716③
✖OSTrackSTB	0.591③	0.790	0.869	0.591①	0.790	0.869	0.680
▶APMT_MR	0.591	0.787	0.877③	0.560	0.768	0.871③	0.686
▲MixFormer	0.587	0.797②	0.874	0.580③	0.796①	0.872②	0.696
■APMT_RT	0.581	0.787	0.877②	0.581②	0.787	0.877①	0.721②
★ADOTstb	0.569	0.775	0.862	0.282	0.672	0.533	0.735①
●SRATransT	0.560	0.764	0.864	0.560	0.764	0.864	0.670
✚Linker_B	0.560	0.789	0.844	0.510	0.766	0.823	0.684
✖TransT_M	0.537	0.765	0.849	0.537	0.765	0.849	0.639
▶vittrack	0.536	0.789	0.818	0.536	0.789	0.818	0.679
▲SuperFus	0.534	0.763	0.828	0.481	0.760	0.782	0.629
■SwinTrack	0.524	0.788	0.803	0.503	0.783	0.791	0.626
★SBT	0.522	0.791③	0.813	0.523	0.791③	0.814	0.641
●TRASFUST	0.514	0.754	0.833	0.401	0.704	0.734	0.674
✚TransT	0.512	0.781	0.800	0.513	0.781	0.800	0.641
✖tomp	0.511	0.752	0.818	0.478	0.728	0.796	0.628
▶oceancycle	0.487	0.702	0.825	0.332	0.658	0.629	0.556
▲GOANET	0.481	0.759	0.772	0.478	0.758	0.768	0.654
■SuperDiMP	0.478	0.742	0.788	0.478	0.736	0.798	0.606
★KYS	0.461	0.688	0.797	0.446	0.690	0.784	0.576
●SiamUSCMix	0.444	0.702	0.773	0.444	0.702	0.773	0.556
✚PrDiMP	0.437	0.725	0.755	0.435	0.723	0.751	0.564
✖ViTCRT	0.433	0.774	0.711	0.434	0.774	0.711	0.609
▶DiMP	0.430	0.689	0.760	0.434	0.689	0.761	0.546
▲SiamVGGpp	0.399	0.719	0.690	0.399	0.719	0.690	0.486
■ATOM	0.386	0.668	0.716	0.391	0.672	0.728	0.505
★NfS	0.384	0.688	0.681	0.384	0.688	0.681	0.456
●TransLL	0.367	0.516	0.859	0.353	0.514	0.842	0.473
✚D3Sv2	0.356	0.521	0.811	0.242	0.486	0.637	0.414
✖DGformer	0.337	0.497	0.831	0.332	0.497	0.824	0.462
▶ReptileFPN	0.295	0.572	0.644	0.180	0.526	0.395	0.393
▲FSC2F	0.279	0.554	0.621	0.263	0.557	0.586	0.327
■DeepTCLCF	0.274	0.550	0.601	0.274	0.550	0.601	0.331
★TCLCFcpp	0.267	0.550	0.598	0.267	0.550	0.598	0.329
●ASMS	0.255	0.526	0.599	0.254	0.526	0.594	0.317
✚SiamFC	0.255	0.562	0.543	0.243	0.565	0.505	0.308
✖CSRDCF	0.251	0.519	0.580	0.250	0.518	0.577	0.300
▶KCF	0.239	0.542	0.532	0.240	0.541	0.533	0.234
▲ANT	0.209	0.492	0.484	0.172	0.471	0.414	0.226
■LGT	0.195	0.461	0.486	0.148	0.422	0.366	0.231

Notably, there is a clear difference between the D-only and RGBD results on the same data (Table 7 vs. Table 8). That confirms that the both modalities, D and RGB, are beneficial for object tracking. For example, the RGB DiMP in Table 7 is clearly better than the depth-only DiMP in Table 8 (EAO 0.534 vs. 0.336), but inferior to the best D-only tracker (MixFormerD 0.600).

The VOT-D2022 Challenge Winner. The three best depth-only trackers, MixFormerD, RSDiMP and OSTrack_D, perform on par, but since MixFormerD obtains the best EAO score, it is selected as the winner.

5   Conclusions

Results of the VOT2022 challenge were presented. The challenge is composed of the following challenges focusing on various tracking aspects and domains: (i) the segmentation-based short-term RGB tracking challenge (VOT-STs2022), (ii) the legacy bounding-box-based short-term RGB tracking challenge (VOT-STb2022), (iii) the realtime counterpart of VOT-STs2022 (VOT-RTs2022), (iv) the realtime counterpart of VOT-STb2022 (VOT-RTb2022), (v) the VOT2022 long-term RGB tracking challenge (VOT-LT2021), (vi) the VOT2022 short-term RGB and depth (D) tracking challenge (VOT-RGBD2022) and its variation (vii) the VOT2022 short-term depth-only tracking challenge (VOT-D2022).

In this VOT edition, new VOT-LT2022, VOT-RGBD2022 and VOT-D2022 datasets were introduced, a legacy bounding-box-based tracking challenge VOT-STb2022 was reintroduced, the VOT-ST2022 public and sequestered datasets were refreshed, and a training dataset has been introduced for VOT-LT2022.

A methodological shift, which was indicated already in the VOT2021 [11], has been made even more aparent this year. Nearly half of the trackers participating in VOT-STs2022 challenge were based on transformers, approximately 40% were using discriminative correlation filters, while only few were based on Siamese correlation trackers (a methodology highly popular in VOT2021). All of the top 9 trackers were based on transformers. Apart from being robust, these trackers are also very fast – 9 of top VOT-STs2022 trackers are among the top trackers on VOT-RTs2022 challenge. Variations of the segmentation trackers were submitted to the legacy bounding-box tracking challenge VOT-STb2022. Seven of the top ten trackers on VOT-STb2022 were modifications of the trackers ranked among the top ten on VOT-STs2022. The winner of the VOT-STs2022 challenge is MS_AOT, while the winner of the VOT-STb2022 challenge in the bounding box ground truth category is OSTrackSTB and the winner in the segmentation mask ground truth category is APMT_MR. The winner of the VOT-RTs2022 challenge is MS_AOT and the winner of the VOT-RTb2022 challenge is OSTrackSTB.

The VOT-LT2022 challenge's top-three performers all apply Transformer-based tracker structure for short-term localization and long-term re-detection. Among all submitted trackers, the dominant methodologies are SuperDiMP [1], STARK [23], KeepTrack [18], and MixFormer [3]. The top perfomer and the winner of the VOT-LT2022 is VITKT_M, which ensembles the results of VitTrack

458 M. Kristan et al.

and KeepTrack. This tracker obtains a significantly better performance than the second-best tracker.

In the VOT-RGBD2022 and VOT-D2022 challenges, the same tracker architecture obtained the best results in all tracking metrics. There are two interesting points in this submission that possibly explain its success as compared to others. At first, the tracker is based on the recent Convolutional Visual Transformer (CvT) model and, secondly, the both RGB and depth representations are learned from data. Since there are no depth-only tracking datasets that are sufficiently large for network training, the existing RGB datasets were converted to pseudo depth map datasets using a monocular depth estimation method. These design choices turned out to be the winning ones this year, and therefore the same authors won the VOT-RGBD2022 and VOT-D2022 challenges with their two trackers adopting the same architecture, MixForRGBD and MixFormerD.

For the last decade, the primary objective of VOT has been to establish a platform for discussion of tracking performance evaluation and contributing to the tracking community with verified annotated datasets, performance measures and evaluation toolkits. The VOT2022 was the tenth effort toward this, following the very successful VOT2013, VOT2014, VOT2015, VOT2016, VOT2017, VOT2018, VOT2019, VOT2020 and VOT2021. Since its beginning, the VOT has successfully identified modern milestone tracking methodologies at their inception, spanning discriminative correlation filters, Siamese trackers and most recently the transformer-based architectures. By pushing the boundaries, presenting ever challenging sequences and opening new challenges, the VOT has been successfully fulfilling its service to community. The effort, however, is joint with the tracking community who continually raises to the challenges and is the one generating the fast pace of tracker architecture development. We thank the community for their collaboration and look forward to future developments in this exciting scientific field.

Acknowledgements. This work was supported in part by the following research programs and projects: Slovenian research agency research program P2-0214 and project J2-2506. The challenge was sponsored by the Faculty of Computer Science, University of Ljubljana, Slovenia. This work was partially supported by the Wallenberg AI, Autonomous Systems and Software Program (WASP), in particular in terms of the Wallenberg research arena for Media and Language, and the Berzelius cluster at NSC, both funded by the Knut and Alice Wallenberg Foundation, as well as by ELLIIT, a strategic research environment funded by the Swedish government. Besides, this work was partially supported by the Fundamental Research Funds for the Central Universities (No. 226-2022-00051). This work has also received funding from the European Union's Horizon 2020 research and innovation program under the Marie Skłodowska-Curie grant agreement no. 899987. Hyung Jin Chang and Aleš Leonardis were supported by the Institute of Information and communications Technology Planning and evaluation (IITP) grant funded by the Korea government (MSIT) (2021-0-00537). Gustavo Fernández was supported by the AIT Strategic Research Program 2022 Visual Surveillance and Insight.

References

1. Bhat, G., Danelljan, M., Gool, L.V., Timofte, R.: Learning discriminative model prediction for tracking. In: Proceedings of the IEEE/CVF International Conference on Computer Vision, pp. 6182–6191 (2019)
2. Chen, X., Yan, B., Zhu, J., Wang, D., Yang, X., Lu, H.: Transformer tracking. In: Proceedings of the IEEE/CVF Conference on Computer Vision and Pattern Recognition, pp. 8126–8135 (2021)
3. Cui, Y., Jiang, C., Wang, L., Wu, G.: Mixformer: End-to-end tracking with iterative mixed attention. In: Proceedings of the IEEE/CVF Conference on Computer Vision and Pattern Recognition, pp. 13608–13618 (2022)
4. Dosovitskiy, A., et al.: An image is worth 16×16 words: Transformers for image recognition at scale. arXiv preprint arXiv:2010.11929 (2020)
5. Kristan, M., et. al.: Appendix of the tenth visual object tracking vot2022 challenge results. In: European Conference on Computer Vision ECCV2022 Workshops (2022)
6. Kristan, M., et al.: The seventh visual object tracking vot2019 challenge results. In: ICCV2019 Workshops, Workshop on Visual Object Tracking Challenge (2019)
7. Kristan, M., et al.: The eighth visual object tracking VOT2020 challenge results. In: Bartoli, A., Fusiello, A. (eds.) ECCV 2020. LNCS, vol. 12539, pp. 547–601. Springer, Cham (2020). https://doi.org/10.1007/978-3-030-68238-5_39
8. Kristan, M., et al.: The visual object tracking vot2018 challenge results. In: ECCV2018 Workshops, Workshop on Visual Object Tracking Challenge (2018)
9. Kristan, M., et al.: The visual object tracking vot2017 challenge results. In: ICCV2017 Workshops, Workshop on Visual Object Tracking Challenge (2017)
10. Kristan, M., et al.: The visual object tracking VOT2016 challenge results. In: Hua, G., Jégou, H. (eds.) ECCV 2016. LNCS, vol. 9914, pp. 777–823. Springer, Cham (2016). https://doi.org/10.1007/978-3-319-48881-3_54
11. Kristan, M., et. al.: The ninth visual object tracking vot2021 challenge results. In: Proceedings of the IEEE/CVF International Conference on Computer Vision ICCV2021 Workshops, Workshop On Visual Object Tracking Challenge, pp. 2711–2738 (2021)
12. Kristan, M., et al.: The visual object tracking vot2015 challenge results. In: ICCV2015 Workshops, Workshop on Visual Object Tracking Challenge (2015)
13. Kristan, M., et al.: The visual object tracking vot2013 challenge results. In: ICCV2013 Workshops, Workshop on Visual Object Tracking Challenge, pp. 98–111 (2013)
14. Kristan, M.: the visual object tracking VOT2014 challenge results. In: Agapito, L., Bronstein, M.M., Rother, C. (eds.) ECCV 2014. LNCS, vol. 8926, pp. 191–217. Springer, Cham (2015). https://doi.org/10.1007/978-3-319-16181-5_14
15. Lukežič, A., Kart, U., Kämäräinen, J., Matas, J., Kristan, M.: CDTB: A Color and Depth Visual Object Tracking Dataset and Benchmark. In: ICCV (2019)
16. Lukežič, A., Čehovin Zajc, L., Vojíř, T., Matas, J., Kristan, M.: Sperformance evaluation methodology for long-term single object tracking. IEEE Trans. Cybern. (2020)
17. Lukežič, A., Matas, J., Kristan, M.: A discriminative single-shot segmentation network for visual object tracking. IEEE Trans. Pattern Anal. Mach. Intell. 1 (2021). https://doi.org/10.1109/TPAMI.2021.3137933
18. Mayer, C., Danelljan, M., Paudel, D.P., Van Gool, L.: Learning target candidate association to keep track of what not to track. In: Proceedings of the IEEE/CVF International Conference on Computer Vision, pp. 13444–13454 (2021)

19. Nam, H., Han, B.: Learning multi-domain convolutional neural networks for visual tracking. In: Proceedings of the IEEE Conference on Computer Vision and Pattern Recognition, pp. 4293–4302 (2016)

20. Touvron, H., Cord, M., Douze, M., Massa, F., Sablayrolles, A., Jégou, H.: Training data-efficient image transformers & distillation through attention. arXiv preprint arXiv:2012.12877 (2020)

21. Čehovin, L.: TraX: the visual tracking exchange protocol and library. Neurocomputing (2017). https://doi.org/10.1016/j.neucom.2017.02.036

22. Wu, Y., Lim, J., Yang, M.H.: Online object tracking: A benchmark. Comp. Vis. Patt. Recogn. (2013)

23. Yan, B., Peng, H., Fu, J., Wang, D., Lu, H.: Learning spatio-temporal transformer for visual tracking. In: Proceedings of the IEEE/CVF International Conference on Computer Vision. pp. 10448–10457 (2021)

24. Yan, B., Zhang, X., Wang, D., Lu, H., Yang, X.: Alpha-refine: Boosting tracking performance by precise bounding box estimation. In: Proceedings of the IEEE/CVF Conference on Computer Vision and Pattern Recognition, pp. 5289–5298 (2021)

25. Yan, S., Yang, J., Käpylä, J., Zheng, F., Leonardis, A., Kämäräinen, J.K.: DepthTrack: Unveiling the power of RGBD tracking. In: Proceedings of the IEEE/CVF International Conference on Computer Vision (ICCV), pp. 10725–10733 (2021)

26. Yan, S., Yang, J., Leonardis, A., Kämäräinen, J.K.: Depth-only object tracking. In: British Machine Vision Conference (BMVC) (2021)

27. Yang, Z., Liu, S., Hu, H., Wang, L., Lin, S.: Reppoints: Point set representation for object detection. In: Proceedings of the IEEE/CVF International Conference on Computer Vision, pp. 9657–9666 (October 2019)

28. Yang, Z., Miao, J., Wang, X., Wei, Y., Yang, Y.: Associating objects with scalable transformers for video object segmentation. arXiv preprint arXiv:2203.11442 (2022)

29. Ye, B., Chang, H., Ma, B., Shan, S.: Joint feature learning and relation modeling for tracking: A one-stream framework. arXiv preprint arXiv:2203.11991 (2022)

Efficient Visual Tracking via Hierarchical Cross-Attention Transformer

Xin Chen[1], Ben Kang[1], Dong Wang[1(✉)], Dongdong Li[2], and Huchuan Lu[1,3]

[1] School of Information and Communication Engineering,
Dalian University of Technology, Dalian, China
wdice@dlut.edu.cn
[2] National Key Laboratory of Science and Technology on Automatic Target
Recognition, National University of Defense Technology, Changsha, China
[3] Peng Cheng Laboratory, Shenzhen, China

Abstract. In recent years, target tracking has made great progress in accuracy. This development is mainly attributed to powerful networks (such as transformers) and additional modules (such as online update and refinement modules). However, less attention has been paid to tracking speed. Most state-of-the-art trackers are satisfied with the real-time speed on powerful GPUs. However, practical applications necessitate higher requirements for tracking speed, especially when edge platforms with limited resources are used. In this work, we present an efficient tracking method via a hierarchical cross-attention transformer named HCAT. Our model runs about 195 fps on GPU, 45 fps on CPU, and 55 fps on the edge AI platform of NVidia Jetson AGX Xavier. Experiments show that our HCAT achieves promising results on LaSOT, GOT-10k, TrackingNet, NFS, OTB100, UAV123, and VOT2020. Code and models are available at https://github.com/chenxin-dlut/HCAT.

Keywords: Efficient tracking · Transformer

1 Introduction

Visual object tracking is a fundamental task in computer vision, in which the aim is to track an arbitrary object in a video given its initial location. It is widely used in drone vision, autonomous driving, surveillance, and other fields. Over the past few years, object tracking has made great progress owing to the development of deep networks [15,19,31,33]. Most trackers aim to achieve high performance on datasets, but they ignore tracking speed and appear satisfied with real-time speed on powerful GPUs. However, real-world application scenarios require trackers to function in real-time or even faster on edge devices, such as CPUs and NVidia Jetson devices. However, most of the recent state-of-the-art trackers cannot achieve real-time speed on edge devices as shown in Fig. 1, thus limiting their real-world applications. In this work, we attempt to develop an accurate and extremely fast tracking algorithm.

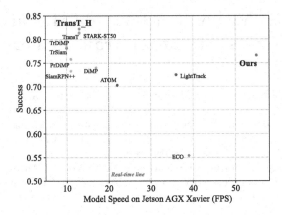

Fig. 1. Speed and performance comparison on TrackingNet [26]. The horizontal axis is the model's speed on the edge AI platform of NVidia Jetson AGX Xavier. The vertical axis is the success (AUC) score. Following the VOT real-time setting [18], we set the real-time line to 20 fps. Our method achieves the best real-time result. We also apply our hierarchical cross-attention layer to TransT [5], namely TransT_H. TransT_H achieves the best result over all compared trackers.

Table 1. Running speed of FFN with different calculation amounts.

	Input size	MFlops	GPU latency (ms)	CPU latency (ms)
FFN1	$8 \times 8 \times 256$	67.1	0.15	0.45
FFN2	$16 \times 16 \times 256$	268.4	0.15	1.22

Recently, transformer [33] has been successfully applied in many vision tasks [4,10,23]. In the tracking field, transformer also boosts the performance [5,35, 40]. However, transformer entails high seriality and its computational amount is proportional to the square of the number of input tokens. In this work, we design a more efficient transformer structure based on the variant transformer in TransT [5]. There are two straightforward ways to speed up the transformer. One approach is to reduce the calculation amount of the operation layers, such as the linear transformation layer, convolution layer, and attention layer. On platforms with limited computing power, the model's speed can be improved in this way. However, reducing the computational amount of operation layers does not always effectively improve the model's speed because devices have the ability of parallel computing. For example, we test the latency of feed-forward network FFN [4,33] commonly used in transformer under different calculation amounts, as shown in Table 1. FFN1 has 4× less calculation amount than FFN2, but its latency is not less on GPU and only 2.7× less on CPU. Another way is to reduce the seriality of the transformer. This approach involves directly reducing the number of layers of the network, which can bring considerably speedup on both edge platforms and powerful GPUs. However, reducing the seriality causes the network to be shallow and it weakens the feature expression ability. So the target is to keep the efficient layers and discard the relatively inefficient layers.

In this work, to speed up the transformer, we first design the **Hierarchical Cross-Attention** (HCA) transformer structure to reduce the seriality and improve the model's representation ability. We employ a full cross-attention structure to improve efficiency and a hierarchical connection method to deepen the network, subsequently enhancing the representation ability. Then, we design the **Feature Sparsification** (FS) module to sparse the template feautures and reduce the computational amount of the transformer. On the basis of these two modules, we propose an efficient visual tracking method named HCAT.

Our contributions can be summarized as follows.

- We propose a novel hierarchical cross-attention transformer that employs a full cross-attention structure to improve efficiency and a hierarchical connection method to enhance the representation ability under a limited number of operation layers.
- We propose a feature sparsification module to sparse the template features and reduce the computational amount of the transformer without affecting performance.
- The feature sparsification module and the hierarchical cross-attention transformer form a new feature fusion network. We combine the feature fusion network with the backbone network and prediction head to develop a new efficient tracker named HCAT.
- Our HCAT has an extremely fast speed. The PyTorch model runs at 195 fps on GPU, 45 fps on CPU, and 55 fps on the edge AI platform of NVidia Jetson AGX Xavier. The ONNX model runs at 589 fps on GPU, 90 fps on CPU, and 127 fps on NVidia Jetson AGX Xavier. Numerous experimental results on many benchmarks show that the proposed tracker performs considerably better than the state-of-the-art high-speed algorithms.

2 Related Work

Visual Object Tracking. In recent years, Siamese trackers have become popular in visual object tracking. SiamFC [1] and SINT [32], the pioneering works, combine naive feature correspondence by using the Siamese framework. Since then, SiamRPN [21] has used the RPN [28] into the Siamese tracking framework for precise bounding box estimation. Many improvements have been achieved to boost tracking performance, such as developing additional branches [36,42], employing deeper architectures [20,44,45], exploiting the anchor-free mechanism [6,14,39,46], and so on. Online trackers [2,7,8,27,46] improve robustness by employing online updating modules. However, due to complex designs, most previous Siamese trackers or online trackers only achieve real-time speed on powerful GPUs, but they hardly achieve real-time speed on edge platforms. The Siamese tracking framework can be divided into three parts: the backbone network, the feature fusion network, and the prediction head. In this work, aiming to develop an extremely fast tracker, we employ a simple backbone network and a simple prediction head network while carefully designing an efficient feature fusion network. Our method can achieve real-time speed on edge platforms.

Transformer in Tracking. Vaswani *et al.* introduced transformer [33] in machine translation. Transformer is composed of attention-based encoders and decoders. Attention mechanism has achieved remarkable results in many tasks because of its ability to integrate global information and less inductive bias. DETR and ViT [4,10] were the first to apply transformer into computer vision field. After that, transformer has been successfully applied in a number of computer vision tasks. In object tracking, transformer has brought great performance improvement. TransT [5] exploits the core idea of transformer and develops a new feature fusion network to fuse template and search region features by using cross-attention rather than correlation operation. TMT [35] uses transformer as a feature enhancement module and combines it with SiamRPN [21] and DiMP [2]. STARK [40] employs the transformer by inputting the concatenated template and search region features. DualTFR and SwinTrack [23,38] use transformer as the backbone network. In this work, we develop an efficient feature fusion network based on transformer. We choose TransT as our baseline because it is relatively simple, does not require additional modules, and has a good speed on GPUs. We develop a feature sparsification module and a hierarchical cross-attention transformer to enable the tracker to achieve real-time speed on edge platforms with good performance.

Efficient Tracking Network. More and more applications that use tracking algorithms have been implemented, including unmanned driving, UAV vision and robot vision. In real-world applications, trackers usually run on the edge platforms with limited computing power. However, most of the current state-of-the-art trackers only run fast on powerful GPUs. The demand for efficient tracking network is urgent. Unfortunately, in recent years, researchers have hardly paid attention to tracker's speed on edge devices. ATOM [8] and ECO [7] can achieve real-time speed on NVidia Jetson AGX Xavier (AGX); however, compared with the popular tracking algorithms (such as PrDiMP and SiamRPN++), their performance is poor. LightTrack [41] is the latest lightweight tracking algorithm that uses NAS to search networks, and it entails low computational amount and relatively high performance. However, a gap exists between the calculation amount and the model's real speed, as described in Sect. 1. LightTrack can achieve real-time speed on CPU and AGX; however, on powerful GPUs, the speed is not extremely fast. We hope our tracker can achieve real-time speed on AGX and CPUs, and acheive extremely fast speed on powerful GPUs. In this work, our tracker's PyTorch model runs at 195 *fps* on GPU, 45 *fps* on CPU, and 55 *fps* on AGX. Our ONNX model runs at 589 *fps* on GPU, 90 *fps* on CPU, and 127 *fps* on AGX.

3   Method

This section describes our method for HCAT. As shown in Fig. 2, our method consists of the feature extraction network, the feature fusion network, and the prediction head. The feature extraction network extracts the features of the search region patch and the template patch. The feature fusion network fuses

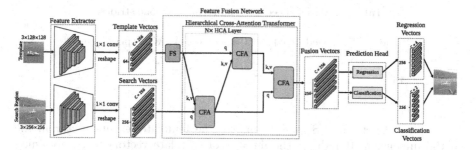

Fig. 2. Architecture of our HCAT framework. This framework contains three components: feature extraction backbone, feature fusion network, and prediction head. The feature fusion network consists of the feature sparsification module (denoted as FS) and the hierarchical cross-attention transformer.

the search region features and the template features by using our hierarchical cross-attention transformer. Before feature fusion, we use the feature sparsification (FS) module to sparse the template feature vectors in the spatial dimension. After feature fusion, we use the prediction head to perform bounding box regression and binary classification on the fusion feature vectors. Finally, we select the best bounding box according to the confidence score. In this section, we introduce the details of each part of HCAT, introduce our hierarchical cross-attention transformer and feature sparsification module, and provide analysis.

3.1 Overall Architecture

Feature Extraction. Similar to most Siamese trackers [1,20,21], we take the template image patch ($\mathbf{Z} \in \mathbb{R}^{3 \times H_{z0} \times W_{z0}}$) and the search region image patch ($\mathbf{X} \in \mathbb{R}^{3 \times H_{x0} \times W_{x0}}$) as the inputs of our HCAT. The template patch is obtained by expanding the initial target bounding box outward by twice the side length, and the search region patch is obtained by expanding the bounding box of the previous frame's target by four times the side length. The search region patch and template patch are reshaped to a square, and then the backbone network extracts their features. We use a modified version of ResNet18 [15] as the backbone network. Specifically, we remove the last stage of ResNet18. In contrast to previous Siamese trackers [5,20], we do not modify the stride; hence, the backbone's stride remains 16 rather than 8. The large backbone stride reduces the feature's resolution, thus further reducing the computational cost. In this manner, the speed gain is not obvious on the powerful GPU, but it is obvious on edge platforms, as shown in Table 2. Finally, we use a 1×1 convolution layer to transform the channel dimension of the backbone features and flatten the features in the spatial dimension. Then, we obtain two sets of feature vectors: the template vectors $\mathbf{F}_z \in \mathbb{R}^{C \times H_z W_z}$ and the search vectors $\mathbf{F}_x \in \mathbb{R}^{C \times H_x W_x}$. $H_z, W_z = \frac{H_{z0}}{16}, \frac{W_{z0}}{16}, H_x, W_x = \frac{H_{x0}}{16}, \frac{W_{x0}}{16}$, and $C = 256$.

Feature Fusion Network. We design a new feature fusion network to fuse the features of the template and the search region. As shown in Fig. 2, the feature fusion network is composed of the FS module and the hierarchical cross-attention

Table 2. Model speed with different backbone strides.

Stride	GPU speed (*fps*)	CPU speed (*fps*)	AGX speed (*fps*)
8	178	20	22
16	195	45	55

transformer. First, the FS module sparses the template feature vector in the spatial dimension. It reduces the number of template vectors to S, obtaining the sparse template vectors $\mathbf{F}_{zs} \in \mathbb{R}^{C \times S}$. Then, the hierarchical cross-attention transformer fuses the sparse template vectors and the search vectors. The hierarchical cross-attention transformer is inspired by the feature fusion network proposed by TransT [5]. We employ the basic unit CFA in TransT, which is a residual structure based on the cross-attention layer and linear layer. CFA can fuse two sets of input features. Two CFAs form a HCA layer, as shown in the yellow dotted box in Fig. 2. In contrast to the method for TransT, we adopt the full cross-attention structure instead of employing self-attention layers. Because the performance gain brought by self-attention layers is relatively small, but it requires extensive running time. In TransT, two CFAs in the same feature fusion layer are juxtaposed but uncorrelated with each other. Instead of juxtaposing two CFAs, we combine them in a hierarchical manner, i.e., the template branch's CFA receives the output of the search branch's CFA of the same layer as the key and value. Hence, the cross-attention utilizes the more accurate features without incurring any additional computational cost. Under the limited number of operation layers, the network is deepened. The HCA layer repeats N times (here, $N = 2$ by default). Then, an additional CFA decodes the final fusion feature vectors $\mathbf{F} \in \mathbb{R}^{C \times H_x W_x}$. The details of the FS module and the hierarchical cross-attention transformer are introduced in Sect. 3.2 and Sect. 3.3.

Prediction Head Network. The prediction head consists of the regression head and classification head, which are both a three-layered perceptron with a hidden dimension ($d = 256$) and the ReLU activation function. The regression head directly outputs the normalized coordinates on each fusion vector, resulting in a total of $H_x W_x$ bounding boxes. The classification head performs binary (foreground/background) classification on each fusion vector. The classification head generates $H_x W_x$ classification scores corresponding to the bounding boxes.

3.2   Feature Sparsification Module

As shown in Fig. 2, before the hierarchical cross-attention transformer, we use the FS module to sparse the template features, thus reducing the calculation amount of subsequent layers, especially the attention layers. The FS module reduces the number of template vectors from $H_z W_z$ to S, and the subsequent linear layer's calculation are reduced accordingly. For attention layer, apart from internal linear transformation, the attention mechanism is calculated as

$$\text{Attention}(\mathbf{Q}, \mathbf{K}, \mathbf{V}) = \text{softmax}(\frac{\mathbf{Q}\mathbf{K}^{\top}}{\sqrt{d}})\mathbf{V}, \tag{1}$$

where $\mathbf{Q} \in \mathbb{R}^{d \times n_q}$ and $\mathbf{K}, \mathbf{V} \in \mathbb{R}^{d \times n_k}$. The calculation amount of attention is given by $2dn_q n_k$; the calculation amount of multi-head attention is processed in the same manner. However, efficiency of attention is criticized because its computational amount is proportional to the product of the two spatial dimensions n_q and n_k. Without the FS module in our method, the attention mechanism takes $\mathbf{F}_z \in \mathbb{R}^{C \times H_z W_z}$ and $\mathbf{F}_x \in \mathbb{R}^{C \times H_x W_x}$ as the inputs, resulting in the calculation amount of $2H_z W_z H_x W_x C$. The direct approach of minimizing the computational amount is to reduce the number of template vectors or search vectors. We find that reducing the number of search vectors has a large impact on performance; however, under certain suitable settings, reducing the number of template vectors has a negligible impact on performance. This finding can be attributed to the network requiring a classification and regression of the target on the search region. The fine-grained appearance information and location information in search vectors are important for accurate predictions. However, template features are relatively redundant as the network only needs to refer to it to know which target to track. Reducing the number of search vectors or template vectors in equal proportions has the same impact on the computational amount of attention; thus we choose to sparse the template vectors. Our FS module reduces the number of template vectors from $H_z W_z$ to S, and the computational amount of attention is reduced to $2SH_x W_x C$, where S is a constant, which is much smaller than $H_z W_z$. With our default settings of $S = 16$ and $H_z W_z = 64$, the computational amount is reduced by a quarter. On powerful GPU, computational amount is not the bottleneck restricting the model's speed. Therefore, this reduction cannot speed up the model on powerful GPUs, but it can substantially speed up the model on edge platforms. The detail structure of the FS module is shown in Fig. 3 (a). The template vectors are the feature vectors $\mathbf{F}_z \in \mathbb{R}^{C \times H_z W_z}$ generated by backbone. The sparse vectors $\mathbf{F}_s \in \mathbb{R}^{C \times S}$ are the general features learned during training. The FS module is a residual structure based on the multi-head cross-attention layer. Multi-head cross-attention takes the sparse vectors as query and the template vectors as key and value. It extracts template vectors with reference to sparse vectors and outputs the sparse feature vectors. Then the sparse feature vectors are added to the sparse vectors $\mathbf{F}_s$ to obtain the final sparse template vectors $\mathbf{F}_{zs} \in \mathbb{R}^{C \times S}$. The mechanism of the FS module can be summarized as

$$\mathbf{F}_{zs} = \mathbf{F}_s + \mathrm{MHCA}(\mathbf{F}_s, \mathbf{F}_z + \mathbf{P}_z, \mathbf{F}_z), \tag{2}$$

where $\mathbf{P}_z \in \mathbb{R}^{C \times H_z W_z}$ denotes the spatial positional encoding, which is generated by a sine function. $\mathrm{MHCA}(.,.,.)$ is the multi-head cross-attention mechanism. We find that such a simple design achieves good results.

3.3 Hierarchical Cross-Attention Transformer

The detailed architecture of our hierarchical cross-attention transformer is shown in Fig. 3 (b). This structure is based on the basic module of the CFA used in TransT. CFA is a residual module based on multi-head cross-attention and FFN,

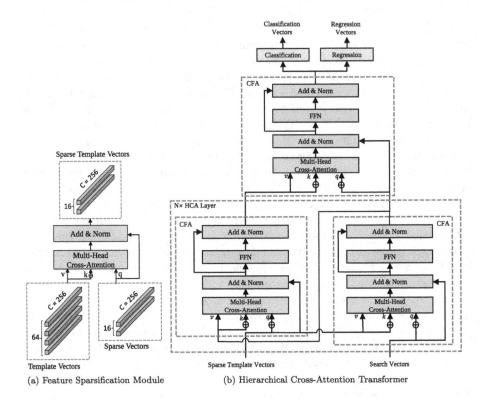

Fig. 3. Detailed architectures of our feature sparsification module and hierarchical cross-attention transformer. ⊕ denotes adding spatial positional encoding.

as shown in the dotted box. FFN is a feed-forward network that consists of two linear transformation layers with a ReLU activation function in between. CFA is formulated as

$$
\begin{aligned}
\mathrm{CFA}(\mathbf{Q}, \mathbf{K}, \mathbf{V}) &= \mathbf{X_{CF}}, \\
\mathbf{X}_{CF} &= \widetilde{\mathbf{X}}_{CF} + \mathrm{FFN}\left(\widetilde{\mathbf{X}}_{CF}\right), \\
\widetilde{\mathbf{X}}_{CF} &= \mathbf{Q} + \mathrm{MHCA}\left(\mathbf{Q} + \mathbf{P}_q, \mathbf{K} + \mathbf{P}_{kv}, \mathbf{V}\right),
\end{aligned}
\tag{3}
$$

where $\mathbf{Q} \in \mathbb{R}^{d \times n_q}$ is the q input, $\mathbf{K}$ and $\mathbf{V} \in \mathbb{R}^{d \times n_{kv}}$ are the k and v inputs, where $\mathbf{K} = \mathbf{V}$ in our method. $\mathbf{P}_q \in \mathbb{R}^{d \times n_q}$ and $\mathbf{P}_{kv} \in \mathbb{R}^{d \times n_{kv}}$ are the spatial positional encodings.

In our method, two CFAs form an HCA layer. The two CFAs in same layer are hierarchical rather than juxtaposed as in TransT. Specifically, the template branch's CFA receives the output of the search branch's CFA of the same layer as the key and value rather than they do not affect each other like TransT. The HCA layer can be formulated as

$$
\begin{aligned}
\mathbf{F}_x^{(l)} &= \mathrm{CFA}(\mathbf{F}_x^{(l-1)}, \mathbf{F}_{zs}^{(l-1)}, \mathbf{F}_{zs}^{(l-1)}), \\
\mathbf{F}_{zs}^{(l)} &= \mathrm{CFA}(\mathbf{F}_{zs}^{(l-1)}, \mathbf{F}_x^{(l)}, \mathbf{F}_x^{(l)}),
\end{aligned}
\tag{4}
$$

where $(l-1)$ represents the previous layer, and (l) represents the current layer. In this manner, the CFA utilizes the more accurate features, which is equivalent to deepening the network without introducing additional layers. Given the different sizes of the inputs, the two CFAs at the same layer cannot be operated in parallel in TransT and our method. Therefore, the hierarchical connection method does not bring any additional running time cost.

We expect our model to achieve excellent speed not only on edge platforms but also powerful GPUs. As discussed in Sect. 1 and Sect. 3.2, owing to the powerful parallel computing capability of the GPU, the calculation amount of operation layers is not the bottleneck restricting the speed. Minimizing the number of serial layers is more effective than reducing the computational amount of operation layers. Therefore, we only use two HCA layers, avoid using any self-attention layer and instead adopt the full hierarchical cross-attention design. In experiments, we find that the self-attention layer brings minimal performance benefit, and it requires a high running time. These settings cause the network shallow, thus affecting the performance, but the hierarchical structure plays a mitigating role.

3.4 Training Loss

The prediction head receives $H_x W_x$ search vectors and generate $H_x W_x$ prediction results. Each result includes a bounding box and a binary classification result. The prediction results whose corresponding positions in the ground-truth bounding box are assigned as positive samples, whereas the others are assigned as negative samples. The class label of the positive samples is foreground and the class label of the negative samples is background. All samples contribute to classification loss. The regression label is the normalized bounding box coordinate. Only the positive samples contribute to regression loss, which enable the feature vectors to predict the targets at their corresponding position. Subsequently, the distractors can be filtered through the position information in online tracking. For classification, we employ the binary cross-entropy loss. To balance the positive and negative samples, we reduce the classification loss contributed by the negative samples by a factor of 16. For regression, we combine the ℓ_1-norm loss and generalized IoU loss [29] linearly.

4 Experiments

4.1 Implementation Details

Offline Training. Our models are trained on the training set of LaSOT [11], GOT-10k [16], COCO [22], and TrackingNet [26]. A search region patch and a template patch form a sampling pair. For the video datasets of LaSOT, GOT-10k, and TrackingNet, we randomly take two frames in the same video sequence with an interval of no more than 100 frames to generate the template patch and search region patch. For the image dataset of COCO, we transform the original image to generate sampling pairs. We employ some normal data augmentation, such as position and scale transformation. The search region patches are resized to $256 \times 256 \times 3$, and the template patches are resized to $128 \times 128 \times 3$. The parameters of the backbone network are initialized with ResNet18 [15] pretrained on ImageNet [30], other parameters are initialized with the Xavier init [13]. The models are trained with the AdamW [24] optimizer. The learning rate of the backbone network is 1e-5, and that of the other parameters is 1e-4. The weight decay is 1e-4. We use 8 NVidia RTX 3090 GPUs to train our network for 500 epochs with a batch size 128. Each epoch contains 60,000 sampling pairs. The learning rate is decayed by $10\times$ at epoch 400.

Online Tracking. In online tracking, the network predicts 16×16 results with classification scores. We employ window penalty [20] to adjust the confidence score. Specifically, we penalize the confidence scores far from the center of the search region by applying a Hanning window. The window penalty is used to filter out the distractors.

4.2   Evaluation on TrackingNet, LaSOT, and GOT-10k Datasets

We compare our HCAT with state-of-the-art trackers on the three large-scale benchmarks of TrackingNet [26], LaSOT [11], and GOT-10k [16]. We test the speed of the trackers' model on three platforms: GPU, CPU, and NVidia Jetson AGX Xavier (AGX). Our GPU is the powerful NVidia Titan RTX, our CPU is Intel(R) Core(TM) i9-9900K @ 3.60 Hz, and the AGX is an edge AI platform. We only test the speeds of the models and exclude the pre-processing and post-processing of the image and results during tracking because the different implementations of these parts affect the tracking speed, and we do not want them to conceal the real speed of the models. For the online update trackers (such as TrDiMP, PrDiMP, ATOM), we turn off their online update module, so their real speed is slower than our test speed. We divide the trackers into two categories according to the running speed of the model. Specifically, following the VOT real-time setting [18], we set the real-time line to 20 fps. The methods that can reach 20 fps on our CPU or AGX are classified as real-time methods, whereas those that cannot reach 20 fps on neither our CPU nor AGX are classified as non-real-time methods. The detailed results are shown in Table 3.

Speed. The speeds on the three platforms are shown in Table 3. AGX indicates NVidia Jetson AGX Xavier. Our model's speed is 195 fps on GPU, which is second only to ECO [7], and it is 52% higher than the recent lightweight tracker LightTrack [41]. On the edge platforms, our model achieves 45 fps on CPU (10% higher than the second fastest tracker LightTrack) and 55 fps on AGX (53% higher than LightTrack). In addition, after using ONNX to accelerate, our model's speed can reach 589 fps on GPU, 90 fps on CPU, and 127 fps on AGX. More results about the ONNX model are introduced in Sect. 4.5.

TrackingNet. TrackingNet [26] is a large-scale dataset containing diverse object categories and scenes. Its test set has 511 video sequences. We upload our

Table 3. State-of-the-art comparison on TrackingNet, LaSOT, and GOT-10k benchmarks. The best real-time results are shown in red fonts, and the best non-real-time results are shown in **blue** fonts.

		Real-time					Non-real-time						
		Ours	E.T. Track [3]	LightTrack [41]	ATOM [8]	ECO [7]	STARK-ST50 [40]	TransT [5]	TrDimp [35]	TrSiam [35]	PrDiMP [9]	DiMP [2]	SiamRPN++ [20]
TrackingNet	AUC	76.6	75.0	72.5	70.3	55.4	81.3	81.4	78.4	78.1	75.8	74.0	73.3
	P_{Norm}	82.6	80.3	77.8	77.1	61.8	86.1	86.7	83.3	82.9	81.6	80.1	80.0
	P	72.9	70.6	69.5	64.8	49.2	–	80.3	73.1	72.7	70.4	68.7	69.4
LaSOT	AUC	59.3	59.1	53.8	51.5	32.4	66.6	64.9	63.9	62.4	59.8	56.9	49.6
	P_{Norm}	68.7	–	–	57.6	33.8	–	73.8	–	–	68.8	65.0	56.9
	P	61.0	–	53.7	50.5	30.1	–	69.0	61.4	60.0	60.8	56.7	49.1
GOT-10k	AO	65.1	–	61.1	55.6	31.6	68.0	72.3	68.8	67.3	63.4	61.1	51.7
	$SR_{0.5}$	76.5	–	71.0	63.4	30.9	77.7	82.4	80.5	78.7	73.8	71.7	61.6
	$SR_{0.75}$	56.7	–	–	40.2	11.1	62.3	68.2	59.7	58.6	54.3	49.2	32.5
Speed (fps)	GPU	195	–	128	83	240	50	63	41	40	47	77	56
	CPU	45	47	41	18	15	7	5	5	5	6	10	4
	AGX	55	–	36	22	39	13	13	10	10	11	17	11

tracker's results to TrackingNet's official evaluation server. The AUC, P_{Norm}, and P are shown in Table 3. Our tracker has the best real-time results, which are 76.6%, 82.6%, and 72.9% for AUC, P_{Norm}, and P, respectively. Compared with the state-of-the-art lightweight tracker LightTrack, our method performs 4.1%, 4.8%, and 3.4% higher in terms of AUC, P_{Norm}, and P, with better speed. Compared with the popular tracker PrDiMP, our method outperforms it by 0.8%, 1% and 2.5% for AUC, P_{Norm}, and P, with 4× higher speed on GPU, 7× higher speed on CPU, and 5× higher speed on AGX.

LaSOT. LaSOT [11] is a large-scale long-term dataset containing 1120 videos for training and 280 videos for testing. We follow the one-pass evaluation to test different tracking methods on the LaSOT test set. The Success (AUC) and Precision (P_{Norm} and P) scores are shown in Table 3. Our tracker has the best real-time results, which are 59.3%, 68.7%, and 61.0% for AUC, P_{Norm}, and P, respectively. Compared with the recent lightweight tracker LightTrack, our method outperforms it by 5.5% and 7.3% for AUC and P, with better speed. Compared with the popular tracker PrDiMP, our tracker offers competitive performance (59.3 *vs.* 59.8 in AUC) but with much faster speed.

GOT-10k. GOT-10k [16] is a large-scale dataset containing a wide range of challenges. We submit HCAT's results to the official server. The obtained AO and SR_T scores are shown in Table 3. Compared with LightTrack, our method outperforms it by 4.0% and 5.5% for AO and $SR_{0.5}$, with better speed. Compared with the popular tracker PrDiMP, our tracker outperforms it by 1.7%, 2.7%, and 2.4% higher for AO, $SR_{0.5}$, and $SR_{0.75}$, with much faster speed.

4.3 Evaluation on VOT2020 Datasets

We compare our tracker with some state-of-the-art trackers on the challenging dataset VOT2020 [18]. VOT2020 contains 60 challenging videos with mask annotation. VOT2020 adopts EAO (expected average overlap) as the final metric to measure the robustness and accuracy of trackers. As our tracker does not generate masks, we only compare trackers that submit bounding boxes. The results are shown in Table 4. Our tracker has the best real-time results. Our tracker's performance is comparable to the powerful tracker STARK-S50, only 0.4% lower in EAO, and 1.9% higher in Robustness, with 4× higher speed on GPU, 5× higher speed on CPU, and 3× higher speed on AGX.

Table 4. State-of-the-art comparison on VOT2020. The best real-time results are shown in red fonts, and the best non-real-time results are shown in **blue** fonts.

	Real-time				Non-real-time		
	Ours	E.T. Track [3]	LightTrack [41]	ATOM [8]	STARK-ST50 [40]	STARK-S50 [40]	DiMP [2]
EAO	27.6	26.7	24.2	27.1	**30.8**	28.0	27.4
Accuracy	45.5	43.2	42.2	46.2	**47.8**	47.7	45.7
Robustness	74.7	74.1	68.9	73.4	**79.9**	72.8	74.0
GPU speed (fps)	195	–	128	83	50	50	77
CPU speed (fps)	45	47	41	18	7	8	10
AGX speed (fps)	55	–	36	22	13	15	17

Table 5. State-of-the-art comparison on OTB100, NFS, and UAV123 in terms of AUC score. The best real-time results are shown in red fonts, and the best non-real-time results are shown in **blue** fonts.

	Real-time					Non-real-time					
	Ours	E.T.Track [3]	LightTrack [41]	ATOM [8]	ECO [7]	TransT [5]	TrDimp [35]	TrSiam [35]	PrDiMP [9]	DiMP [2]	SiamRPN++ [20]
OTB100	68.1	67.8	66.2	66.9	69.1	69.4	**71.1**	70.8	69.6	68.4	69.6
NFS	63.5	59.0	55.3	58.4	46.6	65.7	**66.5**	65.8	63.5	62.0	50.2
UAV123	62.7	62.3	62.5	64.2	53.2	**69.1**	67.5	67.4	68.0	65.3	61.6
GPU speed (fps)	195	–	128	83	240	63	41	40	47	77	56
CPU speed (fps)	45	47	41	18	15	5	5	5	6	10	4
AGX speed (fps)	55	–	36	22	39	13	10	10	11	17	11

4.4 Evaluation on Other Datasets

We evaluate our tracker on some common small-scale datasets, including OTB100, NFS, and UAV123 [17,25,37]. Small-scale datasets easily overfit, and many methods use different hyperparameters for each datasets. Our tracker uses the default hyperparameters without adjustment. We report the AUC scores in Table 5. Our method outperforms LightTrack on these three datasets, with the best real-time results on NFS and the second best results on OTB100 and UAV123.

4.5 Ablation Study and Analysis

Component-Wise Analysis. Table 6 shows the component-wise study results. The model speed after ONNX acceleration (indicated as ONNX speed) is also reported in this table. #4 is our default model. In #2, the baseline is our method without the hierarchical connection structure and the FS module. In #3, after employing the hierarchical connection, the tracker can achieve 0.8% AO improvement on GOT-10k, 2.4% AUC improvement on TrackingNet, and 1.6% AUC improvement on LaSOT. This result verifies the effectiveness of our hierarchical

Table 6. Component-wise study. The best results are shown in red fonts.

#	Component	GOT-10k			TrackingNet			LaSOT			PyTorch Speed (*fps*)			ONNX Speed (*fps*)		
		AO	$SR_{0.5}$	$SR_{0.75}$	AUC	P_{Norm}	P	AUC	P_{Norm}	P	GPU	CPU	AGX	GPU	CPU	AGX
1	Baseline-SA	60.1	71.3	49.1	74.0	79.8	68.8	54.2	62.6	54.0	192	54	63	572	77	119
2	Baseline	64.7	76.6	54.1	74.2	80.6	70.3	57.4	67.2	58.1	197	40	50	575	80	110
3	+Hierarchical	65.5	76.7	57.0	76.6	82.3	72.6	59.0	68.5	60.6	197	40	50	575	80	110
4	+FS Module	65.1	76.5	56.7	76.6	82.6	72.9	59.3	68.7	61.0	195	45	55	589	90	127

Table 7. State-of-the-art comparison of TransT_H. The best two results are shown in red fonts and **blue** fonts.

Method	GOT-10k			TrackingNet			LaSOT		
	AO	$SR_{0.5}$	$SR_{0.75}$	AUC	P_{Norm}	P	AUC	P_{Norm}	P
DiMP [2]	61.1	71.7	49.2	74.0	80.1	68.7	56.9	65.0	56.7
SiamPRN++ [20]	51.7	61.6	32.5	73.3	80.0	69.4	49.6	56.9	49.1
PrDiMP [9]	63.4	73.8	54.3	75.8	81.6	70.4	59.8	68.8	60.8
Ocean [46]	61.1	72.1	47.3	−	−	−	56.0	65.1	56.6
SiamR-CNN [34]	64.9	72.8	59.7	81.2	85.4	80.0	64.8	72.2	−
STMTrack [12]	64.2	73.7	57.5	80.3	85.1	76.7	60.6	69.3	63.3
TrSiam [35]	67.3	78.7	58.6	78.1	82.9	72.7	62.4	−	60.0
TrDiMP [35]	68.8	80.5	59.7	78.4	83.3	73.1	63.9	−	61.4
ARDiMPsuper [42]	70.1	80.0	64.2	80.5	85.6	78.3	65.3	73.2	68.0
DualTFR [38]	73.5	84.8	69.9	80.1	84.9	−	63.5	72.0	66.5
DTT [43]	68.9	79.8	62.2	79.6	85.0	78.9	60.1	−	−
STARK-S50 [40]	67.2	76.1	61.2	80.3	85.1	−	65.8	−	−
STARK-ST50 [40]	68.0	77.7	62.3	81.3	86.1	−	66.6	−	−
TransT [5]	72.3	**82.4**	68.2	**81.4**	**86.7**	**80.3**	64.9	**73.8**	**69.0**
TransT_H	**72.4**	82.0	**68.5**	82.2	87.0	80.4	**66.2**	75.1	70.7

cross-attention manner. In addition, we also apply hierarchical connection structure to the original TransT. We only replace the juxtaposed cross-attention in TransT with our hierarchical cross-attention; the others remain unchanged. The results of TransT with our hierarchical cross-attention are shown in Table 7, denoted as TransT_H. Our method improves TransT to the level of state-of-the-art performance. TransT_H has second best performance on GOT-10k, best performance on TrackingNet, and second best performance on LaSOT. Except for LaSOT, all the results are better than STARK-ST50 with the same backbone network. STARK-ST50 is an online update method, while our TransT_H is a completely offline method.

In #4, after employing the FS module, the change in performance is negligible, but the speedup is noticeable on both CPU and AGX. The FS module reduces the computational amount by sparsing the template features, resulting in 12.5% and 10% speedup of the PyTorch model on CPU and AGX. Given the redundant template feature, sparsification has no negative impact on performance. We notice a slight slowdown on the GPU for the tracker using the FS module. As we discussed in Sect. 1 and Sect. 3, the computational amount of single operation layer is not the bottleneck of the speed on GPU, but the serial layer introduced by FS module slows down the model slightly. However, as the speed is extremely high on the GPU, this speed reduction is negligible. For the ONNX model, the FS module improves the speed on three platforms.

In #1, the baseline-SA is the model that replaces the cross-attention layers in the baseline with self-attention layers. The experimental results show that self-attention is not as effective as cross-attention for our model. Therefore, we discard self-attention and adopt a full cross-attention design.

Number of Sparse Vectors. We explore the effect of the number of sparse vectors. In Table 8, S indicates the number of sparse vectors. As S increases, the performance gradually increases. When $S = 16$, the performance is at par with the performance without the FS module. This result verifies the occurrence of redundancy in the template feature vectors. A total of 16 sparse vectors can achieve performance similar to the original 64 template vectors.

Table 8. Ablation study with different numbers of template queries S. The best results are shown in red fonts.

S	GOT-10k			TrackingNet			LaSOT		
	AO	$SR_{0.5}$	$SR_{0.75}$	AUC	P_{Norm}	P	AUC	P_{Norm}	P
1	62.3	73.7	51.4	74.7	80.5	70.1	57.6	67.0	58.3
4	64.5	75.3	55.7	75.8	81.4	71.4	58.3	67.4	58.8
9	63.9	75.2	55.6	76.2	81.7	71.9	58.3	67.5	59.1
16	65.1	76.5	56.7	76.6	82.6	72.9	59.3	68.7	61.0
25	64.5	75.9	56.1	76.2	82.0	72.2	60.7	70.1	61.9

Table 9. Ablation study with different numbers of feature fusion layers N. The best results are shown in red fonts.

N	GOT-10k			TrackingNet			LaSOT			PyTorch Speed (fps)			ONNX Speed (fps)		
	AO	$SR_{0.5}$	$SR_{0.75}$	AUC	P_{Norm}	P	AUC	P_{Norm}	P	GPU	CPU	AGX	GPU	CPU	AGX
1	62.4	73.8	51.9	74.2	80.0	69.3	57.6	67.2	57.7	240	46	69	702	104	147
2	65.1	76.5	56.7	76.6	82.6	72.9	59.3	68.7	61.0	195	45	55	589	90	127

Number of HCA Layers. We explore the effect of the number of HCA layers. In the results shown in Table 9, N indicates the number of HCA layers. When $N = 1$, the model is extremely fast, the ONNX model can reach 702 fps on GPU, 104 fps on CPU, and 147 fps on AGX, with good performance. We believe that the extremely fast tracker facilitates many vision tasks and real-world applications.

5   Conclusion

In this work, we propose an efficient and accurate tracking framework based on a novel feature fusion network. The feature fusion network is composed of a feature sparsification module and a hierarchical cross-attention transformer. The feature sparsification module sparses the template features to reduce the computational amount of the transformer. The hierarchical cross-attention transformer employs a full cross-attention design and a shallow structure to improve efficiency, and it also employs the hierarchical connection structure to enhance the representation ability. The experimental results on many benchmarks indicate that our tracker outperforms the state-of-the-art high-speed methods. The PyTorch model runs at 195 fps on GPU, 45 fps on CPU, and 55 fps on the edge AI platform NVidia Jetson AGX Xavier, and the ONNX model runs at 589 fps on GPU, 90 fps on CPU, and 127 fps on NVidia Jetson AGX Xavier.

Acknowledgement. This work was supported in part by the National Natural Science Foundation of China (NSFC) under Grant 61902420 and 62022021, in part by Joint Fund of Ministry of Education for Equipment Pre-research under Grant 8091B032155, in part by National Defense Basic Scientific Research Program under Grant WDZC20215250205, in part by the Science and Technology Innovation Foundation of Dalian under Grant no. 2020JJ26GX036, and in part by the Fundamental Research Funds for the Central Universities under Grant DUT21LAB127.

References

1. Bertinetto, L., Valmadre, J., Henriques, J.F., Vedaldi, A., Torr, P.H.S.: Fully-convolutional siamese networks for object tracking. In: ECCVW (2016)
2. Bhat, G., Danelljan, M., Gool, L.V., Timofte, R.: Learning discriminative model prediction for tracking. In: ICCV (2019)
3. Blatter, P., Kanakis, M., Danelljan, M., Van Gool, L.: Efficient visual tracking with exemplar transformers. arXiv preprint arXiv:2112.09686 (2021)
4. Carion, N., Massa, F., Synnaeve, G., Usunier, N., Kirillov, A., Zagoruyko, S.: End-to-end object detection with transformers. In: Vedaldi, A., Bischof, H., Brox, T., Frahm, J.-M. (eds.) ECCV 2020. LNCS, vol. 12346, pp. 213–229. Springer, Cham (2020). https://doi.org/10.1007/978-3-030-58452-8_13
5. Chen, X., Yan, B., Zhu, J., Wang, D., Yang, X., Lu, H.: Transformer tracking. In: CVPR (2021)
6. Chen, Z., Zhong, B., Li, G., Zhang, S., Ji, R.: Siamese box adaptive network for visual tracking. In: CVPR (2020)
7. Danelljan, M., Bhat, G., Khan, F.S., Felsberg, M.: ECO: Efficient convolution operators for tracking. In: CVPR (2017)
8. Danelljan, M., Bhat, G., Khan, F.S., Felsberg, M.: ATOM: Accurate tracking by overlap maximization. In: CVPR (2019)
9. Danelljan, M., Gool, L.V., Timofte, R.: Probabilistic regression for visual tracking. In: CVPR (2020)
10. Dosovitskiy, A., et al.: An image is worth 16 × 16 words: Transformers for image recognition at scale. In: ICLR (2020)

11. Fan, H., et al.: LaSOT: A high-quality benchmark for large-scale single object tracking. In: CVPR (2019)
12. Fu, Z., Liu, Q., Fu, Z., Wang, Y.: Stmtrack: Template-free visual tracking with space-time memory networks. In: CVPR (2021)
13. Glorot, X., Bengio, Y.: Understanding the difficulty of training deep feedforward neural networks. In: ICAIS (2010)
14. Guo, D., Wang, J., Cui, Y., Wang, Z., Chen, S.: SiamCAR: Siamese fully convolutional classification and regression for visual tracking. In: CVPR (2020)
15. He, K., Zhang, X., Ren, S., Sun, J.: Deep residual learning for image recognition. In: CVPR (2016)
16. Huang, L., Zhao, X., Huang, K.: Got-10k: A large high-diversity benchmark for generic object tracking in the wild. In: TPAMI (2019)
17. Kiani Galoogahi, H., Fagg, A., Huang, C., Ramanan, D., Lucey, S.: Need for speed: A benchmark for higher frame rate object tracking. In: ICCV (2017)
18. Kristan, M., et al.: The eighth visual object tracking vot2020 challenge results. In: ECCV (2020)
19. Krizhevsky, A., Sutskever, I., Hinton, G.E.: Imagenet classification with deep convolutional neural networks. In: NIPS (2012)
20. Li, B., Wu, W., Wang, Q., Zhang, F., Xing, J., Yan, J.: SiamRPN++: Evolution of siamese visual tracking with very deep networks. In: CVPR (2019)
21. Li, B., Yan, J., Wu, W., Zhu, Z., Hu, X.: High performance visual tracking with siamese region proposal network. In: CVPR (2018)
22. Lin, T.-Y., et al.: Microsoft COCO: common objects in context. In: Fleet, D., Pajdla, T., Schiele, B., Tuytelaars, T. (eds.) ECCV 2014. LNCS, vol. 8693, pp. 740–755. Springer, Cham (2014). https://doi.org/10.1007/978-3-319-10602-1_48
23. Liu, Z., et al.: Swin transformer: Hierarchical vision transformer using shifted windows. In: ICCV (2021)
24. Loshchilov, I., Hutter, F.: Decoupled weight decay regularization. In: ICLR (2018)
25. Mueller, M., Smith, N., Ghanem, B.: A benchmark and simulator for UAV tracking. In: Leibe, B., Matas, J., Sebe, N., Welling, M. (eds.) ECCV 2016. LNCS, vol. 9905, pp. 445–461. Springer, Cham (2016). https://doi.org/10.1007/978-3-319-46448-0_27
26. Müller, M., Bibi, A., Giancola, S., Alsubaihi, S., Ghanem, B.: TrackingNet: a large-scale dataset and benchmark for object tracking in the wild. In: Ferrari, V., Hebert, M., Sminchisescu, C., Weiss, Y. (eds.) ECCV 2018. LNCS, vol. 11205, pp. 310–327. Springer, Cham (2018). https://doi.org/10.1007/978-3-030-01246-5_19
27. Nam, H., Han, B.: Learning multi-domain convolutional neural networks for visual tracking. In: CVPR (2016)
28. Ren, S., He, K., Girshick, R., Sun, J.: Faster R-CNN: Towards real-time object detection with region proposal networks. In: NIPS (2015)
29. Rezatofighi, H., Tsoi, N., Gwak, J., Sadeghian, A., Reid, I.D., Savarese, S.: Generalized intersection over union: A metric and a loss for bounding box regression. In: CVPR (2019)
30. Russakovsky, O., et al.: ImageNet Large scale visual recognition challenge. In: IJCV (2015)
31. Szegedy, C., et al.: Going deeper with convolutions. In: CVPR (2015)
32. Tao, R., Gavves, E., Smeulders, A.W.M.: Siamese instance search for tracking. In: CVPR (2016)
33. Vaswani, A., et al.: Attention is all you need. In: NIPS (2017)
34. Voigtlaender, P., Luiten, J., Torr, P.H.S., Leibe, B.: Siam R-CNN: Visual tracking by re-detection. In: CVPR (2020)

35. Wang, N., Zhou, W., Wang, J., Li, H.: Transformer meets tracker: Exploiting temporal context for robust visual tracking. In: CVPR (2021)
36. Wang, Q., Zhang, L., Bertinetto, L., Hu, W., Torr, P.H.S.: Fast online object tracking and segmentation: A unifying approach. In: CVPR (2019)
37. Wu, Y., Lim, J., Yang, M.H.: Object tracking benchmark. In: TPAMI (2015)
38. Xie, F., Wang, C., Wang, G., Yang, W., Zeng, W.: Learning tracking representations via dual-branch fully transformer networks. In: ICCV (2021)
39. Xu, Y., Wang, Z., Li, Z., Yuan, Y., Yu, G.: SiamFC++: Towards robust and accurate visual tracking with target estimation guidelines. In: AAAI (2020)
40. Yan, B., Peng, H., Fu, J., Wang, D., Lu, H.: Learning spatio-temporal transformer for visual tracking. In: ICCV (2021)
41. Yan, B., Peng, H., Wu, K., Wang, D., Fu, J., Lu, H.: Lighttrack: Finding lightweight neural networks for object tracking via one-shot architecture search. In: CVPR (2021)
42. Yan, B., Zhang, X., Wang, D., Lu, H., Yang, X.: Alpha-refine: Boosting tracking performance by precise bounding box estimation. In: CVPR (2021)
43. Yu, B., et al.: High-performance discriminative tracking with transformers. In: ICCV (2021)
44. Yu, Y., Xiong, Y., Huang, W., Scott, M.R.: Deformable siamese attention networks for visual object tracking. In: CVPR (2020)
45. Zhang, Z., Peng, H.: Deeper and wider siamese networks for real-time visual tracking. In: CVPR (2019)
46. Zhang, Z., Peng, H., Fu, J., Li, B., Hu, W.: Ocean: object-aware anchor-free tracking. In: Vedaldi, A., Bischof, H., Brox, T., Frahm, J.-M. (eds.) ECCV 2020. LNCS, vol. 12366, pp. 771–787. Springer, Cham (2020). https://doi.org/10.1007/978-3-030-58589-1_46

Learning Dual-Fused Modality-Aware Representations for RGBD Tracking

Shang Gao[1], Jinyu Yang[1,2], Zhe Li[1], Feng Zheng[1(✉)], Aleš Leonardis[2], and Jingkuan Song[3]

[1] Department of Computer Science and Engineering, Southern University of Science and Technology, Shenzhen, China
zfeng02@gmail.com
[2] University of Birmingham, Birmingham, UK
[3] University of Electronic Science and Technology of China, Chengdu, China

Abstract. With the development of depth sensors in recent years, RGBD object tracking has received significant attention. Compared with the traditional RGB object tracking, the addition of the depth modality can effectively solve the target and background interference. However, some existing RGBD trackers use the two modalities separately and thus some particularly useful shared information between them is ignored. On the other hand, some methods attempt to fuse the two modalities by treating them equally, resulting in the missing of modality-specific features. To tackle these limitations, we propose a novel Dual-fused Modality-aware Tracker (termed DMTracker) which aims to learn informative and discriminative representations of the target objects for robust RGBD tracking. The first fusion module focuses on extracting the shared information between modalities based on cross-modal attention. The second aims at integrating the RGB-specific and depth-specific information to enhance the fused features. By fusing both the modality-shared and modality-specific information in a modality-aware scheme, our DMTracker can learn discriminative representations in complex tracking scenes. Experiments show that our proposed tracker achieves very promising results on challenging RGBD benchmarks. Code is available at https://github.com/ShangGaoG/DMTracker.

Keywords: RGBD tracking · Object tracking · Multi-modal learning

1 Introduction

Object tracking is to localize an arbitrary object in a video sequence, given only the object description in the first frame. It can be applied in lots of applications in video surveillance, autonomous driving [18,23,35], and robotics [19]. Recent years witness the development of RGBD (RGB+Depth) object tracking

---

Supplementary Information The online version contains supplementary material available at https://doi.org/10.1007/978-3-031-25085-9_27.

L. Karlinsky et al. (Eds.): ECCV 2022 Workshops, LNCS 13808, pp. 478–494, 2023.
https://doi.org/10.1007/978-3-031-25085-9_27

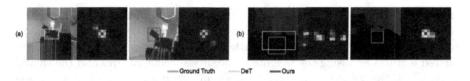

Fig. 1. Visual comparison of our proposed DMTracker and the state-of-the-art DeT [31]. Our tracker is more robust and accurate on handling the tracking challenges, *e.g.*, (a) background clutter, (b) dark scenes.

thanks to the affordable and accurate depth cameras. RGBD tracking aims to track the objects more robustly and accurately with the help of depth information, even in color-failed scenarios, *e.g.*, target occlusion and dark scenes. Compared to conventional RGB-based tracking, the major difficulty of RGBD tracking is the modality discrepancy resulting from intrinsically distinct information obtained from different modalities. For accurate and robust tracking, existing RGBD trackers have many attempts on either RGB and depth fusion [16], or RGB and depth feature extraction [25]. For example, early RGBD trackers fuse RGB and depth HoG features by using different weights [29]. Camplani *et al.* [3] applied the Kernel Correlation Filter (KCF) in RGB and depth maps, respectively. With the boosting of deep learning, researchers combine depth information with pre-trained RGB trackers to improve tracking performance. DAL [24] utilizes depth information to enhance RGB features via depth-aware convolutions. Very recently, Yan *et al.* proposed DeT [31], which utilized the depth colormaps for feature extraction and fused color and depth features via a simple operation like element-wise maximum.

However, on the one hand, a simple modality fusion strategy generally treats the two modalities equally. Thus, some modality-specific features which are strongly complementary cannot be discovered. For example, such fusing operations have side effects on the color modality. As RGB frames are informative, preserving the color and texture information is important for tracking. On the other hand, separately using RGB and depth information may lose the shared information between different modalities. Thus, with simply-fused features only, or using color and depth features separately, the upper bound of the discrimination ability of the feature representation is limited. Up to now, few methods explicitly exploit both intra- and inter-modality characteristics for RGBD tracking. Such deficiencies impede the tracking performance, and thus designing a high-performance RGBD tracker is still challenging.

Existing works in multi-modal learning demonstrate that models can benefit from exploring both the shared information and modality-specific characteristics [11]. Inspired by this, in this paper, we aim to tackle the above deficiencies by proposing **D**ual-fused **M**odality-aware **Tracker** for RGBD object tracking, namely **DMTracker**. The proposed framework consists of dual fusions, which apply the modality similarity and discrepancy for a robust feature representation. The first fusion module is the Cross-Modal Integration Module (CMIM), which is based on cross-attention. The second fusion module is Specificity Preserving Module (SPM). Specifically, CMIM explores the affinities between RGB and depth features to obtain shared information between different modalities. The obtained modality-shared features represent the correlation between RGB and depth channels through cross-modal attention. SPM compensate the modality-shared features by fusing the weighted modality-specific features and the shared ones. Thus, the second fusion tries to make up the missing specific information from the individual modalities. In DMTracker, intra-modality and inter-modality characteristics are both preserved through modality-aware fusions for similarity matching between tracking templates and search regions. This scheme can compensate for the lack of modality-specific information and enhance the modality-shared features, which finally improves the overall representation ability. Experiments on popular RGBD tracking benchmarks demonstrate the effectiveness of the proposed method.

Our contributions are three-fold:

- We propose a new tracking method dedicated to RGBD object tracking. It is capable of end-to-end offline training and real-time applications.
- We develop two novel modality-aware fusion modules that incorporate both modality-shared and modality-specific information for robust tracking.
- The proposed DMTracker achieves state-of-the-art performance on existing challenging RGBD tracking benchmarks with running at real-time speed.

2 Related Work

RGBD Object Tracking. Early RGBD trackers devote to applying RGBD fusion on handcrafted features between color and depth channels. PTB [25] presents a hybrid RGBD tracker composed of HOG features, optical flow, and 3D point clouds. An et al. [1] extends KCF by adding the depth channel. Another idea is to utilize the depth information on specific scenarios, e.g., occlusion handling and target re-detection [13,16], with hand-crafted depth features. In late fused trackers, the decisions from different channels are combined by using weighted summation [29], or multi-step localization [28]. Recently, deep learning-based RGBD trackers appear [24,31,34]. Among them, DAL [24] reformulates a deep discriminative correlation filter (DCF) to embed the depth information into deep features. DeT [31] transfers the depth maps to depth colormaps and fuses the color and depth features via a mixing operation, e.g., maximum. However, these methods ignore the modality discrepancy, which decreases the tracking accuracy. Compared with existing methods, our DMTracker eliminates the bias that heterogeneous features learned from different modalities, and exploits the correlation between them for robust tracking.

Multi-modal Learning. Multi-modal learning has attracted more and more attention since a large amount of data can be collected from various sources or sensors. As events or actions can be described by information from multiple modalities, multi-modal learning aims to understand the correlation between different modalities. Early methods directly concatenate the features from the multiple modalities into a feature vector. However, simple concatenations often fail to exploit the complex correlations among different modalities. Therefore,

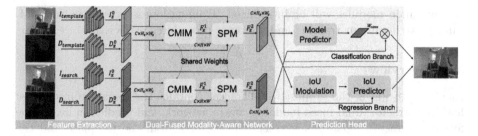

Fig. 2. Overview of the DMTracker pipeline. Our DMTracker consists of three main components: feature extraction, dual-fused modality-aware network, and prediction head. The dual-fused modality-aware network consists of two novel modules, CMIM (cross-modal integration module) and SPM (specificity preserving module), which are introduced in Sect. 3.2 and 3.3.

several multi-modal learning methods have been developed to explicitly fuse the complementary information from different modalities to improve model performance [14,22]. Hu *et al.* [11] present a shareable and individual multi-view learning strategy to explore more properties of multi-modal data. Besides, Zhou *et al.* [36] propose SPNet which improves saliency detection performance by exploring both the shared information and modality-specific properties. Chen *et al.* [7] propose a multi-modal framework with an inter-modal module that learns cross-modal features and an intra-modal module to self-learn feature representations across videos.

Transformer in Tracking. The transformer is originally proposed for natural language processing. With attention-based encoders and decoders, transformers have shown their great potential for vision tasks. Transformers have been applied to generic object tracking with considerable success [4,6,15,27,30,33]. Existing transformers for tracking focus on the correlation between target templates and search regions. Typically, the transformer decoder fuses feature from the template and search region by cross-attention layers to obtain discriminative features. For example, TransT [6] is proposed to use cross-attention instead of the traditional correlation operation. Later, STARK is proposed based on DETR [5], utilizing self-attention to learn the relationship between different inputs. Existing works demonstrate the effectiveness of the transformer on track, while they mostly focus on the RGB tracking area. To the best of our knowledge, there are no transformer architectures focusing on cross-modal RGBD tracking. Our proposed method utilizes an attention mechanism between color and depth modalities and shows promising performance.

3 Methodology

In this section, we present the cross-modal correlation tracker, *i.e.* DMTracker. We first describe our DMTracker pipeline in Sect. 3.1. Next, we present the key modules in the dual-fused modality-aware network in Sect. 3.2 and 3.3, in which the two main modules will be introduced. Training loss is present in Sect. 3.4. Finally, we detail our training and inference procedures in Sect. 3.5.

3.1 Overview

As shown in Fig. 2, the proposed DMTracker contains three main components: feature extractor, dual-fused modality-aware network, and prediction head. The feature extractor separately extracts the features from the template and the search region of both RGB and depth modalities. Then, the features from different modalities are enhanced by the proposed dual-fused modality-aware network to be a robust representation. Finally, the prediction head performs the binary classification and bounding box regression on the fused features to predict the bounding boxes.

Feature Extractor. Firstly, the feature extractor takes template patches from RGB and depth frames Z_{rgb}, $Z_{depth} \in \mathbb{R}^{3 \times H_{z_0} \times W_{z_0}}$ and corresponding search region patches X_{rgb}, $X_{depth} \in \mathbb{R}^{3 \times H_{x_0} \times W_{x_0}}$. With the ResNet-50 backbone, the feature maps of the template and the search patch from the two modalities, *i.e.*, I_z, $D_z \in \mathbb{R}^{C \times H_z \times W_z}$ and $I_x, D_x \in \mathbb{R}^{C \times H_x \times W_x}$, can be obtained. $H_z = \frac{H_{z_0}}{n}$, $W_z = \frac{W_{z_0}}{n}$, $H_x = \frac{H_{x_0}}{n}$, $W_x = \frac{W_{x_0}}{n}$, n is the downsampling factor and C is the dimension of the feature maps.

Dual-Fused Modality-Aware Network. Then, the dual-fused modality-aware network takes the feature maps of the two modalities of RGB and depth under the same branch as input, *i.e.*, I_x, D_x, I_z, and D_z. The whole network consists of two fusion modules, *i.e.*, the Cross-Modal Integration Module (CMIM) and Specificity Preserving Module (SPM). Features are firstly fed into the preliminary fusion in CMIM to obtain the initial fusion features through the modality-aware cross-modal attention. And then the features, and both the RGB and depth features, are fed into the SPM to get the final fused features with preserving the modality-aware specificity. The details of the two modules in the network are introduced in Sect. 3.2 and 3.3.

Prediction Head. After getting the final modality-aware feature representations of templates and search regions, a prediction head [2] is used to obtain the bounding box estimation. For the classification head, the fused template feature maps are input to the model predictor that fully exploits the information from the target and background. The model predictor outputs the weights of the convolution layer, which then performs target classification for the search branch. For the localization head, using the template features and the initial target bounding box, IoU Modulation component computes the modulation vectors carrying the target-specific appearance information. Finally, bounding box estimation is performed by maximizing the IoU prediction based on the modulation vectors and proposal bounding boxed in the search region.

3.2   Cross-Modal Integration Module (CMIM)

The cross-modal integration module is built as shown in Fig. 3. In this module, we fuse the cross-modal features from the RGB and depth modalities to learn their shared representation.

Attention. In the attention mechanism, given queries Q, keys K and values V, the attention function is defined by the scale dot-product attention:

$$Attention(Q, K, V) = Softmax(\frac{QK^\top}{\sqrt{d_k}})V, \tag{1}$$

where d_k is the dimension of the key K.
It can also be extended to multi-head attention (MHA) as:

$$
\begin{aligned}
MHA(Q, K, V) &= Concat(Head_1, Head_h)W^O, \\
Head_i &= Attention(QW_i^Q, KW_i^K, VW_i^V),
\end{aligned}
\tag{2}
$$

where the projections matrices are $W_i^Q, W_i^K \in \mathbb{R}^{d_{model} \times d_k}$, $W_i^V \in \mathbb{R}^{d_{model} \times d_v}$ and $W^O \in \mathbb{R}^{hd_v \times d_{model}}$. By projecting the inputs to different spaces through different linear transformations, the model can learn various data distributions and pay attention to varying levels of information, for more details [26]. In this work, we employ $h = 8$, $d_{model} = 256$ and $d_k = d_v = d_m = 32$ as default values.

Cross-Modal Attention (CMA) Block. To model the global information of the dual modalities and minimize the computation of our real-time tracker, we use the lightweight attention architecture named cross-modal attention (CMA) block as shown in Fig. 5. We simplify the original transformer, remove the self-attention part, but retain the cross-attention, and feed-forward network because we pay more attention to the interactive fusion between the dual modalities, and it can significantly reduce the amount of computation and improve the speed of the tracker. A feed-forward network (FFN) is a fully connected network that consists of two linear functions with a Relu activate function to enhance the fitting ability of the attention. The attention mechanism can not distinguish the position information of the input feature sequence. Thus, we introduce the spatial positional encoding attracting to input following the setting [5,6]. Cross-Modal Attention Block can enhance depth features by finding the most related visual clues from RGB features, and the CMA can be formulated as Eq. (3).

$$
\begin{aligned}
f_t &= Norm(D + MHA(D, I, I)), \\
F &= Norm(f_t + FFN(f_t)),
\end{aligned}
\tag{3}
$$

where $F, D \in \mathbb{R}^{C \times H \times W}$.

Cross-Modal Integration. The cross-modal integration module is designed based on the CMA block. The basic idea of our proposed CMIM is to use the attention mechanism to globally model the correlation information between the dual modalities learning the shared representation. Formally, we use $I^{(0)}, D^{(0)}$

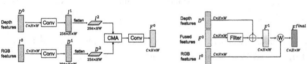

Fig. 3. Cross-Modal Integration Module (CMIM).

Fig. 4. Specificity Preserving Module (SPM).

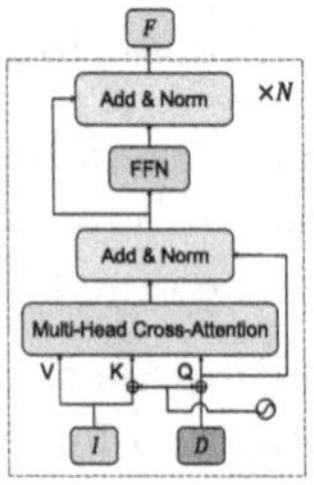

Fig. 5. Cross-Modal Attention (CMA).

to denote the initial RGB feature and depth feature obtained from the backbone network. First, a 1×1 convolution reduces the channel dimension obtaining two lower dimension feature $I^{(1)}$ and $D^{(1)}$:

$$I^{(1)}, D^{(1)} = \mathcal{F}(I^{(0)}, D^{(0)}), \tag{4}$$

where $I^{(0)}, D^{(0)} \in \mathbb{R}^{C \times H \times W}, I^{(1)}, D^{(1)} \in \mathbb{R}^{C_i \times H \times W}$. We employ $C_i = 256$ in our implementation. Since the cross-modal attention (CMA) takes a set of feature vectors as input, we flatten $I^{(1)}$ and $D^{(1)}$ in spatial dimension, obtaining $I^{(2)} \in \mathbb{R}^{C_i \times HW}$ and $D^{(2)} \in \mathbb{R}^{C_i \times HW}$. In our scenario, we have dual modality features which can be used to attend to each other using the Dual-fused Modality-aware Network i.e., using RGB feature to enhance depth feature, or using depth feature to enhance RGB feature. In our method, we adopt the former: D as query, I as key and value, as this essentially puts more emphasis on the RGB feature, if we want to make depth feature to integrate more color or texture features from the RGB feature. After cross-modal attention (CMA) block and a 1×1 convolution, the initial fused RGBD feature $F^{(0)}$ can be obtained as follows:

$$F^{(0)} = \mathcal{F}(CMA(I^{(2)}, D^{(2)})), \quad F^{(0)} \in \mathbb{R}^{C \times H \times W}. \tag{5}$$

3.3 Specificity Preserving Module (SPM)

The specificity preserving module is built as shown in Fig. 4. As we have the shared representations from the fused modality, and the features from the modality-specific feature extractors, we propose SPM that make full use of modality-specific information to further enhance the fusion features. Due to the considerable gap between the depth space information from the depth feature and the visual information from the RGB feature, the initial fused features obtained by the CMIM not only contain the correlation information between the two modalities but are also mixed with a lot of irrelevant interference features. Thus, the initial fused feature first passes through an information filter to weaken the chaotic interference information and enhance the shared characteristics of the dual modalities. Then the specificity of each modality is attached to the filtered features, and the final fused features can be obtained which contain the modality-shared and modality-specific cues. The detailed process is described as follows.

We design a learnable vector $\mathcal{V}$ to filter the interference and enhance the shared cues of the initial fused features. $\odot$ means element-wise multiplication. It is then updated with depth-specific features through residual connections:

$$F^{(1)} = D^{(0)} + \mathcal{V} \odot F^{(0)}, \mathcal{V} \in \mathbb{R}^C, F^{(0)} \in \mathbb{R}^{C \times H \times W}. \qquad (6)$$

Then the obtained features $F^{(1)}$ and initial RGB-specific features $I^{(0)}$ are fused by learnable weights to get the final RGBD feature (6).

$$F^{(final)} = \alpha I^{(0)} + \beta F^{(1)}, \quad F^{(final)} \in \mathbb{R}^{C \times H \times W}. \qquad (7)$$

3.4   Training Loss

We train our network with the following function,

$$\mathcal{L}_{total} = \lambda \mathcal{L}_{cls} + \mathcal{L}_{bbox}. \qquad (8)$$

Classification loss $\mathcal{L}_{cls}$ provides the network the ability to robustly discriminate the target object and background distractors, which is defined as [2].

$$l(s, z) = \begin{cases} s - z & z > T \\ max(0, s) & z \leq T \end{cases} \qquad (9)$$

$$\mathcal{L}_{cls} = \frac{1}{N_{iter}} \sum_{i=0}^{N_{iter}} \sum_{(x,c) \in M_{test}} \|l(x * f, z_c)\|^2. \qquad (10)$$

Here, according to the label confidence z, the threshold T defines the target and background regions. N_{iter} refers to the number of iterations. z_c is the target center label that is set from a Gaussian function. We only penalize positive confidence values for the background if $z \leq T$. If $z > T$, we take the difference between them. Here, f is the wights obtained by the model predictor network.

For bounding box estimation, following [8], it extends the training procedure to image sets by computing the modulation vector on the first frame in M_{train} and sampling candidate boxes from images in M_{test}. The bounding box loss $\mathcal{L}_{bbox}$ is computed as the mean squared error between the predicted BBox defined by B and the groundtruth BBox defined by $\bar{B}$ in M_{test},

$$\mathcal{L}_{bbox} = \underset{i \in M_{test}}{MSE} (B_i, \bar{B}_i). \qquad (11)$$

3.5   Implementation Details

Training. Our tracker is trained in an end-to-end fashion. 1) Template and Search Area. Different from DeT [31], we use RGB images and raw depth images as inputs. To adapt the depth image to the pre-trained feature extraction model,

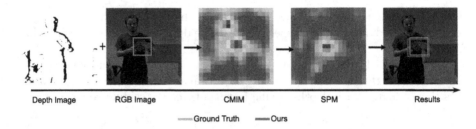

Depth Image   RGB Image   CMIM   SPM   Results

━━━ Ground Truth   ━━━ Ours

Fig. 6. Visualization of the attention maps for a representative frame. From left to right, they are input depth and RGB frames, feature maps in the search region after CMIM and SPM, and the output prediction, respectively.

we copy the raw depth to stack three channels. To incorporate background information, the template and search region is obtained by adding random translation and cropping a square patch centered on the target, which is 5^2 of the target area. Then these regions are resized to 288×288 pixels. Horizontal flip and brightness jittering are used for data augmentations. 2) Initialization. The backbone ResNet50 is initialized by the pre-trained wights on ImageNet. The weights of IoU modulation and IoU predictor in the regression head are initialized using [8]. The model predictor in the classification head is initialized using [2]. $N_{iter} = 5, T = 0.05$. We use $\mathcal{V} = 0.01$ to initialize the information filter. The learnable weights between the two modalities: α and β are initialized to 0.5. The network is optimized using the AdamW optimizer with a learning rate decay of 0.2 every 15th epoch and weight decay $1e - 4$. The target classification loss weight is set to $\lambda = 1e - 2$.

Inference. Different from the training phase, we track the target in a series of subsequent frames giving the target position in the first frame. Once the annotated first frame is given, we use the data enhancement strategy to build an initial set of 15 samples to refine the model predictor. With the subsequent frames with high confidence, the model predictor can continuously refine the classification head and enhance the ability to distinguish the target and background. Bounding box estimation is performed as in the training phase.

3.6 Visualization

To explore how the dual-fused modality-aware network works in our framework, we visualize the attention maps of the fusion modules in a representative frame, as shown in Fig. 6. We visualize the attention maps of the modality shared and specific features from the output of CMIM and SPM, respectively. As shown, we can obtain a very informative feature map after the proposed CMIM. Based on cross-modal attention, the modality-shared feature is generated to make the focus on candidates, *i.e.*, the box and the face. Then the SPM is to use the modality-specific information to suppress unreliable candidates. The combination of two kinds of features in SPM produces high-quality representations. This proves that the shared and specific features can be well fused by our CMIM and SPM, leading to more discriminative feature representations for tracking.

4 Experiments

4.1 Experiment Settings

Datasets. To validate the effectiveness of the proposed tracker, we evaluate it on three public RGBD tracking datasets, including STC [29], CDTB [17] and DepthTrack [31]. Among them, STC is a compact short-term benchmark containing 36 sequences, while CDTB and DepthTrack are long-term ones containing 80 sequences and 50 sequences, respectively. Furthermore, we carry out an attribute-based evaluation to study the performance under different challenges.

Evaluation Metrics. For STC, a representative short-term evaluation protocol, success rate, is used. Given the tracked bounding box b_t and the groundtruth bounding box b_a, the overlap score is defined as $S = \frac{b_t \cap b_a}{b_t \cup b_a}$, which is indeed the intersection and union of two regions. With a series of thresholds from 0 to 1, the area-under-curve (AUC) is calculated as the success rate.

For CDTB and DepthTrack, long-term evaluation, $i.e.$, tracking precision and recall are used [32]. They can be obtained by:

$$ Pr(\tau_\theta) = \frac{1}{N_p} \sum_{A_t(\tau_\theta) \neq \varnothing} \Omega(A_t(\tau_\theta), G_t), \quad Re(\tau_\theta) = \frac{1}{N_g} \sum_{G_t \neq \varnothing} \Omega(A_t(\tau_\theta), G_t), \quad (12) $$

where $Pr(\tau_\theta)$ and $Re(\tau_\theta)$ denote the precision (Pr) and the recall (Re) over all frames, respectively. F-score then can be obtained by $F(\tau_\theta) = \frac{2Re(\tau_\theta)Pr(\tau_\theta)}{Re(\tau_\theta)+Pr(\tau_\theta)}$. N_p denotes the number of frames in which the target is predicted visible, and N_g denotes the number of frames in which the target is indeed visible.

Compared Trackers. We compare the proposed tracker with the following benchmarking RGBD trackers.

- Traditional RGBD trackers: PT [25], OAPF [21], DS-KCF [3], DS-KCF-shape [10], CA3DMS [16], CSR_RGBD++ [12], STC [29] and OTR [13];
- Deep RGBD trackers: DAL [24], TSDM [34], and DeT[1] [31];
- State-of-the-art RGB trackers: ATOM [8], DiMP [2], PrDiMP [9], KeepTrack [20] and TransT [6].

---

[1] For a fair comparison, we use DeT-DiMP50-Max checkpoint in all experiments.

Table 1. Overall performance on Depth-Track test set [31]. The top 3 results are shown in red, blue, and green.

Method	Type	Pr	Re	F-score
DS-KCF [3]	RGBD	0.075	0.077	0.076
DS-KCF-Shape [10]	RGBD	0.070	0.071	0.071
CSR_RGBD++ [12]	RGBD	0.113	0.115	0.114
CA3DMS [16]	RGBD	0.212	0.216	0.214
DAL [24]	RGBD	0.478	0.390	0.421
TSDM [34]	RGBD	0.393	0.376	0.384
DeT [31]	RGBD	0.560	0.500	0.532
DiMP50 [2]	RGB	0.387	0.403	0.395
KeepTrack [20]	RGB	0.503	0.514	0.508
TransT [6]	RGB	0.484	0.494	0.489
DMTracker	RGBD	0.619	0.597	0.608

Table 2. Overall performance on CDTB dataset [17]. The top 3 results are shown in red, blue, and green.

Method	Type	Pr	Re	F-score
DS-KCF [3]	RGBD	0.036	0.039	0.038
DS-KCF-Shape [10]	RGBD	0.042	0.044	0.043
CA3DMS [16]	RGBD	0.271	0.284	0.259
CSR_RGBD++ [12]	RGBD	0.187	0.201	0.194
OTR [13]	RGBD	0.336	0.364	0.312
DAL [24]	RGBD	0.661	0.565	0.592
TSDM [34]	RGBD	0.578	0.541	0.559
DeT [31]	RGBD	0.651	0.633	0.642
ATOM [8]	RGB	0.541	0.537	0.539
DiMP50 [2]	RGB	0.546	0.549	0.547
DMTracker	RGBD	0.662	0.658	0.660

Table 3. Overall performance and attribute-based results on STC [29]. The top 3 results are shown in red, blue, and green.

Method	Type	Success	Attributes									
			IV	DV	SV	CDV	DDV	SDC	SCC	BCC	BSC	PO
OAPF [21]	RGBD	0.26	0.15	0.21	0.15	0.15	0.18	0.24	0.29	0.18	0.23	0.28
DS-KCF [3]	RGBD	0.34	0.26	0.34	0.16	0.07	0.20	0.38	0.39	0.23	0.25	0.29
PT [25]	RGBD	0.35	0.20	0.32	0.13	0.02	0.17	0.32	0.39	0.27	0.27	0.30
DS-KCF-Shape [10]	RGBD	0.39	0.29	0.38	0.21	0.04	0.25	0.38	0.47	0.27	0.31	0.37
STC [29]	RGBD	0.40	0.28	0.36	0.24	0.24	0.36	0.38	0.45	0.32	0.34	0.37
CA3DMS [16]	RGBD	0.43	0.25	0.39	0.29	0.17	0.33	0.41	0.48	0.35	0.39	0.44
CSR_RGBD++ [12]	RGBD	0.45	0.35	0.43	0.30	0.14	0.39	0.40	0.43	0.38	0.40	0.46
OTR [13]	RGBD	0.49	0.39	0.48	0.31	0.19	0.45	0.44	0.46	0.42	0.42	0.50
DAL [24]	RGBD	0.62	0.52	0.60	0.51	0.62	0.46	0.63	0.63	0.55	0.58	0.57
TSDM [34]	RGBD	0.57	0.53	0.42	0.43	0.54	0.37	0.56	0.55	0.41	0.45	0.53
DeT [31]	RGBD	0.60	0.39	0.56	0.48	0.58	0.40	0.59	0.62	0.48	0.46	0.54
DiMP [2]	RGB	0.60	0.49	0.61	0.47	0.56	0.57	0.60	0.64	0.51	0.54	0.57
PrDiMP [9]	RGB	0.60	0.47	0.62	0.47	0.51	0.58	0.64	0.62	0.53	0.54	0.58
KeepTrack [20]	RGB	0.62	0.51	0.63	0.47	0.57	0.56	0.61	0.62	0.53	0.52	0.58
DMTracker	RGBD	0.63	0.52	0.59	0.46	0.67	0.63	0.61	0.65	0.52	0.52	0.60

4.2 Main Results

Quantitative Results. We give the quantitative results of the compared models on DepthTrack, CDTB, and STC in Tables 1, 2, and 3, respectively. For DepthTrack and CDTB, corresponding tracking Precision-Recall and F-score Plots are also given in Fig. 7 and Fig. 8. As shown in Table 1, our method outperforms all of the compared state-of-the-art methods and obtains the best F-score of 0.608 on the DepthTrack dataset. On CDTB, we also obtain the top F-score of 0.660. Compared with the state-of-the-art RGBD tacker DeT, our DMTracker performs 7.6% and 1.8% higher on DepthTrack and CDTB, respectively, according to F-score. DMTracker also outperforms TransT by 11.9% higher on Depth-Track. Overall, the proposed DMTracker provide better accuracy than RGB trackers or advanced RGBD trackers on existing RGBD datasets.

Qualitative Results. Figure 9 shows several representative example results of representative trackers and the proposed tracker on various benchmark datasets. Compared with other RGBD models, we can see that our DMTracker can achieve better tracking performance under multiple difficulties. Sequence 1 shows a scene with background clutter. Our method can accurately distinguish the target object, while other trackers are confused by the clutters. In sequence 2, we show an example with multiple similar objects with fast motion, where it is challenging to accurately track the object. While our DMTracker can track the small-sized target ball robustly. In sequence 3 and 5, we show two examples when there are background clutter or occlusion in outdoor scenarios. Our method locates the objects more accurately compared to other approaches. In sequence 4, under a very dark scene with fast motion of the object, our tracker can robustly track the target object. Finally, It can be seen that some approaches fail to track the rotated object in a dark room, while our DMTracker can produce reliable results. Overall, our tracker can handle various tracking challenges and produce promising results.

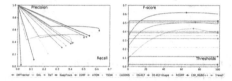

Fig. 7. Overall tracking performance is presented as tracking Precision-Recall and F-score Plots on DepthTrack [31].

Fig. 8. Overall tracking performance is presented as tracking Precision-Recall and F-score Plots on CDTB [17].

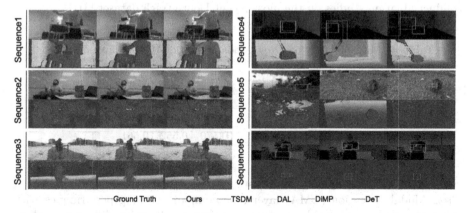

Fig. 9. Qualitative results in different challenging scenarios. Seq 1: background clutter; Seq 2: similar objects and fast motion; Seq 3: outdoor scenario and full occlusion; Seq 4: dark scenes and fast motion; Seq 5: outdoor scenario and camera motion; Seq 6: dark scene and target rotation.

Attribute-Based Analysis. We report attribute-based evaluations of representative state-of-the-art algorithms on STC, DepthTrack, and CDTB for in-depth analysis. Success rates according to different attributes on STC are given in Table 3. We can see that our proposed outperforms other compared models on Color Distribution Variation (CDV), Depth Distribution Variation (DDV), Surrounding Color Clutter (SCC) , and Partial Occlusion (PO). Attribute-specific F-scores for the tested trackers on DepthTrack and CDTB are shown in Fig. 10 and Fig. 11, respectively. As shown, our DMTracker outperforms other trackers on 12 attributes on DepthTrack. Our DMTracker outperforms other trackers on 7 attributes on CDTB. It is obvious that DMTracker can handle the depth-related challenges better, *e.g.*, background clutter (BC), dark scenes (DS), depth change (DC), and similar objects (SO).

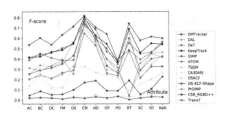

Fig. 10. Attribute-based F-score on DepthTrack [31].

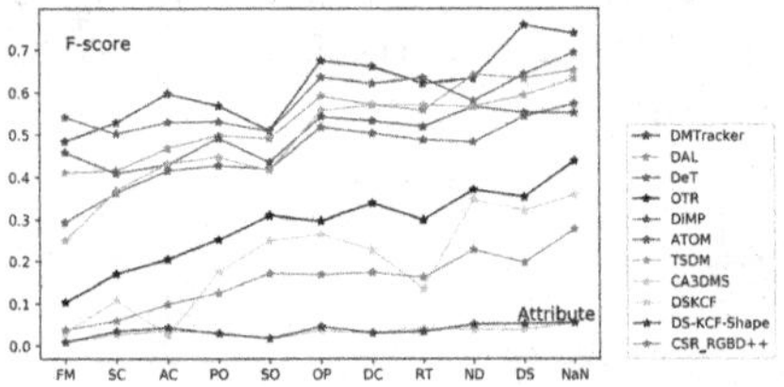

Fig. 11. Attribute-based F-score on CDTB [17].

4.3   Ablation Study

To verify the relative importance of different key components of our tracker, we conduct ablation studies on the most recent DepthTrack dataset [31].

Table 4. Ablation study on Cross-Modal Attention (CMA) layer numbers.

Method	CMA Layer	Pr	Re	F-score
DMTracker	1	0.585	0.571	0.578
DMTracker	2	**0.619**	**0.597**	**0.608**
DMTracker	3	0.570	0.552	0.561

Effects of CMA Layers. To investigate the effects of different numbers of Cross-Modal Attention (CMA) layers, we compare tracking performance with different settings. Table 4 shows the comparison results using different numbers of CMA layers, *i.e.*, 1, 2, and 3. From the results, we can see that our model with 2 CMA layers obtains better performance. This can be explained that only one CMA layer leads to inadequate learning of the target knowledge and cross-modal correlation, while deep layers face the risks of overfitting.

Component-Wise Analysis. Since the proposed DMTracker is used to fuse cross-modal features and learn a unified representation, we ablate each key component for analysis. The base model is constructed by removing the whole dual-fused modality-aware network. The fused features are obtained by simply obtaining the element-wise addition of RGB and depth feature maps. As shown in Table 5, the tracking performance first degrades with only CMIM or SPM and then highly improves with adding the CMIM and SPM. As we analyzed in Sect. 3.6, the CMIM will bring very informative and even redundant features which leads to the sub-optimal performance. With only SPM, the interaction of cross-modal information is lacking, so the model does not fully exploit the modality-aware correlation information. Then, by adding the CMIM and SPM, the redundant information is effectively suppressed, shared and specificity information is enhanced and thus the tracking performance improves.

Table 5. Ablation study on key components in the network.

Base	CMIM	SPM	Pr	Re	F-score
✓			0.517	0.492	0.504
✓	✓		0.522	0.426	0.469
✓		✓	0.584	0.563	0.574
✓	✓	✓	0.619	0.597	0.608

Table 6. Ablation study on different SPM settings.

Method	$\alpha I^0 + \beta F^1$	$\alpha D^0 + \beta F^1$	Pr	Re	F-score
DMTracker		✓	0.576	0.560	0.568
DMTracker	✓		0.619	0.597	0.608

Effects on Modalities Fusion Order in SPM. In the specificity preserving module (SPM) of our DMTracker, We first add depth-specific features to RGBD features and then fuse them with weighted RGB-specific features. When given different orders of specific modalities, the performance of the tracker degrades. This can be explained by the fact that the RGB features are more informative than depth features, and this part of the information is dominant in the tracking process. Assume the spatial information is added in high weight at the end fusion. In that case, the proportion of visual information in the fusion feature is weakened. Mixing excess spatial information confuses the tracker.

5 Conclusions

In this paper, we proposed a novel DMTracker to learn dual-fused modality-aware representations for RGBD tracking. On the one hand, the proposed cross-modal integration module can learn the shared information between RGB and depth channels through cross-modal attention. On the other hand, the following specificity preserving module can adaptively fuse the shared features and RGB and depth features to get discriminative representations. Extensive experiments validate the superior performance of the proposed DMTracker, as well as the effectiveness of each component.

Acknowledgements. This work is supported by the National Natural Science Foundation of China under Grant No. 61972188 and 62122035.

References

1. An, N., Zhao, X.G., Hou, Z.G.: Online rgb-d tracking via detection-learning-segmentation. In: 2016 23rd International Conference on Pattern Recognition, pp. 1231–1236. IEEE (2016)
2. Bhat, G., Danelljan, M., Gool, L.V., Timofte, R.: Learning discriminative model prediction for tracking. In: Proceedings of the IEEE/CVF International Conference on Computer Vision, pp. 6182–6191 (2019)
3. Camplani, M., et al.: Real-time rgb-d tracking with depth scaling kernelised correlation filters and occlusion handling. In: BMVC, vol. 4, p. 5 (2015)
4. Cao, Z., Huang, Z., Pan, L., Zhang, S., Liu, Z., Fu, C.: Tctrack: Temporal contexts for aerial tracking. arXiv preprint arXiv:2203.01885 (2022)
5. Carion, N., Massa, F., Synnaeve, G., Usunier, N., Kirillov, A., Zagoruyko, S.: End-to-end object detection with transformers. In: Vedaldi, A., Bischof, H., Brox, T., Frahm, J.-M. (eds.) ECCV 2020. LNCS, vol. 12346, pp. 213–229. Springer, Cham (2020). https://doi.org/10.1007/978-3-030-58452-8_13
6. Chen, X., Yan, B., Zhu, J., Wang, D., Yang, X., Lu, H.: Transformer tracking. In: Proceedings of the IEEE/CVF Conference on Computer Vision and Pattern Recognition, pp. 8126–8135 (2021)
7. Chen, Y.W., Tsai, Y.H., Yang, M.H.: End-to-end multi-modal video temporal grounding. In: Advances in Neural Information Processing Systems 34 (2021)
8. Danelljan, M., Bhat, G., Khan, F.S., Felsberg, M.: Atom: Accurate tracking by overlap maximization. In: Proceedings of the IEEE/CVF Conference on Computer Vision and Pattern Recognition, pp. 4660–4669 (2019)
9. Danelljan, M., Gool, L.V., Timofte, R.: Probabilistic regression for visual tracking. In: Proceedings of the IEEE/CVF Conference on Computer Vision and Pattern Recognition, pp. 7183–7192 (2020)
10. Hannuna, S., et al.: Ds-kcf: a real-time tracker for rgb-d data. J. Real-Time Image Proc. **16**(5), 1–20 (2016)
11. Hu, J., Lu, J., Tan, Y.P.: Sharable and individual multi-view metric learning. IEEE Trans. Pattern Anal. Mach. Intell. 1 (2017)
12. Kart, U., Kämäräinen, J.K., Matas, J.: How to make an rgbd tracker? In: ECCVW (2018)
13. Kart, U., Lukežič, A., Kristan, M., Kämäräinen, J.K., Matas, J.: Object tracking by reconstruction with view-specific discriminative correlation filters. In: IEEE Conference on Computer Vision and Pattern Recognition (2019)
14. Kim, J., Ma, M., Pham, T., Kim, K., Yoo, C.D.: Modality shifting attention network for multi-modal video question answering. In: Proceedings of the IEEE/CVF Conference on Computer Vision and Pattern Recognition, pp. 10106–10115 (2020)
15. Lin, L., Fan, H., Xu, Y., Ling, H.: Swintrack: A simple and strong baseline for transformer tracking. arXiv preprint arXiv:2112.00995 (2021)
16. Liu, Y., Jing, X.Y., Nie, J., Gao, H., Liu, J., Jiang, G.P.: Context-aware three-dimensional mean-shift with occlusion handling for robust object tracking in rgb-d videos. IEEE Trans. Multimedia **21**(3), 664–677 (2018)
17. Lukezic, A., et al.: Cdtb: A color and depth visual object tracking dataset and benchmark. In: Proceedings of the IEEE/CVF International Conference on Computer Vision, pp. 10013–10022 (2019)

18. Luo, W., Yang, B., Urtasun, R.: Fast and furious: Real time end-to-end 3d detection, tracking and motion forecasting with a single convolutional net. In: Proceedings of the IEEE Conference On Computer Vision and Pattern Recognition, pp. 3569–3577 (2018)
19. Machida, E., Cao, M., Murao, T., Hashimoto, H.: Human motion tracking of mobile robot with kinect 3d sensor. In: 2012 Proceedings of SICE Annual Conference (SICE), pp. 2207–2211. IEEE (2012)
20. Mayer, C., Danelljan, M., Paudel, D.P., Van Gool, L.: Learning target candidate association to keep track of what not to track. In: Proceedings of the IEEE/CVF International Conference on Computer Vision, pp. 13444–13454 (2021)
21. Meshgi, K., ichi Maeda, S., Oba, S., Skibbe, H., zhe Li, Y., Ishii, S.: An occlusion-aware particle filter tracker to handle complex and persistent occlusions. Comput. Vision Image Understand. **150**, 81–94 (2016). https://doi.org/10.1016/j.cviu.2016.05.011, https://www.sciencedirect.com/science/article/pii/S1077314216300649
22. Pan, B., et al.: Spatio-temporal graph for video captioning with knowledge distillation. In: Proceedings of the IEEE/CVF Conference on Computer Vision and Pattern Recognition, pp. 10870–10879 (2020)
23. Qi, H., Feng, C., Cao, Z., Zhao, F., Xiao, Y.: P2b: Point-to-box network for 3d object tracking in point clouds. In: Proceedings of the IEEE/CVF Conference on Computer Vision and Pattern Recognition, pp. 6329–6338 (2020)
24. Qian, Y., Yan, S., Lukežič, A., Kristan, M., Kämäräinen, J.K., Matas, J.: DAL: A deep depth-aware long-term tracker. In: International Conference on Pattern Recognition (2020)
25. Song, S., Xiao, J.: Tracking revisited using rgbd camera: Unified benchmark and baselines. In: Proceedings of the IEEE International Conference On Computer Vision, pp. 233–240 (2013)
26. Vaswani, A., et al.: Attention is all you need. In: Advances in Neural Information Processing Systems 30 (2017)
27. Wang, N., Zhou, W., Wang, J., Li, H.: Transformer meets tracker: Exploiting temporal context for robust visual tracking. In: Proceedings of the IEEE/CVF Conference on Computer Vision and Pattern Recognition, pp. 1571–1580 (2021)
28. Wang, Q., Fang, J., Yuan, Y.: Multi-cue based tracking (2014)
29. Xiao, J., Stolkin, R., Gao, Y., Leonardis, A.: Robust fusion of color and depth data for rgb-d target tracking using adaptive range-invariant depth models and spatio-temporal consistency constraints. IEEE Trans. Cybern. **48**(8), 2485–2499 (2017)
30. Yan, B., Peng, H., Fu, J., Wang, D., Lu, H.: Learning spatio-temporal transformer for visual tracking. In: Proceedings of the IEEE/CVF International Conference on Computer Vision, pp. 10448–10457 (2021)
31. Yan, S., Yang, J., Käpylä, J., Zheng, F., Leonardis, A., Kämäräinen, J.K.: Depthtrack: Unveiling the power of rgbd tracking. In: Proceedings of the IEEE/CVF International Conference on Computer Vision, pp. 10725–10733 (2021)
32. Yang, J., et al.: Rgbd object tracking: An in-depth review. arXiv preprint arXiv:2203.14134 (2022)
33. Yu, B., et al.: High-performance discriminative tracking with transformers. In: Proceedings of the IEEE/CVF International Conference on Computer Vision, pp. 9856–9865 (2021)
34. Zhao, P., Liu, Q., Wang, W., Guo, Q.: Tsdm: Tracking by siamrpn++ with a depth-refiner and a mask-generator. In: 2020 25th International Conference on Pattern Recognition, pp. 670–676. IEEE (2021)

35. Zheng, C., Yan, X., Gao, J., Zhao, W., Zhang, W., Li, Z., Cui, S.: Box-aware feature enhancement for single object tracking on point clouds. In: Proceedings of the IEEE/CVF International Conference on Computer Vision, pp. 13199–13208 (2021)
36. Zhou, T., Fu, H., Chen, G., Zhou, Y., Fan, D.P., Shao, L.: Specificity-preserving rgb-d saliency detection. In: International Conference on Computer Vision (ICCV) (2021)

Author Index

Printed in the United States
by Baker & Taylor Publisher Services